Book of Abstracts of the 63rd Annual Meeting of the European Federation of Animal Science

EAAP - European Federation of Animal Science

The European Federation of Animal Science wishes to express its appreciation to the
Ministero delle Politiche Agricole e Forestali (Italy) and the
Associazione Italiana Allevatori (Italy)
for their valuable support of its activities.

Book of Abstracts of the 63rd Annual Meeting of the European Federation of Animal Science

Bratislava, Slovakia, 27 - 31 August 2012

Proceedings publication and Abstract Submission System (OASES) by

Wageningen Academic
P u b l i s h e r s

ISBN: 978-90-8686-206-1
e-ISBN: 978-90-8686-761-5
DOI: 10.3920/978-90-8686-761-5

ISSN 1382-6077

First published, 2012

© Wageningen Academic Publishers
The Netherlands, 2012

Welcome to EAAP 2012 in Bratislava – Slovakia

On behalf of the Slovak Organising Committee I am pleased to invite you to attend the 63rd Annual Meeting of the EAAP which will take place in Bratislava from August 27 to August 31, 2012. This will be the first time Slovakia is going to host one of the worldwide most important scientific meetings of professionals working in the area of animal production.

One of the main objectives of the meeting is to focus on sustainability and efficiency of the livestock sector. Programme will cover all aspects of scientific achievements withinanimal science dealing with genetics, animal breeding, farm management and technology,product quality, nutrition, physiology, health and animal welfare, and other related topics.

During the meeting scientists and professionals will present how the animal science acquirements and innovations could help to provide effective solutions and long term perspectives needed for a world with resource constraints and environmental limits. You will see the good examples of successful partnerships of international teams consisting of scientists, industry stakeholders and other specialists within Europe and the world. We would also like to pay attention to efficient and faster transfer of knowledge and life education of professionals in agriculture. Ministry of Agriculture and Rural Development and Ministry of Education, Science, Research and Sport of the Slovak Republic agreed to participate in the Meeting.

During the congress we would like to offer you not only an interesting scientific programme, but also present the hospitality and country development during past decades, culture and way of life in Slovakia. It is our pleasure to invite you to participate in the 63rd EAAP Meeting 2012 to help us to make it a highly professional, successful and enjoyable event.

Ľubomír Jahnátek

Minister of Agriculture and Rural Development
President of the Slovak Organising Committee

National organisers of the 63rd EAAP annual meeting

Slovak National Organizing Committee

President
- **Prof. Ľubomír Jahnátek,**
  Minister of Agriculture and Rural Development of the Slovak Republic
  (lubomir.jahnatek@land.gov.sk)

Vice-President
- **Dr. Dana Peškovičová,**
  Animal Production Research Centre Nitra (peskovic@cvzv.sk)

Executive Secretary
- **Dr. Peter Polák,**
  Animal Production Research Centre Nitra (polak@cvzv.sk, eaap2012@cvzv.sk)

Members
- **Dr. Slávka Jánošíková,**
  Ministry of Agriculture and Rural Development of the Slovak Republic
  (slavka.janosikova@land.gov.sk)
- **Mr. Ján Vajs,**
  Ministry of Agriculture and Rural Development of the Slovak Republic
  (jan.vajs@land.gov.sk)
- **Ms. Andrea Hrdá,**
  Ministry of Agriculture and Rural Development of the Slovak Republic
  (andrea.hrda@land.gov.sk)
- Prof. Ladislav Hetényi, National Coordinator EAAP, Animal Production Research Centre Nitra
  (hetenyi@cvzv.sk)
- **Dr. Ján Huba,**
  Animal Production Research Centre Nitra (huba@cvzv.sk)
- **Dr. Štefan Ryba,**
  Breeding Services of Slovak Republic, State Enterprise (stefanryba@pssr.sk)
- **Assoc. Prof. Juraj Candrák,**
  Slovak University of Agriculture Nitra (juraj.candrak@uniag.sk)
- **Prof. Daniel Bíro,**
  Slovak University of Agriculture Nitra (daniel.biro@uniag.sk)
- **Prof. Jozef Bulla,**
  Slovak University of Agriculture Nitra (jozef.bulla@uniag.sk)
- **Prof. Stefan Mihina,**
  Animal Production Research Centre Nitra (mihina@cvzv.sk),
  Slovak University of Agriculture Nitra (stefan.mihina@uniag.sk)
- **Dr. Marián Šurda,**
  Turf Bratislava, State Enterprise, (surda@zavodisko.sk)
- **Dr. Michal Horný,**
  The National Stud Farm in Topoľčianky, State Enterprise, (sekretariat@nztopolcianky.sk)
- **Ms. Alena Péliová,**
  Ministry of Education, Science, Research and Sport of Slovak Republic
  (alena.peliova@minedu.sk)

Slovakian Scientific Committee

President
* **Dr. Dana Peškovičová**,
  Animal Production Research Centre Nitra, (peskovic@cvzv.sk, riaditel@cvzv.sk)

Vice-President
* **Assoc. Prof. Jaroslav Slamečka**,
  Animal Production Research Centre Nitra (slamecka@cvzv.sk)

Secretary
* **Dr. Zuzana Krupová**,
  Animal Production Research Centre Nitra (krupova@cvzv.sk)

Guest members – Coordinators of Study Commissions
Animal Genetics
* **Assoc. Prof. Juraj Candrák**, Slovak Agricultural University Nitra (juraj.candrak@uniag.sk)
* **Prof. Peter Chrenek**, Animal Production Research Centre Nitra (chrenekp@cvzv.sk)

Animal Physiology
* **Assoc. Prof. Vladimír Tančin**, Animal Production Research Centre Nitra (tancin@cvzv.sk)

Animal Nutrition
* **Assoc. Prof. Milan Šimko**, Slovak Agricultural University Nitra (milan.simko@uniag.sk)

Animal Management and Health
* **Assoc. Prof. Peter Strapák**, Slovak Agricultural University, Nitra (peter.strapak@uniag.sk)
* **Dr. Vladimír Foltys**, Animal Production Research Centre Nitra (foltys@cvzv.sk)

Cattle Production
* **Prof. Jozef Bulla**, Slovak Agricultural University Nitra (jozef.bulla@uniag.sk)
* **Dr. Ján Huba**, Animal Production Research Centre Nitra (huba@cvzv.sk)

Sheep and Goat Production
* **Assoc. Prof. Milan Margetín**, Animal Production Research Centre Nitra (margetin@cvzv.sk)

Pig Production
* **Prof. Ladislav Hetényi**, Institute Animal Production Research Centre Nitra (hetenyi@cvzv.sk)
* **Dr. Peter Demo**, Animal Production Research Centre Nitra (demo@cvzv.sk)

Horse Production
* **Prof. Marko Halo**, Slovak Agricultural University Nitra (marko.halo@uniag.sk)

Livestock Farming Systems
* **Prof. Štefan Mihina**, Animal Production Research Centre Nitra (mihina@cvzv.sk)
* **Assoc. Prof. Jan Brouček**, Production Research Centre Nitra (broucek@cvzv.sk)

EAAP Program Foundation

Aims
EAAP aims to bring to our annual meetings, speakers who can present the latest findings and views on developments in the various fields of science relevant to animal production and its allied industries. In order to sustain the quality of the scientific program that will continue to entice the broad interest in EAAP meetings we have created the "EAAP Program Foundation". This Foundation aims to support:
- Invited speakers with a high international profile by funding part or all of registration and travel costs.
- Delegates from less favoured areas by offering scholarships to attend EAAP meetings.
- Young scientists by providing prizes for best presentations.

The "**EAAP Program Foundation**" is an initiative of the Scientific committee (SC) of EAAP. The Foundation aims to stimulate the quality of the scientific program of the EAAP meetings and to ensure that the science meets societal needs. The Foundation Board of Trustees oversees these aims and seeks to recruit sponsors to support its activities.

Sponsorships
1. Meeting sponsor – from 6000 euro
- acknowledgements in the final booklet with contact address and logo
- one page allowance in the final booklet
- advertising/information material inserted in the bags of delegates
- advertising/information material on a stand display
- acknowledgement in the EAAP Newsletter with possibility of a page of publicity
- possibility to add session and speaker support (at additional cost to be negotiated)

2. Session sponsor – from 3000 to 5000 euro
- acknowledgements in the final booklet with contact address and logo
- one page allowance in the final booklet
- advertising/ information material in the delegate bag
- ppt at beginning of session to acknowledge support and recognition by session chair
- acknowledgement in the EAAP Newsletter.

3. Speaker sponsor – from 2500 euro (cost will be defined according to speakers country of origin)
- half page allowance in the final booklet
- advertising/ information material in the delegate bag
- recognition by speaker of the support at session
- acknowledgement in the EAAP Newsletter

4. Registration Sponsor – equivalent to a full registration fee of the Annual Meeting
- acknowledgements in the booklet with contact address and logo
- advertising/information material in the delegate bag

The Association
EAAP (The European Federation of Animal Science) organises every year an international meeting which attracts between 900 and 1500 people. The main aims of EAAP are to promote, by means of active co-operation between its members and other relevant international and national organisations, the advancement of scientific research, sustainable development and systems of production; experimentation, application and extension; to improve the technical and economic conditions of the livestock sector; to promote the welfare of farm animals and the conservation of the rural environment; to control and optimise the use of natural resources in general and animal genetic resources in particular; to encourage the involvement of young scientists and technicians. More information on the organisation and its activities can be found at www.eaap.org.

Contact and further information
If you are interested to become a sponsor of the 'EAAP Program Foundation' or want to have further information, please contact the EAAP Secretariat (eaap@eaap.org, Phone +39 06 44202639).

Acknowledgements

 Animal Health

 animal

 REDNEX

European Federation of Animal Science (EAAP)

President:	Kris Sejrsen
Secretary General:	Andrea Rosati
Address:	Via G.Tomassetti 3, A/I
	I-00161 Rome, Italy
Phone:	+39 06 4420 2639
Fax:	+39 06 4426 6798
E-mail:	eaap@eaap.org
Web:	www.eaap.org

64th EAAP Annual meeting of the European Federation of Animal Science

26 - 30 August 2013, Nantes, France

Organising Committee

President:
• Henri Seegers, INRA – president@nantes.inra.fr

Vice-President:
• Joël Merceron Idele, Institut de l'Elevage – Joel.Merceron@idele.fr

Executive Secretaries:
• Michel Bonneau, INRA – michel.bonneau@rennes.inra.fr
• Bernard Coudurier, INRA – bernard.coudurier@tours.inra.fr

Venue Address:
• La Cité Nantes Events Center, 5 rue de Valmy, 44000 Nantes, France
  Phone: +33 (2) 51 88 20 00 – Fax: +33 (2) 51 88 20 20
  www.lacite-nantes.com/uk/index.html

Conference website: www.eaap2013.org

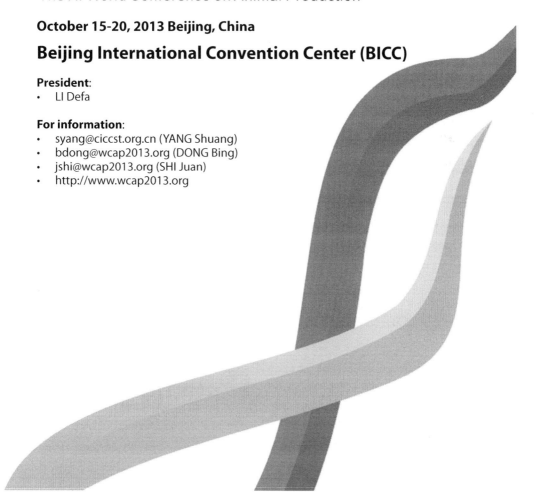

WCAP 2013

The XI World Conference on Animal Production

October 15-20, 2013 Beijing, China

Beijing International Convention Center (BICC)

President:
- LI Defa

For information:
- syang@ciccst.org.cn (YANG Shuang)
- bdong@wcap2013.org (DONG Bing)
- jshi@wcap2013.org (SHI Juan)
- http://www.wcap2013.org

Scientific Programme EAAP 2012

Monday 27 August 8.30 – 12.30	Monday 27 August 14.00 – 18.00	Tuesday 28 August 8.30 – 11.30*	Tuesday 28 August 14.00 – 18.00
Biology of lactation workshop (EAAP/ASAS)		**Plenary session**	**Session 16** Synthesis and secretion of specific constituents into colostrum and mature milk Chair: R. Bruckmaier
Session 1 Mammary gland development and function Chair: K. Singh	**Session 9** Mammary health during the transition period Chair: K Singh	Followed by **Poster Session 1**, 11.30 to 12.30 (Session 1-24)	
Session 2 Opportunities to use genomic information for *in-situ* and *ex-situ* conservation Chair: J. Fernandez	**Session 10a** PRRS eradication or control: what needs to be done? Chair: A. Doeschl-Wilson		**Session 17** Genomic selection Chair: N. Ibanez
			Session 18 Modelling complexity in LFS to address trade-offs and synergies for efficiency Chair: M Tichit
Session 3 Breeding and management for milk and product quality Chair: M. Klopcic	**Session 10b** Breeding and conservation of dog breeds Chair: E. Strandberg		
Session 4a LFS in emerging and developing countries: trends, roles and goals Chair: S. Oosting/I. Stokovic	**Session 11** Breeding and management for meat, milk and product quality Chair: M. Klopcic/P. Polak		**Session 19** Impact of stockmanship on animal welfare and integrity Chair: E. Stasssen
			Session 20 Organic and Low Input Dairy Farming Systems Chair: N. Scollan
Session 4b Labour issues in LFS (gender, lifestyle, workload satisfaction, part-time agriculture, immigration) Chair: K. Eilers	**Session 12** Closing the phenomic gap – Methods, data collection and experiments to select for new traits Chair: R. Veerkamp		**Session 21** Free Communications: Nutrition Chair: J.-E. Lindberg
Session 5 Free Communications in Horse Genetics Chair: I. Cervantes	**Session 13a** Horse genetic resources focused to Central Europe Chair: D Lewzuck		**Session 22** Phenotyping complex traits in dairy and beef cows; applications Chair: M. Coffey
Session 6 Piglet survival through lactation performance, milk composition and management Chair: C. Lauridsen/G. Bee.	**Session 13b** Emerging diseases and challenges in horse production Chair: D. Burger		**Session 23a** AI, fertility and reproductive technologies in sheep and goats Chair: M. Milerski.
	Session 14a Lamb, kid and piglet perinatal survival and vigour Chair: J. Conington		
Session 7 Detection and control of mastitis and lameness Chair: J. Krieter	**Session 14b** Free communications; Sheep and Goats Chair: J. Conington		**Session 23b** The effect of nutrition and metabolic status on follicle, oocyte and embryo development Chair: H. Quesnel
All-day Symposium: Nitrogen Utilization by Ruminants			**Session 24a** Entrepreneurship, farm and herd management Chair: I. Halachmi/A. Kuipers
Session 8 Feeds and feed evaluation for ruminants Chair: G. Broderick	**Session 15** Innovative and practical management strategies to reduce N excretion from dairy farms Chair. A. van Vuuren		**Session 24b** Health and fertility: aspects of breeding, energy balance and transition periods Chair: G. Thaller

* Both the Tuesday morning and Wednesday morning theatre sessions will be from 8.30 to 11.30 and followed by a one-hour poster session.

Wednesday 29 August 8.30 – 11.30*	Wednesday 29 August 14.00 – 18.00	Thursday 30 August 8.30 – 12.30	Thursday 30 August 14.00 – 18.00
Session 25 Using genomic data for management and non-breeding purposes Chair: M. Winters	**Genetics business meeting** Chair: H. Simianer	**Session 41** Industry session: Precision dairy and beef farming Chair: J. Hocquette	**Session 49** What's new about stress, behaviour, physiology and welfare in animals? Chair: I. Veissier
Session 26 Nutritional value of by-products derived from the bio-based economy Chair: L. Bailoni	**Session 32** Free Communications Chair: G. Rose	**Session 42** Genetics Free Communications Chair: H. Simianer	**Session 50** Analysis of complex traits based on genome sequences and high density genotyping Chair: M. Lund
Session 27 Entrepreneurship, farm and herd management Chair: I. Halachmi/A. Kuipers	**Session 33** Cattle business meeting and free communications Chair: A. Kuipers/C. Lazzaroni	**All-day Symposium** **Session 43** Efficiency and optimization in ruminant husbandry Chair: A.Kuipers/S. Mihina	**Session 51** Efficiency, competitiveness and structure of ruminant husbandry in E. Europe Chair: A. Rozstalnyy/A.Svitojus
Session 28 Health and fertility: aspects of breeding, energy balance and transition periods Chair: G. Thaller	**Session 34** Horse business meeting and free communications Chair: N. Miraglia/M. Magistrini	**All-day Symposium**: Industry session **Session 44** Nutrient sensing and signalling in the gastrointestinal tract; mechanisms, physiological significance and impact on animal's performance and health Chair: S.P. Shirazi-Beechey	
Session 29 Animal welfare research and education in enlarged Europe Chair H. Spoolder	**Session 35** Sheep and Goats business meeting and free communications Chair: L. Bodin	**All-day Symposium**: Genomics and horse production: from science to practice **Session 45**	
Session 30 Free communications: Pig nutrition and management Chair: G. Bee	**Session 36** Pigs business meeting and free communications: Pig Genetics Chair: P. Knap	Presented papers Chair: J. Mickelson/A. von Velsen-Zerweck	Round table and discussion Chair: J. Mickelson/A. von Velsen-Zerweck
Followed by **Poster Session 2**, 11.30 to 12.30 (Session 25-54)	**Session 37** Livestock Farming Systems business meeting and free communications Chair: A Bernués	**All-day Symposium**: Lactation and milking of small ruminants	
	Session 38 Animal Physiology free communications and business meeting Chair: M. Vestergaard	**Session 46** Health in small dairy ruminants Chair: N Silanikove	**Session 52** Milk quality, technology and processing in sheep and goats Chair: E. Ugarte Round table discussion
	Session 39 Management and Health business meeting and free communications Chair: C. Fourichon/Gurbus Das	**Session 47** Impact of market requirements and production conditions on pig breeding goals: mainstream and niche Chair P. Knap	**Session 54** Fibre production in Central and Eastern countries Chair: R. Niznikowski
Two-day Symposium: Livestock and climate change: options for mitigation and adaptation			
Session 31 New systems design and technologies to mitigate emissions Chair: I de Boer	**Session 40** Nutrition business meeting. Chair: J-E. Lindberg Nutritional strategies to reduce methane Chair: G. van-Duinkirken	**Session 48** Adaptation of livestock farming systems to climate change and uncertainty Chair: S. Ingrand	**Session 53** Adaptation of livestock to climatic stress Chair: F. Ringdorfer

Commission on Animal Genetics

Prof. Dr Simianer	President	University of Goettingen
	Germany	hsimian@gwdg.de
Dr Baumung	Vice-President	FAO Rome
	Italy	Roswitha.Baumung@fao.org
Dr Meuwissen	Vice-President	Norwegian University of Life Sciences
	Norway	theo.meuwissen@umb.no
Dr Szyda	Vice-President	Agricultural University of Wroclaw
	Poland	szyda@karnet.ar.wroc.pl
Dr Ibañez	Secretary	IRTA
	Spain	noelia.ibanez@irta.es
Dr De Vries	Industry rep.	CRV

Commission on Animal Nutrition

Dr Lindberg	President	Swedish University of Agriculture
	Sweden	jan-eric.lindberg@huv.slu.se
Dr Bailoni	Vice-President	University of Padova
	Italy	Lucia.bailoni@unipd.it
Dr Auclair	Vice President/	
	Industry rep.	LFA Lesaffre
	France	ea@lesaffre.fr
Mrs Tsiplakou	Secretary	Agricultural University of Athens
	Greece	eltisplakou@aua.gr
Mr Van Duinkerken	Secretary	Wageningen University
	Netherlands	gert.vanduinkerken@wur.nl

Commission on Animal Management & Health

Dr Fourichon	President	Oniris INRA
	France	Christine.fourichon@oniris-nantes.fr
Dr Spoolder	Vice-President	ASG-WUR
	Netherlands	Hans.spoolder@wur.nl
Prof. Dr Krieter	Vice-President	University Kiel
	Germany	jkrieter@tierzucht.uni-kiel.de
Mr Pearce	Vice-President/	
	Industry rep.	Pfizer
	United Kingdom	Michael.C.Pearce@pfizer.com
Dr. Boyle	Secretary	Teagasc
	Ireland	Laura.Boyle@teagasc.ie
Mr Das	Secretary	University of Goettingen
	Germany	gdas@gwdg.de

Commission on Animal Physiology

Dr Vestergaard	President	Aarhus University
	Denmark	Mogens.Vestergaard@agrsci.dk
Dr Kuran	Vice-President	Gaziosmanpasa University
	Turkey	mkuran@gop.edu.tr
Dr Driancourt	Vice president/	
	Industry rep.	Intervet
	France	Marc-antoine.driancourt@sp.intervet.com
Dr Bruckmaier	Vice-President	University of Bern
	Switzereland	Rupert.bruckmaier@physio.unibe.ch
Dr Quesnel	Secretary	INRA Saint Gilles
	France	Helene.quesnel@rennes.inra.fr
Dr Scollan	Secretary	Institute of Biological, Environmental and rural sciences
	United Kingdom	ngs@aber.ac.uk

Commission on Livestock Farming Systems

Dr Bernués Jal	President	CITA
	Spain	abernues@aragon.es
Dr Ingrand	Vice-President	INRA
	France	ingrand@clermond.inra.fr
Dr Leroyer	Vice president/	
	industry rep.	ITAB
	France	Joannie.leroyer@itab.asso.fr
Dr Matlova	Vice-President	Res. Institute for Animal Production
	Czech Republic	matlova.vera@vuzv.cz
Dr Tichit	Vice-President	INRA
	France	Muriel.tichit@agroparistech.fr
Dr Eilers	Secretary	Wageningen University
	Netherlands	karen.eilers@wur.nl
Mrs Zehetmeier	Secretary	Institute Agricultural Economics and Farm Management
	Germany	tmonika.zehetmeier@tum.de

Commission on Cattle Production

Dr Kuipers	President	Wageningen UR
	Netherlands	abele.kuipers@wur.nl
Dr Thaller	Vice-President	Animal Breeding and Husbandry
	Germany	Georg.Thaller@tierzucht.uni-kiel.de
Dr Lazzaroni	Vice-president	University of Torino
	Italy	carla.lazzaroni@unito.it
Dr Coffey	Vice president/	
	Industry rep.	SAC, Scotland
	United Kingdom	mike.coffey@sac.ac.uk>
Dr Hocquette	Secretary	INRA
	France	hocquet@clermont.inra.fr
Dr Klopcic	Secretary/	
	Industry rep.	University of Ljublijana
	Slovenia	marija.klopcic@bf.uni-lj.si

Commission on Sheep and Goat Production

Dr Bodin	President	INRA-SAGA
	France	Loys.bodin@toulouse.inra.fr
Dr Ringdorfer	Vice-President	LFZ Raumberg-Gumpenstein
	Austria	ferdinand.ringdorfer@raumberg-gumpenstein.at
Dr Conington	Vice-President	SAC
	United Kingdom	joanne.conington@sac.ac.uk
Dr Papachristoforou	Vice President	Agricultural Research Institute
	Cyprus	Chr.Papachristoforou@arinet.ari.gov.cy
Dr Milerski	Secretary/	
	Industry rep.	Research Institute of Animal Science
	Czech Republic	m.milerski@seznam.cz
Dr Ugarte	Secretary/	
	Industry rep.	NEIKER-Tecnalia
		eugarte@neiker.net

Commission on Pig Production

Dr Knap	President/	
	Industry rep.	PIC International Group
	Germany	Pieter.Knap@genusplc.com
Dr Lauridsen	Vice-President	Aarhus University
	Denmark	charlotte.lauridsen@agrsci.dk
Dr Bee	Vice-President	Agroscope Liebefeld-Posieux ALP
	Switzerland	giuseppe.bee@alp.admin.ch
Dr Pescovicova	Secretary	Research Institute of Animal Production
	Slovak Republic	peskovic@vuzv.sk
Dr Velarde	Secretary	IRTA
	Spain	antonio.velarde@irta.es

Commission on Horse Production

Dr Miraglia	President	Molise University
	Italy	miraglia@unimol.it
Dr Burger	Vice president	Clinic Swiss National Stud
	Switzerland	
		Dominique.burger@mbox.haras.admin.ch
Dr Janssen	Vice president	BIOSYST
	Belgium	Steven.janssens@biw.kuleuven.be
Dr Lewczuk	Vice president	IGABPAS
	Poland	d.lewczuk@ighz.pl
Dr Saastamoinen	Vice president	MTT Agrifood Research Finland
	Finland	markku.saastamoinen@mtt.fi
Dr Palmer	Vice president/	
	Industry rep.	CRYOZOOTECH
	France	ericpalmer@cryozootech.com
Dr Holgersson	Secretary	Swedish University of Agriculture
	Sweden	Anna-Lena.holgersson@hipp.slu.se
Dr Hausberger	Secretary	CNRS University
	France	Martine.hausberger@univ-rennes1.fr

Poster

Session 01. Mammary gland development and function (biology of lactation workshop)

Date: 27 August 2012; 08:30 - 12:30 hours
Chairperson: Singh

Session 02. Opportunities to use genomic information for *in-situ* and *ex-situ* conservation

Date: 27 August 2012; 08:30 - 12:30 hours
Chairperson: Fernandez

Session 03. Breeding and management for milk and product quality

Date: 27 August 2012; 08:30 - 12:30 hours
Chairperson: Klopcic

Session 04a: Livestock farming systems in emerging and developing countries: trends, roles and goals

Date: 27 August 2012; 08:30 - 10:15 hours
Chairperson: Oosting and Stokovic

Session 04b: Labour issues in livestock farming systems (workload, gender, lifestyle, satisfaction, part-time agriculture, immigration)

Date: 27 August 2012; 10:45 - 12:30 hours
Chairperson: Eilers

Session 05. Free communications in horse genetics

Date: 27 August 2012; 08:30 - 12:30 hours
Chairperson: Cervantes

Session 06. Piglet survival through lactation performance, milk composition and management

Date: 27 August 2012; 08:30 - 12:30 hours
Chairperson: Lauridsen and Bee

Session 07. Detection and control of mastitis and lameness

Date: 27 August 2012; 08:30 - 12:30 hours
Chairperson: Fourichon

Session 08. Feeds and feed evaluation for ruminants (nitrogen utilization by ruminants symposium)

Date: 27 August 2012; 08:30 - 12:30 hours
Chairperson: Broderick

Session 09. Mammary health during the transition period (biology of lactation workshop)

Date: 27 August 2012; 14:00 - 18:00 hours
Chairperson: Singh

Session 10a: PRRS eradication or control: what needs to be done?

Date: 27 August 2012; 14:00 - 15:45 hours
Chairperson: Doeschl-Wilson

Session 10b: Breeding and conservation of dog breeds

Date: 27 August 2012; 16:15 - 18:00 hours
Chairperson: Strandberg

Session 11. Breeding and management for meat, milk and product quality

Date: 27 August 2012; 14:00 - 18:00 hours
Chairperson: Klopcic and Polak

Session 12. Closing the phenomic gap: methods, data collection and experiments to select for new traits

Date: 27 August 2012; 14:00 - 18:00 hours
Chairperson: Veerkamp

Poster   **Session 12 no. Page**

Session 13a: Horse genetic resources focused to Central Europe

Date: 27 August 2012; 14:00 - 15:45 hours
Chairperson: Lewczuk

Session 13b: Emerging diseases and challenges in horse production

Date: 27 August 2012; 16:15 - 18:00 hours
Chairperson: Burger

Session 14a: Lamb and kid perinatal survival and vigour

Date: 27 August 2012; 14:00 - 15:45 hours
Chairperson: Conington

Poster   Session 14a no. Page

Lambs behaviour during suckling in the period of creep feeding   6   108
Margetinova, J., Apolen, D. and Broucek, J.

Session 14b: Free communications: sheep and goats

Date: 27 August 2012; 16:15 - 18:00 hours
Chairperson: Conington

Theatre   Session 14b no. Page

Effectiveness of a deactivator on mitigating the impact of endophyte alkaloids in Merino
ewes   1   108
Leury, B.J., Henry, M.L.E., Digiacomo, K., Ng, C., Kemp, S. and Dunshea, F.R.

Mediterranean shrub *Pistacia lentiscus* L. as a potential tool in the control of nematodes
in sheep   2   109
Saric, T., Rogosic, J., Beck, R., Zupan, I., Zjalic, S., Musa, A. and Skobic, D.

Valeric and isovaleric acid concentrations: useful biomarkers for subacute ruminal acidosis?   3   109
Fanning, J., Cockcroft, P. and Hynd, P.

Preliminary assessment of sheep welfare on pasture   4   110
*Mialon, M.M., Robin, C., Verney, A., Brule, A., Pottier, E., Davoine, J.M., Ribaud, D., Boivin, X. and
Boissy, A.*

Curved allometries reveal no constraint in horn length in Rasquera White Goat   5   110
Parés-Casanova, P.M. and Sabaté, J.

The sheep sector in greenhouse gas inventory in Hungary   6   111
Borka, G., Németh, T., Krausz, E. and Kukovics, S.

Poster   Session 14b no. Page

Exploring skull shape and sexual dimorphism in a local goat breed   7   111
Parés-Casanova, P.M., Sabaté, J. and Soto, J.

Session 15. Innovative and practical management strategies to reduce N excretion from dairy farms (nitrogen utilization by ruminants symposium)

Date: 27 August 2012; 14:00 - 18:00 hours
Chairperson: Van Vuuren

Theatre   Session 15 no. Page

invited Constraints to efficient protein utilization in the dairy cow and on the dairy farm   1   112
Broderick, G.A.

The use of antibodies in order to alter bacterial population in the rumen   2   112
Foskolos, A., Cavini, S., Ferret, A. and Calsamiglia, S.

Protein level and forage digestibility interactions on dairy cow production   3   113
Alstrup, L. and Weisbjerg, M.R.

Effect of feeding rumen protected methionine and choline on milk yield and milk composition in early

Soleimani, A., Nourozi Ebdalabadi, M., Kousary Moghaddam, M. and Ahmadzadeh Bazzaz, B.

Effect of total replacement of soybean meal with a sustained-release non-protein nitrogen source

Nollet, L. and Warren, H.

Session 16. Synthesis and secretion of specific constituents into colostrum and mature milk

Date: 28 August 2012; 14:00 - 18:00 hours
Chairperson: Bruckmaier

Session 17. Genomic selection

Date: 28 August 2012; 14:00 - 18:00 hours
Chairperson: Ibañez-Escriche

Session 18. Modelling complexity in LFS to address trade-offs and synergies for efficiency

Date: 28 August 2012; 14:00 - 18:00 hours
Chairperson: Tichit

Session 19. Impact of stockmanship on animal welfare and integrity

Date: 28 August 2012; 14:00 - 18:00 hours
Chairperson: Stassen

Session 20. Organic and low input dairy farming systems

Date: 28 August 2012; 14:00 - 18:00 hours
Chairperson: Scollan

Session 21. Free communications: nutrition

Date: 28 August 2012; 14:00 - 18:00 hours
Chairperson: Lindberg

Session 22. Phenotyping complex traits in dairy and beef cows, applications

Date: 28 August 2012; 14:00 - 18:00 hours
Chairperson: Coffey

Session 23a: AI, reproductive technologies and fertility in sheep and goats

Date: 28 August 2012; 14:00 - 15:45 hours
Chairperson: Milerski

Session 23b: The effect of nutrition and metabolic status on follicle, oocyte and embryo development

Date: 28 August 2012; 16:15 - 18:00 hours
Chairperson: Quesnel

Session 24a: Entrepreneurship, farm and herd management

Date: 28 August 2012; 14:00 - 15:45 hours
Chairperson: Halachmi and Kuipers

Session 24b: Health and fertility: aspects of breeding, energy balance and transition periods

Date: 28 August 2012; 16:15 - 18:00 hours
Chairperson: Thaller

Session 25. Using genomic data for management and non-breeding purposes

Date: 29 August 2012; 08:30 - 11:30 hours
Chairperson: Winters

Session 26. Nutritional value of by-products derived from the bio-based economy

Date: 29 August 2012; 08:30 - 11:30 hours
Chairperson: Bailoni

Session 27. Entrepreneurship, farm and herd management

Date: 29 August 2012; 08:30 - 11:30 hours
Chairperson: Halachmi and Kuipers

Theatre	Session 27 no.	Page
invited · Achieving optimal cow performance with the aid of information systems *Nir (Markusfeld), O.*	1	193
Accuracy and potential of in-line NIR milk composition analysis *Melfsen, A., Haeussermann, A. and Hartung, E.*	2	193
invited · The modern cow bell: activity and rumination sensing collars *Bar, D.*	3	194
Detection of early lactation ketosis by rumination and other sensors *Steensels, M., Bahr, C., Berckmans, D., Antler, A., Maltz, E. and Halachmi, I.*	4	194
Comparison between direct and video observation for locomotion assessment in dairy cow *Schlagater-Tello, A., Lokhorst, C., Bokkers, E.A.M., Koerkamp, P.W.G., Van Hertem, T., Steensels, M., Halachmi, I., Maltz, E., Viazzi, S., Romanini, C.E.B., Bahr, C. and Berckmans, D.*	5	195
Evaluation of potential variables for sensor-based detection of lameness in dairy cattle *Van Hertem, T., Maltz, E., Antler, A., Schlageter Tello, A., Lokhorst, C., Viazzi, S., Romanini, E., Bahr, C., Berckmans, D. and Halachmi, I.*	6	195
Changing conditions require higher level of entrepreneurship for farmers *Beldman, A.C.G., Lakner, D. and Smit, A.B.*	7	196
invited · Interactive strategic management methodology for improvement of entrepreneurship: case of a farmer *Prezelj, K., Klopcic, M. and Beldman, A.*	8	196

Session 28. Health and fertility: aspects of breeding, energy balance and transition periods

Date: 29 August 2012; 08:30 - 11:30 hours
Chairperson: Thaller

Theatre	Session 28 no.	Page
invited · Shortening the dry period for dairy cows: effects on energy balance, health and fertility *Van Knegsel, A.T.M. and Kemp, B.*	1	197
Technical and economical consequences of extended (18 m) calving intervals for dairy cows *Brocard, V., Portier, B., Francois, J. and Tranvoiz, E.*	2	197
The effect of dry period management and nutrition on milk production *Blazkova, K., Cermakova, J., Dolezal, P. and Kudrna, V.*	3	198
Effect of trace mineral supplementation on the reproductive performance of pasture based dairy cows *Watson, H., Evans, A.C.O. and Butler, S.T.*	4	198

Session 29. Animal welfare research and education in enlarged Europe

Date: 29 August 2012; 08:30 - 11:30 hours
Chairperson: Spoolder

Session 30. Free communications: pig nutrition and management

Date: 29 August 2012; 08:30 - 11:30 hours
Chairperson: Bee

Poster   **Session 30 no. Page**

Session 31. New systems design and technologies to mitigate emissions

Date: 29 August 2012; 08:30 - 11:30 hours
Chairperson: De Boer

Session 32. Genetics business meeting and free communications

Date: 29 August 2012; 14:00 - 18:00 hours
Chairperson: Simianer and Rose

Session 33. Cattle business meeting and free communications

Date: 29 August 2012; 14:00 - 18:00 hours
Chairperson: Kuipers and Lazzaroni

Session 34. Horse business meeting and free communications

Date: 29 August 2012; 14:00 - 18:00 hours
Chairperson: Miraglia

Poster **Session 34 no. Page**

Session 35. Sheep and Goats business meeting and free communications

Date: 29 August 2012; 14:00 - 18:00 hours
Chairperson: Bodin

Theatre **Session 35 no. Page**

Poster **Session 35 no. Page**

Session 36. Pigs business meeting and free communications: pig genetics

Date: 29 August 2012; 14:00 - 18:00 hours
Chairperson: Knap

Session 37. Livestock Farming Systems business meeting and free communications

Date: 29 August 2012; 14:00 - 18:00 hours
Chairperson: Bernués

Session 38. Animal Physiology business meeting and free communications

Date: 29 August 2012; 14:00 - 18:00 hours
Chairperson: Vestergaard

Session 39. Management and health business meeting and free communications

Date: 29 August 2012; 14:00 - 18:00 hours
Chairperson: Das

Session 40. Nutritional strategies to reduce methane

Date: 29 August 2012; 14:00 - 18:00 hours
Chairperson: Van Duinkerken

Theatre	Session 40 no.	Page
invited Nutritional strategies to reduce enteric methane emissions in ruminants Newbold, C.J.	1	294
Greenhouse gas emissions from feed production and enteric fermentation of rations for dairy cows Mogensen, L., Kristensen, T., Nguyen, T.L.T., Knudsen, M.T., Brask, M., Hellwing, A.L.F., Lund, P. and Weisbjerg, M.R.	2	294
Role of the nature of forages on methane emission in cattle Doreau, M., Nguyen, T.T.H., Van Der Werf, H.M.G. and Martin, C.	3	295
Assessment of archaeol as a molecular proxy for methane production in cattle Mccartney, C.A., Bull, I.D., Yan, T. and Dewhurst, R.J.	4	295
invited Trade-offs between methane emission reduction and nitrogen losses Bannink, A., Ellis, J.L., Sebek, L.B.J., France, J. and Dijkstra, J.	5	296

Poster	Session 40 no.	Page
Effect of supplementation of whole crushed rapeseed on methane emission from heifers Hellwing, A.L.F., Sørensen, M.T., Weisbjerg, M.R., Vestergaard, M. and Alstrup, L.	6	296
Milk production and carbon footprint in two samples of Italian dairy cattle and buffalo farm Carè, S., Terzano, M.G. and Pirlo, G.	7	297
Methane generating potential of *Lotus uliginosus* var. Maku harvested in three consecutive dates Marichal, M.D.E.J., Crespi, R., Arias, G., Furtado, S., Guerra, M.H. and Piaggio, L.	8	297

Session 41. Industry session: Precision dairy and beef farming

Date: 30 August 2012; 08:30 - 12:30 hours
Chairperson: Hocquette

Theatre	Session 41 no.	Page
invited New perspectives and risks of precision livestock farming systems Faverdin, P.	1	298
Innovation in animal feeding, a key driver in the concept of sustainable precision livestock farming Den Hartog, L. and Sijtsma, R.	2	298
invited Precision livestock farming: review and case studies Halachmi, I.	3	299
ATOL: an ontology for livestock Meunier-Salaün, M.C., Bugeon, J., Dameron, O., Fatet, A., Hue, I., Hurtaud, C., Nedellec, C., Reichstadt, M., Vernet, J., Reecy, J., Park, C. and Le Bail, P.Y.	4	299

Poster   Session 41 no. Page

Session 42. Genetics free communications

Date: 30 August 2012; 08:30 - 12:30 hours
Chairperson: Simianer

Theatre   Session 42 no. Page

Session 43. Efficiency and optimization in ruminant husbandry

Date: 30 August 2012; 08:30 - 12:30 hours
Chairperson: Kuipers and Mihina

Poster   Session 43 no. Page

Session 44. Nutrient sensing and signalling in the gastrointestinal tract; mechanisms, physiological significance and impact on animal's performance and health

Date: 30 August 2012; 08:30 - 12:30 hours
Chairperson: Shirazi-Beechey

Theatre   Session 44 no. Page

Poster   Session 44 no. Page

Effect of dietary free or protein-bound Lys, Thr, and Met on expression of b0,+ and myosin in pigs   9   322
Grageola, F., García, H., Morales, A., Araiza, A., Arce, N. and Cervantes, M.

Session 45. Genomics and horse production: from science to practice

Date: 30 August 2012; 08:30 - 12:30 hours
Chairperson: Mickelson and Von Velsen-Zerweck

Session 46. Health in small dairy ruminants (lactation and milking of small ruminants symposium)

Date: 30 August 2012; 08:30 - 12:30 hours
Chairperson: Marnet

Session 47. Impact of market requirements and production conditions on pig breeding goals: mainstream and niche

Date: 30 August 2012; 08:30 - 12:30 hours
Chairperson: Knap

Session 48. Adaptation of livestock farming systems to climate change and uncertainty

Date: 30 August 2012; 08:30 - 12:30 hours
Chairperson: Ingrand

Session 49. What's new about stress, behaviour, physiology and welfare in animals?

Date: 30 August 2012; 14:00 - 18:00 hours
Chairperson: Veissier

Session 50. Analysis of complex traits based on genome sequences and high density genotyping

Date: 30 August 2012; 14:00 - 18:00 hours
Chairperson: Lund

Session 51. Efficiency, competitiveness and structure of ruminant husbandry in Eastern Europe

Date: 30 August 2012; 14:00 - 18:00 hours
Chairperson: Rozstalnyy and Svitojus

Session 52. Milk quality, technology and processing in S&G (lactation and milking of small ruminants symposium)

Date: 30 August 2012; 14:00 - 18:00 hours
Chairperson: Ugarte Sagastizabal

Session 53. Adaptation of livestock to climatic stress

Date: 30 August 2012; 14:00 - 18:00 hours
Chairperson: Ringdorfer

Session 54. Fibre production in Central and Eastern countries

Date: 30 August 2012; 14:00 - 18:00 hours
Chairperson: Niznikowski

Something about mammary development

Sejrsen, K., Aarhus University – Foulum, Animal Science, Blicher Allé 20, P.O. Box 50, 8830, Tjele, Denmark; Kr.Sejrsen@agrsci.dk

Over the last decades there has been considerable progress in the knowledge of mammary gland development. Much of the increase in knowledge has been recorded in the 'Red books' published after each of the Biology of Lactation in Farm Animal (BOLFA) Workshops. Nevertheless, there are still many unanswered questions that remain to be solved. One such question is: what is the mechanism behind long term effects, such as the effect of nutrition before and during the pubertal period on subsequent milk yield and the carry-over effect of milking frequency in early lactation on the milk yield later in lactation? The effect of photoperiod during puberty and pregnancy falls into the same category. Are these effects all due alteration of mammary tissue development or are other mechanisms involved? New studies on the role and importance of mammary stem cells and epigenetic regulation of mammary development will no doubt help shed light over these questions as will the studies on mammary gene transcription, especially when studies on individual mammary cell types become more common place. Even if the late H. Allen Tucker, at one of the first BOLFA meetings stated, that 'there are too damned many growth factors', the studies on gene transcription will most likely reveal that even more growth factors (or transcription factors) than we know now are involved in regulation of mammary development.

Role of ovarian secretions in mammary gland development and function in ruminants

Yart, L.[1,2], Lollivier, V.[1,2] and Dessauge, F.[1,2], [1]Agrocampus Ouest, UMR 1348 Pegase, 66 Rue de Saint Brieuc, 35000 Rennes, France, [2]INRA, UMR 1348 Pegase, Domaine de la PRISE, 35590 Saint Gilles, France; frederic.dessauge@rennes.inra.fr

The milk yield potential in dairy ruminants is highly dependent of mammary gland development. Fifty years ago, data suggested that the ovaries play a critical role during mammogenesis in cattle and reported a strong decrease in parenchymal weight in the mammary glands from ovariectomized heifers. Through this review, we will focus on the effects of ovarian hormones on mammary gland development and function *in vivo* (before puberty and during lactation) and *in vitro* (on MAC-T cells). Before puberty a rapid period of allometric growth of mammary parenchyma occurs and persists through the first estrous cycles. We will highlight that the prepubertal period is decisive for the next step of lobulo-alveolar mammogenesis, which takes place during gestation and continues during the future milk yield in goats and cows. In lactating cows, the decline in milk yield after peak of lactation is associated with a regression of the secretory tissue, a reduction in the alveolar size and a loss of mammary epithelial cells (MEC). Lactation persistency is characterized by the rate of decline in milk yield after peak of lactation. During this period, MEC undergo apoptosis to decrease cell number within mammary tissue. Ovarian hormones, such as estrogen, seem to negatively affect milk yield during lactation. Through a model of ovariectomized lactating cows, we will present the effects of ovarian secretions on MEC and milk production during lactation. Estrogens and progesterone are known to be the main ovarian hormones, and it is now well established that both are essential for MEC proliferation and differentiation. Therefore, we will present *in vitro* results describing the apoptotic effect of estrogen on bovine MEC. Discovery of the ovary-mediated MEC number and activity modulation will permit novel management strategies to optimize mammary growth and function.

Milking frequency modifies DNA methylation at a CSN1 regulatory region in the bovine mammary gland
Nguyen, M.[1], Boutinaud, M.[2], Dessauge, F.[2], Charlier, M.[1], Gabory, A.[1], Galio, L.[1], Jammes, H.[1], Kress, C.[1] and Devinoy, E.[1], [1]UR1196 GPL; UMR1198 BDR, INRA, 78352 Jouy-en-Josas, France, [2]UMR1348 Pegase, INRA, 35590 Saint-Gilles, France; eve.devinoy@jouy.inra.fr

Milking frequency influences milk production in cows by regulating milk synthesis and secretion. This in turn involves a number of genes that are differentially expressed. In this study, we hypothesized that milking frequency changes the milk yield through altered gene expression by modifying the methylation at CpGs dinucleotides located within regulatory regions. We focused on a candidate regulatory region of one casein gene (CSN1). Eight heifers were studied over three periods. Period 1: twice daily milking for 5 days of both udders of each heifer. Period 2: differential milking for 7 days: right and left udders were milked twice and once a day, respectively. Period 3: both udders returned to twice daily milking. Milk yield was recorded every day over the three periods. Biopsies from left & right udders were taken at the end of period 2. Global DNA methylation was evaluated using the LUMA assay. DNA methylation of three CpGs located within a distal regulatory region CSN1 was analyzed using pyrosequencing after bisulfite treatment of genomic DNA. The mRNA expression levels have been quantified by RNASeq. Significance of results was studied with the Wilcoxon test. Global methylation of the once daily milking group (75.9%) is higher compared with the twice daily milking group (74.6%) but only significant if one of the eight heifers, which behaved differently from the others, was excluded (P=0.04). The DNA methylation of all three CpG in the once daily milking group is higher than that of twice daily milking group (P=0.02, P=0.07, and P=0.02, respectively). Once daily milking increases DNA methylation at a regulatory region of the CSN1 gene and at the global level in the bovine mammary gland, to a certain extent. These modifications are related to transcription of the casein genes and to some ncRNA.

Suckling effects in sows: does teat use in first lactation affect its yield in second lactation?
Farmer, C., Agriculture and Agri-Food Canada, Dairy and Swine R & D Centre, 2000 College St., Sherbrooke, QC, J1M 0C8, Canada; farmerc@agr.gc.ca

Sows cannot produce enough milk to sustain optimal growth of their litters and it is essential to develop management strategies to increase sow milk yield. The number of mammary cells present at the onset of lactation affects milk yield. At each parity, the mammary gland undergoes a cycle of rapid accretion in late gestation and lactation, followed by involution at weaning. Yet, the potential impact of a teat not being used in one lactation on its milk yield in the next lactation is not known. In a recent study, first-parity sows were divided into 2 groups: (1) same teats used in 2 subsequent lactations (controls, CTL); and (2) different teats used in 2 subsequent lactations (treated, TRT). In the first lactation, teats from both sides of the udder were blinded so that 6 functional teats remained. During the next lactation, CTL sows had the same teats blinded as in the first lactation and the reverse was done for TRT sows. In both parities, litters were standardized to 6 piglets. Behavioural measures were obtained on d 3 and 10 to evaluate satiety of piglets based on aggressiveness and nursing behaviour. At weaning on the second lactation, sows were slaughtered and mammary glands collected for compositional analyses and measurement of mRNA abundance for selected genes. Piglets from CTL sows weighed 1.12 kg more than piglets from TRT sows (P<0.05) on d 56, and functional mammary glands from CTL sows contained more parenchymal tissue, DNA and RNA (P<0.01) than those from TRT sows. Relative mRNA abundance of prolactin in parenchymal tissue tended to be greater in CTL than TRT sows (P<0.10). Behavioural measures indicated a greater hunger level for piglets using teats that were not previously used. These findings clearly show that teats which were suckled in first lactation produce more milk and have a greater development in the second lactation, than unused teats. This could be partly due to changes in mRNA abundance for prolactin at the level of the mammary gland.

Alternative animal models to better understand mammary gland development and function

Nicholas, K.R., Sharp, J.A. and Lefevre, C.M., Deakin University, Pigdons Road, Geelong VIC 3016, Australia; kevin.nicholas@deakin.edu.au

Hormones and extracellular matrix regulate mammary development and function during the lactation cycle. However, it is becoming apparent that milk plays a central role not only in providing nutrition and programming signals to the young but also in regulating the development of the mammary gland. With the availability of comparative genomics and bioinformatics, research is increasingly turning to alternative models to better understand the relationships between these regulatory processes. Animal models with extreme adaptation to lactation allow researchers to more easily identify mechanisms that control mammary function and that are present, but not readily apparent, in eutherian species. Lactation has evolved in monotremes, marsupials and eutherians by exploiting a diverse range of strategies. For example the monotremes are egg laying mammals and the mother begins to produce milk when the immature young hatch. Marsupials have a very short gestation and a relatively long lactation. They give birth to an altricial young and change the composition of milk progressively during lactation to regulate growth and development of the young. Therefore there is greater postnatal investment in development of the young in both marsupials and monotremes. In contrast, eutherians have a long gestation relative to lactation and the majority of development of the young occurs in utero. However, there is considerable adaption to lactation in eutherians. For example, the fur seal has a lactation characterised by a repeated cycle of long at-sea foraging trips (up to 28 days) alternating with short suckling periods of 2-3 days ashore. Lactation almost ceases while the seal is off shore but the mammary gland does not progress to apoptosis and involution. Exploiting the comparative genomics of lactation in these species provides new opportunities to identify key genes and mechanisms regulating mammary gland development, milk production and composition, and the subsequent role of milk bioactives in signalling growth in the suckled young.

Molecular profiling of putative bovine mammary stem cells and identification of potential biomarkers

Choudhary, R.K.[1], Li, R.W.[2] and Capuco, A.V.[2], [1]University of Maryland, ANSC, Bulding 142, 20740 College Park, USA, [2]USDA-ARS, BFGL, Powder Mill Rd, Bldg 200, 20705 Beltsville, USA; vetdrrkc@gmail.com

Knowledge of mammary stem cell (MaSC) biology is critical for understanding mammary epithelial growth, homeostasis and tissue repair. Our objective was to evaluate the molecular profiles of putative MaSC and control cells and to indentify potential MaSC biomarkers, based upon genes that are differentially expressed by these cells. Putative MaSC were identified by their ability to retain bromodeoxyuridine (BrdU) for an extended period. Five Holstein calves were injected with BrdU, tissue was harvested 45 d later and label retaining epithelial cells (LREC) were identified in mammary cryosections by immunostaining. Using laser microdissection, LREC from basal (LRECb) and from embedded (LRECe) layers of mammary epithelium were isolated along with adjacent control epithelial cells (EC). Cells in each of the four categories were lysed, cDNA synthesized, amplified and labeled for microarray hybridization. Data analysis revealed genes that were differentially expressed ($P \leq 0.05$; ≥ 2-fold change) between LREC and EC. The LRECb displayed increased expression of genes that were associated with stem cell attributes and enrichment of WNT, TGF-β and MAPK pathways of self-renewal and proliferation. Comparison between LRECb and LRECe indicated that LRECb possessed more stem cell attributes and LRECe more differentiation markers. Data support the hypothesis that BrdU label retention identifies stem and progenitor cells, wherein MaSC (LRECb) are located in the basal region of the mammary epithelium and committed progenitor cells (LRECe) are localized in more apical layers. This study also identified potential novel biomarkers of MaSC/progenitor cells, including NR5A2, NUP153, FNDC3B, HNF4A and USP15. Abundance and localization of cells that were immunologically positive for these markers were similar to that of LRECs. Insight into the biology of MaSC will be gained by confirmation and utilization of such biomarkers.

Effect of milking frequency and nutrition in pasture-based dairy cows during an extended lactation
Rius, A.G., Phyn, C.V.C., Kay, J.K. and Roche, J.R., DairyNZ, Cnr Ruakura & Morrinsville Roads, Newstead, Private Bag 3221, Hamilton 3240, New Zealand; Agustin.Rius@dairynz.co.nz

Extended lactations (EL) of >600 DIM could benefit seasonal dairy systems by improving animal welfare and reducing breeding-related costs; however, lower milk yields on an annualized basis limits farmer adoption. This study determined if temporary increases in milking frequency (MF) and nutrition increased milk production during an EL. Non-pregnant, multiparous Holstein-Friesian cows (333 DIM, n=120) were randomly allocated to one of four treatments (n=30 cows) for 68 d in a 2×2 factorial arrangement which included two MF and two diets. Treatments were cows (1) milked twice daily (2X) and fed pasture only (PAS); (2) milked 2X and fed pasture plus 6 kg DM/d of concentrate (CON); (3) milked thrice daily (3X) and fed PAS; and (4) milked 3X and fed CON. Cows were offered a pasture allowance of 23 kg DM/d throughout and milked 2X post-treatment. Statistical analyses were conducted on milk production and body weight (BW) during treatments and an 84-d carry-over period. There were no interactions between MF and diet for each trait. During treatment, cows milked 3X had increased yields of milk (13.3 vs. 12.0 kg/d; $P<0.003$) and lactose (0.602 vs. 0.565 kg/d; $P<0.05$), but not energy-corrected milk (14.7 vs. 14.1 kg/d; $P>0.18$), relative to 2X. Cows fed CON increased yields of milk (13.1 vs. 12.1 kg/d; $P<0.01$), protein (0.545 vs. 0.503 kg/d; $P<0.02$), and lactose (0.612 vs. 0.554 kg/d; $P<0.002$), and tended to increase energy-corrected milk (14.8 vs. 14.0 kg/d; $P=0.08$), relative to PAS. Diet and MF did not affect production during the carry-over period or cumulative production from d 1 of treatment to dry-off. Cows fed CON gained more BW (544 vs. 523 kg; $P<0.001$) and the difference remained for 5 weeks post-treatment. In conclusion, short-term 3X milking and CON increased milk production but these responses did not carry-over to improve total EL yields. Imposing treatments longer than 68 d could increase subsequent production responses.

Effect of processing on the immunomodulatory activity of milk proteins
Wheeler, T.T.[1], Gupta, S.[1] and Seyfert, H.-M.[2], [1]AgResearch, Ruakura Research Centre, Private Bag 3123, Hamilton, New Zealand, [2]Leibniz Institute for Farm Animal Biology (FBN), Molecular Biology, Wilhelm-Stahl Allee, 18196 Dummerstorf, Germany; tom.wheeler@agresearch.co.nz

Milk contributes to the development and health of the digestive tract of the newborn through a range of minor milk proteins. Cows' milk has been processed to ensure it does not contain viable microbial pathogens. However, this has the potential to modify the activity of the minor host-defence associated proteins in milk. To address this, we tested the activity of bulk skimmed cows' milk, obtained either before or after factory processing, to modulate *E. coli*-stimulated innate immune responses in cultured bovine mammary epithelial cells. The presence of raw milk, but not processed milk, stimulated the expression of key cytokines and chemokines including TNF-α, IL1A, IL6, IL8 and CCL20 in the absence of heat-killed *E. coli*, indicating that raw milk contains pro-inflammatory substances. Raw milk obtained from cows with mastitis resulted in stimulation of a wider range of cytokines and effector proteins, including TNF-α, IL1A, IL6, IL8, CCL20, serum amyloid A3, and β-defensin, suggesting that infection of the mammary gland results in an enhanced pro-inflammatory activity in milk. In order to investigate the effects of milk processing on the activity of individual milk proteins, lactoferrin was purified from both raw and processed milk, and its immunomodulatory activity assessed. The presence of lactoferrin purified from raw milk resulted in a 2 to 3 fold enhancement of the expression of TNF-α, IL1A, IL6, IL8 and CCL20 in response to heat-killed *E. coli* over a 1 hour period, contradicting some previous reports suggesting that lactoferrin suppresses inflammatory signalling. However, the presence of lactoferrin purified from processed milk resulted in no enhancement, providing a possible explanation for the apparently contradictory reports and suggest that processing of milk could influence the activity of lactoferrin as well as other milk components.

Analysis of primary cilia in the lactating bovine mammary gland
Millier, M.[1], Poole, C.A.[1] and Singh, K.[2], [1]Dunedin School of Medicine, University of Otago, Dunedin, New Zealand, [2]AgResearch Ltd, Ruakura Research Centre, Hamilton, New Zealand; kuljeet.singh@agresearch.co.nz

Primary cilia are multifunctional cellular receptors found on virtually all vertebrate cells, but their role in bovine lactation and involution have not been investigated. To understand the potential role of primary cilia mechanosensation and/or chemoreception in the milk production process, our objective was to analyse primary cilia distribution and morphology in the bovine mammary gland during active lactation and throughout early involution. Mammary gland involution was induced by abrupt termination of milking in non-pregnant Friesian dairy cows at mid-lactation. Alveolar tissue was obtained following slaughter at 6, 12, 18, 24, 36, 72 and 192 h after the last milking (n=6 per group) and was fixed, wax embedded, sectioned at 5µm. Immunohistochemistry was performed to define primary cilia (anti-acetylated tubulin), centrioles (anti-gamma tubulin), myoepithelial cells (anti-smooth muscle actin, (SMA)) and nuclei (Hoechst) (n=3 per group). Fluorescently labelled sections were examined by confocal microscopy and the images analysed for primary cilia distribution and morphology. Primary cilia were identified on the apical surface of mammary epithelia, in SMA-positive myoepithelia and in stromal connective tissue cells at 6 and 12 hours after milking. However, as involution advanced and alveolar remodelling accelerated, fewer primary cilia were identified in any cell type. Epithelial cilia always projected into the alveolar lumen, were generally short, tapered at the distal tip, and were often bent. In myoepithelial and stromal cells, the cilia were of similar morphology but did not show a preferred orientation. Results show that all cell types involved with bovine lactation can express a primary cilium, and that ciliary bending is consistent with mechanotransduction. Apparent differences exist in the ciliary orientation of epithelial and myoepithelial cells, suggesting potential differences in ciliary function for these two critical cells.

Study of the sn-2 fatty acid composition in ovine milk fat enriched with CLA and omega-3
Mele, M., Serra, A., Conte, G. and Secchiari, P., University of Pisa, Dipartimento di Agronomia e Gestione dell'Agroecosistema, Via S. Michele degli Scalzi 2, 56124 Pisa, Italy; mmele@agr.unipi.it

In the last years, several studies have succeeded to enrich milk fat from dairy cattle, sheep and goats with conjugated linoleic acid (CLA) and omega-3 fatty acids (FA), by using unsaturated vegetable or marine oils. However, the effect of the enrichment on the triglyceride (TG) composition has been poorly studied. Since in human intestine FA in the sn-2 position are preferentially absorbed than FA in the sn-1 and sn-3 positions, it is of interest to study the position of CLA and omega-3 FA in the triglycerides (TG) of enriched milk fat, in order to better evaluate the bio-availability of beneficial FA from dairy products. With this aim, individual milk samples from twenty dairy ewes were analyzed in order to determine the total and sn-2 FA composition. Milk samples were obtained from a previous trial where the diet of dairy sheep was supplemented or not with extruded linseed, in order to achieve an enrichment of milk fat with CLA and C18:3 n-3. Ten individual milk samples for each group were collected after 5 weeks of treatment. Monoglycerides (MG) were obtained by incubation of milk TG with porcine pancreatic lipase and the FA composition of total and sn-2 position was obtained by gas-chromatography. The preferential incorporation of individual FA in the sn-2 position was evaluated by the ratio between percentage composition of MG and TG. Data were analyzed by ANOVA taking into consideration diet as a fixed factor (linseed or control diet). In the milk from the linseed diet, the triglyceride contents of CLA and C18:3 n-3 were enriched by 290% and 250%, respectively, whereas saturated FA content decreased by 30%. The content of these FA in the sn-2 MG reflected the TG composition, but the content of SFA in sn-2 MG from enriched milk decreased more than 40%. In the control milk the FA preferentially incorporated in sn-2 were C14:1, C16:1 C14:0, whereas in the enriched milk the FA were C18:2 n-6, C14:1, C14:0.

Production and application of a polyclonal antibody against purified bovine adiponectin
Mielenz, M., Mielenz, B., Kopp, C., Heinz, J., Häussler, S. and Sauerwein, H., Institute of Animal Science, Physiology and Hygiene, Katzenburgweg 7-9, 53115 Bonn, Germany; mmielenz@uni-bonn.de

The highly abundant multimeric adipose-tissue derived protein adiponectin (AdipoQ) is related to glucose- and fatty acid metabolism and exhibits anti-inflammatory properties. Data about AdipoQ in human milk revealed a correlation with early postnatal growth of the infant. Less information is available about this protein in dairy cattle. Therefore, we aimed to produce and apply a polyclonal antibody against bovine AdipoQ. The protein used for immunization was purified from newborn calf serum. Its purity was approved by SDS-PAGE. The specificity of the antibody was checked by Western blot using reducing and denaturing conditions. Using the purified antibody, AdipoQ expression was analyzed by Western blot in serum of different species, in bovine primary preadipocyte culture supernatant, in milk, as well as by immunohistochemistry (IHC) in bovine adipose tissue. Semi-quantitatively, dairy cows' sera (n=6) were analyzed from day 21 ante partum up week 36 of lactation. The mixed model was used for data analysis (repeated measures). Parturition was considered via nested periods. Reducing and/or denaturing conditions substantiated the specificity of the antibody based on the breakdown of the multimeric protein to dimeric (~56 kDa) and monomeric (~28 kDa) AdipoQ. As expected, we observed AdipoQ only in culture supernatant of differentiated but not undifferentiated preadipocytes. Using the produced antibody we were able to show bands of comparable size to the monomeric AdipoQ in serum of several ungulate species as well as of mice, and humans. By applying IHC, a characteristic adipocyte specific staining pattern was observed. Serum AdipoQ values of dairy cows were lower ante than post partum (P=0.001) indicating a relation to the reduced insulin sensitivity and inflammatory conditions during early lactation. The presence of AdipoQ in bovine milk as demonstrated herein supports its role as a bioactive component of cow milk.

mTOR signalling in the developing ovine fetal mammary gland
Sciascia, Q.[1,2], Blair, H.[1], Senna Salerno, M.[3], Pacheco, D.[2] and Mccoard, S.[2], [1]Massey University, International Sheep Research Centre, Private Bag 11222, Palmerston North 4442, New Zealand, [2]AgResearch, Grasslands, Animal Nutrition Team, Animal Nutrition & Health Group, Private Bag 11008, Palmerston North 4472, New Zealand, [3]AgResearch, Ruakura, Growth & Lactation Team, Animal Productivity Group, Private Bag 3123, Hamilton 3240, New Zealand; quentin.sciascia@agresearch.co.nz

The objective of this study was to determine if changes in fetal mammary gland development are associated with changes in biochemical indices and mechanistic target of rapamycin (mTOR) pathway signaling. Archived fetal mammary tissue from day-140 twin fetuses born to dams fed *ad libitum* (AD) or maintenance (M) levels of nutrition from day-21 to day-140 of pregnancy were used. Previous research showed ewes born to M-dams had greater lactose percentage, milk and accumulated lactose yields compared to AD in their first lactation. Total DNA, RNA and protein was extracted from fetal mammary tissue and relative ratios used to determine cell size, number and translational efficiency and capacity. The abundance and phosphorylation status of mTOR signalling targets were evaluated using Western blotting. Statistical analysis was performed using ANOVA procedure in SAS. Fetal mammary glands from AD-dams were heavier (P=0.06) compared to M, and biochemical indices showed they have increased DNA content (P<0.05) but reduced protein synthetic capacity (P<0.05), while protein synthetic efficiency and cell size were unchanged. Western blotting results showed decreased abundance of all targets analysed, eIF4E (P<0.05), eIF4E^{Ser209} (P<0.01), 4EBP1 (P<0.05), 4EBP1^{Ser65} (P=0.05), RPS6 (P<0.01) and RPS6$^{Ser235/236}$ (P<0.05), in mammary tissue of fetuses from AD-dams compared to M. Overall, these results indicate that the mTOR pathway may play a role in fetal mammary gland development by regulating ribosome biogenesis and the abundance and activation state of downstream targets involved in translational, developmental and cell cycle regulation, with possible links to future lactation potential.

Behavior, cortisol concentrations, ewes and lambs performance after ACTH administration and weaning

Negrao, J.A., Rodrigues, A.D. and Stradiotto, M.M., USP/FZEA, University of Sao Paulo, Basic Science Department, Av. Duque de Caxias norte 225, Pirassununga, SP, 13630-970, Brazil; jnegrao@usp.br

There are several different managements that can cause stress and influence behavior. However, it is possible to identify adequate behavior at parturition to improve the performance of the ewe and lamb after weaning. In this context, the objective of this study was to investigate the behavioural changes and cortisol release in sixty Santa Ines ewes and their lambs. Consequently, ewes and lambs were submitted to stress imposed by ACTH administration and weaning (experimental stress). Cortisol concentration was measured before, during and after both experimental stress. Ewes and lambs were in full view throughout the experiment and recordings started immediately after parturition and experimental stress. During this period general activity patterns of each ewes and lamb were observed. Statistical analysis was performed using mixed models, Student´s t-test and Newman-Keuls tests. The relationship between behavior and cortisol was evaluated by Pearson correlation coefficients. Significance was set as $P<0.05$. Regarding ewes and lamb behavior and their interactions after parturition, ewes were classified as calm, agitated and very agitated. Although, the relationships between behavior and cortisol were small for ewes and their lambs, cortisol concentrations in the ewes classified as very agitated were higher when compared with calm ewes. When exposed to ACTH administration and weaning, there was a direct relationship between cortisol and maternal behavior. This indicates that relationships between behavior and cortisol during early lactation may help improve ewes and lamb performance after weaning.

Cortisol and insulin like growth factor I after ACTH administration, weaning and first milking

Negrao, J.A., Delgado, T.F.G. and Gaiato, A.P.R., USP/FZEA, University of Sao Paulo, Basic Science Department, Av. Duque de Caxias norte 225, Pirassununga, SP, 13630-970, Brazil; jnegrao@usp.br

Environmental and management can cause stress and influence behavior and milk production. However, it is possible to identify biomarkers of stress and milk production in dairy goats. In this context, the objective of this study was to investigate the behavioural changes, cortisol and insulin like growth factor I (IGF-I) release in twenty four Saanen goats submitted to stress imposed by: ACTH administration, weaning and first milking (experimental stress). Cortisol concentration was measured before, during and after experimental stress. Goats were in full view throughout the experiment and recordings started immediately after the parturition and experimental stress. During this period general activity patterns of each goat were observed in the milking parlor. Statistical analysis was performed using mixed models, Student´s t-test and Newman-Keuls tests. Significance was set as $P<0.05$. Regarding goats behavior during milking, goats were classified as calm, agitated and very agitated. There was a direct relationship between these classifications and cortisol concentrations during weaning and first milking. Although, there was no significant influence of ACTH administration in IGF-I concentration, high concentrations of IGF-I was positively associated with high milk production. This indicates that IGF-I can act a marker for milk production in goats and cortisol can act as a marker for behavior in milking parlor.

Discovering the unique value of indigenous livestock populations: the opportunities of genomics

Hanotte, O., Ndila, M., Wragg, D. and Mwacharo, J., The University of Nottingham, School of Biology, University park, NG7 2RD Nottingham, United Kingdom; olivier.hanotte@nottingham.ac.uk

The advent of next generation sequencing and associated high throughput genome screening technologies are providing new opportunities in livestock breeding improvements. Their potential application in the areas of genetic characterization in connection to the conservation of indigenous livestock genetic resources is yet poorly documented. Human and natural selection have shaped the genetic make-up of these populations. Besides a fine understanding of haplotypes and nucleotides diversities, at individuals and populations levels, these technologies allow us unravelling the unique adaptive genetic make-up of local livestock populations to their environments and production systems. The detection of signatures of selection across the full genome is offering to stamp out the genetic uniqueness of these populations while valuing them as adaptive traits. It is applicable to reproductively isolated livestock populations, often already characterized intensively; but as well for the larger number of the non-descript indigenous livestock populations for which their genetic uniqueness remained largely hidden behind a mosaics of phenotypic diversity. We will illustrate these points through example from our work in African indigenous cattle, fancy and village chicken, ending by an advocacy of the unique value of indigenous livestock genetic diversity as reservoir of adaptations and unique research models for the mapping of the genetic control of adaptive traits.

Using genomic information provided by selection schemes to assess French dairy breeds diversity

Laloë, D.[1], Allais, S.[2], Baloche, G.[3], Barillet, F.[3], Raoul, J.[4,5] and Danchin-Burge, C.[4,5], [1]INRA, UMR1313 GABI, Domaine de Vilvert, 78352 Jouy en Josas, France, [2]UNCEIA, 149 rue de Bercy, 75595 Paris 12, France, [3]INRA SAGA, BP 52627, 31326 Castanet-Tolosan, France, [4]Institut de l'Elevage, 149 rue de Bercy, 75595 Paris 12, France, [5]Institut de l'Elevage, BP 42118, 31321 Castanet-Tolosan, France; coralie.danchin@idele.fr

In France, selection programs in ruminant species are extremely efficient and a major contributor to the proficiency of the meat and dairy industries. These programs are characterized by the selection of few elite breeding animals. The drawbackis a loss of genetic variability in most breeds, which means that selection programs should take into account this parameter. There is, therefore, a need to provide genetic variability indicators, on a regular basis, so that breed associations can adjust their management accordingly. The aim of the project VARUME (Genetic Variability of RUMinants and Equine species) is to set up an observatory of the genetic variability of the French ruminant and equine species, based on pedigree and molecular data. In dairy cattle and sheep, there are now numerous molecular data generated for the needs of selection programs. The project evaluates the feasibility of setting up a genetic variability observatory based on these SNP data. A first step is to define which type of indicators can be generated from SNP data in order to characterize a breed's diversity. An inventory of all the molecular (SNP) data available to build up the observatory is done, in three cattle breeds (Holstein, Montbéliarde and Normande) and two sheep breeds (Lacaune and Manech Tête Rousse). A list of the best indicators to monitor a breed's diversity is defined by testing them with the various molecular data available and evaluating their usability depending on various contexts. The indicators are also used to target males with outstanding indicators so that their semen can be transferred to the French Cryobank. Finally, the indicators are compared with the ones obtained with pedigree data.

Comparing linkage disequilibrium between taurine and indicine cattle with a high density SNP chip

Pérez O'brien, A.M.[1], Garcia, J.F.[2], Carvalheiro, R.[3], Neves, H.[2], Vantassell, C.[4], Sonstegard, T.[4], Utsunomiya, Y.T.[2], Mcewan, J.C.[5] and Sölkner, J.[1], [1]University of Natural Resources and Applied Life Sciences – BOKU, Gregor-Mendel-Strasse 33, 1180 Vienna, Austria, [2]São Paulo State University – UNESP, Clovis Pestana 793, 16050680 Aracatuba, Brazil, [3]GenSys Consultores Associados, Rio Grande do Sul, 90080000 Porto Alegre, Brazil, [4]USDA, ARS, Beltsville Agricultural Research Center, 10300 Baltimore Ave., 20705 Beltsville, MD, USA, [5]AgResearch, Invermay Agricultural Centre, 55034 Mosgiel, New Zealand; anita_op@students.boku.ac.at

Linkage Disequilibrium (LD) patterns are the basis for the application of Genomic Selection (GS) and Genome-wide association studies (GWAS). This study focused on comparing the extent of LD across different cattle breeds of the Bos indicus and Bos taurus subspecies, using a high density single nucleotide polymorphism (SNP) array (Illumina BovineHD). The extent of LD was characterized in a specific population including analysis on effective population size (Ne) as well as variation and size of haplotype blocks identified for each breed. Four taurine breeds (Angus, Brown Swiss, Holstein Friesian and Fleckvieh) and three indicine breeds (Gir, Nelore, and Brahman) were used in the analysis. Breed representation ranged from 30 to 100 animals. LD decay analysis of HD genotypes for different distances (100kb, 1Mb and 10Mb) and haplotype block construction were based on the r^2 correlation measure. Current and past Ne were estimated using the chromosome segment homozygosity methodology. Levels of LD were lower in indicine breeds over short distances, but stabilized at long distances (>100 kb). These results are consistent with both the larger Ne found for Bos indicus cattle, and the differences found in the length and variation of haplotype blocks. Overall, indicine cattle have dramatically lower levels of LD at distances less than 100 kb. These results indicate the need to use higher density genotyping platforms for GS and GWAS studies in indicine cattle breeds.

Screening of indigenous Czech cattle breeds for polymorphism in innate immunity genes

Novák, K., Institute of Animal Science, Department of Molecular Genetics, Přátelství 815, 104 00 Prague 22, Czech Republic; novak100@centrum.cz

The historical Czech cattle breeds Czech (Bohemian) Red and Czech Pied are included into the national program of Conservation and Use of Farm Animal Genetic Resources. Like other historical breeds of farm animals, they should be considered as a source of variability for the improvement of modern genotypes via introgression of the desirable genes. Both traditional breeds play an important role in enhancing the reservoir of disease resistance genes. The value of additional resistance resources is highlighted by the continually changing microbial scene, limits on the antibiotic use, and the decreasing genetic diversity of modern cattle breeds. Therefore, a screening has been initiated in both the national breeds for the structural polymorphism of genes controlling innate immunity response. The survey is confined to 50 individuals of each breed. In the first phase, the work is concentrated at the genes coding for those TOLL-like receptors that participate in the interactions with the bacterial pathogens, partly, genes TLR1, TLR2, TL4, TLR5 and TLR6. Their exon regions were amplified in 3, 6, 9, 10 and 4 PCR reactions, respectively. The resulting fragments of 400-800 bp are probed for polymorphism using three approaches: direct sequencing with capillary sequencer, heteroduplex detection with endonuclease CEL1 (TILLING), and fluorescent SSCP. The results allow to evaluate the level of polymorphism in the monitored genes and the potential role of historical breeds for resistance breeding. In parallel, the work enables to estimate the efficiency of the cleavage by CEL1 and the SSCP method for SNP discovery compared to the error-free direct sequencing. There is a hope that the structural polymorphism will be linked to health traits in the studied breeds.

Ascertainment bias in the estimation of the effective population size from genome-wide SNP data

Ober, U.[1], Malinowski, A.[2], Schlather, M.[2,3] and Simianer, H.[1], [1]Georg-August-University Goettingen, Department of Animal Sciences, Albrecht-Thaer-Weg 3, 37075 Goettingen, Germany, [2]Georg-August-University Goettingen, Institute for Mathematical Stochastics, Goldschmidtstraße 7, 37077 Goettingen, Germany, [3]University of Mannheim, Institute for Mathematics, B6, 26, 68131 Mannheim, Germany; hsimian@gwdg.de

Effective population size N_e is an important parameter in population genetics and has a direct impact on the achievable accuracy of genomic predictions. With the availability of high throughput genotyping data or whole genome sequences, N_e can be estimated from the linkage disequilibrium (r^2) between pairs of markers that are c Morgan units apart, usually based on the equation $E(r^2) = (1+4N_e c)^{-1}$ suggested by Sved. Although this expression lacks a sound mathematical justification, it was found empirically to provide reasonable results in simulation studies. In practical applications $E(r^2)$ is estimated by the empirical mean of observed r^2 values, which can be shown to be strongly affected by the allele frequency spectrum of the marker panel used. While the true minor allele frequency (MAF) distribution, as e.g. obtained in sequencing studies, usually is skewed with a substantial excess of extreme MAF values, commercial SNP chips are constructed such that the MAF distribution is uniform, so that alleles with extreme MAF are systematically underrepresented. We studied the impact of this practice on N_e estimation using sequence data of the human HAPMAP project, where we sampled alleles from the true skewed distribution or an enforced uniform distribution. Estimates of N_e obtained with an improved version of Sved's underlying recursion formula and using a uniformly distributed SNP panel were found to be up to 30% lower than those obtained from the skewed SNP distribution. This suggests that by establishing uniform MAF distributions, which is common practice in commercial SNP chip construction, a systematic and substantial downward bias is introduced, especially in the estimation of historical effective population sizes.

Genomic management strategies for conservation of endangered livestock populations

Villanueva, B., De Cara, A., Gomez-Romano, F., Fernandez, J., Saura, M., Fernandez, A., Barragan, C. and Rodriguez, M.C., INIA, Crta. de La Coruña km 7.5, 28040 Madrid, Spain; villanueva.beatriz@inia.es

While the application of genome-wide markers in selection programmes of farm animals has received much attention in the last years, few studies have considered this information in conservation programmes where the aim is to maintain the highest possible levels of genetic diversity and limit the increase in inbreeding. Here, we evaluate the benefits of using genomic information for managing endangered populations both through computer simulations and through data analysis of a strain of Iberian pigs (Guadyerbas) facing now a high risk of extinction. Simulations included a set of populations varying in effective population size (N). Two sets of loci were generated (SNP markers used for managing the populations and non-markers used for computing diversity). Management was carried out for 10 generations and implied optimizing contributions of potential parents by minimizing the global coancestry computed either from SNP genotypes or from genealogies. Different genome sizes and marker densities were investigated. Results show that with high marker densities the heterozygosity maintained when using molecular coancestry is always higher than when using genealogical coancestry although differences are small for large genomes. The density required for obtaining a benefit from markers ranged from about 100 to 500 SNP/Morgan for n=20 and n=160, respectively. We also compared molecular and genealogical coancestry for the Guadyerbas strain that has been maintained as a closed herd since 1944 (average n=17). Data available included the complete genealogy and SNP genotypes from the PorcineSNP60 BeadChip for 200 animals. Although molecular coancestry values were higher than genealogical values the correlation between both measures of coancestry was very high (>0.9).

Conserving a single gene versus overall genetic diversity with the help of optimal contributions

Windig, J.J., Hulsegge, I. and Engelsma, K.A., Livestock Research, Wageningen UR, Animal Breeding and Genomics Centre, P.O. Box 65, 8200 AB Lelystad, Netherlands; jack.windig@wur.nl

Frequently specific genes are targeted when livestock breeds are conserved in a gene bank. Examples are the eradication of genetic defects, preservation of a characteristic gene for a breed such as alleles for coat colour or blood groups, or storing an allele that is threatened to be lost in the breed. However, focussing on a single gene may be at the expense of overall genetic diversity. To optimise overall genetic diversity while storing a single allele in a gene bank we extended the optimal contribution method. Now a group of animals can be chosen so that overall genetic diversity is maximised while the frequency of a specific allele is fixed. We studied effects of this method in a small Holstein population genotyped with the 50K SNP chip, with simulations. SNPs with different minor allele frequency were chosen at random across the genome, and stored at a specific frequency in the gene bank while maximising genetic diversity. These simulations show that the amount of genetic diversity that can be conserved, depends on the allele frequency chosen. Particularly when the allele frequency chosen for the gene bank differs widely from the allele frequency in the population the overall genetic diversity can be reduced. Loss of diversity was studied in a practical example, storage of the B19 blood group in the Groningen White headed cattle, as well.

Use of inbred matings for population management in conservation programmes

De Cara, M.A.R.[1], Villanueva, B.[1], Toro, M.A.[2] and Fernández, J.[1], [1]INIA, Dept. Mejora Genética Animal, Ctra Coruña km 7.5, 28040 Madrid, Spain, [2]ETSIA, UPM, 28040 Madrid, Spain; miguel.toro@upm.es

Conservation programmes try to maximise the survival probability of the populations of interest. Their goals are to minimise the loss of genetic diversity, which allows populations to adapt to changes, and to reduce the increase in inbreeding, which can lead to a decreased fitness. An optimal strategy to achieve these goals is to optimise the number of offspring every individual contributes to the next generation by minimising global coancestry. However, this strategy may maintain deleterious mutations in the population, compromising its long-term viability. In order to avoid this, optimal contributions can be combined with inbred matings, to expose and eliminate recessive deleterious mutations by natural selection in a process known as purging. Whether purging by inbred matings is efficient in conservation programmes depends on the balance between the loss of diversity, the increase in fitness and the reduction in mutational load. Some studies have concluded that purged populations experience reduced inbreeding depression, but others have found that purging in small populations increased the probability of extinction. Thus, purging is not commonly used, as its positive benefits are unclear. In this study, we perform computer simulations to determine whether managing a population with inbred matings can help to improve its long-term viability while keeping reasonable levels of diversity. Our results are highly dependent on the genetic architecture and the mutational model assumed for the selected trait, with very different outcomes if this trait is controlled by many loci of small effect compared with a trait controlled by less loci of larger effect. Using molecular coancestry in the management of the population can maintain a larger genetic diversity but leads to a lower fitness than using genealogical coancestry. While this pattern is clear for species with large reproductive rates, it is much less so for species with small reproductive rates.

Estimation of genomic inbreeding in cattle: the impact of density

Ferencakovic, M.[1], Curik, I.[1] and Soelkner, J.[2], [1]Faculty of Agriculture, University of Zagreb, Department of Animal Science, Svetosimunska 25, 10000 Zagreb, Croatia, [2]University of Natural Resources and Life Sciences Vienna, Department of Sustainable Agricultural Systems, Division of Livestock Sciences, Gregor Mendel Str. 33, 1180 Vienna, Austria; mferencakovic@agr.hr

Runs of homozygosity (ROH) were recently proposed as genomic estimate of inbreeding that accounts for stochastic variation. In addition ROHs provide inference of more remote inbreeding than estimates obtained from pedigree. We analyzed impact of SNP chip density [Illumina BovineSNP50 Genotyping BeadChip (54k chip) and BovineHD Genotyping BeadChip (777k chip)] on the inbreeding estimates calculated for 115 Tyrolean Grey bulls. Inbreeding was calculated for whole pedigree (F_{pedT}), for five generation pedigree (F_{ped5}), and from five ROH lengths (k>1Mb, k>2Mb, k>4Mb, k>8Mb, k>16Mb) for both SNP chips (F_{ROHk_54} & F_{ROHk_777}). ROH greater than 1Mb representing overall inbreeding of individual cover (on average) 7.7%, 15.5%, 5.4%, 6.2%, 6.8% and 7.7% of genome for $F_{ROH1\_54\_0}$, $F_{ROH1\_54\_1}$, $F_{ROH1\_777\_1}$, $F_{ROH1\_777\_1}$, $F_{ROH1\_777\_2}$ and $F_{ROH1\_777\_3}$, respectively. Pedigree inbreeding coefficients F_{pedT} (2.2%) and F_{ped5} (1.4%) indicates much lower levels. F_{ROH}s from same lengths, regardless to the SNP chip density, were showing high correlations ranging from 0.90 to 0.99. Correlations of F_{ped} and F_{ROH} were in range between 0.73 to 0.81. We conclude that correlations between pedigree and genomic inbreeding coefficients with respect to density were approximately the same. F_{ROH} estimates of inbreeding coefficients from 777k chip were lower. The total length of ROHs >1Mb is much shorter when considering 777k chip data compared to 54k chip data. Allowing one SNP in a ROH to be heterozyous for 777k chip data results in similar levels of inbreeding as derived from 54k chip data. Given the chance of genotyping errors, it seems advisable to allow one heterozyous SNP in a ROH.

Mitochondrial genetic variation in two Podolian cattle breeds in Croatia

Ivanković, A.[1], Ramljak, J.[1] and Dovč, P.[2], [1]Faculty of Agriculture, Department of animal production and technology, Svetošimunska cesta 25, 10000 Zagreb, Croatia, [2]Biotechnical faculty, Deapartment of animal genetics, Groblje 3, 1230 Rodica, Slovenia; aivankovic@agr.hr

Two Croatian endangered autochthonous cattle breeds, Istrian cattle and Slavonian-Syrmian podolian cattle are included in conservation program and insight in their genetic structure is a basic prerequisite of successful conservation strategies. Mitochondrial DNA represents an informative genetic marker in knowing the genetic structure of the populations. The aim of research was evaluation of genetic variability in two local Podolian cattle breeds based on analysis of the mtDNA D-loop region from 25 animals. Sequencing the 785-bp fragment of mtDNA D-loop region revealed 23 polymorphic sites representing 15 different haplotypes which were clustered into five haplogroups. In sequences of Slavonian-Syrmian Podolian cattle six polymorphic sites were found and in sequences of Istrian cattle 21. The nucleotide sequence diversity in Slavonian-Syrmian Podolian cattle was lower than in population of Istrian cattle, 1.78% and 3.03%, respectively. In Slavonian-Syrmian Podolian population were observed three haplogroups of which two are observed in the population of Istrian cattle. Overall mean of genetic distances in Slavonian-Syrmian Podolian cattle was lower (0.002) than in population Istrian cattle (0.004). Coefficient of differentiation in entire population was 0.081. The high numbers of identified haplotypes in population of Istrian cattle indicate a high genetic variability of the maternal component, while in population of the Slavonian-Syrmian Podolian cattle lower genetic variability was observed. Our results show a relatively moderate genotypic diversity within and between the two analyzed cattle breeds. The genetic diversity information based on mtDNA is important baseline data for the future breed conservation strategy.

Genetic diversity in buffalo populations of Iraq using microsatellite markers
Jaayid, T.A.[1] and Dragh, M.A.[2], [1]College of Agriculture, Animal Production, Basrah University, Basrah, Iraq, [2]College Education, Biology, Amara University, Maysam, Iraq; taleb1968@yahoo.com

This study was conducted in the College of Agriculture, Basrah University, Iraq and Huazhong university, China. Divided Iraq into three main regions included certain provinces: southern area included Basrah, Missan, and Dhi-Qar, middle area included Al-Qadisiyah, Babil, Karbala and Baghdad and northern area included Diyala, Kirkuk and Mosul. The aim of the study was to measure Genetic structure of Iraqi buffalo population using microsatellites techniques. Sixty nine blood and hair samples collected from unrelated animals in the three areas mentioned above. Six microsatellite markers used (ETH125, CSSM060, BM1706, ETH02, ETH225 and INRA005). Polymerase Chain reaction (PCR) was done using specific bovine primers. The PCR products was run on the Genetic analyzer (ABI-3730) including specific programmes for results analysis such as GENEPOP, FSTAT, microsatellite toolkit and MEGA program for genetic map construction. Seventy allele were detected across the 6 loci. Total number of alleles per locus (TNA) varied from 3 (INRA005 locus) to 16 (ETH152 locus). The mean number of allele (MNA) across the 6 loci in Iraqi indigenous buffalo was (11.4). The locus ETH152 was the most polymorphic marker a according to its number of alleles (16), the expected heterozygosity was 0.86 and polymorphism information content 0.80, number of alleles 3, expected heterozygosity 0.1-0.2 and polymorphism information content 0.1- 0.2. Results showed that these markers were suitable in population genetics researches. It was concluded that a high degree of genetic diversity exist in the Iraqi buffalo populations.

Introduction to cattle sector in the Slovak republic
Chovan, V., Slovak Holstein Association, Nadrazna 36, 900 28 Ivanka pri Dunaji, Slovakia; holstein@holstein.sk

After 1989 there was a significant drop in number of cattle in the Slovak republic. Total number in that year was 1,559,395 heads, of which 536,568 cows and 542,092 calves. For comparison, in year 2011 total number of cattle was 467,125, out of it 201,285 cows. Average drop per year from 1989 is 49,820, of which 15,240 cows. As for breed composition, there was a substantial increase in Holstein cows, which in 1990 in pure breed represented only 4.92%. The numbers for 2011 are as follows: Holstein 62.0%, Slovak Simmental 33.92%, Slovak Pinzgau 1.06% and 3.02% Beef cattle. In the system of suckling cows there are 31,461 cows. A remarkable progress was recorded in the average milk recorded production: in 1989 3,602 kg and in 2011 6,769 kg of milk. So the average production increase per year represents 144 kg. The highest milk production was achieved in the pure bred Holsteins (8,087 kg). In 2011 393,275 inseminations were performed, of which 168,365 first service ones. Conception rate after all inseminations was in heifers 55.4%, cows 35.0% and for all females 38.6%. National milk quota for the quota year 2010/11 was 1,082,841 tons of milk. For the whole Slovak republic 818,010 tons of milk was delivered including direct sales, which represents 76.4% of the total capacity. In 2011, 59.8% of the delivered milk was classified in the Q quality, 37.6% in first class quality and substandard quality was represented by 2.6%. Concerning somatic cell count 602,081 inspections were performed. This number represents 66% of all inspections. Somatic cells count was always below 400,000. Concerning the housing technologies, 85.7% of milking cows are housed in free stall premises with milking parlours, while 12.8% of cows are in tie stalls without milking parlours and 1.5% in tie stalls with milking parlours. Nowadays in the Slovak republic 15 milking robots are in operation at 4 farms.

Modern dairy needs an extensive management system
Kiljunen, J., Valio Ltd dairy company, P.O. Box 10, 00039 Valio, Finland; Jaana.Kiljunen@valio.fi

Co-operative dairy industry operates on a wide scale on production and quality management processes to fulfill the owners milk prize as well as consumers needs. The industry is interested in milk quality i.e. somatic cells, bacteria, temperature, spores, milk contents etc. and medicine residues like antibiotics. Management control is based often on programs on farm level, prizing, agreements, education, evaluations and control samples in several points on farm as well as in industrial processes (HACCP). Recent discussions in quality management are dealing with sustainability including animal welfare, environment, food safety, nutrition, local welfare, different footprints and fare trade.

Genetic selection for optimal milk yield and quality in US grazing systems
Schutz, M.M., Purdue University, Department of Animal Sciences, 125 S. Russell St. #105, West Lafayette, IN 47907, USA; mschutz@purdue.edu

Grazing has increased in popularity in the US. However, grazing herds (GH) often forego US Artificial Insemination (AI) sires, in favor of sires proven in other grazing countries or novel breeds. The objective was to investigate the possible existence of genotype-by-environment interaction (GxE) in GH versus confinement herds (CH) in the US and to develop a more suitable selection index based on economic merit in pastoral systems. Estimates of genetic correlations for the traits in GH versus CH were 0.89, 0.88, and 0.91 for milk, fat, and protein, respectively. Within-quartile analyses revealed a lower correlation for milk and protein between the upper and lower GH quartiles, but not for upper and lower CH quartiles. Rank correlations of sire PTA for somatic cell score based on GH versus CH was 0.70. Genetic correlations less than 1.0 suggest re-ranking among sires in environments, while the rank correlations indicate sire re-ranking at extremes within management system. Differences were not explained by adjustment for heterogeneous variance. While evidence for GxE was limited, separate selection indices to optimize merit within production system are warranted, although separate progeny testing programs are not. A Grazing Merit index (GM$) parallel to the US NM$ index was constructed using costs and revenues of US grazing systems. Spearman rank correlation of AI bulls between GM$ and NM$ was 0.93 (P<0.0001). Traits in GM$ (and their percentage of weight) include: milk volume (24%), fat yield (16%), protein yield (4%), productive life (7%), somatic cell count (-8%), Feet and Leg Composite (4%), Body Size Composite (-3%), Udder Composite (7%), daughter pregnancy rate (18%), calving ability (3%), and dairy form (6%). Weights in NM$ were 0, 19, 16, 22, 10, 4, 6, 7, 11, 5, and 0% for the same traits, respectively. It appears that NM$ may provide guidance to select for milk yield and quality in pasture systems, however a GM$ index based on appropriate costs and revenues, is more beneficial.

Session 03

Theatre 4

Proper antibiotics use in animals a social responsibility: case dairy cows

Kuipers, A., Wageningen UR, Expertise Centre for Farm Management and Knowledge Transfer, P.O Box 35, 6700 AA Wageningen, Netherlands; abele.kuipers@wur.nl

Quality schemes deal with measurement of hygiene, cell count, bacteria count, etc., while milk is tested for contaminants, like antibiotics. These days, a careful use of antibiotics is considered a social responsibility. It is linked to human health, because of growing resistance against antibiotics in humans. Especially 3rd and 4th generation antibiotics are needed in hospitals, but are also used in animals. To gain insight, antibiotic use was registered on 114 farms, including 31 FADN farms during 5-6 years. Use was collected by invoices from 50 Veterinary practices serving these farmers. Antibiotic use on farm level is expressed in daily dosages/cow/year. The average over 5 years was 5,80 dosages per cow/year with a spread from 2-12. Daily dosages were split up in contributions to dry cow therapy (43%), mastitis (22%), calves (3%), uterine and after birth treatments (3%) and other diseases (29%). 2/3 of antibiotics is going to the udder. Part of the farmers were guided by means of study groups. These farms showed a decline in antibiotic use in contrast to the others. Farm and herd factors were studied affecting antibiotics use, practising a step-wise regression procedure. Variation in total antibiotics use was explained for 39% (R^2) by such factors, and dry cow therapy use even by 46%. The most determining factors appeared to be quota size, milk amount sold per cow, health status, cell count, and calving interval. Interesting is the significant negative correlation between cell count and level of antibiotic use (-0,55). The question has arisen: which of both is more important in the long run to be reduced for the dairy sector? Analysis showed that the 'more successful' farmers tend to use on average somewhat more antibiotics than the other colleagues. 'More successful' reflects a lower cell count, higher health status of herd and a larger quota. New wishes of society ask the judgement of being a 'more successful' farmer to be redefined. A change in mindset is needed of stakeholders, farmers and veterinaries.

Session 03

Theatre 5

Genetic variance in environmental sensitivity for milk and milk quality in Walloon Holstein cattle

Vandenplas, J.[1,2], Bastin, C.[2], Gengler, N.[2] and Mulder, H.A.[3,4], [1]National Fund for Scientific Research, FNRS, 1000 Brussels, Belgium, [2]University of Liege Gembloux Agro-Bio Tech, Animal Science Unit, 5030 Gembloux, Belgium, [3]Wageningen University, Animal Breeding and Genomics Centre, 6700 AH Wageningen, Netherlands, [4]Wageningen UR Livestock Research, Animal Breeding and Genomics Centre, 8200 AB Lelystad, Netherlands; jvandenplas@ulg.ac.be

Animals that are robust to environmental changes are desirable in the current dairy industry. This difference in environmental sensitivity can be studied through the heterogeneity of residual variance while homogeneous residual variance between animals is usually assumed homogeneous in traditional genetic evaluations. The aim of this study was to study genetic heterogeneity of residual variance by the estimation of variance components in residual variance for 5 milk and milk quality traits. 146,027 test-day records from 26,887 Walloon Holstein first-parity cows in 747 herds were available. All cows had at least 3 records and had a known sire. These sires had at least 10 cows with records and each herd × test-day had at least 5 cows. Five traits, milk yield, somatic cell score, and content in milk (g/dl) of oleic acid, monounsaturated and unsaturated fatty acids, were analyzed separately. Estimation of variance components was performed by running iteratively Expectation Maximization-Restricted Maximum Likelihood algorithm by the implementation of double hierarchical generalized linear models. For all traits, the genetic standard deviation in residual variance (i.e. approximately the genetic coefficient of variation of residual variance) was low and ranged between 0.12 and 0.17. The standard deviations due to herd × test day and permanent environment in residual variance ranged between 0.35 and 0.44 for herd × test-day effect and between 0.55 and 0.96 for permanent environmental effect. This study shows the existence of heterogeneity of residual variance and the existence of some genetic variance in environmental sensitivity for all studied traits in the Walloon Holstein cattle.

A dynamic artificial neural network for the prediction of milk yields from dairy cattle

Murphy, M.D.[1,2], Upton, J.[2], O'Mahony, M.J.[1] and French, P.[2], [1]Cork Institute of Technology, Mechanical Engineering Department, Bishopstown, Cork, Ireland, [2]Teagasc Moorepark, Animal & Grassland Research & Innovation Centre, Kilworth, Fermoy, Co.Cork, Ireland; michael.murphy@teagasc.ie

Accurately predicting daily milk yields from dairy cattle is an inherently difficult undertaking. Pasture based dairy cattle are subject to a host of stochastic conditions (weather, grazing conditions, grass growth, etc...) which dictate the levels of milk production during a typical season. In this study, a dynamic artificial neural network model is proposed to predict annual daily milk yield using daily lactation records pertaining to Holstein Friesian dairy cattle. The network architecture selected was the non-linear auto regressive model with exogenous input (NARX). This model's main advantage over other time series analyses tools is its ability to actively learn and take into account exogenous data that affects the future values of the time series. The daily milk yield is treated as a time series, the cow number is feed through as an exogenous input and a number is chronologically assigned to the day of milking dependent on its position during the cycle. The NARX learns the characteristics of the yield curve relative to time of the year, by doing this, seasonal dependent external inputs such as weather conditions and cow metabolic conditions are indirectly taken into account during the training of the network and implemented through the NARX generalisation abilities. Several model configurations were tested, the quality of each network was measured using coefficients of determination, sum of the squared errors, and root mean square errors. Three previous lactation cycles with corresponding cow numbers were used to train the network. The model was then used to predict milk yield for both an entire annual lactation cycle and a continuous 50 day rolling cycle. The model was capable of predicting daily milk yield with a mean average percentage error of 11.48 and 6.11 for a full lactation cycle and a 50 day rolling period respectively.

Differences in bovine milk fat composition among dairy breeds in the Netherlands

Maurice - Van Eijndhoven, M.H.T.[1,2], Bovenhuis, H.[2], Soyeurt, H.[3,4] and Calus, M.P.L.[1], [1]Wageningen Livestock Research, Animal Breeding and Genomics Centre, P.O. Box 65, 8200 AB Lelystad, Netherlands, [2]Wageningen University, Animal Breeding and Genomics Centre, P.O. Box 338, 6700 AH Wageningen, Netherlands, [3]National Fund for Scientific Research, rue d'Egmont 5, 1000 Brussels, Belgium, [4]Gembloux Agro-Bio Tech, University of Liège, Animal Science Unit, Passage des Déportés 2, 5030 Gembloux, Belgium; myrthe.maurice-vaneijndhoven@wur.nl

Knowledge on variation in detailed milk fat composition among cows is of interest for the dairy industry because fat composition is associated with processability, human health and also methane emission. The aim of our study is to quantify breed differences in fat composition among different cattle breeds in the Netherlands. We included milk samples of Holstein Friesian (HF), Meuse-Rhine-Yssel (MRY), Dutch Friesian (DF) Groningen White Headed (G), and Jersey (JER) cows. In total 159,437 records of 99,250 purebred and crossbred cows were included. The detailed fat composition in g/dl milk was predicted using MIR profiles from routine milk recordings based on calibration equations developed in the EU project RobustMilk. Breed effects were estimated by calculating predicted means for fatty acids (FA) using an animal model. For FA which arise in milk through de novo synthesis (short chain FA, C12:0, C14:0, and partly C16:0) differences in breed effects were found for JER and G. Also breed differences were found for FA that arise in milk directly from cows diet (C18:2cis9,12 and C18:3cis9,12,15) and FA in which Δ^9-desaturases plays a role (C14:1, C16:1, and C18:1). In general differences between HF, MRY and DF in fat composition were not significant. JER tended to produce more saturated FA, while G tended to produce relatively less saturated FA. Adjusting for differences in total fat content in the model showed that differences in detailed fat composition can to a large extent be explained by differences in total fat content.

Genetic relation between composition of bovine milk fat in winter and summer

Duchemin, S.I.[1], Bovenhuis, H.[1], Stoop, W.M.[2], Bouwman, A.C.[1], Van Arendonk, J.A.M.[1] and Visker, M.H.P.W.[1], [1]Wageningen University, Animal Breeding and Genomics Centre, De Elst 1, Building 122, 6708 WD Wageningen, Netherlands, [2]Cooperative Cattle Improvement Organization – CRV, P.O. Box 454, 6800 AL Arnhem, Netherlands; sandrine.duchemin@wur.nl

Milk fat composition shows substantial seasonal variation. In this study we investigated if milk fat composition in winter is regulated by the same genes as milk fat composition in summer. For this purpose, we: (1) estimated genetic correlations between winter and summer milk fat composition; and (2) tested for genotype by season interactions of DGAT1 and SCD1 polymorphisms. Milk samples were obtained from about 2,000 first lactation Dutch Holstein Friesian cows, most of which with records in both winter and summer. Milk fat composition was measured by gas chromatography. Summer milk contained higher amounts of unsaturated fatty acids (FA) and lower amounts of saturated FA compared to winter milk. Heritability estimates were comparable between seasons: moderate to high for short and medium chain FA (0.35 to 0.72) and moderate for long chain FA (0.18 to 0.42) in both seasons. Genetic correlations between winter and summer milk were high (0.70 to 1.00), indicating that milk fat composition in winter and in summer is genetically the same trait. For most FA effects of DGAT1 K232A and SCD1 A293V polymorphisms were similar across seasons. Significant DGAT1 by season interaction was found for some FA and SCD1 by season interaction was only found for C18:1trans-11. Genotype by season interactions were due to scaling rather than to re-ranking, probably caused by differences in the diets of the dairy cows between seasons.

Mid-infrared prediction of milk titratable acidity and its genetic variability in first-parity cows

Colinet, F.G.[1], Vanlierde, A.[1,2], Vanden Bossche, S.[2], Sindic, M.[2], Dehareng, F.[3], Sinnaeve, G.[3], Vandenplas, J.[1,4], Soyeurt, H.[1,4], Bastin, C.[1] and Gengler, N.[1], [1]University of Liege, Gembloux Agro-Bio Tech, Animal Science Unit, 5030 Gembloux, Belgium, [2]University of Liege, Gembloux Agro-Bio Tech, Analysis, Quality and Risk Unit, 5030 Gembloux, Belgium, [3]Walloon Agricultural Research Centre, Valorisation of Agricultural Products Department, 5030 Gembloux, Belgium, [4]National Fund for Scientific Research, FNRS, 1000 Brussels, Belgium; Frederic.Colinet@ulg.ac.be

Coagulation of milk has a direct effect on cheese yield. Among several parameters, titratable acidity of milk (TA) influences all the phases of milk coagulation. In order to study the genetic variability of this trait on a large scale, mid-infrared (MIR) chemometric methods were used to predict TA. A total of 507 milk samples collected in the Walloon Region of Belgium from individual cows (Holstein, Red-Holstein, Dual Purpose Belgian Blue, Montbeliarde, and Jersey) were analyzed using a MIR spectrometer. TA was recorded as Dornic degree. An equation to predict TA from milk MIR spectrum was developed using PLS regression after a first derivative pre-treatment applied to the spectra. During the calibration process, 45 outliers were detected and removed. The TA mean of the final calibration set was 16.62 (standard deviation (SD)=1.80). The coefficient of determination (R^2) was 0.82 for the calibration with a standard error (SE) of 0.76. A cross-validation (cv) was performed (R^2_{cv}=0.81 with SE_{cv}=0.80). This equation was then applied on the spectral database generated during the Walloon routine milk recording. The variances components were estimated by REML using single-trait random regression animal test-day model. The dataset included 33,717 records from 9,191 Holstein first-parity cows; the TA mean was 17.05 (SD=1.35) and TA ranged from 12.83 to 20.87. Estimated daily heritabilities ranged from 0.43 at 5th day in milk to 0.59 at 215th day in milk indicating potential of selection. Further research will study phenotypic and genetic correlations between TA and milk production traits.

Genetics of the mineral contents in bovine milk predicted by mid-infrared spectrometry

Soyeurt, H.[1,2], Dehareng, F.[3], Romnee, J.-M.[3], Gengler, N.[2] and Dardenne, P.[3], [1]National Fund for Scientific Research, Rue d'Egmont 5, 1000 Brussels, Belgium, [2]University of Liège, Gembloux Agro-Bio Tech, Animal Science Unit, Passage des déportés 2, 5030 Gembloux, Belgium, [3]Agricultural Walloon Research Centre, Valorisation of Agricultural Products Department, Chaussée de Namur 24, 5030 Gembloux, Belgium; hsoyeurt@ulg.ac.be

Knowing the contents of minerals in milk like Ca or Na could be interesting to improve the nutritional quality of milk and to assess the animal health status. This study had two aims: (1) development of mid-infrared equations for mineral contents in milk by using an approach combining multiple countries, breeds, and production systems; and (2) study of the genetic variability of these traits in the Walloon Holstein dairy cattle. Samples included in the calibration set were collected in Belgium, Luxembourg and France over 5 years. The calibration set included at least 400 samples analyzed by coupled plasma atomic emission spectrometry to quantify the contents of Na, Ca, Mg, P and K. The calibration coefficient of determination ranged between 0.69 for K and 0.93 for Na. The standard error of cross-validation was 63.35, 49.24, 64.33, 7.04, and 93.22 mg/kg of milk for Na, Ca, P, Mg and K. From these results, the quantification of milk minerals by mid-infrared is feasible. These equations were applied to more than 140,000 spectral records collected from 43,797 first parity Holstein cows in 1,233 herds. The variance components were estimated using Gibbs Sampling using single trait random regression models derived from the one used for the Walloon genetic evaluation of milk production traits. First results gave a daily heritability of 0.26 for Na, 0.45 for Ca, 0.48 for P, 0.46 for Mg, and 0.41 for K. Moderate negative genetic correlations were found between Na and the other studied traits. The highest correlation (0.69) was observed between P and Mg. These results confirmed the genetic variability of these traits. Further studies will be conducted to study the relationship between these traits other traits (e.g., production, health).

Genetics of the mid-infrared predicted lactoferrin content in milk of dairy cows

Bastin, C.[1], Leclercq, G.[1], Soyeurt, H.[1,2] and Gengler, N.[1], [1]University of Liège, Gembloux Agro-Bio Tech, Animal Science Unit, 5030, Gembloux, Belgium, [2]National Fund for Scientific Research (F.R.S.-FNRS), 1000, Brussels, Belgium; catherine.bastin@ulg.ac.be

Lactoferrin is an iron-binding glycoprotein found in bovine milk that presents therapeutic properties. It can be interesting for humans, as a biologically active food component. Also lactoferrin concentration was shown to be higher in milk from (sub)clinical mastitic cows and it could be considered as an indicator of udder health. The objectives of this study were to estimate genetic parameters of the mid-infrared prediction of lactoferrin content in milk (MIRLf; mg/l) and to estimate its genetic correlations with milk, fat, and protein yields, somatic cell count (SCS), and contents in milk (g/dl) of 7 groups of fatty acids (FA) predicted by mid-infrared spectrometry (saturated, monounsaturated, polyunsaturated, unsaturated, short chain, medium chain, and long chain). Variance components were estimated using a series of 11 two-trait random regression models. Dataset included records from 4928 first-parity Walloon Holstein cows and contained 88,270 milk, fat, and protein records; 85,664 SCS records; 62,628 MIRLf records; and from 61,834 to 61,974 FA records according to the trait. Heritability estimates for MIRLf ranged from 0.20 at 5 days in milk (DIM) to 0.40 at 250 DIM. Genetic correlation of MIRLf with yields decreased along the lactation; from 0.50 to -0.50 for milk, from 0.40 to -0.10 for fat, and from 0.55 to -0.25 for protein. Genetic correlation between MIRLf and SCS increased along the lactation from 0.10 at 5 DIM to 0.30 at 305 DIM. This positive correlation substantiated the potential interest of MIRLf as an indicator of udder health. Average daily genetic correlations between MIRLf and FA ranged from 0.25 to 0.35 and were higher for unsaturated FA. Concomitant selection of MIRLf and unsaturated FA is therefore feasible and this would be beneficial for the genetic improvement of nutraceutical properties of milk.

Genetic analysis of vitamin B12 content of bovine milk

Visker, M.H.P.W.[1], Rutten, M.J.M.[1], Sprong, R.C.[2], Bouwman, A.C.[1] and Van Arendonk, J.A.M.[1], [1]Wageningen University, Animal Breeding and Genomics Centre, De Elst 1, 6708 WD Wageningen, Netherlands, [2]NIZO food research, Kernhemseweg 2, 6718 ZB Ede, Netherlands; marleen.visker@wur.nl

Vitamin B12 is an essential component of the human diet, and milk and dairy products are important sources of this vitamin. Current intake levels of vitamin B12 are suboptimal, therefore, we have established whether vitamin B12 content of milk can be improved through selective breeding. Vitamin B12 content was measured in milk samples of 544 Dutch Holstein Friesian cows: content was 4.4 µg/l on average, and varied between 1.0 and 12.9 µg/l. This variation between cows could to a large extend be attributed to genetic factors (h^2=0.37). The presence of genetic variation suggests that improvement of milk vitamin B12 content through breeding is possible. Presence of genetic variation also justified more in-depth genetic analysis. Therefore, an association analysis was performed in which the bovine genome was screened in order to identify chromosomal regions that affect vitamin B12 content of milk. This association analysis comprised 487 animals with phenotypes for milk vitamin B12content and genotypes for 49,994 molecular (SNP) markers. Significant associations with milk vitamin B12 content ($P<0.001$) were found on 22 Bos taurus (BTA) chromosomes. Clusters of significantly associated SNP were found on BTA 10, 13, 14, and 26. These associations suggest that genes that affect vitamin B12 content of milk are located on these chromosomes. However, confirmation of these results is needed. Our genetic analysis shows that vitamin B12 content of bovine milk can be improved through selective breeding. Our association analysis is a first step towards identification of genes that affect vitamin B12 content of milk. Such genes may be located on chromosomes 10, 13, 14, and 26.

Milk production black and white cows' in Lithuania

Sileika, A., Juozaitiene, V., Zostautiene, V., Muzikevicius, A. and Jurgaitis, A., Lithuanian University of Health Sciences, Veterinary Academy, Tilzes 18, 47181 Kaunas, Lithuania; arunas@zum.lt

Dairy production in Lithuania is one of the most important traditional agricultural industries. After Lithuania's accession to the EU cow under control productivity every year increased on average 169 kg of milk, kg (R^2=0.97). Much attention is paid to breeding programs for milk composition and to longevity. Lithuanian dairy cattle population is dominated by Black and White cattle, which make up about 74% cows under control in the country. To improve this breed is used Holstein cattle from USA, Canada, EU and other countries. We examined 61,214 Black and White cows' genealogy data and identified, that the Black and White cattle population of Lithuania has an average of 64.5±0.09% of Holstein genes. Cows were stratified by Holstein breed gene content to classes with class interval 12.5%. Linear regression analysis showed that increasing amount of Holstein breed genes of one class (12.5%), improves Black and White cows productivity (y = 139.72x + 5331.2; R^2=0.852), milk fat (y = 5.2298x + 235.2; R^2=0.7342) and milk protein production (y = 4.275x + 177.78; R^2=0.7799). Also found a quantity of Holstein breed genes with negative dependence on the duration of economic use of cow's productivity live in months (y = -1.6745x + 38.513; R^2=0.6926).

Feed as a risk factor for raw milk contamination with *Listeria monocytogenes*

Konosonoka, I.H., Jemeljanovs, A., Ikauniece, D. and Gulbe, G., Latvia University of Agriculture, Research Institute of Biotechnology and Veterinary Medicine Sigra, Instituta street 1, 2150 Sigulda, Latvia; biolab.sigra@lis.lv

Dairy farming is the leading sector of the Latvia agriculture. Enhanced nutrition qualities, taste, and health benefits are the reasons for increase interest in raw milk consumption. Unfortunately, milk is a good source of nutrients not only for humans but also for numerous microorganisms, which thus can grow in milk. Feed is a risk factor for poisoning the farm environment thus also raw milk with pathogenic microorganisms of Listeria genus species. The objective of this study was to clarify incidence of bacteria from the genus *Listeria* (L.) int.al. foodborne pathogen *L. monocytogenes* in the feed and raw milk from one organic and three conventional dairy farms in Latvia. In total, 167 feed samples and 244 bulk tank milk samples were analyzed. Feed and milk samples were collected randomly in all seasons of year. *L. monocytogenes* from samples were isolated in accordance with standard LVS EN ISO 11290-1+A1. Presumptive *L. monocytogenes* isolates were purified and confirmed by Fourier transform infrared spectroscopy technique. *L. ivanovii*, *L. innocua* and *L. seeligeri* were isolated from 12.0%, but *L. monocytogenes* from 19.8% of feed samples. Most often feed concentrates (7.8%) and silage (6.0%) were contaminated with *L. monocytogenes*. *Listeria* genus species were isolated more often from feed prepared and used in organic dairy farm than from that used in conventional dairy farms, correspondingly 44.4% and 18.3%. The obtained results show that both *Listeria* spp. and *L. monocytogenes* are found more often in feed prepared in organic farm (correspondingly 14.8% and 29.6%) than in feed used in conventional farms (correspondingly 5.2% and 13.1%). Although different feed samples contained *L. monocytogenes*, no such bacteria were found in samples of bulk milk from organic farm (n=33), but in samples of bulk milk from conventional farms *L. monocytogenes* found three times or in 1.4% of all cases (n=211).

Relation of somatic cell score and natural antibodies with functional longevity in Holstein-Friesian

Xue, P., Van Der Poel, J.J., Heuven, H.C.M., Ducro, B.J. and Van Arendonk, J.A.M., Wageningen University, Animal Breeding and Genomics Centre, P.O. Box 338, 6700AH, Wageningen, Netherlands; peng.xue@wur.nl

Innate immunity plays an important role in preventing and combating infection. Natural antibody (NAb) is an important humoral component of innate immunity. Within early predictors, immune related factors have the highest correlation with longevity. In this study the relation of NAbs and SCS in milk with functional longevity of Holstein-Friesian heifers was investigated, as well as the influence of the cows' sub-mastitic status on this relationship. Milk samples were collected from 1,939 heifers in 398 commercial herds in February/March 2005. SCS, titers of total and isotypes (IgA, IgM and IgG1) of NAb binding keyhole limpet hemocyanin (KLH), lipoteichoic acid (LTA), lipopolysaccharide (LPS) and peptidoglycan (PGN), were measured in milk samples. Cows were divided into a healthy and a sub-mastitic group based on SCC threshold of 150,000 cells/ml in their milk samples. Longevity status was based on the cows' presence in the herd at the end of February 2010. Hazard models were applied to get hazard ratios (HR)of immune-related parameters on risk of being culled. Results showed that high SCS significantly increased the risk of being culled, and nearly all NAb titers had hazard ratio(HR)>1 of which 3 were statistically significant. In the healthy cows group SCS had no significant effect on longevity, and IgA binding KLH was the only NAb which significantly increased HR. In the subclinical mastitis group, only average SCS of the whole lactation had a significant higher HR, and IgG1-LTA had a significant lower HR. In general, most immune-related parameters had negative effect on functional longevity. Among the studied predictors for functional longevity, SCS-related parameters were better predictors than NAbs, and multiple measured SCS had better predictive ability. The sub-mastitic status of the heifer influenced the effects of SCS and NAbs on functional longevity.

Effect of dietary organic zinc (Zn-Methionine) supplementation in dairy cows

Fantuz, F.[1], Beghelli, D.[1], Mocchegiani, E.[2], Malavolta, M.[2], Lebboroni, G.[1] and Renieri, C.[1], [1]University of Camerino, Dept. of Environmental Sciences, via Gentile III da Varano, 62032 Camerino, Italy, [2]Italian National Research Center on Aging (INRCA), Immunology Center, Research Department, via Birarelli 8, 60100 Ancona, Italy; francesco.fantuz@unicam.it

In human nutrition dietary zinc, whose intake is often inadequate in elderly, may promote healthy aging. The aim of this study was to increase the milk zinc (Zn) content by dietary supplementation with organic zinc (zinc-methionine), also investigating the effect of such supplementation on milk yield and composition and on milk and blood antioxidant status. In a 3 month study, 12 Simmenthal cows (4th month from parturition) were used to provide milk and blood samples. Animals were divided into 2 groups, control (CTL) and treated (Zn-Meth): control cows received a total mixed ratio (TMR) including 3 kg of commercial concentrate added with 330 mg inorganic Zn (Zn sulphate)/kg. Treated cows received the same TMR but commercial concentrate was added with 165 mg inorganic Zn (Zn sulphate) and 165 mg organic Zn (zn-methionine)/kg. Milk production was recorded every 2 weeks and individual milk and blood samples were collected and analysed for Zn concentration (ICP-MS) and redox status (colorimetric) by total antioxidant power (AOP) and reactive oxygen metabolites (ROMs) evaluation. ROMs were measured in blood serum, whereas milk samples were also analysed for total solids, protein, fat, ash content and AOP. Data were analysed by analysis of variance for repeated measures. Supplementation of 50% inorganic and 50% organic Zn did not affect the Zn concentration in milk (CTL 3240.4 vs Zn-Meth 3266.3 µg/l, SE=131.1) and blood serum (CTL 705.1 vs Zn-Meth 692.2 µg/l, SE=25.5). Milk yield and composition, serum ROMs (CTL 71.0 vs Zn-Meth 68.4 Carratelli Units, SE 2.7), and milk (CTL 146.1 vs Zn-Meth 141.8 µmol HClO/ml) and serum (CTL 431.1 vs Zn-Meth 444.0 µmol HClO/ml, SE=3.9) AOP, were not affected by the dietary treatment.

Effect of different selenium sources on blood and milk selenium levels in dairy cows

Brucker, L.[1], Schenkel, H.[1] and Warren, H.[2], [1]University of Hohenheim, Landesanstalt für Landwirtschaftliche Chemie, 70599 Stuttgart, Germany, [2]Alltech Biotechnology Centre, Dunboyne, Co Meath, Ireland; hwarren@alltech.com

Research has highlighted the potential positive effects of feeding organic compared with inorganic selenium (Se) sources. The present work is a field study, involving 15 farms from Bavaria, was carried out to investigate the influence of differing sources of dietary Se on whole blood, plasma and milk Se contents, as well as GSHPx activity. Following an initial synchronisation phase, during which all animals (n=150) were offered 0.3 mg/kg of Se as sodium selenite (NaSe), animals were allocated to one of three dietary treatments differing in their Se source: NaSe, rumen-stable NaSe or Se yeast (Sel-Plex®, Alltech Inc., KY) provided at a rate of 0.3 mg/kg of Se via the mineral feed. All diets were analysed monthly for Se level. Initial whole blood, plasma and milk Se levels were determined in 10 animals per farm and animals remained on treatment for 105d after which final whole blood, plasma and milk Se levels were determined. All animals had high blood Se levels prior to the experimental (93% of animals were between 80-110 µg/l). Despite this, there were significant differences between all sources of Se for whole blood and plasma. Mean whole blood Se levels differed significantly ($P<0.05$) at 190, 232 and 255 µg/l for NaSe, rumen-stable NaSe and Se yeast, respectively. Mean plasma Se contents were significantly ($P<0.001$) different at 74, 89 and 108 µg/l for NaSe, rumen-stable NaSe and Se yeast, respectively. Significant ($P<0.05$) differences were also noted between the NaSe and both treatment groups for GSHPx activity at 280, 312 and 322 U/g Hb for NaSe, rumen-stable NaSe and Se yeast, respectively. With regards to milk Se, the two inorganic Se sources had a significantly ($P<0.001$) lower mean of 0.027 compared with 0.055 µg/ml for the Se yeast. These data suggest that significant enrichment of milk Se can only be achieved via organic Se sources.

Preliminary results on zinc, copper, selenium and manganese concentration in donkey milk
Fantuz, F.[1], Ferraro, S.[1], Todini, L.[1], Mariani, P.[1], Piloni, R.[1] and Salimei, E.[2], [1]Università di Camerino, Dipartimento di Scienze ambientali, via Gentile III da Varano, 62032 Camerino, Italy, [2]Università degli Studi del Molise, Dipartimento di Scienze e Tecnologie Agroalimentari, Ambientali e Microbiologiche, via De Sanctis, 86100 Campobasso, Italy; francesco.fantuz@unicam.it

Donkey milk is well tolerated by infants with cow milk protein allergy. Only few or no data are available on trace elements content of donkey milk. Aim of this study was to determine the content of Zinc (Zn), Copper (Cu), Selenium (Se) and Manganese (Mn) in donkey milk. Individual milk samples were obtained from 16 lactating donkeys (46-72 days from parturition) fed 8 kg of coarse hay and 2.5 kg of commercial concentrate daily. During the experimental period (3 months), foals were separated from the dams 3h before mechanical milking (11:00 a.m.). Individual milk samples were collected every 2 weeks (6 sampling time). Milk samples were analyzed for Zn, Cu, Se, and Mn concentration by inductively coupled plasma-Mass Spectrometry. Average Zn, Cu, Se and Mn milk concentration (mean±SE) was respectively 2246.92±8.12 μg/l, 97.57±2.98 μg/l, 4.48±0.15 μg/l and 3.57±0.11 μg/l. The concentration of all elements was significantly affected by the stage of lactation. A clear decrease of approximately 38% was observed for Cu (from 122 to 73 μg/l) and Se (from 5.31 to 3.3 μg/l) with the advancing of lactation. On the contrary, Zn and Mn tended to increase during the study. Milk Zn was positively correlated with Mn (r=0.40) and inversely correlated with Cu (r=-0.28), and Se (r=-0.24). Copper was positively correlated with Se (r=0.63) and negatively correlated with Mn (r=-0.40). Correlations were also significant between Mn and Se (r=-0.27).

Polymorphisam of kappa kasein and milk procesing to cheese
Nemeš, Ž.[1], Vidović, V.[2], Lukač, D.[2], Stojčević Maletić, J.[3] and Stupar, M.[2], [1]Pik Bečej, Dairy production, Topolski put bb, 21220 Bečej, Serbia, [2]Faculty of Agriculture, Animal science, Trg Dositeja Obradovića 8, 21000 Novi Sad, Serbia, [3]Faculty of Medicine, Biochemistry, Hajduk Veljkova 3, 21000 Novi Sad, Serbia; vidovic.vitomir@gmail.com

The aim of this study was to perform identification of alleles and genotypes of k-casein on the DNA level and to study the relationship of different genotypes of k-casein with milk processing to cheese. K-casein genotyping was performed on 54 cows Holstein breed and based on the analysis of DNA from blood using PCR – RFLP techniques. Determining the impact of different factors on the speed of coagulation of milk has been investigated by the method of least square, LSMLMW. Using the identified alleles (A and B) and genotypes (AA, AB and BB) k-casein evaluated their frequency in the examined population of cows. Frequency k-casein genotype was: 0.447 AA +0.497 AB + 0.074 BB = 1.00. Furthermore, the frequency of alleles A and B which is derived from frequencies of genotypes was 0.68 for allele A and 0.32 for allele B. This relationship expresses the preliminary information about presence of different genotypes k-casein in Holstein cattle. Of all examined factors (lactation number, calving season, stage of lactation, the amount of calcium) k-casein genotype and total casein had the highest and highly significant impact on the speed of coagulation of milk. Casein BB genotype had the fastest time of coagulation (about 3 times faster than the AA genotype, and about two times faster than the AB genotype). Cheeses obtained from milk k-BB genotype were classified into groups of semi-hard cheeses (% water ranged from 55.63 to 52.60), while the cheeses obtained from milk AA and AB k-casein genotypes classified in the group of hard cheese (54.72 to 58.57% of water). The greatest amount of milk needed to produce one kg of cheese was from the cow k- AA genotype (4.69 kg). The highest cheese yield of 10.43% was achieved with the k-AB genotype.

Comparison between automated and manual systems of sheep milk collection in Tuscany (Italy)
Lombardo, A.[1], Dal Pra, A.[1], Bozzi, R.[2], Gradassi, S.[1], Amatiste, S.[1], Piazza, A.[1] and Brajon, G.[1], [1]Istituto Zooprofilattico Sperimentale delle Regioni Lazio e Toscana, Via di Castelpulci, 50010 San Martino alla Palma, Scandicci, Florence, Italy, [2]University of Florence, Dept. Agricultural Biotechnology, Via delle Cascine 5, 50100 Florence, Italy; lombardo.andrea@yahoo.it

In order to guarantee traceability in the sheep dairy production chain, the measurement of the milk produced in each farm represents a critical and troublesome phase. In Tuscany (Italy), the most common system to measure milk production in sheep farms is the manual one, by means of calibrated bars installed at farm tanks. Recently a fully-automated system consisting of pumps, installed on tank trucks, enable to measure milk quantity, pH and temperature, and also to collect samples for quality analysis. In the present study, the two measurement systems for milk collection (manual and automated) have been compared on 46 sheep dairy farms in Tuscany. Data concerning milk quantity (litres) and quality parameters (fat, protein, lactose, somatic cell count and total bacterial count) that were measured automatically and manually, were then submitted to statistical analysis (Pearson correlation coefficient). Results showed an excellent concordance ($P<0.0001$) for almost each of the analysed parameters; the diffusion of automated systems for sheep milk collection may therefore represent an important tool to guarantee traceability and to obtain the certification under the regulation UNI EN ISO 22005/2008.

A quality-based system for differentiated payment of sheep milk in Tuscany
Lombardo, A.[1], Bozzi, R.[2], Dal Pra, A.[1], Ragona, G.[1], Paladini, I.[1] and Brajon, G.[1], [1]Istituto Zooprofilattico Sperimentale delle Regioni Lazio e Toscana, Via di Castelpulci, 50010 San Martino alla Palma, Scandicci, Florence, Italy, [2]University of Florence, Dept. Agricultural Biotechnology, Viale delle Cascine 5, 50144 Florence, Italy; lombardo.andrea@yahoo.it

In Tuscany (Italy) the sheep dairy sector is broadly developed; milk price fluctuations represent a critical point in the difficult relationships between farmers and cheese factories. The object of the study was to define a quality-based system for differentiated payment of sheep milk in Tuscany. Fat and protein content (%) values, collected from 2008 to 2011 and belonged to 11 dairy factories were employed for a total of 19,782 analysed samples. The weighted average was computed for fat (X) and protein (Y), on the litres of milk in the farm tank at the sampling time. For both parameters (fat and protein), starting from the mean X and Y values, one neutral class and 4 awarding and 4 penalising classes have been hypothesized with an increasing/decreasing of 0.3% for fat and 0.2% for protein. The application of such a system all over Tuscany could represent a challenge for farmers to increase their breeding and productive standards.

Changes in work in livestock farms: from general trends to the implications on LFS approaches
Dedieu, B.[1], Cournut, S.[2], Fiorelli, C.[1], Hostiou, N.[1] and Madelrieux, S.[3], [1]INRA, SAD, Metafort Theix, 63122 Saint Genes Champanelle, France, [2]Vetagro Sup, Metafort Marmilhat, 63370 Lempdes, France, [3]Irstea, DTM, BP 76, 38402 Saint Martin D'hères, France; dedieu@clermont.inra.fr

Changes in work in livestock farms are firstly demographic, with the high decrease of farms' number and agricultural workers' most of the European countries have faced, and the related huge labor productivity gains. Whereas the couple form of farming is still dominant, associative forms (in France for example) and entrepreneurial units with wage earners are now on the scope and rapidly increasing. Changes are also technical in relation with the different long term pathways farms are engaged in ('get big, special or diversified'), and with the European regulations pressures. Changes are at least social, because farmers aim to have better-working conditions, as the other socio-professional categories and express a desire to separate their private and working lives. In France, a research, education and extension networking has been developed in order to include work prospects into scenarios for the future of livestock farming and to initiate a common understanding about the interactions between farmers as work organizer, the whole workforce and animal production management. Several work models, which underlie research and operational tools for 'work advice in livestock farms', coexist in approaches to analyze how livestock farming systems (LFS) operate. We describe three models: 'resource work', 'organized work', 'work and subjectivity' and the operational indicators they built. We specify the way in which LFS research has mobilized social science disciplines (ergonomy, economics, sociology) to construct them. We illustrate how they complement each other to examine the changes in livestock farming systems, the ways in which the farming activity is carried out and the different levers farmers can use to improve their situations, in different countries including Southern. Consequences on extensionists skills and mode of actions will be presented.

Implications of two dynamics of changes (specialization and diversification) on work organization
Hostiou, N.[1], Cialdella, N.[2,3], Le Gal, P.Y.[3], Vazquez, V.[3,4] and Muller, A.G.[2], [1]Umr Métafort, INRA, 63122 Saint Genès Champanelle, France, [2]Embrapa Cerrados, BR 020 Km 18, CEP 73310-970 Planaltina, Brazil, [3]Cirad, UMR Innovation, 73 Avenue Jean-François Breton, 34398 Montpellier, France, [4]Montpellier SupAgro, 2, place Pierre Viala 34060 Montpellier Cedex 2, 34060 Montpellier Cedex 2, France; nhostiou@clermont.inra.fr

Development's opportunities for family farms often present the intensification of a specialized production and diversification of livestock and crop systems as two opposed pathways to enhance agricultural viability. In the Brazilian Cerrados, family agriculture is an exemplar case of these two dynamics. On the one hand, a structured and growing milk chain supply offers a real opportunity for smallholder farms. More and more farms, from agrarian reform settlements, specialize and intensify in dairy production. On the other hand, opportunities for diversification exist too, thanks to the support policy of family farms of the Brazilian government. We studied work organization of 15 dairy farmers, in an agrarian reform settlement in the municipality of Unaí, using a work method assessment developed in France (QuaeWork). We supposed that the dynamics of changes tending to specialization or diversification have different implications on work organization during the year. Our results show that the dynamics of development have direct implications on the work organization, as work durations, work distribution over the year and the calculated time available for the farmers (room for manoeuvre in time). For a given family, depending on workforce availability and equipment, one or another dynamic of change could be possible. Work organization induces possible evolution of family farms. These results could help research and development organizations to support family farms dynamics.

Work organisation in livestock farms and farm liveability

Cournut, S.[1] and Chauvat, S.[2], [1]VetAgro Sup, UMR 1273 Métafort, Campus agronomique, 89 boulevard de l'Europe, BP 35, 63370 Lempdes, France, [2]Institut de l'Elevage, Supagro, 2 place Pierre Viala, 34060 Montpellier cedex 1, France; sylvie.cournut@vetagro-sup.fr

Qualifying the liveability of livestock farms from the work viewpoint is becoming an important issue in the present context where the renewal of breeders is highly questioned. We analysed the diversity of work organisation in 630 French livestock farms from varied production sectors (dairy and beef cattle, dairy and meat sheep, goats, pig and poultry) specialized or mixed. We used the Work Assessment method, which quantifies the work duration and evaluates room for manoeuvre time for farmers. The objectives of the study was to describe the diversity of work organisation and put in light the differences or proximities between production sectors. Results showed that the production sector highly orientates work organization in livestock farms, and that every production sector had its specificities linked to different technical but also socio-cultural models. Some proximities have been founded between herbivore and monogastric sectors. The evaluation of the room for manoeuvre showed that for 30% of the farms surveyed, the liveability is questioned. By describing the diversity of logics of work organisation, we showed that there are different ways to decrease the routine work load and get some available time. We identified four logics of work organisation leading to contrasted room for manoeuvre which result from complex interactions between dimension and combination of activities, and collective workforce configuration. We discussed the contribution of such results for Extension and Development.

Labour and attitude, important factors in development strategies of Dutch pig and poultry farmers

Eilers, C.H.A.M., Oosting, S.J., Boumans, I.J.M. and Cornelissen, L., Wageningen University, Animal Production Systems, Department of Animal Sciences, De Elst 1, 6708 WD Wageningen, Netherlands; karen.eilers@wur.nl

Pig and poultry farmers choose often for a modernization development strategy, which consists mainly of scale enlargement and specialization. A multifunctional agriculture strategy (MFA), in which activities such as recreation, home sales and care are developed are less common among those farmers. In the last 10 years, however, Dutch pig and poultry farmers are changing their strategy, following an earlier trend of multifunctional activities on crop and dairy farms. To understand the factors that influence the choice of a pig or poultry farmer for a certain development strategy 22 pig farmers and 20 laying hen farmers are interviewed, half of which had home sales and thus followed the MFA strategy. Creating labour opportunities for family members, mainly the partner of the farmer, was a strong incentive to start a multifunctional activity. The total number of persons working on MFA farms and the involvement of the partner was significantly higher than on conventional farms. Most important factor influencing the choice for a certain strategy was the attitude of the farmer. MFA farmers agreed significantly more with the 'social attitude' than conventional farmers, i.e. they found the preservation of nature, sustainable production, bringing citizens and pig/poultry farming together, regional selling and regional cooperation with other farmers, processors etc. very important on their farms. Farmers that chose for MFA strategy are still small in number supplying for a niche market. The whole pig and poultry sector, however, profits from the positive image that is created among citizens visiting those MFA farms.

Work efficiency and work flexibility in organic sheep farms differentiated by reproduction rhythm
Hostiou, N., INRA, UMR Métafort, 63122 Saint Genès Champanelle, France; nhostiou@clermont.inra.fr

Changes affecting livestock farming systems have made labour a central concern for both the sector and the farmers. The challenge lies in increasing labour productivity to improve work conditions. Farmers voice new work expectations particularly a demand to separate their private lives from their work as controlling working hours and being able to have free weekends and holidays. In order to provide viable livestock farming systems, work productivity and work flexibility can be analysed together. A work organization survey was carried on 15 organic sheep farms in the Massif Central region in France. The sample was characterized by different reproductions rhythms (one to three lambings/year). Routine work (work to be done every day) varied from 78 to 4,343 hours/year. This variability can be explained by the number of workers, the sheep breeding and the level of equipment. 96% of the routine work was carried out by farmers themselves. Only one farmer employed a permanent wage earner. The efficiency of the routine work was also diverse (1 to 7.8 hours/ewe/y). It was better in large herds. The calculated time available (room for manoeuvre in time) varied from 454 to 1,199 hours/farm/year. Three annual distributions of the calculated time available were identified according to the reproduction rhythm of the herd (periods of lambing) and to the production of forage (silage, hay making). Adaptive capacities were assessed by adjustments of forms of work organization. The results highlighted diverse adaptive capacities: the stability or the constant adjustments over the year to diverse events (climatic, workforce, livestock management, etc.). The reproduction rhythm can be a lever of work flexibility for farmers. Other factors are also important: other activities on the farm (transformation, direct selling, etc.), workforce composition and level of equipment. These results could help research and development organizations to design farming systems that are better aligned to the quality-of-life expectations voiced by farmers.

Livestock farming systems in emerging and developing countries: trends, roles and goals
Zjalic, M., EAAP, European federation of Animal Scientists, Via G. Tomassetti 3, 00161 Rome, Italy; zjalic@libero.it

Farming systems in Central and Eastern European countries (CEECs) are characterized by a dual structure: there are large corporate farms (CF) and small family farms (FF). The size of FF varies between 0.38 ha (household plots in Russian Federation: RF) to 42 ha (Slovakia) and 58 ha (private farms in RF). The proportion of agricultural land owned by FF varies between 12% (Slovakia) and 94% (Slovenia). The share of FF in total animal production in majority of CEECs (including the Russian Federation) is higher then in plant production and in land ownership. The number of small and medium size FF with livestock production in EU MS from the region is declining parallel with an increase in livestock farm size and milk and meat yield. In the period 1989-1999 the total per caput agricultural production declined in CEEC and CIS by 16 and 40% respectively while the animal production marked 21 and 50% decline. In 2010, meat production was 20% higher then in 2000, but still under the level of 1989. Milk production remained stable. Sustainability of small to medium (SM) livestock farms will depend on a number of factors, such as growing per caput consumption of animal products, potential of niche markets of local, organic and specific products, availability of up and down-stream services, technology development including the use of genomics and trends in agrarian policies. Higher costs of production of SM farms in mountain and less advantaged areas in the EU MS could be off-set by income support, rural development policies and recognition of additional non-productive roles such as protection and management of the environment and farm animal biodiversity. The market share of SM farms will further decline. New technologies, use of genomics and innovations in corporate management will contribute to an increase in the volume of production and the market share of the CF sector.

War and transition: livestock farming systems trends, roles and goals in Croatia
Stokovic, I.[1], Kostelic, A.[2] and Matkovic, K.[1], [1]University of Zagreb Faculty for Veterinary Medicine, Heinzelova 55, 10000 Zagreb, Croatia, [2]University of Zagreb Faculty of Agriculture, Svetosimunska 25, 10000 Zagreb, Croatia; igor.stokovic@vef.hr

Croatian livestock sector endured two huge turnovers – war (1990-1995) and post-war transition (1996-2010). The biggest drop in livestock numbers happened in 1992 and continuously dropping until 1998 that could be contributed to the war events. In 1991 National cattle breeding program was established (still in force). Number of cattle was still dropping until 2002 after that increase was obvious till 2006 with continuous fall after which could be contributed to transition. The similar trends could be observed for sheep, pig and poultry populations. The fact is we never reached pre-war numbers again. Agriculture and livestock production had very important role in pre-war Croatia and was supposed to play important role further more. Instead of that they almost collapsed. In 2000 there was 3.16 million ha of arable land in use but in 2010 there was1.33 million ha still in use. In spite of proclaimed goals of livestock sector intensification and concentration and in spite of huge amount of money invested, Croatian livestock sector stayed mostly in small households (almost 80% of cattle on farms with less than 11 cows). In last decade we also witnessed expected fall in number of households involved in agricultural production (in 2003 we had almost 60,000 households delivering milk and in 2010 only 17,000 left). Considering global trends of intensification and concentration and in the same time viability of production we came to the crossroad (crux mortis of agriculture). Less people is involved in agricultural production earning less money for their hard work and more people living in the city's earning more money wanting to pay less for food but forgetting how hard it is to produce it. Some authors came to conclusion that smaller production units are more flexible because of their lover fixed costs but still forcing concentration in this turbulent time.

Dynamics of livestock farming in extensive livestock territories: what processes are going on?
Cournut, S.[1], Madelrieux, S.[2], Rapey, H.[3], Nozières, M.O.[4], Poccard-Chapuis, R.[5], Corniaux, C.[6], Choisis, J.P.[7] and Ryshawy, J.[7], [1]VetAgro Sup, UMR 1273 Métafort, BP 35, 63370 Lempdes, France, [2]Irstea, DTM, BP 76, 38402 Saint-Martin d'Hères, France, [3]Irstea, UMR Métafort, BP50085, 63172 Aubière, France, [4]INRA, UMR Selmet, 2 place Viala, 34060 Montpellier, France, [5]CIRAD, ES, Embrapa Amazônia Oriental, Belem, Brazil, [6]CIRAD, ES, BP 1813, Bmako, Mali, [7]INRA, UMR 1201 Dynafor, BP 52 627, 31 326 Castanet-Tolosan, France; sylvie.cournut@vetagro-sup.fr

The dynamics of family livestock farmiing in extensive livestock farming territories are crucial for the future of these territories. Indeed livestock farming is an activity anchored in a society, with its market sectors and its local environment. It provides products and multiple services: social and economic dynamics, desirable landscape, biodiversity. Family livestock farming has to reshape itself, and face up to local factors of change such as demographic evolution or land pressure and global factors like climate or market. The purpose of our communication is to shed light on the processes of transformation of family livestock farming, based on a comparative analysis carried out on 8 extensive livestock farming territories: 2 in South America, 1 in Sub-Saharan Africa, 2 in the Mediterranean area and 3 in the French mountains. We propose a transversal reading of these processes based on three complementary considerations: the link between family and livestock farming, the link of the farm to space and finally the link of the livestock farm to market sectors. This reading informs us on the diversity of adaptation dynamics of family livestock farms. In all the territories we can demonstrate the influence of two driving forces for change on the construction of this diversity of family livestock farm adaptation dynamics. These driving forces are globalisation and territorialisation. We propose a stylised representation of this and discuss the tensions and/or complementarities which this diversity of dynamics creates within the territories.

The LIVCAF framework to analyse constraints for livestock production development in the tropics

Oosting, S.J., Wageningen University, Animal Production Systems group, P.O. Box 338, 6700 AH Wageningen, Netherlands; simon.oosting@wur.nl

To analyse constraints for development of livestock production in the tropics from the perspective of the farming system the so-called 'livestock value chain analytical framework (LIVCAF)' was developed. The framework distinguishes 4 major value chains for livestock products (the rural-rural, the rural-urban, the urban-urban and the import-urban) each predominated by a specific livestock production system, with in turn, different roles and functions of livestock, different constraints for development and different stakeholders for each of these chain-specific livestock farming systems. On the basis of the framework it is hypothesised that important constraints for development of livestock production from mixed crop-livestock farms is the competition between crops and livestock, the availability of land and the competition between value chains.

Crossbreeding as innovation for dairy systems in the tropics: case study Amhara region, Ethiopia

Roschinsky, R.[1], Sölkner, J.[1], Puskur, R.[2] and Wurzinger, M.[1], [1]BOKU – University of Natural Resources and Life Sciences Vienna, Gregor Mendel Strasse 33, 1180 Vienna, Austria, [2]ILRI-International Livestock Research Institute, P.O. Box 5689, Addis Ababa, Ethiopia; romana.roschinsky@boku.ac.at

Livelihoods of many resource-poor farmers in Ethiopia depend on livestock. The availability of productive breeds is a major development constraint. Crossbreeding of local, well adapted breeds with high-yielding dairy breeds from temperate zones is encouraged by the government to ensure productivity growth of the dairy sector within a short period of time. A slow up-take rate has been reported and a substantial information gap has been identified. The aim of the study is to create empirical knowledge on the development of crossbreeding by smallholder dairy cattle keepers in Amhara Region, Ethiopia. The results contribute insights necessary to ensure sustainability of crossbreeding programs for dairy cattle by means of analysing adoption, adaptation and farmers' breeding experiments. Pretested, semi-structured questionnaires have been used to interview 62 farmers in 4 districts. Socioeconomic data has been collected to ascertain factors influencing change of production systems at farm level. Data on herd structure and history have been collected to depict the development of breeding over time. In addition, in-depth interviews were carried out with non-farming actors of the crossbreeding network. First results show that the provision of crossbred animals alone is not enough, but also training has to be provided to farmers to improve the herd management (feeding strategies, health interventions, housing). Farmers are only likely to adopt crossbreeding if there is a secure market for their products and if necessary production inputs are available. The access to AI service has been identified as one crucial, but often weak, point and therefore farmers seek alternative sources and use crossbred bulls from the vicinity. The highly unreliable support of extension services and veterinarians is also seen as a hampering factor to the adoption of crossbreeding.

Effect of genetic improvement of sheep in Ethiopia: systems analysis
Gebre, K.T.[1], Wurzinger, M.[1], Gizaw, S.[2], Haile, A.[3], Rischkowsky, B.[3] and Sölkner, J.[1], [1]BOKU-University of Natural Resources and Life Sciences, Gregor-Mendel-Strasse 33, 1180 Vienna, Austria, [2]ARARI-Amhara Reginal Agricultural Research Institute, P.O. Box, Debre Berhan, Ethiopia, [3]ICARDA-International Centre for Agricultural Research in the Dry Areas, P.O. Box, Aleppo, Syrian Arab Republic; maria.wurzinger@boku.ac.at

In the cool highlands of Menz area the predominant production system is a mixed sheep-crop production system. Due to frost and unreliable rainfalls crop production is limited and therefore many families depend on food aid. These effects are becoming more severe as climate changes and human population increases. Consequently, some farmers tend to shift from the sheep-crop production system to sole sheep production. Thus, to recommend the appropriate management system in this agro-ecological zone, understanding of the current production system and the interrelationship between system components are required. This study compares the the the ecological sustainability and economic viability of alternative production systems. System dynamics considers dynamic interaction between the elements of the studied system and can help to understand their behaviour over time with the use of stocks and flows. Dynamic system modelling software STELLA 9.0.2. (High performance Systems, Inc., Hanover, New Hampshire) is used. The system is rainfall driven, the model includes sheep herd, range and crop components as well as an economical module. Data are collected via questionnaires, interviews and physical mapping of two reference villages (Mehal-Meda and Molale) that are part of a community based sheep breeding program. Reliable information about sheep herd structure and performance data is available from this program. Meteorological data are available from weather stations. Ecological and economic analyses are carried out for the reference model and for each scenario; this will also lead to develop policy recommendations for the region.

Adoption of exotic chickens in rural areas of Ethiopia: implication for breed introduction
Woldegiorgiss, W.E.[1,2], Tadelle, D.[3], Van Der Waaij, E.H.[2] and Van Arendonk, J.A.M.[2], [1]Debre Zeit Agricultural Research Centre, Poultry Research Case Team, Debre Zeit, P.O. Box 32, Ethiopia, [2]Wageningen University, Animal Breeding and Genomics Centre, Wageningen, De Elst 1, 6708 WD, Netherlands, [3]International Livestock Research Institute, Biotech theme, Addis Ababa, P.O. Box 5689, Ethiopia; wondmeneh.esatu@wur.nl

This study examines factors that affect the probability and intensity of adoption of exotic chicken breeds among rural poultry producers in Ethiopia. Primary data were collected from 240 respondents using a structured questionnaire after the households were selected by systematic random sampling. A Heckman selection two-step model was used to identify factors that affect the probability and intensity of adoption. Statistically significant differences (P<0.001) were observed between adopters and non-adopters for social contact, income from livestock, off-farm participation, credit access and compatibility(easiness to manage). Probability of adoption was positively affected by off-farm participation and negatively by income from livestock (P<0.05) and gender (P<0.001). Farm size and income from livestock negatively affected the intensity of adoption (P<0.05). With this study we have generated insight in factors that determine a successful adoption of an improved poultry breed. The results of this study will be used in the set-up of a project which aims at successful adoption of a genetically improved indigenous poultry breed.

Winter daily live weight gain in females calves: effects on productive and reproductive parameters
Quintans, G.[1], Scarsi, A.[1], López-Mazz, C.[2] and Velazco, J.I.[1], [1]National Institute for Agricultural Research, Beef and Wool, Ruta 8, Km 281, Treinta y Tres, 33000, Uruguay, [2]School of Agronomy, Garzón 780, 12900, Uruguay; gquintans@tyt.inia.org.uy

Winter daily weight gain (DLWG) is a good predictor of fertility in heifers that are first mated at 15 or 18 month old. The aim of this experiment was to evaluate two different winter DLWG in females calves and its effect on their performace. Twenty eight females calves (AAxHH) of 8m and weighing 158 ± 5.6kg, were assigned to 2 treatments during winter (120d) to achieve low and moderate DLWG: (1) grazed *Lotus* Maku at a low forage allowance, 5% = 5 kg DM/100kg LW (LFA, n=14); (2) grazed *Lotus* Maku at a high forage allowance, 15% (HFA, n=14). Before and after the differential management, all animals grazed together L. Maku at 7%. Animals were weighed biweekly from 8 to 26m old. Hip height (HH) and Rump Fat Thickness (RFT) were measured at 8, 12, 15, 18 and 24 m. and presence of corpus luteum was assessed by ultrasonography at 12, 15, 18, 24 and 26 m. BW, HH and RFT were analyzed by repeated measures using the MIXED procedure with time as the repeated effect. Probability of cycling cows was fitted using the GENMOD procedure with the binomial distribution. Heifers DLWG during winter was -0.128 ± 0.03 and 0.140 ± 0.03 kg/a/d for LFA and HFA, respectively (P<0.001). From 12 until 26m, heifers in LFA tended (P=0.06) to present higher DLWG than those in HFA (0.534 vs. 0.495 kg/a/d). At 12m female calves in HFA weighed more (P<0.0001) than those in LFA (183 ± 5.6 vs. 146 ± 5.6 kg), but no differences were found at 15, 18, 24 and 26m. HH was similar between heifers at 8 and 24 m, but HFA heifers had greater HH at 12, 15, and 18 m. RFT was greater in HFA only at 24 m (2.7 vs. 2.4 mm). No differences in ovarian cyclicity were found between heifers of both groups until 26 m when there was a higher probability (P=0.056) of cycling heifers in HFA than in LFA (36 vs. 7%).Winter management of young female calves would affect their future productivity and reproductive performance.

The role of subsidies in beef cattle farms in SW Spain: a comparison of organic and extensive farms
Escribano, A.J., Gaspar, P., Pulido, A.F., Mesías, F.J., Escribano, M., Pulido, F. and Rodríguez De Ledesma, A., Universidad de Extremadura, Escuela de Ingenierías Agrarias, Ctra. Cáceres s/n, 06071 Badajoz, Spain; pgaspar@unex.es

Livestock farming systems of the Spanish rangelands (dehesas) are traditional systems characteristic of the Iberian Peninsula where native herbaceous vegetation and evergreen species of Quercus provide the foundation for extensive beef farming enterprises. Dehesas are considered to be the most extensive, diverse, and low-intensity land use systems in Europe. Nevertheless, in the last twenty years, the Common Agricultural Policy (CAP) has produced some important effects in these systems, specially an increase of the stocking rate. Although CAP policies contributed to stabilizing income in dehesa farms, they also produced a loss of sustainability. Nevertheless, EU subsidies have become a key element in these enterprises, accounting for as much as 30% of the farms' revenues. In recent years, many of these traditional beef farms have turned to organic production trying to take advantage both of new CAP subsidies and of new market trends. Theoretically, once these farms are organically certified, they can increase their profits by accessing niche markets with higher prices. Often things do not happen this way, and farms are forced to higher expenses on input purchases and greater reliance on subsidies if they want to remain in operation. In the present context, in which the maintenance of agricultural subsidies in their current state is not clear –though with some confidence a reduction can be expected in the medium term – these farms may face serious problems. Therefore, this paper examines the economic indicators of 90 conventional and organic dehesa beef farms in Extremadura (SW Spain), developing a comparative study. Specifically, the role of subsidies in both types of farms is reported, and strategies for the future proposed based on alternative CAP scenarios.

Cattle systems in Misaje area, Cameroon: biomass resource pressure and decentralisation challenges
Bessong Ojong, W. and Piasentier, E., University of Udine, Agricultural and Environmental Sciences Dpt., Via Sondrio 2, 33100, Italy; willington@hygienecameroon.org

Misaje Subdivision is a Sudan savannah grassland of Cameroon at its Northwest frontiers with Nigeria (lat. 6°59' long. 10°55') and an important hotspot for beef-type cattle production; providing livelihood for over 70% of the population. The aim of the study was to audit the prevailing cattle farming systems and profile the challenges on their sustainability. Data set for analysis was obtained from field visits and a questionnaire survey conducted in early 2012. Of 164 farmers surveyed, who managed 213 herds with 17,000 cattle, 88% are landless and predominantly of the minority Mbororo cultural decent, while 12% are indigenous farmers including a 38,000 hectare ranch, breeding 6,200 Goudali cattle, owned by a parastatal; SODEPA. Except for SODEPA ranch, transhumance is the main pastoral system. Three main zebu breeds were identified in transhumant herds. One hundred and forty one herds were of homogeneous breeds (Goudali 27.2%, Aku 22.1%, Djafun 16.9%) while 72 herds where of mixed breeds. Minimum, mean and maximum herd size were 23, 79 and 270 respectively, while most of the herds (55%) fell within 40-80 class size. Shortage of pasture, agro-pastoral conflicts, cattle rustling, and reduced fertility were the determinants for migration during the dry season. Cultivation of cereals; notably maize, was intended for domestic use. Apart from micro-mineral licks, feed supplementation with farm residues is not practiced. The only pasture maintenance action was off-season bush fires. However, a combination of natural pasture rotation and bush fires was noted in SODEPA ranch. Access to land by the Mbororo cattle-rearing minority, in a situation where natural resource management is decentralised, may not be in favour of production. In the absence of specialised seed-stock breeders and organised pasture development and maintenance plans, the prevailing situation is chaotic and unsustainable; lacking essential elements for management and genetic progress.

Evaluation of two different artificial pastures as alternative source for livestock production
Turk, M.[1], Albayrak, S.[1] and Bozkurt, Y.[2], [1]Suleyman Demirel University, Faculty of Agriculture, Department of Field Crops Science, Isparta, 32260, Turkey, [2]Suleyman Demirel University, Faculty of Agriculture, Department of Animal Science, Isparta, 32260, Turkey; yalcinbozkurt@sdu.edu.tr

This research was conducted to evaluate the pastures artificially established in the Mediterranean region in terms of stage of maturity, quality, yield and botanical composition during the 2010 and 2011. For this purpose, a-3ha pasture land was chosen adjacent to the university farm located in Isparta province and divided into two pasture lands (AP1 and AP2 each 1.5ha), and cultivated in March 2010 with two different botanical compositions, AP1, mixtures of *Medicago sativa* L.(20%) + *Bromus inermis* L. (40%) + *Agropyron cristatum* L. (30%) + *Poterium sanguisorba* (10%); and AP2, mixtures of *M. sativa* L. (15%) + *Onobrychis sativa* Lam. (15%) + *A. cristatum* L. (35%) + *B. inermis* L. (35%), respectively. Seven harvesting times starting from the 1st May to the 1st August 2011 were applied and grass samples were collected by using 1m² quadrats fortnightly. Chemical analysis were performed to determine fresh biomass (FB) yield, dry matter (DM) yield, crude protein (CP), acid detergent fibre (ADF) and neutral detergent fibre (NDF) contents. Besides, the changes in dry weight ratios of botanical compositions were also determined at each sample collection. There were significant (P<0.05) differences in FB (24.3 v. 31.2t/ha) and DM yield (10.6 v. 13.8t/ha) between pastures while no significant differences were observed for CP, ADF and NDF. The significant changes were also observed in botanical composition of the pastures. While dry weight ratios of *B. inermis* L. was increased gradually those of the rest of mixtures were decreased. The results indicated that there was a reduction in forage quality at the late stages of vegetation period. Contents of CP ratios decreased with advancing growth while DM yield, ADF and NDF increased in both pastures. It may be concluded that the establishment of AP2 is recommended for the region.

Assessment of morphological and gait scores given by judges in the Spanish Purebred horses' shows
Sánchez, M.J., Gómez, M.D., Azor, P.J., Horcada, A. and Valera, M., ETSIA, University of Seville, Dpt Agro-Forestal Sciences, Ctra. Utrera km 1, 41013 Seville, Spain; v32sagum@gmail.com

The Spanish Purebred (SPB) horse is the most recognized horse breed in Spain. In morphological shows, assessment of morphology is made in a subjective way determined by the experience of the show-judge, who explains most of the phenotypic variance of the 8 morphological scores (MS) and 2 gait score (GS). Moreover, the breeders use MS to achieve high prices and to select future reproducers by morphological criteria. The main aim of this study was to analyze the factors that influence in the MS obtained by the animals in the horse-shows. The quality of the scores given by different judges in 69 horse-shows held between 2002-2010 was analyzed. Information is available for 7,288 horses (3,604 males and 3,684 females), aged between 1-19 years old. They belonged to 1,042 different studs and had 8 different coat colours. Basic statistics for MS and GS were made for the horse shows with 2, 3 and 5 judges, 1,687, 4,482 and 119 horses, respectively. A General lineal model (GLM) was also applied to ascertain the influence of the different environmental factors (sex, colour coat, stud, age and judge) on MS (head-neck, shoulder-withers, chest-thorax, back-loins, croup, hindlimb, forelimb and harmony) and GS (walk and trot). Finally, the Indexes of disagreement (ID) between judges were estimated. Our results showed that the judges did not use all the scale range (from 1 to 10). The maximum score used was 5.5 and 6.0 for GS and MS respectively; the minimum score used was 3.5 and 2.5 for GS and MS, respectively. The mean of the ID was 20.67. The morphological shows with 2, 3 and 5 judges had an ID of 36.18, 14.52 and 22.30 respectively. Almost every environmental factor was significant in the GLM for MS and GS. For example, the judge was significant for all GS and MS traits except for the croup in the shows with 2 judges, for all the GS and MS traits in the shows with 3 judges, and finally, in the case of 5 Judges shows, it was significant only for head-neck and back-loins.

Influence of age and training level on biokinematic and morphometric traits in Menorca horses
Solé, M.[1], Gómez, M.D.[2,3], Galisteo, A.M.[1] and Valera, M.[2], [1]University of Córdoba, Córdoba, 14071, Spain, [2]University of Seville, 41013 Seville, Spain, [3]A.C.P.C. Raza Menorquina, 07760 Ciutadella de Menorca, Spain; ge2sobem@uco.es

The Menorca Horse (PRMe) is an endangered breed mainly located in the Balearic Island used as a sportive and leisure horse. The main aims of this work were to analyze the morphometric and kinematic traits at walk and trot and to determine the influence of age and training level on them. A total of 29 kinematic and 40 morphometric traits were analyzed from 35 PRMe: 17 young (3-5 years old) and 18 adult horses (6-10 years old); 16 trained for dressage and 19 not trained horses. Video records were processed by using a semi-automatic movement analysis system (SOMCAM3D). Basic statistics, Pearson correlations and an ANOVA were carried out to detect differences between age groups and training levels. Significant correlations were attained between morphometric and kinematic traits. At walk, 80% of them were positive (0.3-0.58) and 64% at trot (0.34-0.96), mainly those related with angular traits. Significant differences were also found between training levels and morphometric traits, mainly those related with shoulder and hindlimbs, at walk and trot. In this sense, the age and the combination of both effects had a great influence on the hindlimb angulations at trot. Moreover, significant differences were obtained between age groups, training levels and the combination of both effects, mainly at walk (50%). In this way, young animals showed larger stride duration at both gaits, and a greater forelimb retraction at trot than adult horses. Additionally, trained animals had lower range of movement of scapula, lower hindlimb retraction at walk and higher overtracking and forelimb protraction at trot. Temporal traits were also influenced by training effect at trot. In conclusion, changes over age groups and training levels for morphometric traits and gait quality at walk and trot were detected, which could be useful to define selection procedures in this breed.

The effect of international show jumping performances on the genetic evaluation of Belgian Warmblood
Janssens, S.[1], Aerts, M.[1,2], Volckaert, F.[2] and Buys, N.[1], [1]KU Leuven, Livestock Genetics, Dep Biosystems, Kasteelpark Arenberg 30, 3001 Heverlee, Belgium, [2]KU Leuven, Lab Biodiversity and evolutionary genomics, Dep Biology, De Beriotstraat 32, 3000 Leuven, Belgium; steven.janssens@biw.kuleuven.be

Belgian horses are very successful in show jumping at international level but until now, these international performances have not used for the estimation of breeding values. Current estimated breeding values are based on single performances in national competitions only, both from recreational (LRV) and more advanced level (KBRSF) competitions. The lack of international performances is considered a deficiency by many breeders so a study was conducted to quantify the effect of international performances on the current breeding value estimations for show jumping. Pedigree data on Belgian Warmbloods (n=44,757) and national performance data (707,221 rankings in individual contests) were complemented with a manually constructed dataset containing individual results of BWP horses in international competitions. This manual collection was first compared with a dataset obtained from the FEI but the latter turned out to be largely incomplete. In order to express performances on the same 'scale', competitions were classified according to difficulty (based on obstacle height and competition level) and the 'highest level' achieved by a horse (HLH) was determined. Genetic parameters were estimated in multiple trait models in different datasets to evaluate the effect of the international data on heritabilities, genetic correlations and the ranking of stallions. Inclusion of international performances did increase the estimated heritabilities for performing at national level from 0.055 (LRV) and 0.082 (KBRSF) to 0.082 and 0.090. The heritability for HLH was estimated at 0.399. The genetic correlations between national performances and HLH vary from 0.55 to 0.80 and depending on the data and model, rather large changes were observed on the ebv's of stallions.

Genetic analyses of eventing data of Swedish Warmblood horses
Viklund, Å., Ray, B. and Philipsson, J., Swedish University of Agricultural Sciences, Animal Breeding and Genetics, P.O. Box 7023, 75007 Uppsala, Sweden; asa.viklund@slu.se

Eventing performance is expressed in the breeding objective of the Swedish Warmblood horse (SWB), even though the main objectives are show jumping and dressage. In the SWB population 75% of the horses compete in show jumping, 40% in dressage and 8% in eventing, thus about 20% compete in more than one discipline. Breeding values (BVs) have been estimated for show jumping and dressage with an animal model since 1986, but no genetic analyses had earlier been conducted with eventing data. Data included 3798 horses competing in eventing since the 1970s. 85% of the horses had also competed in show jumping or dressage, and one third had results in all three disciplines. 22% of the horses had been tested at Riding Horse Quality Test (RHQT) for 4-year-old horses. The analyses were performed with uni- and bivariate animal models. The heritability was low (0.12) for lifetime eventing performance. The genetic correlation between eventing and show jumping was 0.44, whereas the correlation between eventing and dressage was zero. For traits tested at RHQT the highest correlations were found between eventing and jumping technique (0.34), jumping temperament (0.41) and canter under rider (0.31), respectively. Conformation traits, walk and trot were only weakly correlated to eventing (-0.15-0.26). The genetic trend has been slow for eventing due to the low heritability in combination with no organized selection for eventing performance. However, a slight positive trend has been observed in recent years, most likely due to a correlated response from selection for show jumping, where the trend has been strongly positive in the last 20-30 years. Among the top stallions 20-40% were Thoroughbreds. Inclusion of breeding strategies for eventing in the breeding program, by publishing BVs for eventing, and use of stallions with high BVs for jumping, could improve the eventing performance in the SWB.

The use of a Tobit-like-Threshold-Model for genetic evaluation of German thoroughbreds
Bugislaus, A.-E.[1] and Reinsch, N.[2], [1]University of Rostock, Justus-von-Liebig Weg 8, 18059 Rostock, Germany, [2]FBN, Wilhelm-Stahl-Allee 2, 18196 Dummerstorf, Germany; antke-elsabe.bugislaus@uni-rostock.de

In previous genetic evaluations of German thoroughbreds, racing performance was described by the trait square root of rank at finish using individual race results from all starting thoroughbreds. In most gallop races in Germany, only the first four ranks receive earnings. Against this background, the question may raise if only thoroughbreds with earnings show their real racing potential, whereas thoroughbreds without earnings are not ridden to their limit at the end of the race. The first objective of this study was to analyze the suitability of a univariate Tobit-like-Threshold-model for genetic evaluation of German thoroughbreds. For this reason, the threshold trait placing status was developed in a problem-oriented manner and genetically analyzed using a Gibbs Sampling algorithm implemented in LMMG. The first, second, third and fourth ranked thoroughbreds obtained values of 4, 3, 2 and 1 for the placing status; all other observations were lumped together into a common class with trait value of 0, thereby treating them as censored observations with the only information of having been slower than their competitors with higher values. Significant fixed effects were sex (3 classes), age (9 classes), year-season (12 seasons), distance of race (4 distance classes), trainer (706 classes), jockey (686 classes), race (6,524 classes) and a fixed linear regression of carried weights. Estimated heritability for the trait placing status was 0.19 (0.03). In comparison, variance components were estimated for the trait square root of rank at finish including results from all starting thoroughbreds in races. Heritability was lower by half with a value of 0.09 (0.01). The rankings of breeding values for the traits placing status and square root of rank at finish showed indeed a high correlation (r=0.95), but differ also significantly. Similar results have previously been obtained in analyses of trotter data.

Modelling the rider effect in the genetic evaluation of dressage in Spanish Purebred Horses
Sánchez, M.J.[1], Valera, M.[1], Gómez, M.D.[1], Cervantes, I.[2] and Gutiérrez, J.P.[2], [1]ETSIA, University of Seville, Dpt Agro-Forestal Sciences, Ctra. Utrera km 1, 41013 Seville, Spain, [2]University of Madrid, Dpt Animal Production, Avda. Puerta de Hierro s/n, 28040 Madrid, Spain; v32sagum@gmail.com

Dressage competition is the most popular sport use of the Spanish Purebred Horse. Tests for young sport horses were established in 2003 in Spain in order to collect data to develop a genetic evaluation for this performance in this breed. The aim of this study was to compare four different models to find out the most appropriate way to include the rider in the genetic evaluation of dressage performance. A total of 8,867 performance records from 1,234 horses, aged between 4-6 years old were used. They were collected between 2003-2011 in the official dressage test for young horses in Spain. The final score in the test was used as performance trait. The pedigree matrix contained 8,487 individuals. A BLUP animal model was applied using a Bayesian approach with TM software. The age (3 levels), sex (3), level of stress (46), training level (41), stud of birth (330), and event (179) were included as fixed effects in all models. Permanent animal and residual random effects were fitted in all models. Different models were compared adding as random effects the following: (1) the genetic effect of the animal; (2) the rider and the animal; (3) the combination rider*animal and the animal; and (4) the combination rider*animal, the rider and the animal. The correlations between the estimated scores under each model and the real scores and other criteria were used to compare models. The results showed that the best model was that fitting the rider and the animal. The heritability under this model was 0.39 and the ratio of rider component was 0.26.

Genetic correlations between indications of imbalance and performance patterns in Warmblood horses

Stock, K.F.[1], Becker, A.-C.[2], Schulze-Schleppinghoff, W.[3], Hahn, A.[3] and Distl, O.[2], [1]Vereinigte Informationssysteme Tierhaltung w.V., Heideweg 1, 27283 Verden / Aller, Germany, [2]University of Veterinary Medicine Hannover, Institute for Animal Breeding and Genetics, Buenteweg 17, 30559 Hannover, Germany, [3]Oldenburg horse breeding society, Grafenhorststrasse 5, 49377 Vechta, Germany; friederike.katharina.stock@vit.de

Routine foal and mare evaluations provide valuable sources of information on movement characteristics in the riding horse. New movement traits have been defined on a descriptive basis using records from routine inspections of the Oldenburg Societies in 2009 and 2010. Genetic analyses implied opportunities to select for improved movement using the new traits and the need of refined correlation analyses with performance traits. To make maximum use of available performance information, the results of the integrated genetic evaluation for riding horses in Germany 2011 were considered for 2,518 foals and 2,748 mares with detailed movement evaluations (DME) and 1,057 sires with DME offspring. Breeding values (EBV) for indications of imbalance (IMB) in foals and mares were tested for their correlations with Integrated EBV for performance traits and derived measures of performance patterns. Differences between EBV for trot and walk (dTW) and trot and canter (dTC) and variance of EBV for gaits (vWTC) served as measures of gait heterogeneity, differences between EBV for performance in young horse and regular competitions (dYC) as measures of performance stability. In 246 sires with ≥ 5 DME offspring, significantly negative correlations were found between dTC and EBV for IMB in foals ($P<0.05$) and mares ($P<0.001$) and between dYC dressage and IMB in mares ($P<0.05$), indicating increasing risk of IMB with increasing difference between EBV for trot and canter and lower genetic potential for dressage performance stability in horses with IMB disposition. Continuation and extension of DME by the Oldenburg Societies will allow inclusion of DME information in future breeding programs to improve performance patterns in the dressage horse.

Mathematical measurement variables for the evaluation of durability of Hungarian show-jumping horses

Posta, J., Mezei, A. and Mihók, S., University of Debrecen, Department of Animal Husbandry, Böszörményi str. 138, 4032 Debrecen, Hungary; postaj@agr.unideb.hu

The main aim of this study was the comparison of different mathematical transformations of the 'number of years in competition' to measure of durability in the genetic evaluation of Hungarian show-jumping horses. Competition results recorded between 1996 and 2009 in show-jumping competitions were used to estimate genetic parameters for number of years in competition. The results were collected by the Show-jumping Group of the Hungarian Horse Breeder's Society. For the estimation of genetic parameters for number of years in competition different linear mixed models were tested with two measurement variables. Measurement variables were constructed used logarithmic transformation of number of years in competition and the fourth-root of number of years in competition. In the first model sex and birth year were taken into account as fixed effects. In the second model an adjustment was made for age at first result in competition. The model fit was compared based on their determination coefficient. Variance components were estimated with VCE-6 software package. The determination coefficients of the models for logarithmic transformation were 0.09 and 0.20, the estimated heritability were 0.27 (0.025) and 0.23 (0.025). The determination coefficients of the models for fourth-root transformation were 0.10 and 0.21, the estimated heritability were 0.26 (0.025) and 0.23 (0.024). Inclusion of age at first result in competition in the evaluation models is suggested and both mathematical functions can be used for transformation of number of years in competition.

Genetic analysis of health status of 4-year-old riding horses

Jönsson, L.[1], Roepstorff, L.[2], Näsholm, A.[1], Egenvall, A.[3], Dalin, G.[2] and Philipsson, J.[1], [1]Swedish University of Agricultural Sciences, Department of Animal Breeding and Genetics, P.O. Box 7023, 75007 Uppsala, Sweden, [2]Swedish University of Agricultural Sciences, Department of Equine Sciences, P.O. Box 7043, 75007 Uppsala, Sweden, [3]Swedish University of Agricultural Sciences, Department of Clinical Sciences, P.O. Box 7054, 75007 Uppsala, Sweden; Lina.Jonsson@slu.se

Musculoskeletal disorders are the most common cause of days lost to training and culling of riding horses. Due to lack of systematic health recording there is lack of knowledge on horse population health status and related heritabilities. The objective was to evaluate the health of young riding horses, where such recording had taken place, and to estimate the genetic predisposition to unsoundness. The Riding Horse Quality Test (RHQT) of 4-year-old Swedish Warmblood horses includes a systematic health examination that until now has not been available for evaluation. Out of 8281 studied horses 74% had one or several remarks on palpatory orthopaedic health, mainly effusions or swellings in locations associated to joints. Flexion test reactions were reported in 21% of the horses, and out of these 20% also had remarks at initial movements and 83% had palpatory orthopaedic health remarks. Mares had significantly more remarks compared to males. Event, including examining veterinarian, had a significant effect on the prevalence indicating that the veterinary examinations could be further standardised. The heritability of palpatory remarks (sum of all remarks) was estimated at 0.12 (s.e. 0.02), using a linear mixed animal model, where remarks of effusion (0.14, s.e. 0.02) and remarks located to joints (0.13, s.e. 0.02) had the highest heritabilities when distinguishing between type of remarks. The heritability of hoof quality remarks was 0.10 (s.e. 0.02). The results suggest large differences in health status between progeny groups and a possibility for improvement of these traits in the population through continued health recording and selection following genetic evaluations.

Variation factors of the use of foal heat in warm- and cold-blooded horse

Langlois, B. and Blouin, C., INRA, Animal Genetics, Domaine de vilvert, 78352 Jouy-en-josas, France; bertrand.langlois@jouy.inra.fr

The French administrative data of the breeding season 1994-2008 were analyzed to study the variation factors of the use of foal heat in breeding warm and coldblooded horses. Dataset consisted of 306,540 intervals foaling-first mating for warm-blood and 204,329 for cold-blood. GLM from SAS was used for computing. The factors of variation were: year, age of the mare, month of mating, breeding area and breed. The same analysis was conducted trough LOGISTIC from SAS on a bimodal variable taking the value 1 if the interval was ≤ 14 days and 0 if > 14 days. The two methods gave exactly the same results. Cold-blooded breeds are using foal heat more intensively and clear differences appear between breeds. However intra-breed genetic analyses using ASREML through a maternal animal model splitting the mare's effect in a additive genetic part and an environmental part showed low heritabilities of the interval foaling first mating (between 0 and 0.10) and non neglectable effects of environment common to the same mare (between 0.03 and 0.11). Seasonal effect and age of the mare (mainly for warm blood) reveals some physiological influences on the occurrence or not of the foal heat. However, the most important factor of variation seems to be the breeding management revealed by the effects of the breeding region and that of the year. The declining use of foal heat with time is worrying and should be reversed. Indeed, this tendency impacts negatively the productivity rate of our breeds by reducing the number of opportunities getting the mare in foal.

Genetic analysis of dystocia in French draft horses

Sabbagh, M.[1], Danvy, S.[1] and Ricard, A.[1,2], [1]Institut français du cheval et de l'équitation (IFCE), Recherche et Innovation, La Jumenterie du Pin, 61310 Exmes, France, [2]INRA, GABI, UMR1313, 78350 Jouy-en-Josas, France; margot.sabbagh@ifce.fr

As in many species, dystocia (foaling difficulty) is a significant cause of female and neonatal death in horse breeding. The aim of this paper was to calculate genetic parameters of dystocia for 4 main populations of draft horses: Bretons (B), Comtois (C), Percheron (P), and as a whole Ardennais, Trait du Nord and Auxois (A). Data included 10,610 (A), 13,274 (P), 35,758 (C) and 38,868 (B) births over 13 years performed by 3,828 (A), 4,928 (P), 12,954 (C) and 12,781 (B) mares. Dystocia was recorded at birth by the breeder in 3 categories: 'easy or without help', 'difficult', 'with veterinarian'. A threshold mixed model was used including fixed effects of year, month of birth, region of birth, age of mare, sex of foal, and 3 random effects: direct genetic effect of the foal, maternal genetic effect of the mare and permanent environmental effect common to the different foalings of the mare. The variance covariance matrix between the two genetic effects involved the relationship matrix calculated with the whole pedigree available so that 18,519 (A) to 56,285 (B) animals were included in the analysis. To estimate the variances and the effects a MCMC and a Gibbs Sampling were used. Main fixed effects were age of mare and sex of foal with a negative effect of young mares (3 and 4 years old) and male foals. Repeatability of dystocia ranged from 0.26 (P) to 0.33 (A). Heritability of direct effect ranged from 0.12 (B) to 0.31 (A), heritability of maternal effect ranged from 0.11 (C) to 0.22 (A), and was always lower except for (B) (0.21). The genetic correlation was unfavorable for 3 populations from -0.42 (P) to -0.27 (A) and neutral for C (-0.02). Estimated breeding values for stallions in activity for the two traits are now available.

The German Riding Pony: a genealogical study and a genetic analysis

Schöpke, K. and Swalve, H.H., Martin-Luther-University Halle-Wittenberg, Institute of Agricultural and Nutritional Sciences, Theodor-Lieser-Str. 11, 06120 Halle, Germany; kati.schoepke@landw.uni-halle.de

The objective of this study was the analysis of a subpopulation of the German Riding Pony. Evaluations were based on performance information from foal and mare inspections as well as from mare performance tests from 1990 to 2010. Data included 2,680 (foal inspection), 1,927 (mare inspection) and 236 (mare performance test) ponies. The outcome of this was 4,092 animals with performance records and a pedigree totaling 12,638 horses. For all animals with performance information, the first parental generation was inspected with respect to the influence of other breeds. Stallions of the breed German Riding Pony were dominating. Welsh-sires (16.7%) – especially Welsh B (8.7%) – and Arabian (4.5%) as well as Thoroughbred (3.7%) stallions also have some impact. For the ponies registered as 'active breeding ponies' for 2011, a backtracking up to the great-grandparents was carried out. The majority of the investigated stallions (36.6%) and mares (43.5) had a fraction of 50-75% Riding Pony genes, followed by animals with at least 50% Welsh-participation (in mares) and 87.5- 100% Riding-Pony-participation (in stallions), respectively. A descriptive analysis of all traits showed a right-skewed distribution and a rather small standard deviation. Genetic parameters for foal inspection ranged from $h^2=0.16$ (Conformation) over $h^2=0.31$ (Gait) to $h^2=0.61$ (Type). Estimates of the heritability for traits from mare inspection showed marginal (e.g. correctness of gaits) as well as medium (e.g. frame) and high (e.g. breed and sex type) values. Breeding objectives and performance recording of German Riding Ponies are similar to the system of the German Warmblood Horse. Additionally, the present results show an analogy in the distribution of traits and genetic parameters estimated. It is therefore obvious to use existing synergies and to transfer knowledge and experience from German Warmblood Horses to German Riding Ponies.

The relationship between genetic diversity and phenotypic characteristics in the Irish Draught Horse
Brady, K.M.[1], Corbally, A.[1], Harty, D.[1] and Fahey, A.G.[2], [1]Horse Sport Ireland, Beech House, Millennium Park, Osberstown, Naas, Co. Kildare, Ireland, [2]UCD School of Agriculture & Food Science, Belfield, Dublin 4, Ireland, Ireland; kbrady@horsesportireland.ie

The Irish Draught is a native Irish horse breed that was traditionally used for both draught and riding purposes. The breed has previously been characterised using both mtDNA and pedigree analyses using PEDIG software to assess its genetic diversity. Breeders are actively encouraged and supported in including genetic diversity as part of their mating plans. For example, mean kinship information on all purebred Irish Draught horses has been made available to all breeders and has been incorporated into the breeding programme of stallion and mare selection processes. The main market for the breed is in the production of crossbred Irish Draught progeny suitable for showjumping. Although this has been a key component in the survival of the breed, it has affected the numbers of purebred foals produced and has raised concerns about the erosion of the traditional phenotypic characteristics of the breed. In order to assess this, the linear scoring results for 529 mares inspected in 2010 and 2011 and conformation characteristics for 506 stallions born between 1961 and 2008 were analysed using SAS software (GLM and chi-square analyses). The analysis of the conformation of stallions passed at inspection showed a significant decrease in bone circumference ($P \leq 0.01$) from 23.7 to 22.9cm over this time period although no significant change for observed in either height at withers or girth. The analysis of the mares' linear profiles showed that mares with a mean kinship value of less than 2% were significantly less likely to meet the required standard for breed type. Conservation of the breed including its traditional type will require mares to be considered on a number of criteria and not solely for their contribution to genetic diversity. Research into the effect of the use of AI and sales results for both purebred and crossbreed ID horses will also be presented.

Differences in judging of young horse free jumping
Lewczuk, D., PAS Institute of Genetics and Animal Breeding, ul. Postepu 1, Jastrzebiec, 05-552 Magdalenka, Poland; d.lewczuk@ighz.pl

The aim of presented study was to investigate the judging agreement on horse jumping skills and relations of judges' notes with measured jumping parameters based on video image analysis. In the end of 100 days performance test, the group of 32 warmblood stallions was judged in free jumping by six experienced judges in following traits: 'willingness to jump', 'easiness of jump', 'work of front', 'work of hind' and 'work of trunk, head and neck'. Simultaneously, horses were filmed during jumping and following linear parameters were measured: taking off, landing, lifting of limbs and elevation of bascule points above the obstacle. The statistical analysis consisted of calculation of judges' notes repeatability using procedure Mixed of the SAS program, calculation of Pearson correlations between notes of individual judges as well as calculation of correlations corrected for the height of the jump and successive number of jump between judges' notes and jumping parameters. The statistical model for repeatability calculation included fixed effect of judge and random effect of the horse. Correlations between the individual notes of different judges were not equal. Notes for the specific traits were in some cases more correlated with other traits like with notes for the same trait. Repeatability of judges notes was 0.33-0.48. Correlations between notes for jumping and measured jumping parameters were low and medium. Received results showed that definition of the traits evaluated by judges is not the same for all of them.

Opportunities of canalizing the rank in the Endurance competitions in Arab Horses

Cervantes, I.[1], Pérez-Cabal, M.A.[1], Pun, A.[1], Valera, M.[2], Molina, A.[3] and Gutiérrez, J.P.[1], [1]Department of Animal Production, University Complutense of Madrid, 28040, Spain, [2]Department of Agro-forestry Sciences, University of Sevilla, 41013, Spain, [3]Department of Genetics, University of Córdoba, 14071, Spain; icervantes@vet.ucm.es

The control of the environmental variability by genetic selection represents a new alternative for selecting productive traits. This methodology aims to modify the homogeneity in productive traits and has been applied in birth weight, which is a very important trait in several species where the homogeneity of the animals is profitable. Here we estimated the genetic parameters concerning environmental variability to the rank trait in endurance competitions in Spanish Arab Horses to analyze the horse's capability of maintaining the desired ranking in each endurance race. This type of competition is carried out across country; it consists of long-distance races where not only speed is important, but the endurance and the adapted physical condition of the horse are essential. The endurance ride data belonged to Arab horses totalling 1,905 records. The pedigree matrix contained 2,754 animals. The rank trait was used as original or as normalized variable. The model included sex, age, and competition as systematic effects, and the rider o rider-horse interaction as well as additive genetic effects as random effects besides the residual. This model assumes that the environmental variance is heterogeneous and partially under genetic control. It was solved by using the GSEVM program. The genetic variance for the variability of the trait ranged between 0.25 and 0.48, leading the higher values of this parameter to more opportunities to select animals for canalization purposes. The correlation between the genetic variance of the trait and its environmental variability ranged between -0.48 and 0.29. Further analyses are needed to establish a canalization process in endurance of horses.

Aplication of multitrait animal models to evaluate the plasticity in horse show jumping performance

Bartolomé, E.[1], Menéndez-Buxadera, A.[2], Valera, M.[1], Cervantes, I.[3] and Molina, A.[2], [1]Department of Agro-forestry Sciences, University of Sevilla, 41013, Spain, [2]Department of Genetics, University of Córdoba, 14071, Spain, [3]Department of Animal Production, University Complutense of Madrid, 28040, Spain; icervantes@vet.ucm.es

Repeatability Animal models have been often used in animal breeding because of simplicity as they assume a unique genetic trait for all age stages. But several studies have shown changes in horse performance due to age suggesting the need to consider every age stage as a different genetic trait. The objective of the present study was to study a phenotypic plasticity effect due to differences among ages for Show Jumping performance. For this study, 11,352 participations (87 Show Jumping competitions and 405 different riders), belonging to 1,085 horses grouped regarding to the age (4, 5 or 6 years old), were analyzed. Repeatability Animal Model (RAM) and a Multiple Trait Animal Model (MTAM) were applied. MTAM considered the results of every animal at every age stage as different (but correlated) traits. The age, sex, starting order and training level were included as fixed effect. The random effects were the animal, individual permanent environment, the competition and the rider. Different models were compared adding as random effects the interaction rider*animal and, only for MTAM, considering or not a heterogeneous residual variance. To study the presence of an age-genotype interaction and, therefore, an environmental underlying effect, the genetic correlations between age groups were estimated. The genetic correlations between age groups showed values of 0.299, 0.397 and 0.335 between 4-5, 4-6 and 5-6 age levels, respectively. These values could be due to the existence of a phenotypic plasticity. This study could allow the selection of horses with a consistent response, increasing their genetic potential as they got older, and horses with a precocious response, loosing genetic potential with age.

Genetic parameters estimation for linear type traits described in the Czech draft breeds of horses
Vostry, L., Hofmanova, B., Capkova, Z. and Majzlik, I., Czech University of Life Sciences Prague, Faculty of Agrobiology, Food and Natural Resources, Kamycka 129, 165 21 Prague 6, Czech Republic; hofmanova@af.czu.cz

Genetic parameters and breeding values for 22 linear type descriptions of conformation and type characters and 4 body measurements were evaluated in 1,744 horses of three original cold-blooded breeds included among the genetic resources in the Czech Republic in a period of 18 years (1990-2007). The genetic parameters were estimated using the restricted maximum likelihood method with a multiple trait animal model. Based on the values of Akaike's information criterion, residual variance and heritability coefficient, a model with fixed effects (sex, year of description, breed, and classifier) and with random effect (animal) was selected. Heritability coefficients for the particular traits were in the range of 0.11 to 0.55 and genetic correlations ranged from -0.63 to 0.97. Standard deviations of the breeding values for linear type description of conformation and type characters were in the range of 0.30 to 0.72 and 0.62 to 6.18 for body measurement traits. The breeding values will be used for more effective selection of typical conformation traits in the Czech cold-blooded breeds of horses. Supported by the project No. MSM 6046070901.

Genetic connections between dressage and show jumping in Dutch Warmblood horses
Rovere, G.[1], Madsen, P.[2], Ducro, B.J.[1], Norberg, E.[2], Arts, D.[3] and Van Arendonk, J.A.M.[1], [1]Wageningen University, Animal Breeding and Genomics Centre, De Elst 1, 6700 AH Wageningen, Netherlands, [2]Aarhus University, Department of Genetics and Biotechnology, Blichers Alle 20, 8830 Tjele, Denmark, [3]KWPN, P.O. Box 156, 3840 AD Harderwijk, Netherlands; gabriel.rovere@wur.nl

During the last decades a process of specialization has occurred into show jumping (JH) and dressage (DH) in the Dutch Warmblood studbook (KWPN). As a consequence, the genetic base might become stratified. The objective of this study was to estimate the connectedness between JH and DH subpopulations of the current generation. The material comprised horses that participated in the studbook entry inspections between 2005 and 2010. KWPN is registering jumping and dressage horses using different codes since 2005. Ancestors were traced back as far as possible in the pedigree to define the base generation. Subsequently the base generation was divided into 3 base groups, according to having only descendants in JH, in DH or in both subpopulations. Subsequently, the genetic contribution of the 3 base groups to the subpopulations was estimated. DMUTRACE software was used to make the pedigrees files and to calculate the genetic contributions. The base generation comprised 8,205 horses, 2,857 only had descendants in JH, 1893 only in DH and 3455 in both subpopulations. The latter base-group also had the largest contributions to both subpopulations: 92.6% to JH and 96.0% to DH. The 'specialist' base groups had smaller genetic contributions but 47.8% of horses in JH and 34.1% in DH receive genes from them. Observed by year of birth, the number of horses that received genes from specialist groups increased from 39.8 to 56.0% in JH and from 31.9 to 39.5% in DH. Results from this study show that JH and DH subpopulations have a large part of their genetic base in common. However, an increasing percentage of animals receive genetic contributions from 'specialist' genetic groups within the genetic base. This trend is more profound in the jumping discipline.

Influence of foreign breeds in the Show Jumping performance of the Spanish Sport Horse
Bartolomé, E.[1], Valera, M.[1], Cervantes, I.[2] and Ducro, B.J.[3], [1]ETSIA. Universidad de Sevilla., Dp Agroforestales, Ctra. Utrera km 1, 41013, Spain, [2]Universidad Complutense de Madrid, Dp Producción Animal, Avda. Puerta de Hierro s/n, 28040, Spain, [3]Wageningen Institute of Animal Sciences, Wageningen University, P.O. Box 338, 6700 AH Wageningen, Netherlands; ebartolome@us.es

The Spanish Sport Horse (SSH) is a composite breed designed in the search for an animal with a first-rate performance at any of the Olympic equestrian disciplines. The aim of this study was to ascertain the influence of foreign breeds on the SSH sport performance in Show Jumping competitions. For this analysis, 1,086 animals (82.9% SSH), with a total of 5,674 participations (87.3% SSH) in 87 Show Jumping competitions, were used. The pedigree matrix included 5,443 animals (24.1% SSH). In order to assess the breed influence, 17 genetic groups were defined according to the breed of the SSH ascendants (Anglo-Arab, BWP, SSH, SSH inscribed on the Auxiliary Registry Book (SSHaux), Hannover, Holsteiner, KWPN, Oldenburg, Spanish Arab, Spanish Purebred, Thoroughbred, Selle Français, Spanish Troter Horse, Trakehner, Westphalian, Zangersheide and Other Minority Foreign Breeds). For the genetic evaluation, a BLUP Animal model was used. The final ranking weighted and subsequently standardized with a BLOM statistical method, was used as trait; the competition and the age were included as fixed factors and the rider-horse interaction, the trainer and the animal as random factors. Results highlighted that significant genetic groups could be identified of which some were beneficial to SSH sport performance.

Pedigree and molecular data to perform optimum contribution selection in a Menorca horse population
Solé, M.[1], Gómez, M.D.[2,3], Valera, M.[3], Cervantes, I.[4] and Fernández, J.[5], [1]University of Córdoba, 14071 Córdoba, Spain, [2]A.C.P.C. Raza Menorquina, Ciutadella de Menorca, 07760, Spain, [3]University of Seville, 41013 Seville, Spain, [4]University Complutense of Madrid, 28040 Madrid, Spain, [5]INIA, 28040 Madrid, Spain; ge2sobem@uco.es

Breeding programs are usually designed to maximize genetic gain for some traits of economic interest in a population and to limit the rate of inbreeding through the so called Optimum Contribution selection (OC). This is especially important in small populations of local breeds where selection progress must be carefully balanced against the conservation of genetic variation to ensure the probability of survival. The Menorca Horse (MH) is a Spanish breed under special protection, but where the performance for some traits may be important to increase the profitability of the breed and, thus, the probability of survival. Consequently, MH is a suitable candidate to implement OC. The aim of this study was to assess if a breeding program aiming for improved Menorca Dressage performance is feasible in such a small local breed. To perform the analysis, animals that were currently available for breeding (3-20 years) were included. Pedigree of 1646 animals (837 males and 809 females) were used for the calculation of different parameters, totalling 3735 individuals. Results for OC selection were studied based on EBVs for Menorca Dressage and coancestries calculated from pedigree and molecular data (separately or in combination). Whatever the considered coancestry (genealogical, molecular or mixed) it was possible to find solutions with low levels of expected inbreeding in the next generation without losing much response in the trait of interest. Consequently, it seems that Menorca Horse is a feasible candidate to implement artificial selection even being a vulnerable breed. The use of both types of information in the calculation of coancestry may be the recommended option to account for differences within families.

Formation of new riding horse population in Lithuania

Sveistiene, R.[1,2], [1]Lithuanian Endangered Farm Animal Breeders Association, R.Zebenkos 12, 82317, Lithuania, [2]Institute of Animal Science of Lithuanian University of Health Sciences, Animal Breedig and Genetics, R.Zebenkos 12, 82317, Lithuania; ruta@lgi.lt

The aim of the study was to evaluate population of Lithuanian riding horses (LJ) for the development of a new breed. In Lithuania riding horses are bred as an open population. In 1999, after selection of the present stallions and mares of riding breeds, a subpopulation of new Lithuanian riding horses was started. The first volume of studbook was prepared in 2000 and is supplemented every year. The average pedigree completeness for animals identified as LJ and born within the last 10 years was: 1 generations deep = 87.7%, 2 generations deep = 44.8%, 3 generations deep = 29.9%, 4 generations deep = 22.4%, 5 generations deep = 17.9%, 6 generations deep = 14.9%. The population of riding horses in Lithuania can be divided into six main related groups according to the distribution by genotype, i.e. Hanoverian – 38.7%, Trakehner – 27.4%, Holstein – 7.3%, Thoroughbred – 5.4%, Budenny – 11.9%, Arab- 3.2% and other -6%. The analysis of the studbook indicates that out of 86 stallions listed in the Studbook, 73.2% were used for breeding. 48.3% of stallions born in Lithuania and 51.7% of imported ones were used for breeding. The founders of the saddle horse population in Lithuania are the stallions of the classic riding hose lines: Pilger IC00383, Forpost T44, Agronavt 2192, Cordelabryere 210398168, Langraf I 210391966, Espri 311100481. The analysis of the progeny shows that 18% of breeding horses were inbreed animals, being the average inbreeding coefficient 2.2%. It is suggested to breed the horses using rational level of relatedness, thus consolidated their eligible showjumping horse type.

Session 06   Theatre 1

Importance of sow colostrum in relation to piglet survival

Quesnel, H., INRA, UMR 1348 Pegase, 35590 Saint-Gilles, France; helene.quesnel@rennes.inra.fr

Over the past few decades, selection to improve prolificacy rate and carcass merit has been accompanied by a substantial increase in piglet mortality before weaning. The most critical time is the first 3 days after birth, and this early mortality is mainly due to low colostrum consumption. Because piglets are born with low energy reserves, colostrum plays an essential role in survival by providing energy for heat production and metabolism. Colostrum also provides growth factors and hormones that stimulate the development and maturation of the piglet's gastrointestinal tract. Moreover, like most farm animals, piglets are born devoid of immunoglobulins (Ig) and rely on Ig, mainly IgG, and immune cells provided by colostrum to acquire immune protection. Piglet performance before weaning largely depends on the amount of colostrum ingested. The mortality rate before weaning sharply decreases when colostrum intake increases. Low mortality rate has been reported when piglets ingest more than 200 g of colostrum. Piglet body-weight gain increases concomitantly with colostrum intake during the first 24 h after birth, provided that intake is above 110-120 g/kg BW. With regard to immunity, IgG concentrations in piglet plasma at 24 h of age are strongly correlated with colostrum intake. However, plasma IgG concentrations have been shown to reach a plateau when colostrum intake increases beyond 200-250 g. Long-term effects of colostrum intake have also been reported on piglet immune status at weaning and on piglet growth from 3 weeks of age until after weaning. Ways to increase piglet colostrum intake must be considered, such as increasing the ability of piglets to suckle, reducing within-litter variation in birth weight, and increasing the quantity of colostrum that sows produce. Sow nutrition during gestation may increase piglet vitality at birth and the acquisition of passive immunity. Ways to increase litter uniformity at birth and colostrum production by the sow need to be further investigated.

Maternal and postnatal dietary n-3 fatty acids improve piglet performance and health

Gabler, N.K.[1], Mani, V.[1], Boddicker, R.L.[1], Ross, J.W.[1], Spurlock, M.E.[1] and Spencer, J.D.[2], [1]Iowa State University, Ames, IA 50011, USA, [2]JBS Unites Inc., Sheridan, IN 46069, USA; ngabler@iastate.edu

Modifying the maternal diet and early postnatal nutrition provides an excellent opportunity for swine producers to manipulate offspring performance and health to enhance production and feed efficiencies. In particular, enhancing the uptake and tissue enrichment of long chain n-3 polyunsaturated fatty acids such as docosahexaenoic acid (DHA) and eicosapentaenoic acid (EPA) has proven to be a viable approach. These two n-3 fatty acids are important for cell growth, metabolism, communication, and regulation of expression and function of genes and proteins. In pigs, higher n-3:n-6 ratios due to increased consumption of DHA and EPA, can increase litter size, reduce mortalities, advance piglet transition at weaning, and attenuate inflammatory processes in pigs. Specifically, we have shown that DHA and EPA improve piglet intestinal nutrient transport through the activation of jejunum AMP-activated protein kinase, Na+, K+ -ATPase activity and increased translocation of GLUT2 to the brush border membrane. Further, we have evidence that these n-3 fatty acids improve intestinal integrity and attenuate endotoxin transport and signaling. In pig peripheral tissues, DHA and EPA block the inflammatory response to endotoxin via antagonizing membrane recruitment of the innate immune receptor, toll like receptor 4, and downstream signaling processes. Interestingly, feeding n-3 fatty acids to sows in gestation attenuated the innate immune response of their offspring at 10 weeks of age. This may be explained by several treatment-specific DNA methylation patterns that are impacting the epigenetic code of these offspring. Altogether, this presentation will discuss the role of maternal and postnatal dietary n-3 fatty acids on sow and piglet performance and health.

Improving piglet survivability by altered maternal nutrient supply for transition sows

Theil, P. and Lauridsen, C., Aarhus University, Animal Science, P.O. Box 50, 8830 Tjele, Denmark; Peter.Theil@agrsci.dk

Neonatal piglet survival is a major challenge these years and a major determinant of the sow productivity. The total yield of colostrum and milk of lactating sows are important to ensure survival of suckling piglets in the neonatal period. Furthermore, newborn piglets also rely upon glycogen depots in liver and muscles and upon early and efficient initiation of milk synthesis in the udder. Two studies have been carried out to investigate aspects related with sow lactation and piglet survival. In study 1, colostrum yield, milk yield and glycogen depots of newborn piglets, as well as plasma and milk contents of immunoglobulins and fatty acids, were studied in 40 sows fed one of four gestation diets until d 108 of gestation, and then sows were fed one of six transition diets until weaning. Three gestation diets were formulated to contain 35% DF (mainly from sugar beet pulp, pectin residue or potato pulp) and a low fiber control diet contained 17.5% DF from mainly wheat and barley. Transition diets contained either 3% supplemented animal fat or 8% supplemented fat originating from coconut oil, sunflower oil, fish oil or fish oil/octanoic acid mixture (4+4%). Sows fed octanoic acid had a higher colostrum production and a higher piglet survivability than control sows. Colostrum contained more fatty acids and immunoglobulins than milk, and dietary treatment of the sow influenced the contents. In the second experiment, 23 sows were divided in two groups and fed +/- 1.3% CLA (trans 10, cis12). Colostrum yield was lower but milk yield in early lactation was higher in sows fed 1.3% CLA, whereas the time for initiating milk production was unaffected. This presentation review the current knowledge on aspects related to piglet survival and how these traits may be improved by dietary manipulations of sows during the transition period from gestation to lactation.

Effect of parity and number of suckling piglets on milk production of sows

Dourmad, J.Y.[1,2], Quiniou, N.[3], Heugebaert, S.[4], Paboeuf, F.[4] and Ngo, T.T.[1,2], [1]Agrocampus Ouest, UMR1348 Pegase, 35000 Rennes, France, [2]INRA, UMR1348 Pegase, 35590 Saint-Gilles, France, [3]IFIP-Institut du Porc, BP 35104, 35651 Le Rheu Cedex, France, [4]Chambres d'agriculture de Bretagne, CS 14226, 35000 Rennes, France; jean-yves.dourmad@rennes.inra.fr

Different factors are known to affect milk production in sows, especially their parity and litter size (LS). In practice, milk production or energy and protein output in milk can be estimated from litter growth rate and LS, according to prediction equations available in the literature. This approach is often used for the determination of nutrient requirements of lactating sows. However, information on factors affecting litter performance is scarce. Data from 3,500 litters from three experimental units were used to quantify the effect of LS and sow's parity on the amount of milk estimated from piglet's growth rate. The number of piglets weaned per litter, at 27.8 days of age, was 11.0 on average with an average live weight of 8.6 kg. Average piglets and litter weight gains were 256 and 2,790 g/d. Calculated milk production reached 10.7 kg/d on average, with a higher value of 11.4 kg/d for the two best performing farms. Total milk production increased linearly with LS, by about 0.75 kg/d per extra piglet, up to 12 piglets per litter and then tended to plateau. The amount of milk per piglet was the highest for LS with 6 to 8 piglets. Above 8 piglets, the amount of milk available per piglet decreased, especially when LS exceeded 12 piglets. Milk production increased from parity 1 to 2, was the highest in parity 2 to 4 and decreased thereafter. The amount of milk per piglet was the lowest for parity 1 (925 g/d) and the highest for parities 2-4 (1,015 g/d).These results will contribute to a better estimation of milk production in sows. They will also contribute to a better evaluation of the variability of milk production within a batch and, consequently, the variability of nutritional requirements in lactating sows.

Genetic impact of Hampshire sires on litter size and piglet survival

Lundeheim, N.[1] and Serenius, T.[2], [1]Swedish University of Agricultural Sciences, Dept.of Animal Breeding and Genetics, P.O. Box 7023, 75007 Uppsala, Sweden, [2]Nordic Genetics, Råby 2003, 24292 Hörby, Sweden; nils.lundeheim@slu.se

Most Swedish pigs are offspring of Hampshire boars and Yorkshire*Landrace crossbred sows. For the mother lines, litter size as well as productions traits are included in the breeding goal, but for the Hampshire breed, only production traits are included. Since half of the genes in the crossbred piglets comes from the sire, the present study focused on the genetic impact of the sire of the litter on litter size and piglet mortality (both stillborn and those lost during nursing period). Data on purebred Hampshire litters in 5 nucleus herds were provided by the breeding organization Nordic Genetics. Data comprised of 9,881 litters (920 boars), born in the period 2000 to 2011. The genetic analyses [trivariate analysis: litter size (total born, TB); number of stillborn piglets (SB); number of piglets lost during nursing period (Mort)] were performed using the DMU software. The statistical model included the fixed effects of herd (5 herds), parity number (1, 2, 3, 4+), farrowing year, and the random effects of herd-year-month combination, sow (without pedigree) and boar (with pedigree). The estimated heritabilities were all very low (<0.03), the genetic correlations between the traits were all positive and significant. The estimated σ_A were 0.17 (TB); 0.02 (SB) and 0.03 (Mort). In spite of the low heritabilities, the distribution of the estimated breeding values shows that some genetic improvement in piglet mortality can be achieved by including paternal fertility in the breeding goal for the Hampshire breed. 5% of the boars had EBV for SB<-0.08, and 5% of the boars had EBV for Mort<-0.12.

Causes of variation in piglet mortality

Engblom, L. and Lundeheim, N., Swedish University of Agricultural Sciences, Department of Animal Breeding and Genetics, P.O. Box 7023, 75007 Uppsala, Sweden; linda.engblom@slu.se

Piglet mortality is a major issue in commercial piglet producing farms in Sweden as well as in other countries. Recent years' selection for increased litter size has been efficient, but unfortunately also piglet mortality has increased. Larger litter size results in piglets with lower birth weight, which reduce their chance to survive. Present study analyzed loss of piglets, both before birth (stillborn; SB) and of the live born piglets until weaning (mortality during lactation; LMORT). Data was collected from 17 commercial piglet farms in the southern part of Sweden using the herd monitoring program PigWin Sugg. The analyzed data included 40,751 farrowings (by 9,939 crossbred L*Y sows) from 2008 to 2010. Analyses were performed with the SAS Software; GLM, MIXED and GLIMMIX procedures. On average each litter had 0.87 SB piglet and 17.7% LMORT (average 33 days lactation). The effect of farm, parity, gestation length and total number born were significant ($P<0.001$) in SB analysis. The effect of farm, parity, gestation length, piglets born alive, transferred piglets and lactation length were significant ($P<0.001$) in LMORT analysis. The effect of year and season were not significant ($P>0.05$) for either of the analyzed traits. There was large variation between farms in SB (0.49-1.44) and LMORT (12.6-22.2%). Both SB and LMORT were lowest in parity two and increased thereafter with increasing parity number. Gestation length less than 114 days had highest SB and less than 113 days had highest LMORT. Number of SB increased in litters with increasing total born and LMORT increased in litters with increasing live born. Litters which received additional piglets had higher LMORT (24.5%) compared with intact litters (18.2%) and litters which had piglets removed (10.8%). Litters weaned before 28 days or after 37 days had highest LMORT. High mortality litters with few live piglets got more piglets added and/or was weaned early. Identified variation between farms can be used to find superior management for reduction of piglet mortality.

Preventing lactation failure: associations between genotype and postpartum dysgalactia syndrome

Bardehle, D.[1], Preissler, R.[1], Lehmann, J.[1], Looft, H.[2] and Kemper, N.[1], [1]Institute of Agricultural and Nutritional Science, Martin-Luther-University Halle-Wittenberg, Animal Hygiene and Reproduction, Theodor-Lieser-Straße 11, 06120 Halle (Saale), Germany, [2]PIC Germany, Ratsteich 31, 24837 Schleswig, Germany; danilo.bardehle@landw.uni-halle.de

Piglet losses after birth are often caused by sows' lactation failure with the postpartum dysgalactia syndrome (PDS) as the most important disease complex. In the analysis of data from 992 litters, we showed that PDS-affected sows had less weaned piglets (10.4 vs. 10.6) although the litters of these sows were characterized by a higher number of piglets born alive (11.8 vs. 11.6). In addition, the number of stillborn piglets increased by the occurrence of PDS (1.3 vs. 1.1). Known as a multifactorial disease, the genetic susceptibility was investigated to get a comprehensive picture of this complex disease. Heritability of about 10% was estimated and loci associated with PDS were detected through a genome-wide association study with 590 sows using the commercially available Illumina PorcineSNP60 BeadChip. The further aim of our study was to analyse these genome-wide associated SNPs in a custom panel of low-multiplex genotyping using the BeadXpress system. Therefore, further 1000 sows are genotyped at 48 selected SNPs. Sows were selected in a family based case-control-design with matched sampling of affected sows and healthy half- or full sib control sows on six farms. Sows were defined as affected by PDS when they showed rectal temperatures above 39.5 °C and/or clinical signs of mastitis such as reddening, swelling or hardening and/or changes in piglets' behaviour within the 12 to 48 h post partum. The phenotype PDS was used in statistical analysis as a binary trait (affected/not affected). Genotype calling and cluster analysis was performed with GenomeStudio. The statistical environment R (www.r-project.org) and several specific R-packages such as GenABEL were used for data control and further statistical analysis.

Echium or linseed oil in the maternal diet affects fatty acids, but not oxidative status of piglets

Tanghe, S.[1], Millet, S.[2] and De Smet, S.[1], [1]Ghent University, Laboratory for Animal Production and Animal Product Quality, Department of Animal Production, Proefhoevestraat 10, 9090 Melle, Belgium, [2]Institute for Agricultural and Fisheries Research (ILVO), Animal Sciences Unit, Scheldeweg 68, 9090 Melle, Belgium; stanghe.tanghe@ugent.be

Polyunsaturated fatty acids (PUFA) and mainly docosahexaenoic acid (DHA) are essential for the development of the foetus and dietary supplementation may be beneficial. DHA can be directly supplied from the diet by adding fish oil, or it may result from the conversion of dietary precursors. As fish oil becomes a scarce resource, research for sustainable alternatives is needed. This study aimed to evaluate linseed oil (LO) as a source of α-linolenic acid (ALA, C18:3n-3) and echium oil (EO) as a source of stearidonic acid (SDA, C18:4n-3). As SDA is one step further than ALA in the n-3 pathway, the hypothesis was that EO may yield a higher DHA content than LO. Because higher levels of dietary PUFA could induce oxidative stress, the oxidative status of the piglets was also studied. Two groups of sixteen sows were fed a diet containing 1% LO or 1% EO from d 73 of gestation and during lactation. Feeds contained similar amounts of vitamin E (75 mg/kg feed). At birth, two piglets per sow were sacrificed and blood, liver and Longissimus dorsi (LD) samples were taken for analyses of fatty acids and oxidative parameters. The transfer of the dietary n-3 PUFA from sows to piglets differed according to the source. All tissues had a higher content of ALA when LO was fed. When EO was fed, plasma and LD but not red blood cells and liver showed a higher content of SDA. No difference in DHA response was found between LO and EO in any tissue. As EO contains less SDA than LO contains ALA, feeding richer sources of SDA should be investigated. No effect of feed was seen on oxidative status parameters. Hence, LO or EO can be added to the sows' diet at a concentration of 1%, without compromising oxidative status of the newborn piglets at standard antioxidant levels.

Natural or synthetic vitamin E supplementation of lactating sows on fatty acid profile of milk

Rey, A.I.[1], Amazan, D.[1], Lopez-Bote, C.[1], Cordero, G.[2] and Piñeiro, C.[2], [1]Facultad de Veterinaria. UCM, Dpto. Producción Animal, Avda. Puerta de Hierro s/n, 28040, Spain, [2]PigCHAMP Pro Europa, Segovia, 40003 Segovia, Spain

The influence of natural vitamin E supplementation (D-α-tocopherol) in drinking water vs. the synthetic form (Dl-α-tocopheryl acetate) in feed to sows during the gestation and lactation periods on tocopherol concentration and the fatty acid profile in colostrum and milk at 7 and 28 days of lactation was studied. From day 45 of gestation to 28 day of lactation sows (n=24) were divided in two dietary treatments: (1) natural vitamin E was administered at 30 ppm in drinking water; or (2) in feed (30 ppm of the synthetic form). In both treatments the vitamin E intake was 150 mg/day. Extraction of α-tocopherol concentration in colostrum and milk samples was carried out by saponification and analysis was made by reverse phase HPLC. Fatty acid composition of colostrum and milk was determined by gas chromatography. The statistical model included the fixed effect of vitamin E source. A repeated measurement test was used to study time and treatment effects and its interactions. α-Tocopherol in colostrum and milk from sows increased with the supplementation time (P=0.0001) and the increase tended to be higher in the group that received the natural vitamin E source in water. Fatty acid proportion of calostrum and milk was affected by the type of vitamin E. Those sows supplemented with natural α-tocopherol in water had lower saturated fatty acids (P=0.0334) and higher polyunsaturated proportion and unsaturation index (P=0.044) in colostrum than the group supplemented with the synthetic form of α-tocopherol. Significant differences were not detected at day 7 of lactation. However, milk produced at day 28 of lactation had lower C18:2 (P=0.048) and polyunsaturated fatty acids (P=0.054) in group supplemented with the natural form. In conclusion, the form of vitamin E supplementation of sows during gestation and lactation affects on the α-tocopherol concentration and fatty acid profile of colostrum and milk.

Milk composition of sows from 0 to 14 days postpartum

Nitrayová, S., Kirchnerová, K., Brestenský, M. and Patráš, P., Animal production research centre Nitra, Nutrition, Hlohovecká 2, 951 41 Lužianky, Slovakia; kirchnerova@cvzv.sk

The porcine mammary gland starts to produce colostrum before parturition and it is gradually replaced by mature milk from around 24 to 36 hours after partum. The selected qualitative milk parameters were determined in 48 samples of sow's milk with the aim to measure their changes during lactation. Through the experiment 12 lactating sows (second to fourth lactation, no health problems) were fed by feedstuff for lactating sows in daily amount 3 kg/ day prior to partum and 4.8-7.2 kg/day postpartum depending on the number of piglets in the litter. The samples of milk were collected by hand-milking on the day of partum, and on the 2nd, 7th and 14th day postpartum. Average crude protein feed intake during the lactation was 1,029.8 g/day. The mean litter size was 12.3 piglets.The dry matter content includes the sum of the individual components of milk. The amount of dry matter in colostrum was 21.6%. From the 2nd day of lactation this figure remained stable at around 18%. The lowest mean fat content was 4.62% in colostrum, this then gradually increased to 6.64% on the 14th day. The mean protein content in early lactation was 11.93% and in the period between the 2nd and 14th day protein levels decreased to 4.59%. The lactose content in milk during the monitored 14 days of lactation increased from 3.59% to 6.18%. The amount of urea in milk is an indicator of diet balances in crude protein and energy. The highest content of urea was recorded on the second day after birth, when it increased to 5.21 mg/100 ml. The active acidity of milk had gradually increased from the initial value of 6.17 to 6.92.Increased knowledge regarding the quality of the sow's milk has contributed to the overall understanding of piglet survival and growth.This article was written during realization of the project 'BELNUZ č. 26220120052' supported by the Operational Programme Research and Development funded from the European Regional Development Fund.

Session 07   Theatre 1

Detection of mastitis during milking: current solutions and prospective ideas

Haeussermann, A. and Hartung, E., Christian-Albrechts-University, Institute of Agricultural Engineering, Max-Eyth-Str. 6, 24098 Kiel, Germany; ahaeussermann@ilv.uni-kiel.de

Mastitis is primarily caused by an infection of the mammary gland and concomitant immune reaction. Clinical mastitis is in general one of the most important factors in cow replacement. In addition, milk yield is severely reduced in mastitic cows, and fertility and immune status can be adversely affected. Discharge of milk with organoleptic or physico-chemical abnormalities is required in Regulation (EC) 853/2004. In automatic milking systems (AMS), utilization of suited sensors for detection of abnormal milk is mandatory. Early indication of subclinical mastitis is important in herd health management in all types of milking systems. The objective of the presentation is to provide an overview on the current state of mastitis detection in AMS and to review budding sensor techniques. Up-to-date, indicators for mastitis detection in AMS are: milk yield, conductivity, colour, milk temperature, and SCC. Sensitivity and specificity of systems which use at least one of the first four above named indicators for mastitis detection varied between 14-100% and 8-96%, respectively. The results were influenced by system, farm, health status, and gold standard. Measurement of SCC during milking improved mastitis detection. Upcoming solutions should either increase detection accuracy or informative value, e.g. by indicating kind and severity of infection. In addition, use of chemical reagents and test sticks versus common non-destructive in-line methods is a topic. Indicators with currently minor or prospective utilization are, e.g. flakes, minerals, lactose, proteins, and enzymes in milk, but also metabolites or DNA of pathogens. Sensor systems can have advantages if more than one indicator is measured, and if they can be applied in more than one field. Although several sensors are already commercially available, recent investigations underline that detection of abnormal milk and mastitis during milking is not yet sufficiently solved.

Principal component analysis for early detection of mastitis and lameness

Miekley, B., Traulsen, I. and Krieter, J., Institute of Animal Breeding and Husbandry, Hermann-Rodewald-Straße 6, 24118 Kiel, Germany; bmiekley@tierzucht.uni-kiel.de

This investigation analyses the applicability of principal component analysis (PCA), a latent variable method, for early detection of mastitis and lameness. Data used was recorded on the dairy research farm Karkendamm between August 2008 and December 2010. For mastitis and lameness detection, data of 328 and 315 cows in their first 200 days in milk were analysed, respectively. Mastitis as well as lameness was specified according to veterinary treatments. Diseases were defined as disease blocks. The different definitions used (two for mastitis, three for lameness) varied solely in the sequence length of the blocks. Only the days before the treatment were included in the blocks. For recognition of mastitis, milk electrical conductivity, milk yield and feeding patterns (feed intake, number of feeding visits and feeding time) were used. Pedometer activity and feeding patterns were utilised for lameness detection. To develop and verify the PCA model, the mastitis as well as the lameness dataset was divided into training and test datasets. PCA extracted uncorrelated principle components (PC) by linear transformations of the original variables so that the first few PCs captured most of the variations in the original dataset. For process monitoring and fault detection, these resulting PCs were applied to the Hotelling's T2 chart and to the residual control chart. First results show that block-sensitivity of mastitis detection ranged from 38.7% to 74.2%, whilst specificity was between 86.4% and 97.5%. The error rates varied between 97.5% and 98.3%. In the case of lameness detection, the block-sensitivity ranged from 33.7% to 89.2% while the obtained specificity was between 68.2% and 91.9%. The error rates varied from 87.6% to 89.1%. In conclusion, PCA seems to be appropriate for disease detection in dairy cows. Results could probably be enhanced if different traits and further sensor data are included in the analysis.

Automated detection of lameness in dairy cows based on day-to-day variation in behaviour

De Mol, R.M., André, G., Bleumer, E.J.B., Van Der Werf, J.T.N. and Van Reenen, C.G., Wageningen UR Livestock Research, Edelhertweg 15, 8219 PH Lelystad, Netherlands; rudi.demol@wur.nl

Lameness is a major problem in dairy husbandry. The behaviour of dairy cows is influenced by lameness. Therefore the perspectives of lameness detection based on day-to-day variation in behaviour were examined. Icetag sensors were used to record the behaviour per day: lying time, number of lying bouts, maximum and average length of lying bout, maximum and average length of standing bout and number of steps. These behavioural data were combined with milking and concentrates feeding data. Data from up to 100 cows collected at an experimental farm in 2010 and 2011 were available for model development. Monthly locomotion scores were available as reference data for lameness. Behaviour is cow-dependent. Per cow, quadratic trend models were fitted on-line with a Dynamic Linear Model (DLM) for seven behavioural variables and two other variables (milk yield per day and concentrates remainder per day). It is assumed that lameness comes up gradually, therefore a lameness alert was given when the trend in two or more of the nine models differed significantly from zero for two or more days in a direction that corresponds with lameness symptoms: higher lying time, change in number of lying bouts, longer maximum/average lying bout, shorter maximum/average standing bouts, fewer steps, lower milk yield or more concentrate remainder. Lameness cases were defined as cows with a locomotion score of 3 or higher (on a scale from 1 to 5). The number of detected cases depends on the type of alert. The sensitivity was 79% when the alerts were only based on trend changes in one or more of the variables lying time, number of lying bouts and number of steps. The sensitivity was higher when more behavioural variables were included. The specificity was not yet analysed. These results are valid for the data set that was also used for model development. The developed model was tested on new data during the first months of 2012 on the same farm by generating lameness alerts each week.

Walking impulses of sound and lame dairy cows
Thorup, V.M.[1], Skjøth, F.[2], Nascimento, O.F.D.[3], Voigt, M.[3], Rasmussen, M.D.[1], Bennedsgaard, T.W.[1] and Ingvartsen, K.L.[1], [1]Aarhus University, Animal Science, Blichers Alle 20, 8830 Tjele, Denmark, [2]AgroTech, Agro Food Park 15, 8200 Aarhus, Denmark, [3]Aalborg University, Health Science and Technology, Fredrik Bajers Vej 7, 9220 Aalborg, Denmark; vivim.thorup@agrsci.dk

Automated lameness detection is objective and fast contrary to visual scoring. Here semi-automatically obtained walking impulses of lame and sound cows before and after claw trimming are compared. Four primiparous and 5 multiparous cows walked individually across 2 parallel 3D force plates repeatedly on 2 consecutive days before trimming (time=before); 2 consecutive days after (time=early); and 1 week after trimming (time=late). Cows were visually lameness scored on a scale from 1 (sound) to 5 (severely lame). Before trimming 4 cows were moderately lame (score=3), and 5 walked normally (score<2), 1 week after trimming all walked normally. Daily means per cow per leg (n=180) were calculated. Symmetry between sides for front and hind ends was calculated, where 100% is full symmetry. Parameters were mixed model analysed with score, end, time, and walking speed as fixed and cow as random effect. Lame cows walked slower than sound cows (t-test: 1.25 (0.19) vs. 1.37 (0.15) m/s; P<0.01). Front end vertical impulse (I_v) was 3.19 (0.09) N×s/kg, hind was 2.94 (0.11) N×s/kg, speed lowered I_v (-1.11 (0.17) N×s/kg; P<0.001). Front I_v symmetry was 88.5 (2.4)%, hind 85.3 (2.4)%. Front horizontal braking impulse (I_b) was -0.25 (0.01) N×s/kg, hind -0.12 (0.01) N×s/kg, speed diminished I_b (0.06 (0.02) N×s/kg; P<0.01). Front I_b symmetry was 78.2 (4.4)%, hind was lower (63.4 (4.4)%; P<0.001). Front horizontal accelerating impulse (I_a) was 0.11 (0.01) N×s/kg, hind 0.24 (0.01) N×s/kg. Front I_a symmetry was 65.0 (4.5)%, hind was higher (73.6 (4.5)%; P<0.05). Briefly, sound cows did not produce 100% symmetric impulses. Walking speed lowered impulses, except I_a. Trimming and lameness did not affect impulses, but symmetry differed between ends. Including speed in the model probably removed some effect of lameness.

Claw health diagnoses in the routine health monitoring system of Austrian Fleckvieh cattle
Fuerst-Waltl, B.[1], Fuerst, C.[2] and Egger-Danner, C.[2], [1]University of Natural Resources and Life Sciences, Vienna, Department of Sustainable Agricultural Systems, Division of Livestock Sciences, Gregor Mendel-Str.33, 1180 Vienna, Austria, [2]ZuchtData EDV-Dienstleistungen GmbH, Dresdner Str. 89/19, 1200 Vienna, Austria; birgit.fuerst-waltl@boku.ac.at

In Austrian cattle, veterinary treated claw diseases are recorded within the nation-wide health monitoring system. However, a routine genetic evaluation of health traits is currently only available for mastitis, ovarian cysts, early reproductive disorders and milk fever. For this analysis, the diagnoses panaritium, mortellaro, sole ulcer and white line disease of Austrian Fleckvieh were considered. The trait claw disorder was defined as a binary trait; whether or not a cow was at least once treated for either of these diseases in the interval of 10 days before to 300 days after calving. In total, 109,239 records of 57,898 cows were available for the years 2007 to 2010. For the genetic analysis, the single trait linear BLUP animal model from the routine genetic evaluation of health traits was applied. The fixed effects parity*age, year*month and type of recording*year as well as the random herd*year, random genetic animal and the random permanent environmental effects were included in the model. Breeding values were calculated with higher values being desirable as for all traits in the routine genetic evaluation. The frequency of claw disorders was 4.18% in this data set. The estimated heritability was 0.018±0.003. Significant positive breeding value correlations for bulls with a minimum of 20 records were found between claw disorders and functional longevity (0.22), the conformation traits feet and legs score (0.26), hocks (0.17) and pasterns (0.19) and the health traits milk fever (0.20) and ovarian cysts (0.12). As breeders tend to seek advice of veterinarians in severe cases of claw disorders only, further work should focus on the inclusion of claw trimmers' records.

Genetic parameters of hoof lesions in French Holstein dairy cattle

Van Der Spek, D.[1], Bovenhuis, H.[1], Vallée, A.A.A.[1,2] and Van Arendonk, J.A.M.[1], [1]Animal Breeding and Genomics Centre, Wageningen University, P.O. Box 338, 6700 AH, Netherlands, [2]Genes Diffusion, Route de Tournai 3595, 59501 Douai, France; dianne.vanderspek@wur.nl

Hoof lesions in dairy cattle are painful disorders reducing cow welfare. Further, hoof lesions are considered the third most costly disease after mastitis and fertility problems. Genetic selection provides a means to improve hoof health, but requires genetic variation. Limited information on heritabilities of detailed scores for hoof lesions is available. The objective of this research was to estimate the prevalence and heritability for nine different hoof lesions. Data on 10,679 Holstein cows was collected in France from 2007 until 2010 by three trained hoof trimmers. The scored hoof lesions were haemorrhages, white line separation, sole ulcers, digital dermatitis, heel erosion, interdigital hyperplasia, interdigital phlegmon, detachment of the sole, and hoof wall cracks. Prevalences of the different lesions ranged from 2.2% (sole ulcers) to 15% (heel erosion) of the cows having at least one hind leg affected. Each lesion was scored as a binary trait; zero if the lesion was absent and one if at least one hind leg was affected. The analyses were performed using a linear animal model in ASReml. The model accounted for the fixed effect of herd-year-season, parity and lactation stage and included a random permanent environmental effect to account for repeated observations. Heritabilities ranged from 0.00 (detachment of the sole, interdigital phlegmon, and hoof wall cracks) to 0.08 (interdigital hyperplasia). When all lesions were combined into one binary trait, zero if all lesions were absent and one if at least one lesion was present, a heritability of 0.04 was found. Despite rather low heritabilities, the present study detected genetic variation in most lesions.

Laminitis in dairy cows: relationships with production and functional traits

Schöpke, K.[1], Weidling, S.[1], Rosner, F.[1], Pijl, R.[2] and Swalve, H.H.[1], [1]Martin-Luther-University Halle-Wittenberg, Institute of Agricultural and Nutritional Sciences, Theodor-Lieser-Str. 11, 06120 Halle, Germany, [2]Claw Health GmbH, Fischershäuser 1, 26441 Jever, Germany; kati.schoepke@landw.uni-halle.de

In modern intensive dairy production, laminitis, also known as sole hemorrhage, is a common disease which has a multifactorial etiology and pathogenesis. Aim of the present study was the identification of some of these factors and their interactions by using a defined experimental design in a field study and a complex data collection. Claw health records taken at time of trimming from 1,962 Holstein heifers were merged with different body condition traits, conformation traits, treatment data for various diseases, functional heard life as well as test-day data and insemination records. All cows were from seven large herds equipped with standard slatted flooring. All trimmings and diagnoses were done by the same hoof trimmer in years 2008 and 2009. 70% of the inspected heifers exhibited at least one claw disorder, while for 57.3% subclinical or clinical cases of laminitis was observed. The magnitude of this incidence rate mostly was a function of the time period of trimming, i.e. recording focused on the critical period between day 50 and day 150 pp. For the conformation traits Rear leg set side view, Rear leg set rear view, Hock quality and Locomotion significant relationships with the health status for laminitis were found. The retrospective examination of the test-day criteria showed differences between laminitis-positive and –negative heifers for test-day milk fat percentage (1 month preceding diagnosis), fat-to-protein-ratio (1 month preceding diagnosis) and test-day milk protein percentage (2 to 3 months preceding diagnosis). As of July 2011, 53% of the cows examined were culled. The culling rate for cows with no claw findings was 43.5% compared to 61.5% for heifers with at least one claw disease.

Factors affecting incidence of digital dermatitis in dairy farms
Relun, A.[1,2], Bareille, N.[2], Lehébel, A.[2], Fourichon, C.[2] and Guatteo, R.[2], [1]Institut de l'Elevage, Atlanpole la chantrerie, BP 40706, 44307 Nantes, France, [2]LUNAM Université, ONIRIS, INRA, UMR 1300, Biology, Epidemiology and Risk Analysis, Atlanpole la chantrerie, BP 40706, 44307 Nantes, France; christine.fourichon@oniris-nantes.fr

Digital dermatitis is a widespread cause of lameness in dairy cows. A controlled field trial was conducted in 52 French dairy farms endemically affected by digital dermatitis (DD) to estimate the effect of treatment and management practices on DD incidence. The farms were quasi-randomly allocated to 1 of 4 treatment regimens for 6 months: no collective treatment (Control), walk-through footbath during 4 consecutive milkings every 4 weeks (FB/4W) or every 2 weeks (FB/2W) and collective spraying during 2 milkings fortnightly (CS/2W). DD was scored on all lactating cows during milking 7 times every 4 weeks by 14 trained investigators. During these farm visits, data related to cow characteristics and farm management also were collected. A cox frailty model including time-varying covariates was used to estimate, on the 5,598 feet free of lesions, the effect of potential risk factors and treatment practices on the time until the first case of an active DD lesion. A high initial DD prevalence strongly increased the hazard for DD occurrence (RR=2.1), as well as absence of hoof-trimming (RR=1.8), and poor leg cleanliness (RR=2.4). Prim'Holstein (RR=2.0) and high-productive cows (RR=1.3) were found to be more likely to develop DD. Collective treatments tended to decreased hazard for DD occurrence only when applied at least every fortnight (RR=0.7), compared to the control group. These results confirm the multi-factorial character of DD and provide useful data to design prevention programs.

Analysis of management practices associated with major contagious mastitis pathogens in BTM
Pinho, L.[1], Carvalheira, J.[1,2] and Thompson, G.[1,2], [1]ICBAS, Universidade do Porto, Largo Abel Salazar, 4099-003 Porto, Portugal, [2]ICETA/CIBIO, Universidade do Porto, Rua Padre Armando Quintas, 4485-661 Vairão, Portugal; gat1@mail.icav.up.pt

Determining possible associations between management practices and prevalence of mastitis pathogens is important for designing and developing specific mastitis-prevention and control programs. Major directives may be common to dairy farms, but specific procedures may influence the true prevalence in certain regions. The main objective of this study was to estimate the association level of some of these practices and the isolation of contagious mastitis pathogens in Portuguese dairy farms. In total, 164 dairy farms where randomly selected and analysis of bulk tank milk was performed in all farms. Information regarding management practices and other characteristics were also obtained through a questionnaire. The three major pathogens studied were *Mycoplasma* spp., *S. aureus* and *S. agalactiae*, with a prevalence of 3.0%, 37.5% and 6.6%, respectively. Univariate analysis found 10 risk factors that were associated (P≤0.25) with the prevalence of those pathogens as a group. The use of other systems than free-stalls (P=0.004), no bedding on stalls (P=0.005) and use of industrial fans for ventilation (P=0.082), were associated with higher probability of isolating contagious mastitis pathogens in the bulk tank milk. Having the therapy defined by a veterinarian (P=0.244) and the number of dry quarters (P=0.203) also showed strong association. On the biosecurity category, importing replacement animals (P=0.053) increased the risk of contagious mastitis. Using gloves (P=0.059), individual use of towels (P=0.087), and avoiding cloth towels (P=0.001), was associated with a lower risk. Finally, maintaining mastitis treatment records (P=0.048) was inversely associated with the odds of contagious mastitis, probably the consequence of a cause-effect reversal. The multivariate analysis kept four predictors (housing, replacements, towels and record of treatments) significantly associated (P<0.05).

Genomic regions affecting innate health of Dutch dairy cows

Wijga, S.[1], Bastiaansen, J.W.M.[1], Van Arendonk, J.A.M.[1], Ploegaert, T.C.W.[2] and Van Der Poel, J.J.[1], [1]Animal Breeding and Genomics Centre, Wageningen University, P.O. Box 338, 6700 AH Wageningen, Netherlands, [2]Cell Biology and Immunology Group, Wageningen University, P.O. Box 338, 6700 AH Wageningen, Netherlands; susan.wijga@wur.nl

Natural antibodies (NAb), part of the innate immune system, are considered to provide a barrier at the onset of infection and to act as a stimulus for activation of the adaptive immune system. Variation in NAb levels can result from genetic and environmental influences. This study aimed to detect regions on the cow genome that affect NAb levels in cow's milk. Generally, NAb bind antigens shared by classes of pathogens. This study included NAb that bind *Staphylococcus aureus*-derived lipopolysaccharide (LPS), *Escherichia coli*-derived lipoteichoic acid (LTA), *S. aureus*-derived peptidoglycan (PGN), and the model antigen keyhole limpet hemocyanin (KLH). Natural antibody levels were pre-corrected for the effect of herd, season, lactation stage, calving age and sire; residuals from the model with these effects were used as phenotypes for the marker analyses. Cows were genotyped for about 50,000 markers. Marker genotypes and NAb levels from 1,939 cows, part of the Dutch Milk Genomics Initiative, were analysed for associations using the PLINK software. Each individual marker was tested for its ability to detect variation in NAb levels. Findings were considered significant at a 0.10 false discovery rate, i.e. 90% chance that the finding was a true positive. Analyses did not show consistent regions affecting total NAb (making no isotype class distinction) and NAb isotypes immunoglobulin (Ig)A and IgG1 binding KLH, LPS, LTA, and PGN. A region on chromosome 23, however, was consistently associated with variation in levels of NAb isotype IgM that bind either KLH, LPS, LTA or PGN. Genes located near this region include the bovine major histocompatibility complex, making this a region of candidate gene(s) involved in NAb expression in dairy cows both from a functional and positional perspective.

Minimal nitrogen loss by dairy cattle, theory versus practice

Van Vuuren, A.M.[1], Dijkstra, J.[2] and Reynolds, C.K.[3], [1]Wageningen UR Livestock Research, P.O. Box 60, 8200 AB Lelystad, Netherlands, [2]Wageningen University, Animal Nutrition, P.O. Box 338, 6700 AH, Wageningen, Netherlands, [3]Reading University, Whiteknights, School of Agriculture, Policy and Development, P.O. Box 237, Reading RG6 6AR, United Kingdom; ad.vanvuuren@wur.nl

The efficiency of nitrogen (N) utilization by ruminants is relatively low and consequently a relatively large proportion of consumed N will be excreted. In regions with a dense ruminant population high N excretion may impair air and water quality. In ruminants inevitable N losses are related to both ruminal as well as intermediary (protein) N metabolism. In feed evaluation systems for dairy cattle it is assumed that ca. 40% of MP available to the mammary gland will not be secreted as milk protein. Inevitable metabolic losses can be recycled in ruminants by the entrapment of urea-N in microbial biomass in the rumen. Tamminga *et al.* estimated MP requirement for maintenance at 110 g/d. Accounting for N losses in hair and scurf the losses via urine would be ca. 14 g/d. Although rumen-degraded feed N and urea N secreted into the rumen will be recycled in microbial biomass, only ca. 55% of the entrapped N will be incorporated in microbial amino acids (AA). Consequently, a 600-kg cow producing 1.75 kg of milk protein/d, will excrete at minimum 90 g N/d as nucleic acid metabolites. Effect of digestibility on N efficiency will be included in the final paper, as well as variations In N efficiencies from literature studies in the revised, final version.

A technique to quantify metabolites of the purine and also of the pyrimidine metabolism
Stentoft, C. and Vestergaard, M., Aarhus University, Department of Animal Science, Tjele, 8830, Denmark; CharlotteS.Nielsen@agrsci.dk

The objective of the study was to develop a rapid, specific and reliable method for determination of the quantitative absorption, turnover and excretion of metabolites of purine/pyrimidine metabolism. Samples were obtained from feeding experiments with Holstein cows at different lactation stages fitted with ruminal cannulas and permanent indwelling catheters in major splanchnic blood vessels. Milk, urine and blood (arterial and portal, hepatic and ruminal vein) samples will be used for nutritional-physiological mapping of nucleic acid turnover in dairy cows to improve their utilization of nitrogen and/or to find new ways to monitor their nutritional status. A HPLC electrospray ionization tandem mass spectrometry-based technique was developed that was able to quantify 23 purine and pyrimidine bases, nucleosides and degradation products of the purine/pyrimidine metabolism in a single run based on calibration standards and isotope-labeled internal standards. A simple and repeatable pre-treatment featuring ethanol precipitation, filtration and evaporation/re-solution was developed and optimized. The analysis can distinguish between the components and quantify these individually. The blood plasma samples have been found to contain large amounts of uric acid as well as smaller amounts of some nucleosides such as guanosine, inosine, 2'-deoxyuridine, cytidine and uridine. The urine samples on the other hand have been found to contain large amounts of primarily allantoin and uric acid but also bases such as guanine, adenine, thymine and uracil. Also, degradation products of pyrimidine turnover such as β-alanine, β-ureidopropionic acid and β-aminoisobutyric acid are found in urine in considerable amounts. In conclusion, even though only preliminary results have been obtained from single plasma and urine samples, we have found it possible with this newly developed technique to quantify components not only from the purine metabolism but also the pyrimidine metabolism in lactating dairy cows.

Group separation of feedstuffs improves prediction of degradability parameters by NIRS
Foskolos, A.[1], Albanell, E.[1], Chrenkova, M.[2], Calsamiglia, S.[1] and Weisbjerg, M.R.[3], [1]Univ. Autónoma Barcelona, Ciéncia Animal i dels Aliments, Bellaterra, 08193, Bellaterra, Spain, [2]Animal Production Research Center Nitra, Hlohovecka 2, 951 41 Luzianky, Slovakia, [3]University of Aarhus, Animal Science, AU-Foulum, P.O. Box 50, 8830 Tjele, Denmark; andreas.foskolos@uab.cat

Previous work by our group defined the potential and limitations of NIRS to predict degradability parameters using a wide range of feedstuffs. Our objective was to improve predictions by separating feedstuffs in groups according to their use in ruminant nutrition. Degradability of DM, CP and NDF was determined using the in situ technique. Feedstuffs were grouped as forages (FF; n=254) and non-forages (NF; n=555). Degradability was described in terms of immediately rumen soluble fraction (a); the degradable but not soluble faction (b); and its rate of degradation (c). Overall effective degradability (ED) of DM, CP (5%/h passage rate) and NDF (2%/h passage rate) were calculated according to the equation of Ørskov and McDonald. All samples were scanned using a NIR Systems 5000 scanning monochromator. A total of 20% of samples were used as external validation set. The ED, a and b fractions of DM and CP were well predicted and improved by group separation. Group separation did not improve the prediction DM_ED of NF but it did for FF and CP_ED of FF and NF. DM_c and CP_c were not satisfactorily predicted when all samples were included ($r^2<0.7$). However separating samples improved the prediction of DM ($r^2>0.7$). Moreover, group separation improved the prediction of CP_c, especially for FF. For NDF, the number of feedstuffs was lower and the majority was grouped as FF. Equations obtained satisfactorily predicted NDF_ED and NDF_b and group separation further improved predictions. When all feedstuffs were included NDF_c was not well predicted, but when samples were grouped prediction of c_NDF in FF was acceptable. In conclusion, group separation into FF and NF improved NIRS equations especially for prediction of degradation rate.

Prediction of CP content and rumen degradability by Fournier transform infrared spectroscopy

Belanche, A.[1], Allison, G.G.[1], Newbold, C.J.[1], Weisbjerg, M.R.[2] and Moorby, J.M.[1], [1]Institute of Biological, Environmental and Rural Sciences, Aberystwyth University, SY23 3EE, United Kingdom, [2]Animal Science, Aarhus University, P.O. Box 50, 8830 Tjele, Denmark; aib@aber.ac.uk

This study examined whether Fournier Transform Infrared (FTIR) could be used to develop a universal equation that enables a low-cost determination of the crude protein (CP) content and rumen CP degradability for most feeds used in ruminant nutrition. A total of 664 feeds samples were used (180 forages, 484 concentrates). In situ CP degradability was described in terms of immediately available rumen soluble fraction a, the degradable but not soluble fraction b, and the rate of degradation of the b fraction c. Concentrations of CP, water soluble CP (CPWS), total tract mobile bag CP digestibility (CPTTD), potential CP degradability (CPPD) and effective degradability (CPED) considering 5%/h passage rate were determined. For FTIR analysis samples were dried and ground at 1 mm. Infrared spectra were collected by attenuated total reflectance from 4,000-600 cm^{-1} using a Equinox 55 FTIR spectrometer (Bruker Optik GmbH, Germany) fitted with a Golden Gate ATR accessory (Specac, UK). Duplicate spectra were averaged, derivatized, normalized and mean centred. Models were developed by PLS regression (MatLab) using a randomly selected subset of samples (570). The precision of the equations was confirmed by an external validation set (94 samples). Feeds varied in the content of CP (5-55%), CPED (27-96%), CPPD (52-99%), CPWS (0-77%), CPTTD (62-99%) and a (1-96%), b (4-100%) and c (1-37%/h). Contents of CP (R^2=0.93, SECV=3.4%) and CPWS (R^2=0.83, SECV=6.5%) were accurately predicted by FTIR. The concentrations of CPED and CPTTD and a and b were predicted with a moderate accuracy (R^2 between 0.67 and 0.77), whilst the c was poorly predicted. In conclusion, this work demonstrates the potential and also the limitations of FTIR to universally predict the protein values of feeds. The development of predictions for particular feedstuff groups might substantially improve FTIR equations.

Application of modelling to improve N utilization in cattle

Ellis, J.L.[1,2], France, J.[1], Bannink, A.[3] and Dijkstra, J.[2], [1]University of Guelph, N1G 2W1, Guelph, Canada, [2]Wageningen University, 6708 WD Wageningen, Netherlands, [3]Wageningen UR Livestock Research, 8219 PH Lelystad, Netherlands; jennifer.st-pierre@wur.nl

One of the current challenges in the agriculture sector is growing concern over the environmental impact of farming, and pressure has been applied through government and international agreements to introduce ceilings to N emissions whilst maintaining animal production. Variation in N utilization efficiency (NUE) in ruminants is high (0.10 to 0.40), and such variation suggests there is opportunity to manipulate the system. Meta-analysis of the data shows that NUE is higher in cows with a higher DMI and milk yield, but also when CP intake is low. More detailed relationships become evident when a closer look is taken. Meta-analysis allows the development of equations to predict animal responses from a collection of data spanning multiple experiments or studies. This form of empirical modelling can be useful to distill out relationships within biological data and suggest future feeding strategies. However, significant variation exists even around these regression lines meant to describe the data. Dynamic mechanistic modelling is a tool that can help elucidate sources of underlying variation based on an understanding of the system. While front-end development is significantly slower and initially less gratifying, mechanistic models have the capability of integrating historic and state of the art knowledge such that predictions of animal response, given a lattice of input conditions, can be made. This type of model is already in use for enteric methane emission estimation in the Dutch Tier 3 methodology. Several mechanistic rumen models exist which can predict N delivery to the small intestine but appreciably less work has been done post-absorptively, where significant variation in metabolism exists. Advancement of mechanistic models will lead to greater understanding of variation in post-absorptive N metabolism and NUE, and they should eventually replace current empirical systems.

A revised amino acid model for the liver of the lactating dairy cow

*Ellis, J.L.[1,2], France, J.[1], Bannink, A.[3] and Dijkstra, J.[2], [1]University of Guelph, N1G 2W1, Guelph, Canada,
[2]Wageningen University, 6708 WD Wageningen, Netherlands, [3]Wageningen UR Livestock Research, 8219
PH Lelystad, Netherlands; jennifer.st-pierre@wur.nl*

The ability of current feed rationing systems to predict effects of metabolisable protein supply on milk protein production and N excretion to the environment by lactating dairy cows is limited by oversimplification of post-absorptive metabolism. A more detailed quantification of amino acid (AA) metabolism within the splanchnic tissues is needed to improve on current empirical models of the efficiency of AA utilization for production. Significant clearance of AA from circulation by the liver occurs on a net daily basis, with significant changes to the profile of AA in the peripheral circulation compared to that absorbed from the GI tract. Several liver metabolism models exist in the literature, and it was the purpose of this project to take the most advanced model a step further. The liver model of Hanigan *et al.* was selected, as this model represents individual AA fluxes across the liver while also representing energy metabolism pathways. The original model performed well on an independent database which examined ruminal urea infusion with a low crude protein diet, but the model failed on some data points with the highest urea infusion rates or water infusion. As a result, initial modifications to the model included replacing linear kinetics with Michaelis-Menten saturation kinetics. Michaelis-Menten kinetics add stability to the model by not allowing fluxes to become infinitely large or small, and this is particularly important for diets that may not have been covered in the developmental database. These modifications allowed realistic values to be obtained for all data points. Other model modifications included simplification of the energy metabolism intermediates represented which behaviour and sensitivity analysis deemed unnecessary. The modified model performed well on the independent database and represents an improvement in prediction.

A protein binding agent decreases the *in vitro* rate of fermentation of protein meals

*Dunshea, F.R., Russo, V.M. and Leury, B.J., The University of Melbourne, The Department of Agriculture
and Food Systems, Royal Parade, Parkville, 3030 Victoria, Australia; fdunshea@unimelb.edu.au*

The rapid rumen fermentation of dietary protein can reduce the amount of rumen undegradable protein (UDP) and metabolisable protein (MP). The aim of the present study was to investigate the effect of a protein binding agent (Bioprotect™, RealisticAgri, Ironbridge, UK) on the rate of *in vitro* gas production from protein meals. The active ingredient in Bioprotect™ is a stable non-volatile organic salt that complexes with the 1^0 and 2^0 amino groups of protein and the hydroxyl groups of starch at neutral or slightly acidic conditions (pH 6 to 7), as observed in the rumen. These complexes decompose under more acidic (pH 2 to 3) conditions as in the abomasum and duodenum, making the protein and starch available for enzymatic digestion. Lupins and canola meal were ground and passed through a 1 mm sieve and samples of the ground meals were treated with Bioprotect™ (0, 15 and 100 ml/kg for lupins and 0 and 15 ml/kg for canola meal). Samples (1.0 g) of the meals (n=6 for each grain) were added to flasks containing buffered rumen fluid obtained from lactating dairy cows. The flasks were purged with CO_2 and maintained in a shaking water bath at 39 °C. Gas production was monitored every 5 min for 48 h using the ANKOM™ wireless *in vitro* gas production system. Gas production was modeled using a Gompetz equation to determine the maximum amount of gas production (Rmax) and the rate constant (β). The Rmax (140 vs 134 and 118 ml/g for 0, 15 and 100 g/kg Bioprotect™, P<0.05) and β (0.127 vs 0.105 and 0.059 min^{-1}, P<0.001) were decreased in a dose-dependent manner for lupins. The Rmax (91 vs 80 control and treated canola meal, P<0.05) but not β (0.166 vs 0.155 min^{-1}, P>0.10) were decreased by treatment of canola meal. These data demonstrate that Bioprotect™ can slow the *in vitro* fermentation of lupin and canola meal presunmably through increasing UDP. If more UDP can pass through the rumen it may improve MP yield through allowing greater post-ruminal enzymatic digestion of UDP.

Influence of hydrothermic processing of cereals on their nutritive value in ruminants

Chrenková, M.[1], Formelová, Z.[1], Poláčiková, M.[1], Čerešňáková, Z.[1], Weisbjerg, M.R.[2] and Chrastinová, Ľ.[1],
[1]Animal Production Research Centre Nitra, Hlohovecká 2, 951 41 Lužianky, Slovakia, [2]Aarhus University,
Blichers Allé 20, Postboks 50, 8830 Tjele, Denmark; chrenkova@cvzv.sk

For optimal utilization of cereals additional processing is required prior to feeding. Wheat is high in rapidly degradable starch. Up to 93% of wheat starch is degraded already after 9 h incubation in the rumen. Therefore, it can be appropriate to reduce its degradability. The effect of steam flaking on wheat (SFW) and steam flaking with addition of urea (SFWU) was determined in situ. Wheat was steam flaked at 90 °C for 30 min. Degradability of crude protein (CP) and starch (S) was measured in situ in the rumen of non-lactating cows. Intestinal digestibility of CP was determined by the mobile bag method. Nylon bags were incubated in the rumen for 16 h and then introduced into the small intestine through a duodenal cannula. We found some differences between untreated wheat and mainly SFWU in the content of CP (130.2 and 175.4 g/kg DM), nitrogen-free extract (809.7 and 767.3 g/kg DM), NDF (114.8 and 109.6 g/kg DM) and in buffer soluble N (25.0 and 30.0% N of total N). Treatment by steam flaking resulted in a marked decrease in effective degradability of CP (from 74% to 48%). Mainly the rate of CP degradability (untreated wheat 11.9%/h, SFW 3.25%/h and SFWU 2.6%/h) was affected. Urea treatment affected also soluble CP fraction 'a' (22.9 and 42.0%). Increased undegraded CP after 16 h rumen incubation subsequently increased the intestinal CP digestibility. The results demonstrated a positive effect of heat treatment on intestinal digestibility of rumen escape CP (87% untreated wheat, 96% treated). Also effective starch degradability decreased after treatment from 86.2% in wheat to 59.7% in SFWU. Proper steam flaking of wheat reduced ruminal protein degradation and increased the PDIN value from 81.6 to 109.9 g/kg DM and PDIE value from 99.9 to 107.7 g/kg DM for untreated wheat and SFWU, respectively.

Postruminal digestibility of crude protein in maize and grass silages in dairy cows

Ali, M.[1,2], Van Duinkerken, G.[1], Weisbjerg, M.R.[3], Cone, J.W.[2] and Hendriks, W.H.[2,4], [1]Wageningen UR
Livestock Research, 8200 AB, Lelystad, Netherlands, [2]Wageningen University, Animal Nutrition Group, 6700
AH Wageningen, Netherlands, [3]Aarhus University, Department of Animal Science, 8830 Tjele, Denmark,
[4]Utrecht University, Faculty of Veterinary Medicine, 3508 TD Utrecht, Netherlands; mubarak.ali@wur.nl

Various protein evaluation systems for ruminants, such as the DVE/OEB$_{2010}$ system in the Netherlands and the PDI system in France, use regression equations and simple models to estimate the intestinal digestibility of crude protein (CP). These equations and models are based on old databases obtained 20 to 50 years ago. The objective of this study was to investigate the relationship between the chemical composition and the determined intestinal digestibility of CP in maize and grass silages, in order to develop new regression equations. Twenty samples of maize silage and 20 samples of grass silage were selected to represent a broad range in chemical composition. The selected samples were incubated in the rumen of three lactating Holstein Friesian cows for 12 h using the nylon bag technique. Rumen incubated residues were transferred to mobile nylon bags and inserted in the duodenum through a cannula. Half of the bags were collected from the ileal cannula and the remaining half of the bags from the faeces. Regression equations were derived with significant predictors (P<0.1) to estimate the rumen degradable (12 h), the intestinal digestible and the undegradable fractions of CP using PROC REG backward procedure in SAS 9.2. There was a large variation in rumen degradability (12 h) and intestinal digestibility (small and large intestinal) of CP. Regression analysis showed that the rumen degradability, the intestinal digestibility and the undegradable fraction were influenced by the chemical composition of the maize and grass silages. The regression also showed a strong linear relationship between the calculated (DVE/OEB$_{2010}$) and determined (present study) values of intestinal digestion of rumen undegraded protein.

Relationship between chemical composition and rumen degradation characteristics of grass silages

Ali, M.[1,2], Cone, J.W.[2], Van Duinkerken, G.[1] and Hendriks, W.H.[2,3], [1]Wageningen UR Livestock Research, P.O. Box 65, 8200 AB Lelystad, Netherlands, [2]Wageningen University, Animal Nutrition Group, 6700 AH Wageningen, Netherlands, [3]Utrecht University, Faculty of Veterinary Medicine, 3508 TD Utrecht, Netherlands; mubarak.ali@wur.nl

Feed evaluation systems for ruminants use regression equations to estimate the rumen degradability of dry matter (DM), crude protein (CP) and neutral detergent fibre (NDF). The objective of this study was to improve the feed evaluation of grass silages by developing new regression equations to estimate the relationship between the chemical composition and the in situ rumen degradability of DM, CP and NDF of grass silages. The variation of the in situ rumen degradability between the cows was also investigated for DM, CP, NDF and organic matter (OM). Fifty-nine samples of grass silage with a broad range in chemical composition were selected. The selected samples were incubated in the rumen of three lactating Holstein Friesian cows for 0, 2, 4, 8, 16, 32, 72 and 336 h using nylon bag technique. Regression equations with significant predictors ($P<0.05$) were derived to determine the relationship between the rumen degradability and the chemical composition of grass silages using PROC REG backward stepwise procedure of SAS 9.2. The variation between the cows was checked by PROC MIX procedure of SAS 9.2. There was a large variation in the rumen degradability of DM, CP and NDF of the grass silage samples at each rumen incubation period. Regression analysis showed that there was a relationship between the rumen degradability of DM, CP and NDF and the chemical composition of grass silages. The regression analysis also showed the relationship between the rumen escape protein and the chemical composition of grass silages. The effect of cow variation on the rumen degradability of DM, CP, NDF and OM was not significant compared to the effect of the variation in chemical composition of grass silage samples.

Applying knowledge of AA metabolism to maximize N-efficiency of the dairy cow: the case of histidine

Lapierre, H.[1], Hristov, A.N.[2], Lee, C.[2] and Ouellet, D.R.[1], [1]Agriculture and Agri-Food Canada, Sherbrooke, QC, J1M 0C8, Canada, [2]Dpt Dairy and Animal Science, Pennsylvania State University, PA, 16802, USA; Helene.Lapierre@agr.gc.ca

Feeding low crude protein diets balanced for metabolizable (M) essential AA (EAA) would represent the best alternative to decrease N intake without altering milk and milk protein yields, assuming sufficient supply of rumen degradable N to support growth of microflora. This approach necessitates determination of EAA requirements (rqt). Amongst EAA, His is the one with the largest variation in the estimations of its rqt, varying from 2.4 to 3.2% of M protein (MP) supply. Therefore, His will be used as an example to illustrate how current knowledge can be used. His behaves as a Group 1 EAA, i.e. has a large liver extraction and a post-liver supply equal to mammary uptake, itself similar to output in milk protein. Therefore, the major net fates of His are milk protein secretion, endogenous fecal protein excretion, and liver removal. Altogether that would suggest that His should have similar rqt to Met, also a Group 1 EAA with a similar concentration in milk protein. The large variation in the estimated rqt for His may be related to carnosine. As this important muscular pool decreases with His deficiency, this depletion can contribute to His supply, thereby hiding a dietary deficiency. Based on the variable proportion of total EAA relative to total AA in MP, it is proposed that rqt of individual EAA would be better expressed as a proportion of total EAA rather than MP. Based on the EAA pattern of the fractions contributing to MP, His might become limiting in situations where microbial protein represents a large portion of MP, because of its relatively low His concentration. This has been proposed for grass silage-based diets and has recently been shown with low crude protein, corn silage-based diets. This example indicates how knowledge of the metabolism of EAA can help to refine estimations of rqt and even to predict under which type of feeding situations an EAA might become limiting.

Effects of Brown-midrib corn silage on dry matter intake, milk yield and composition in dairy cows
Gorniak, T., Meyer, U., Lebzien, P. and Dänicke, S., Friedrich-Loeffler-Institute (FLI), Institute of Animal Nutrition, Bundesallee 50, 38116 Braunschweig, Germany; tobias.gorniak@fli.bund.de

Brown-midrib (Bm) mutations in Zea mays were discovered in the 1920s and are known to improve fibre digestibility and dry matter intake (DMI) in ruminants which were most likely due to decreased contents of lignin and changes in lignin composition. Two trials with 64 lactating German Holstein cows (trial 1 and 2) and one trial with 4 wethers (trial 3) were conducted to examine the influence of an experimental Bm-hybrid compared to a common corn hybrid (Con) on DMI, milk yield, and milk composition in dairy cows and on digestibility in wethers. In trial 1, dairy cows were fed a total mixed ration of silage, Con or Bm respectively and concentrate (50:50 on dry matter (DM) basis). In trial 2, the cows received the respective silages for *ad libitum* intake and 6 kg of concentrate per animal per day. In trial 3, wethers received 1 kg DM of either silage, whereas animals were adapted to the silage for 14 days followed by 8 days of total collection of faeces. Treatment effects could be found for DMI only in trial 1 (Con: 22.5 kg/day; Bm: 21.5 kg/day) and for milk yield and fat corrected milk (FCM) only in trial 2 (milk yield Con: 25.8 kg/day; Bm: 29.4 kg/day; FCM Con: 27.2 kg/day; Bm: 29.6 kg/day). Milk fat content was decreased significantly by Bm-treatment in both trials (trial 1 Con: 3.8%; Bm: 3.3%; trial 2 Con: 4.4%; Bm: 4.0%). Milk protein content was not influenced. In trial 3, silages revealed an increased NDF digestibility (Con: 56.8%; Bm: 64.8%) but similar energy contents (Con: 6.4 MJ NEL/kg DM; Bm: 6.3 MJ NEL/kg DM). It needs to be clarified to what extent the influences of Bm on DMI, milk yield, and milk composition are related to alterations in ruminal fermentation associated with the improved fibre digestibility and changes in lignin content and lignin structure. Analysis of lignin structure and further trials with emphasis on ruminal fibre degradation will be needed.

Effect of deactivation of tannins of pistachio hull by PolyethyleneGlycol on gas production *in vitro*
Rahimi, A.[1], Naserian, A.[1], Shahdadi, A.[2] and Saremi, B.[3], [1]Agriculture, Animal Science, Ferdowsi University of Mashhad, P. O. Box- 91775-1163, Iran, [2]Agriculture, Animal Science, Agriculture Sciences and Natural Resources University of Gorgan, 49138, Iran, [3]Animal Science, Physiology & Hygiene Unit, University of Bonn, 53115, Germany; behnamsaremi@yahoo.com

The objective of this study was to evaluate effects of deactivation of tannin content of pistachio hull (PH) by PEG (as a tannin-binding agent) and to investigate its effects on gas production parameters. Treatments were: Control with 30% alfalfa, 20% straw and 50% concentration (0.42% tannin), T1) 15% alfalfa, 15% PH, 20% straw and 50% concentration (1.06% tannin) and T2) 30% PH, 20% straw and 50% concentration (1.83% tannin). All treatments were processed with 100 or 200 mg PEG in a 3×2 factorial experiment. Gas production was measured with 4 replicates for each treatment. The pressure and volume of gas was recorded before incubation (0) and 2, 4, 6, 8, 12, 24, 36, 48, 72, 96 and 120 hours after incubation. Gas production kinetics was estimated using the equation of Ørskov and McDonald. Statistical analysis was performed using the MIXED procedure of SAS (2000). Results showed that extent of gas production in treatments including 100 mg PEG weren't significantly affected, the highest extent was observed in T2 (213.515 ml/g DM). Extent of gas production in treatments including of 200 mg PEG in T2 (222.436) relative to control (209.159) was significantly increased (P=0.015), but relative to T1 (211.849) was not significantly different. Increasing PEG from 100 to 200 mg significantly increased rate of gas production in T1 (from 0.067 to 0.081 ml/h/g) (P=0.015), although there was no significant difference between rate of gas production in T2 and control. Increasing PEG from 100 to 200 mg in three experimental treatments led to an increment in extent and rate of gas production. Therefore, it seems that PEG could neutralize tannin negative effects based on gas production technique.

Effect of cellooligosaccharide feeding on growth performance and hormone concentrations in calves
Kushibiki, S., Shingu, H., Moriya, N., Kobayashi, H. and Yamaji, K., National Institute of Livestock and Grassland Science, Tsukuba, Ibaraki, 305-0031, Japan; mendoza@affrc.go.jp

Cellooligosaccharide (CE) is considered an indigestible oligosaccharide (NDO) because of its beta-1-4 linkages. NDOs prevent diarrhea and improve growth performance in calves. It has been reported that CE feeding improves daily body weight gain (DG) in weanling pigs. In an *in vitro* study, CE affected the fermentation of organic acids by mixed ruminal bacteria. The objective of this study was to evaluate the effects of oral CE administration on growth performance, feed intake, and plasma hormone concentrations in Holstein calves. Thirty-four female Holstein calves were allocated to a CE feeding group (n=17) and a control group (n=17). All calves were separated from their dams and given colostrum within 3 h. They were housed individually in pens and received their dams' milk for 2 days. The calves had *ad libitum* access to fresh water and were fed warm Holstein whole milk daily at 09:00 h and 16:00 h until weaning (42 days of age). The experimental period lasted 90 days (0-90 days of age). The CE feeding group was fed CE (Nippon Paper Chemicals Co., Ltd., Tokyo, Japan) dissolved in whole milk at 5 g/day. Following weaning, CE was dissolved in warm water. Body weight at 56-90 days of age was greater (P<0.05) in the CE feeding group than in the control group. Ninety-day body weight, DG, and feed efficiency in the postweaning period were also greater (P<0.05) in the CE feeding group than in the control group. There was no significant differences in the number of days to weaning or in plasma growth hormone concentrations between the groups. However, plasma insulin and insulin-like growth factor-I concentrations were higher (P<0.05) in the CE feeding group than in the control group at 90 days of age. These results suggest that CE feeding of calves improves DG and feeding efficiency during the postweaning period.

The effects of some essential oils on rumen total bacteria and protozoa numbers
Boga, M.[1], Kilic, U.[2] and Gorgulu, M.[3], [1]University of Nigde, Crop and Animal Production, Bor Vocational School, 51700 Nigde, Turkey, [2]University of Ondokuz Mayis, Department of Animal Science, Faculty of Agriculture, 55139 Samsun, Turkey, [3]University of Cukurova, Department of Animal Science, Faculty of Agriculture, 01130 Adana, Turkey; mboga@nigde.edu.tr

The aim of this study was to determine the effect of essential oils (EO) of Oregano (ORE, *Origanum vulgare*), black seed (BSD, *Nigella sativa*), Laurel (LAU, *Laurus nobilis*), cummin (CUM, *Cumminum cyminum*), garlic (GAR, *Allium sativum*), anise (ANI, *Pinpinella anisum*), cinnamon, (CIN, *Cinnamomum verum*), oleaster (OLE, *Eleagnus angustifolia*), mint (MNT, *Mentha longifolia*), rosemary (ROS, *Rosmarinus officinalis*), coriander (COR, *Coriandrum sativum*), grape seed (GRA, *Vitis vinifera*), orange peel (ORA, (*Citrus cinensis*) and fennel (FEN, *Foenicum vulgare*) on the number of rumen microorganisms. Rumen microorganisms were determined by using residual rumen liquor from *in vitro* gas production technique (IVGPT). Barley, soybean meal and wheat straw were incubated as substrate in IVGPT with doses of 0, 50, 100, 150 ppm being tested for all essential oils. The bacterial and protozoa contents of BSD and MNT treatments for barley were found higher compared to other plant extracts. Total bacteria counts with the MNT treatment for wheat straw were higher compared to other extracts. Bacteria counts for MNT were higher than those for other extracts. BSD ranked first in terms of protozoa counts. It was concluded that EO had significant effects on bacteria and protozoa counts (P<0.05), but EO doses had no significant effect (P>0.05). It is known that there is a symbiotic relationship between protozoa and methanogenic bacteria. In conclusion the highest total protozoal counts were determined with BSD for barley, wheat straw and soybean meals. For this reason, BSD may increase methanogenesis in the rumen. *In vivo* experimentation is required to determine the effects of BSD and other EOs on methanogenesis.

Use of odd-chain fatty acids to estimate the microbial protein synthesis

Chizzotti, M.L.[1,2], Lopes, L.S.[2], Busato, K.C.[2], Rodrigues, R.T.S.[1], Chizzotti, F.H.M.[3] and Ladeira, M.M.[2], [1]Univasf, Zootecnia, 56304-917 Petrolina PE, Brazil, [2]Universidade Federal de Lavras, Depto Zootecnia, UFLA, cx postal 3037, 37200-000 Lavras MG, Brazil, [3]Universidade Federal de Viçosa, DZO, Campus UFV, 36571-000 Viçosa MG, Brazil; mariochizzotti@dzo.ufla.br

It was aimed to evaluate the correlation of body odd-chain fatty acids (OCFA) with the digestible energy intake and microbial protein synthesis. Sixty goats from different breed groups (20 Moxotó, 20 Canindé and 20 F1 Boer × non-descript goats), averaging 15 kg of initial BW, were alocated to different feeding levels: *ad libitum*, restricted fed (75% of *ad libitum*), and fed at maintenance level, with five animals from each breed group at each level of feeding. The experimental design provided ranges in ME intake, and as consequence on the mass of empty body and of fatty acids. A baseline group of 15 randomly selected kids (five from each breed group) was used to assess the initial body composition. The diets consisted of 60% of elephant grass and 40% of concentrate, on dry matter basis. The feeding period lasted 90 days, after a 30 day adaptation period. After slaughter the carcasses were dissected and ground and total body meat (muscle and carcass fat) and offal were quantified and sampled. The fatty acid content in meat and offal were determined by gas chromatography. The total fat was determined by ether extraction. The mass of body odd-chain fatty acids ($C_{15:0}$, iso $C_{15:0}$, anteiso $C_{15:0}$, $C_{17:0}$, iso $C_{17:0}$, anteiso $C_{17:0}$, and $C_{17:1}$) and total fatty acids were determined. The total fatty acid and metabolizable energy intake were not affected by breed (P>0.14) but there was an effect of the level of feeding on those variables. The total and odd-chain fatty acids were correlated (P<0.01) with the metabolizable energy intake. The estimated microbial protein synthesis (MP) was correlated with total OCFA: MP (g/day) = 37.54 + 2.113 x OCFA (r^2=0.780). We conclude that the mass of odd-chain fatty acid in the body can be used to estimate the microbial protein synthesis. Funded by FAPEMIG and INCT-CA, CNPq.

Diurnal excretion patterns of milk urea, urinary urea and urinary creatinine in dairy cows

Eriksson, T., Swedish University of Agricultural Sciences, Department of Animal Nutrition and Management, Kungsangen Research Centre, 75323 Uppsala, Sweden; torsten.eriksson@slu.se

Milk urea concentration and urinary urea excretion are common estimates of dairy cow ration balancing and nitrogen excretion. Measurements of urinary urea excretion could either be based upon total urine collection or volume estimates from creatinine concentration. Diurnal patterns in the concentration and excretion of milk urea, urinary urea and urinary creatinine were evaluated in four change-over experiments with 85 cow × period observations from nine different diets. In all experiments, grass-legume silage constituted 65-70% of total ration, fed either *ad libitum* or at a semi-restricted level. Ration crude protein proportion was 14.6-17.7% of dry matter. Silage was fed in three or four equal meals, starting at 05:00 to 05:45 h, morning milking started 30-45 min thereafter and afternoon milking at 15:30 h (10 h interval). Quantitative urine collection was performed during 72 h with container renewal and sampling at 06:00 (nighttime excretion) and 18:00 h (daytime excretion). Time was used as the sole fixed factor in a mixed model of SAS together with random effects of trial and its interactions with time, cow, period and diet. Daytime concentrations were 25% (P=0.02), 17% (P=0.03) and 7% (P=0.07) higher than at nighttime for milk urea, urinary urea and urinary creatinine, respectively. Diurnal proportions excreted at daytime and probability for difference relative to nighttime were 46% (P=0.14), 51% (P=0.38), 55% (P=0.01) and 52% (P=0.04) for milk urea, urine volume, urinary urea and urinary creatinine, respectively. Experiments were then analysed separately for the abovementioned response variables with a GLM procedure that included effects of diet, cow and period and using either only daytime values (12 h/d) or values from the entire period (24 h/d). Diet effects were ranked numerically similarly in both cases although variation was increased with daytime values for 80% of all possible pairwise comparisons.

Effect of ruminally protected methionine on some blood serum metabolites in lactating Holstein cows
Kousary Moghaddam, M.[1], Soleimani, A.[1], Norouzi Ebdalabadi, M.[2], Ahmadzadeh Bazzaz, B.[1] and Khajeh Ghiassi, P.[1], [1]Islamic Azad University, Kashmar Branch, Animal Science, Kashmar, seyyed Morteza Blvd, 9197965816 Mashhad, Iran, [2]Khorasan Razavi Agricultural and Natural Resources Research Center, Animal Science, Mashhad, Toroq, 9185745896, Iran; mehdimoghaddam98@yahoo.com

An experiment was conducted to study the effect of rumen-protected methionine (RPMet; Metilock™) on some blood metabolites in dairy cows. Eighteen multiparous (parities 2-6) Holstein dairy cows were randomly assigned to 1 of 3 treatments: (1) 0 g daily RPMet (control); (2) 15 g daily RPMet; and (3) 30 g daily RPMet from calving to 42 days in milk. All cows were fed a routine farm ration (as a total mixed ration) three times a day and had free access to water at all times. Blood samples were collected weekly via venipuncture of coccygeal vessels before the morning feeding. The data were analyzed using the PROC MIXED model of SAS for a completely randomized design with repeated measures. The blood urea nitrogen (P=0.30; 21.58, 23.66, 24.30±1.27 mg/dl, respectively), glucose (P=0.06; 55.11, 49.12, 52.24±1.70 mg/dl, respectively), cholesterol (P=0.10; 132.11, 105.75, 129.16±9.14 mg/dl, respectively), non esterified fatty acids (P=0.55; 0.70, 0.54, 0.6372±0.10 mmol/l, respectively) and beta hydroxy butyrate acid (P=0.09; 0.87, 0.53, 0.62±0.11 mmol/l, respectively) were all similar among the groups. There are some trends (P<0.10) for effects of RPMet on glucose, cholesterol and beta hydroxy butyrate. Cholesterol was significant over time (P<0.01). Results of this study showed that the supplementation of ruminally protected methionine had little effect on these blood serum metabolites and metabolic responses.

Different rumen environments can cause different DM degradation profile of sainfoin
Khalilvandi Behroozyar, H., Rezayazdi, K. and Dehghan Banadaki, M., University of Tehran, Department of Animal Science, Karaj, Tehran, 3158777871, Iran; Khalilvandi@ut.ac.ir

Conventional in situ degradability has some problems in the case of tanniferous feeds because of small sample weights in nylon bags and dilution of antinutrients in rumen fluid. Thus, we used another method to assess tannin effects on DM degradability. Second cut sainfoin was dried and chopped (3-5 cm length), and exposed to a 5% (w/v) solution of polyethylene glycol (PEG 6000 MW) and water that was mixed with forage with v/w ratio of 1:1 and 4:1, respectively. Treatments were carried out at ambient temperature for 20 min with hand shaking for water and overnight for PEG. Extractable condensed tannin content was determined using butanol-HCl reagent. An in situ trial was done with a 3×3 change-over design. Three ruminally fistulated Holstein multiparous cows (680±20kg of BW) were used with 10 days for adaptation and 7 days for nylon bag incubations in each period. Forages were fed as sole diet (equal amount fed at 0800 and 1600) to meet 110% of maintenance. The CT content of control was 21.3, which was decreased by the water and PEG treatments to 1.7 and 0.3 g/kg DM, respectively. Untreated forage samples were ground to pass 2 mm screen. Five g of samples were weighed into nylon bags (10×20 cm, 50 μm pore size). Duplicates were incubated up to 96 h in ventral rumen. Degradation profiles were calculated by the nonlinear model of Ørskov and McDonald using NEWAY package. PROC MIXED of SAS 9.1 (2002) was used for statistical analysis at 0.05 probability level. Degradation rate increased from 0.0626[b]h[-1](control) to 0.0751[b]h[-1]and 0.0961[a] h[-1] and lag time decreased from 2.4[b] h to 1.33[a] and 0.8[a] h for water and PEG, respectively. PEG significantly increased ED from 52.4% to 55.9% and from 47.9% to 51.6% at outflow rates of 0.05h[-1]and 0.08 h[-1], respectively. We concluded that PEG and water treatments diminished phenolic compound effects in the rumen environment and increased nutrient availability in situ due to increased microbial activity and nutrient availability.

Effect of rumen protected choline on some blood serum metabolites in early lactating Holstein cows

Ahmadzadeh Bazzaz, B.[1], Soleimani, A.[2], Norouzi Ebdalabadi, M.[3] and Kousary Moghaddam, M.[4], [1]Islamic Azad University, Kashmar Branch, Animal Sience, kashmar, Seyyed mortezza Blvd, 4548484552-kashmar, Iran, [2]Islamic Azad University, Kashmar Branch, Animal sience, Kashmar, Seyyed mortezza Blvd, 4548484552-kashmar, Iran, [3]KHorasan Razavi Agricultural and Natural Resources Research Center, Animal sience, Mashhad, Torog, 4548484552 Kashmar, Iran, [4]Islamic Azad University, Kashmar Branch, Animal sience, Kashmar, Seyyed mortezza Blvd, 4548484552 kashmar, Iran; babak.ahmadzadeh@gmail.com

An experiment was conducted to evaluate the effect of rumen-protected choline (RPC; Vicomb™) on some blood metabolites in dairy cows. Eighteen multiparous (parities 2-6) Holstein dairy cows were randomly assigned to 1 of 3 treatments: (1) 0 g daily RPC (control); (2) 50 g daily RPC; and (3) 100 g daily RPC from calving to 42 days in milk. All cows were fed a routine farm ration (as total mixed ration) three times a day and had free access to water at all times. Blood samples were collected weekly via venipuncture of coccygeal vessels before the morning feeding. The data were analyzed using the PROC MIXED model of SAS for a completely randomized design with repeated measures. The blood urea nitrogen (P=0.89; 21.58, 22.02, 22.47±1.34 mg/dl, respectively), glucose (P=0.23; 55.11, 50.30, 51.63±2.00 mg/dl, respectively), cholesterol (P=0.41; 132.11, 119.97, 113.47±9.83 mg/dl, respectively), non-esterified fatty acid (P=0.97; 0.70, 0.71, 0.67±0.13 mmol/l, respectively) and beta hydroxy butyrate acid (P=0.58; 0.87, 0.64, 0.81±0.16 mmol/l, respectively) were all similar among the groups. Cholesterol and non-esterified fatty acids changed significantly over time (P<0.01). Results of this study showed that the supplementation of ruminally protected choline had no significant effect on these blood serum metabolites and metabolic responses.

Effects of dietary protein concentration on composition of exhaled air in dairy cows

Rustas, B.O.[1], Öberg, M.[2] and Lersten, K.[2], [1]Swedish University of Agricultural Sciences, Kungsängen Research Center, 753 23 Uppsala, Sweden, [2]Tibria, Vera Sandbergs allé 8, 412 96 Göteborg, Sweden; bengt-ove.rustas@slu.se

The objective of this study was to evaluate how dietary protein concentration affects the composition of exhaled air in dairy cows. Six mid- to late-lactation dairy cows (Swedish red breed) were offered grass silage at *ad libitum* intake supplemented with different proportions of rolled barley and a commercial protein concentrate. Cows were randomly assigned to one of six sequences in a balanced 3×3 latin-square design with three 4-day periods and three treatments. The treatments were 12, 16 and 20% crude protein in the diet. Measurements and sample collection were made on day 3 and 4 in each period. Samples of exhaled air were collected by face mask in Tedlar bags, transferred to adsorbent tubes and analysed with a gas chromatography–mass spectrometry technique. Urine was collected by spot samples. Results were analysed by the MIXED procedure in SAS. No carryover effects were detected (P>0.05). Milk production averaged 28.5 kg/day and did not differ (P=0.36) among treatments. Urea concentration in urine increased with increasing dietary protein concentration (P<0.0001). Dimetylsulphide, methylcyclopentane and hexane in exhaled air all increased with increasing dietary protein concentration (P<0.0001). Urinary urea concentration was linearly related to dimethylsulphide (P<0.0001) and curvilinearly related to methylcyclopentane (P<0.0001) and hexane (P<0.0001) in exhaled air. The results indicate a potential for volatile organic compounds in exhaled air from dairy cows to serve as indicators of dietary protein concentration and nitrogen excretion in the urine.

Deactivation of tannins affects metabolizable protein profile of sainfoin (*Onobrychis viciifolia*)

Khalilvandi Behroozyar, H., Dehghan Banadaki, M. and Rezayazdi, K., University of Tehran, Department of Animal Science, Daneshkadeh street, Karaj/Tehran, 3158777781, Iran; Khalilvandi@ut.ac.ir

Sainfoin (*Onobrychis viciifolia*) is a tanniferous legume forage with high hay quality. There are few reports about effects of tannin deactivation on metabolizable protein profile of sainfoin. Second cut forage was shade dried and chopped (3-5 cm length), and then exposed to solutions of KMnO4 (0.03 M, pH 12.04), NaOH (0.05 M, pH 12.28) and water with forage to reagent ratio of 1:4 (W/V). Treatments were carried out in triplicate, at 25 °C, for 20 min, with hand shaking. Condensed tannin (CT) concentration was determined using a Butanol-HCl reagent. Samples were analyzed for CP degradability using 3 ruminally fistulated Holstein cows, fed balanced rations with forage:concentrate ratio of 60:40. Samples were ground to pass a 2 mm screen and 5 g weighed in duplicate into nylon bags with 50 micron pore size and incubated in the ventral rumen up to 96 h. Effective degradability (ED) was calculated with NEWAY computer package. GLM PROC of SAS 9.1 and Duncan's Test were used for data analysis. The CT concentration of the control forage was 21.3±0.4 g/kg dry matter. Treatments decreased CT more than 90% and increased degradability at all rumen incubation times. The rapidly degradable fraction increased with treatments, but NaOH was most effective. The slowly degradable fraction was not affected by treatments, but rate of degradation of this fraction was significantly increased by water and NaOH. Treatments improved ED at different rumen outflow rates. Sodium hydroxide increased RDP more. Treatments improved protein metabolizablity of sainfoin. Also, digestible undegradable protein/undegradable protein ratio was improved by water and KMnO4. This study revealed that deactivation of CT can improve metabolizability of protein from tanniferous forages. Treatment of sainfoin with water, which is available on-farm and has no environmental hazard regarding it's use, greatly improved protein metabolism characteristics of sainfoin.

Peripartal tissue-specific transcriptome dynamics: a step towards integrative systems biology

Loor, J.J. and Bionaz, M., Department of Animal Sciences and Division of Nutritional Sciences, University of Illinois, Urbana, IL 61801, USA; jloor@illinois.edu

The advent and application of genome-enabled technologies (e.g., microarrays, next-generation sequencing) constituted a setback to the widespread use of the reductionist approach in livestock research. Those tools along with bioinformatics analyses of the resulting data are the foundation of modern systems biology. Systems biology is a field of study widely-used in model organisms to enhance understanding of the complex biological interactions occurring within cells and tissues. Application of systems biology concepts is ideal for the study of interactions between nutrition and physiological state with tissue/cell metabolism/function during key life stages of mammalian organisms such as the transition from pregnancy to lactation (the peripartal period). The nature of the physiologic and metabolic adaptations during this period is multifaceted and involves key tissues and cell types. Within that framework, the use of a single time point is reductive and insufficient to capture the dynamism of the underlying biological adaptations; thus, implementation of time-course experiments must be undertaken. We have developed and validated a bioinformatics approach for 'omics' data termed Dynamic Impact Approach (DIA) to help interpret longitudinal biological adaptations of liver, adipose, and mammary tissue to lactation using large transcriptomics datasets. Results demonstrate that the DIA is a suitable tool for systems analysis of complex genome-wide and tissue-specific datasets. Furthermore, the systems approach allowed simultaneous visualization of the complex inter-tissue adaptations to physiological state and nutrition. The knowledge generated from this integrative approach provided a more holistic understanding of the complex dynamic biological adaptations of tissues, and in the future may prove useful for fine-tuning nutritional interventions. An important goal during that process is to uncover key molecular players involved in the tissue's adaptations to physiological state or nutrition.

Bovine neutrophils' oscillation and transition period
Mehrzad, J., Faculty Veterinary Medicine, Pathobiology, Faculty of Veterinary Medicine, 9177948974-1793, Iran; mehrzad@um.ac.ir

As a pivotally cellular and molecular arms of the circulating innate immune system, neutrophils are the most vital primary mobile phagocytes; their appropriate function is pivotal to enhance udder performance in high yielding dairy cows. Despite substantial advances in biomedicine, the severity and incidence of periparturien-related diseases are unstoppably rising in high yielding dairy. Transition from parturition to lactation is highly stressful in dairy cows, peculiarly leading to oscillation of neutrophils' functions; this impairment can be cumulative in post-diapedetic udder neutrophils. Clearly, neutrophils' oscillation happens to many sensing-and-effector arms of the neutrophils' cytoskeleton machinery like membrane receptors, chemotaxis, diapedesis, phagocytosis, phagolysosome formation, release of enzymes, granules and free radicals into phagolysosome, eventually compromising udder immunity. Here I address the oscillatory events affecting the effectiveness of particularly neutrophils' effector arms of innate molecules in peripartum cows; I also discuss recent advances and future prospects in the area. Aspects such as cellular and molecular innate immunity, biology, biochemistry and biophysics of blood and milk neutrophils and neutrophil-pathogen interactions as well as increasingly environmental bovine granulotoxins will be addressed to gain more insight into the oscillatory events of blood and milk neutrophils' functions in peripartum cows. In short, occurrence of neutrophils' oscillation in the udder of periparturient dairy cows is inevitable, and is considered as a central host-attributable factor to the peripartum-related diseases especially mastitis. The sustainable solution for mastitis would be strengthening the sensing-and-effector arms of blood and milk neutrophils by means of attainable physio-immunological approaches.

Adiponectin and leptin system: long term physiological and conjugated linoleic acid induced changes
Saremi, B.[1], Winand, S.[1], Friedrichs, P.[1], Sauerwein, H.[1], Dänicke, S.[2] and Mielenz, M.[1], [1]University of Bonn, Katzenburgweg 7-9, 53115 Bonn, Germany, [2]Friedrich-Loeffler-Institute (FLI), Bundesallee 50, 38116 Braunschweig, Germany; bsaremi@uni-bonn.de

In two trials, Holstein cows were allocated to a conjugated linoleic acid (CLA) or control fat supplement at 100 g/day from d 1 to 182 post partum (p.p.) in trial 1 and d 1 to 105 p.p. in trial 2. Trial 1: 21 multiparous cows were biopsied from tail head adipose tissue (AT) and liver at d 21 ante partum (a.p.) and d 1, 21, 70, 105, 182, 196, 224, 252 p.p. In trial 2, 25 primiparous cows were slaughtered on d 1 (n=5; control), 42 and 105 p.p. (n=5 per each group) and 3 visceral (v.c.) and 3 subcutaneous (s.c.) AT depots, liver, mammary gland, and muscle tissues were sampled. The mRNA abundance (Ab) was quantified by qPCR. Data were analyzed using the mixed model (trial 1) and GLM or non parametric test (trial 2) (SPSS 17; $P<0.05$). The Ab of leptin and its receptor (ObR), adiponectin (Ad) and its receptors (AdR1 and AdR2) were reduced when comparing p.p. vs. a.p. in s.c.AT from tail head. Adiponectin mRNA Ab in retroperitoneal and omental AT a.p. increased from d 1 and 42 to d 105 p.p. in heifers whereas no such changes were observed s.c.AT. From d 1 to 105 p.p. AdR1 Ab decreased in muscle and increased in the liver and mammary gland tissues. AdR2 Ab was stably expressed in liver and muscle tissues but was increased in mammary gland tissue from d 42 to 105 p.p. Leptin remained stable p.p. regardless of parity and fat depot. Cows stably expressed ObR or ObRb in s.c.AT p.p. while ObRb Ab decreased p.p. in v.c.AT. Ad, AdR1, and AdR2 Ab were lower in omental AT, muscle, and in retroperitoneal fat from CLA treated cows, indicating insulin desensitizing effects of CLA supplementation. The relevance for the entire organism remains to be clarified. The observed shifts in the mRNA expression of the Ad and the leptin system in different tissues of cows were in line with the reduction in insulin sensitivity after parturition and its increase thereafter.

Mammary gland metabolism during manipulated insulin and glucose concentrations in dairy cows

Gross, J.J., Vernay, M.C.M.B., Kreipe, L., Wellnitz, O., Van Dorland, H.A. and Bruckmaier, R.M., Veterinary Physiology, Vetsuisse Faculty, University of Bern, Bremgartenstrasse 109a, 3001 Bern, Switzerland; rupert.bruckmaier@vetsuisse.unibe.ch

The regulation of mammary mRNA abundance of GLUT and other genes involved in mammary gland metabolism have been investigated during manipulated concentrations of insulin and glucose for 48 h in 18 diestric mid-lactating dairy cows. Six animals each were assigned to a hyperinsulinemic hypoglycemic clamp (HypoG), a hyperinsulinemic euglycemic clamp (EuG) and a control (NaCl). Blood samples were collected before and hourly during the infusions for analysis of glucose and insulin. Mammary gland biopsies were taken before and after 48 h of infusion and mRNA abundance of genes involved in the mammary gland metabolism were measured by RT-qPCR. Changes in the measured parameters, and the area under the curve of plasma parameters from 24 to 48 h of infusion were evaluated by analysis of variance with the treatment as fixed effect. From 24 to 48 h of infusion, HypoG cows had a lower glucose concentration (2.25±0.05 mmol/l) in comparison to EuG (3.80±0.16 mmol/l) and NaCl (4.17±0.10 mmol/l) cows (P<0.05), and HypoG and EuG cows had higher insulin concentrations than NaCl cows (41.9±8.1, 57.8±7.8, and 12.2±2.8 mU/l, resp.; P<0.05). In mammary tissue, the mRNA abundance of GLUT4 was decreased in EuG compared to HypoG, and that of αS1-Casein and INSR was decreased compared to HypoG and NaCl, while mRNA abundance of INSIG1 (insulin induced gene 1) and UGP2 (UDP-glucose pyrophosphorylase) was upregulated compared to NaCl and HypoG (P<0.05), while mRNA abundance of κ-Casein and ACoC were increased in EuG by trend (P=0.06). No differences between groups were found for mRNA abundance of GLUT1, α-Lactalbumin, SREBF1 and FASN. In conclusion, expression of GLUT1 is independent of glucose and insulin concentration, whereas GLUT4 is regulated at a status of increased glucose turnover. HypoG has a clear influence on the mRNA expression of milk proteins associated with glucose metabolism (α-Lactalbumin, κ-Casein).

Dendritic cells, neutrophils, and immune enhancement in the battle against mastitis

Kanevsky Mullarky, I., Virginia Tech, Dairy Science, 2050 Litton Reaves Hall, Blacksburg, VA 24061, USA; mullarky@vt.edu

The suppression of innate immunity during the periparturient period and the resulting increased incidence of infectious mastitis are of significant economic impact to the dairy industry. In addition, the mucosal immune system can be enhanced and suppressed by the antigens in the mammary gland. To date, research has focused on the role of neutrophils in defending the mammary gland against invading pathogens. Previous literature indicates that increased glucocorticoid levels at parturition suppress neutrophil function resulting in increased incidence of mastitis. Recently, we have described the dendritic cell population of the mammary gland and are delineating antigenic responses. Development of both nutritional and immunological therapies that target neutrophils and dendritic cells, will enable researchers to overcome immune suppression, enhance immune function, and thereby prevent mastitis.

Milk vesicles: evidence of their Golgi apparatus origin and outlining their physiological roles

Silanikove, N., ARO, Biol. of Lactation, P.O. Box 6, Bet Dagan 50250, Israel; nsilanik@agri.huji.ac.il

There are two sources of lipoprotein membrane in milk: the rather well characterized membrane that surrounds the milk fat droplets (MFGM) and nanosized vesicles floating in the milk serum in the range of 50 to 60 nM that account for 40-60% of the total membranous phospholipids content in milk. These vesicles are scarcely studied and their source is of debate. Before, they were considered as secondly derived from MFGM after secretion, or originating from mammary epithelial cells (MEC). Here, we provide compelling evidence that these vesicles originate from the golgi apparatus (GA) of MEC and are secreted intact into the milk; therefore, we named them golgisomes (Gsom). The most powerful evidence for this idea is the ability of isolated Gsom to synthesize ^{12}C-lactose from its precursors, glucose and ^{12}C-UTP-galactose. Both components of lactose synthase, galactosyltranferase and α-lactalbumin, are present in the golgisomes lumen. Phlorrihizin inhibited lactose synthesis in intact Gsom, indicating the presence of the glucose transport system in their membrane. The internal composition of Gsom lumen in terms of organic acid (citrate), carbohydrate (lactose), proteins (presence of casein in a colloidal form, α-lactalbumin, albumin) and mineral content is uniquely different from what could be expected to be found in the cell cytosol, or exosomes, and similar to what is expected to be found in secretory vesicles. The membrane composition is complex and as in exsomes, it contains many proteins such as TLRs, lactadherin and butyrophilin that may endow them with various regulatory and immune function. Where the lactose synthesis capacity of Gsoms is probably a superfluous feature reflecting their GA origin, our data suggest that they have important physiological roles in the regulation of milk secretion, innate and acquired immunology of the mammary gland and secretion of a toxic mineral that may poison MEC.

Can new probiont isolate of bovine milk reverse aflatoxin M1-induced neutrophil oscillation?

Mehrzad, J., Mahmoudi, M., Schneider, M., Klein, G., Vahabi, A., Mohammadi, A., Tabasi, N., Rad, M. and Bassami, M., Ferdowsi University of Mashhad, Pathobiology, Faculty of Veterinary Medicine, 9177948974-1793, Iran; mehrzad@um.ac.ir

As a metabolite of aflatoxin (AF) B_1, AFM1 is a potential granulotoxin, and in AFB_1-exposed cows the metabolite appears everywhere, especially in milk. Neutrophils are pivotal for defense against mastitis; they have enormous potentials to finally eliminate engulfed pathogens. We have recently observed the antagonistic properties of a *Lactobacillus fermentum*, isolated from milk of healthy dairy cows, against *Staphylococcus aureus*. To investigate the effect of AFM_1 on neutrophils killing activity against *S. aureus*, and to determine whether *L. fermentum* reverses the diminished effects of the AFM_1 on neutrophils, healthy dairy cows were used as a source of neutrophils. The isolated blood neutrophils were exposed with: (1) only 25 ng/ml of AFM_1; (2) AFM_1 plus overnight grown *L. fermentum*; (3) only overnight grown *L. fermentum*; and (4) none of them for 3 hours; their capacity to kill was then monitored by a bactericidal assay, accordingly, using *in vitro* challenge of *S. aureus* with above mentioned neutrophil groups for 1 h and then cfu counting of the *S. aureus*. Results from the bactericidal assay were expressed and compared as the percentage of killed *S. aureus*. The killing of *S. aureus* by AFM_1-treated neutrophils was markedly minimal and improved significantly with application of the *L. fermentum*. Observed improved phagocytic activity of neutrophils sheds fresh light on the application of this lactobacillus as a good probiotic to prevent infection in immunocompromised periparturient dairy cows.

The effect of choline and methionine on oxidative stress in a bovine mammary epithelial cell line

Baldi, A.[1], Skrivanova, E.[2], Rebucci, R.[1], Fusi, E.[1], Cheli, F.[1] and Pinotti, L.[1], [1]Università degli Studi di Milano, Veterinary Medicine Faculty, Milan, Italy, [2]Institute of Animal Science, Prague, 10401, Czech Republic; antonella.baldi@unimi.it

The aim of the present study was to evaluate the effect of hydrogen peroxide exposure on the survival and viability of bovine mammary epithelial cells in presence of choline (Cho) and methionine (Met). The BME-UV1 cell line has been used as the *in vitro* model of the bovine mammary epithelium. Cells were incubated with Cho and Met at two different concentrations: 500 μM and 715 μM for Cho and Met or 1000 μM and 1,430 μM for Cho and Met, respectively. The ratio between Cho and Met has been established on molar basis and insulin (1 μg/ml) has been included in order to support their uptake. First of all, the cells were exposed to increasing concentration of hydrogen peroxide (from 15.62 up to 890μM) for the following 24, 48 and 72h in order to calculate the half lethal concentrations (LC50) by MTT test. LC50 were 376.5μM at 24h, 249.9μM at 48h and 244.9μM of hydrogen peroxide at 72h. Subsequently, the cells were firstly exposed for 2h to Cho and Met and then treated with hydrogen peroxide for the following 24, 48 and 72h. After 24h of incubation, Cho and Met did not significantly influenced the viability of cells exposed to hydrogen peroxide. By contrast, after 48 and 72h of incubation, the cell proliferation was significantly ($P<0.05$) enhanced on average by 21% and 25.8% at the lowest range of hydrogen peroxide concentration tested (15.62 up to 62.5 μM), while at the highest concentrations (83.2 up to 333μM) by 14% and 15%. When apoptosis (TUNEL assay) was considered, presence of supplemental Cho and Met in the medium exerted a dose dependent effect on BME-UV1 cells, reducing the cell death according to the nutrients concentration. Our results indicate that Cho and Met could play a crucial role in counteracting oxidative damage induced by hydrogen peroxide in bovine mammary epithelial cells, even though the real mechanism merit further investigations.

Prioritising health in pedigree dog breeding

Lewis, T.W., Animal Health Trust, Lanwades Park, Kentford, Newmarket, Suffolk CB8 8SW, United Kingdom; tom.lewis@aht.org.uk

The genetic challenge facing pedigree dog breeders is the same as those facing livestock breeders: to achieve a satisfactory response to selection while maintaining sufficient genetic diversity within a breed, herd or population. It is simply the traits under selection that differ. For livestock breeders the objectives are to improve yields and efficiencies and to cut costs. For pedigree dog breeders primary selection objectives have often been defined by what is successful in the show ring. Effective selection relies on the collection of data pertaining to the objective traits, and genetic progress on the application of sufficient selective pressure to elicit a response. This entails possessing the motivation for change and the degree of control to enact it and is relatively easy to achieve in livestock industries where few individuals control a large number of animals, and where improvement is quantitatively evaluated and financially rewarded. The motivation to improve health in pedigree dogs is weaker since such traits are still sub-ordinate to 'type' traits in many overall objectives. Furthermore, the competitive nature of showing and the large number of individuals each controlling relatively few animals makes universal co-operation much harder to achieve. Improving selection for health in pedigree dogs therefore relies on fostering the motivation to include health in breeding objectives and in developing the opportunities for co-operation among breeders necessary to gather information and exert sufficient selective pressure throughout the breed population. This presentation aims to highlight some of the current initiatives in the UK intended to motivate breeders to put a higher priority on health in pedigree dog breeding objectives. Potential international collaboration both in calculation of genetic parameters and better use of generated screening data will also be explored, as will the differences in optimal strategies to effect selection given the differing industry and control structures in livestock and pedigree dog breeding.

Progressive retinal atrophy in dogs, its causal mutations in several dog breeds in Czech Republic

Hrdlicova, A., Majzlik, I., Hofmanova, B. and Vostry, L., Czech University of Life Sci., Animal Science and Ethology, Kamycka 129, 165 21 Prague 6 Suchdol, Czech Republic; majzlik@af.czu.cz

The aims of this study were to asses the occurrence of the Progressive retinal atrophy /gPRE/ and other eye diseases / rod cone dysplasia type2 rcd2, Central Stationary Night Blindness CSNB, Progressive Rod Cone Degeneration PRCD, Collie eye anomally CEA/ in 29 different dog breeds kept in Czech Republic at present. The laboratory investigation analysed 2,050 of blood samples from 29 dog breeds. We found in 27 breeds one cause mutations whereas in two breeds were 2 mutations disclosed. The different occurrence of mutations in breeds showed statistical significance. Mutation cord1 in sample of 552 Dachshund was found in 73 heterozygotes and 5 recessive homozygotes which were normal in phenotype. CSNB was tested in 193 Briards – among them we found 36 carriers. Testing the gPRA presence in 36 Sloughi we found 1 carrier only. CEA is widely spread in Collie population – of 46 animals are 43 recessive homozygotes and 3 carriers. The results of study showed the necessity of testing on all known mutation of eye diseases in dogs. We adopted a simplified PCR analysis using PICO system omitting DNA isolation for routine testing. This project was supported by grant MSM 604 607 09 01.

Management of inbreeding rates and relatedness in pedigreed dogs in the Netherlands

Windig, J.J.[1] and Oldenbroek, J.K.[2], [1]Wageningen Livestock Research, P.O. Box 65, 6500 AB Lelystad, Netherlands, [2]Centre for Genetic Resources the Netherlands, P.O. Box 16, 6700 AA Wageningen, Netherlands; jack.windig@wur.nl

Pedigreed dogs can suffer from effects associated with excessive inbreeding such as genetic defects. A project with the objective to prevent a further increase of inbreeding rates in dog breeds in the Netherlands was started. Two approaches are used. First a handbook is written for dog breeders on management of relatedness and inbreeding, translating the scientific knowledge to a practical level. In addition to the handbook, software is developed that can (1) monitor breeds with respect to inbreeding, relatedness and demography; (2) evaluate measures taken to restrict future inbreeding; and (3) provide information on relatedness of individual dogs. The handbook and software are integrated and will become available for all dog breeds and breeders associated with the Dutch kennel club. Three dog breeds participate in a pilot project to implement the software and handbook. Initial results show that the three breeds vary both in population size and inbreeding levels. Inbreeding rates up to 10% per generation were found. However, measures to restrict inbreeding can be effective for all three breeds.

Conservation of genetic diversity within dog breeds: what solutions in practice?

Leroy, G. and Rognon, X., INRA/AgroParisTech, UMR 1313 Génétique Animale et Biologie Intégrative, 16 rue Claude Bernard, 75231 Paris 05, France; gregoire.leroy@agroparistech.fr

Management of genetic diversity, in relation to dissemination of genetic disorders, constitutes a growing concern, for dog owners, breeders and kennel clubs. As a consequence, Fédération Cynologique Internationale (FCI) has set recently several recommendations concerning either the studs to be used, limitation of popular sire effect (through restriction of number of puppies produced per reproducers), or cross breeding use. We propose here to investigate the impact of some of those breeding strategies on genetic diversity, combining pedigree and simulation scenarios, based on different breeds with various demographic situations. We will analyse the consequence of close-breeding practice and popular sire phenomenon on diversity and breed health. The amount of genetic diversity that can be saved, through reproducer use restrictions will also be investigated. We will finally study to which extent cross-breeding may contribute to the gene pool of a given population, using the example of Barbet, a small French dog breed. Barbet breed has known several episodes of cross breeding from different origins, other breeds contributing to at least 45% of its gene pool. These results will be discussed in relation to considerations about genetic disorders, as well as breeders own choices.

The domestic livestock resources of Turkey: breed descriptions and status of guard and hunting dogs

Yilmaz, O., Ertugrul, M. and Wilson, R.T., Bartridge Partners, Bartridge House, EX37 9AS Umberleigh, United Kingdom; trevorbart@aol.com

The present day inhabitants of modern Turkey arrived in the country with the expansion of the Turkic Empire out from Centra Asia in the middle of the eleventh century. They travelled with their herds and flocks and with the guard and hunting dogs as part of their array of domestic animals. In the one thousand years since their arrival several specialized dog breeds have developed. This paper describes ten such, five of which are molossers, one is a sighthound, one is a scenthound and one is a small Spitz type. Two of the molossers (Kangal and Akbash) have local breed societies or associations and are well known and have breed societies internationally but are not recognized by the Fédération Cynoloqique Internationale (FCI). One molosser (Kars) is registered by the Turkish Standards Institute and another (Koyun) has been recently identified. The sighthound (Tazi) is similar to other Near and Middle East greyhounds. The scenthound (Tarsus Catalburun also known in English as Fork-nose and Turkish Pointer) is little known outside Turkey but is celebrated in its home area for its skills and is finding employment as a sniffer dog for narcotics, explosives and live and dead people. The Spitz-type (Dikkulak) is employed mainly as a household guard dog as are two other breeds of indeterminate type. The Cynology Federation of Turkey was formed in 2006 and is a contract partner of the FCI (and considers there may be as many as twenty dog breeds as opposed to the ten here described). A Turkish NGO known as Let's Adopt tries to place street dogs. Turkey's Animal Welfare Act No. 5199 of 2004 seeks to protect animals from torture, abuse and maltreatment but with regard to dogs is mainly concerned with capture-neutering-return of stray street dogs.

Genetic differentiation among three populations of Iranian guardian dogs

Asgarijafarabadi, G. and Allahyarkhankhorasani, D., Islamic Azad University, Varamin-Pishva Branch, Department of Animal Science, 7489-33817 Varamin, Iran; gh.asgari@iauvaramin.ac.ir

Several breeds of dogs can be found in Iran. Some of these breeds are completely different in their phenotype while others are similar. These animals are kept in rural families as guard animals for their herds. One of the indicators of genetic diversity within and between populations is Y sex chromosome haplotype which is inherited from sires to their male offspring. In order to study the genetic diversity of a part of Y sex chromosome between and within the populations of Sarabi, Sangsari and Afshari dogs, total number of 21 non inbred male dogs were sampled from these populations. Blood samples collected from these animals were transferred into laboratory by EDTA containers. DNA extraction was applied using extraction kits and forward and reverse primers were designed using Oligo5 software. After sequencing, results were analyzed by Finch and Mega4 softwares. Analyzing the replicated sequences showed that all of the sequences were conserved and having repeated sequences caused the existence of slip pages. In this study, both neighbor joint approach and unweighted pair-group method of the arithmetic average (UPGMA) were utilized to draw phylogenetic tree. Results showed that these three populations of dogs, based on their genetics, are completely different from each other which is very important in conservation of pure breeds and avoiding crossbreeding between distinct populations.

Analysis of the origin of sires and dams used in Polish breeding of Hovawart dogs

Głażewska, I., University of Gdańsk, Department of Plant Taxonomy and Nature Conservation, Al. Piłsudskiego 46, 81-378 Gdynia, Poland; i.glazewska@ug.edu.pl

Hovawart dogs are a German guard breed from the FCI 2 group. They are working dogs, however, the majority of Hovawarts are kept as family dogs. One of the important factors in the breeding of working breeds is a country of dog origin because of differences in the qualification criteria for breeding between particular kennel federations. This concerns mainly requirements as to dog shows, behaviour traits and health. The aim of the analysis was to evaluate the origin of sires and dams used in Polish breeding of Hovawarts in the 1988-2010 period. A pedigree analysis was performed on a total of 230 litters born in 87 kennels. The parents of the litters were 136 sires and 113 dams. Great differences were found in regard to the country of the parents' origin. The majority of the sires were of foreign origin (77.9%), mainly German (48.5%). The foreign sires originated also from the Czech Republic and Slovakia (21.3%), Austria, Sweden, Finland, Belgium, France and Great Britain. Contrary to the sires, the group of dams was predominated by Polish bitches (75.2%). From among 28 imported dams originating from five countries, 57.1% dams were born in Czech and Slovakian kennels, and 25.0% were from Germany. The difference regarding the choice of the country of the dams' purchase (mainly the Czech Reublic and Slovakia), and the country the dams were mated in (mainly Germany) resulted mainly from economic reasons (a significant difference in prices for puppies). The breeder's wishes to avoid mating between relatives and to use German sires, which were recognised as dogs of a higher utility value than the Polish dogs, are worth notice.

Breeding successes and defeats: gene flow in dam lines of Polish Hounds and Hovawarts

Głażewska, I.[1] and Prusak, B.[2], [1]University of Gdańsk, Department of Plant Taxonomy and Nature Conservation, Al. Piłsudskiego 46, 81-378 Gdynia, Poland, [2]Institute of Genetics and Animal Breeding PAS, Jastrzębiec, 05-552 Magdalenka, Poland; i.glazewska@ug.edu.pl

The aim of the analysis was the observation of gene flow in dam lines of two dog breeds bred in Poland: Hovawart and Polish Hound (ogar polski). Great differences in the effectiveness of transmitting genes of dam line founders to descendant generations were found in the both breeds. The lines differed regarding to a number of generations and a number of litters born in particular lines. In the Polish Hounds, 15 female founders of unknown origin were used in breeding between 1960 and 2008. From among lines established by these dams, six lines had only by one descendant generation and merely four lines had produced three or more generations. In the Polish population of Hovawarts bred between 1988 and 2008, which was descended from 23 imported dams, only 11 lines had more than one descendant generation. In both breeds the highest breeding success was achieved by the oldest line. In the Polish Hounds, 83.5% of the total number of litters represented the first dam line, and this number was equal to 37.0% in the Hovawarts. The breeding success of the first lines was mainly connected to the decisions of dog buyers as to the choice of a kennel. Persons with precise plans as to future breeding looked for puppies at established kennels. Meanwhile, persons interested in acquiring a family dog bought puppies from kennels located closer to their homes or from those that were selling at lower prices. This stratification in buyer intention negatively impacted the dynamics of change in the gene pool of the population. The second reason of the breeding defeat of new lines in the Polish Hounds was the worst quality of their representatives connected to a mixed breed origin of the founder dams which could be observed on the mtDNA level.

A regional approach for stable PRRS virus control and elimination

Rowland, R.R.R., Kansas State University, DMP, College of Veterinary Medicine, 1800 Denison Ave, Manhattan, KS 66506, USA; browland@vet.k-state.edu

PRRSV is a stealthy agent that can efficiently enter a swine unit at any stage of production and is efficiently spread by horizontal and vertical transmission. A subclinical, long-lived infection in immune competent pigs results in undetected circulation of the virus. Modified live virus (MLV) vaccines are available and offer some control over disease, but have not fully met the conditions needed for effective elimination (eradication) of the virus. High frequencies of genetic mutation and recombination result in the continuous evolution of several viral genes, which contribute to weak cross-protection offered by vaccination. Other deficiencies include vaccine virus shedding, persistent infection, inability to distinguish infected from vaccinated animals (DIVA), and a potential for reversion to virulence. However, PRRSV does possess an Achilles' Heel that makes virus control and elimination possible, even without an effective vaccine. Diagnostic tools, such as RT-PCR and serology, are sufficient to identify infected pigs. Using the available tools, it is relatively easy and cost-effective to eliminate PRRSV from a single herd; however, an outbreak with a new virus is all but inevitable. Therefore, the best approach for the stable control and elimination of PRRSV is the implementation of herd control methods that simultaneously target multiple farms within a region. The rationale for a regional approach is to lessen the risk of re-introduction of virus into any single farm. In 2004, the first regional elimination project was initiated in Minnesota, which included 87 pig sites and 164,000 pigs. By 2010, the region was free of PRRSV and the project was expanded to cover all of Northern Minnesota. As of 2012, more than 20 regional projects have been initiated or planned in the U.S. The estimated benefits of an effective elimination program include decreased costs of production and increased profits.

The role of vaccines for PRRS control and PRRSV eradication

Nauwynck, H.J., Trus, I., Christiaens, I., Karniychuk, U., Frydas, I., Cao, J. and Van Breedam, W., Ghent University, Department of Virology, Parasitology and Immunology, Salisburylaan 133, 9820 Merelbeke, Belgium; hans.nauwynck@ugent.be

After its first appearance in the late eighties, porcine reproductive and respiratory syndrome virus (PRRSV) caused for several years huge financial losses worldwide. Vaccines were developed shortly after, with big success. However, due to genetic drift, PRRSV was escaping more and more from vaccination immunity. At present, the original vaccines are no longer giving a full control of PRRS. Therefore, there is a need for adaptable vaccines that can be regularly updated. One should even think in the direction of auto-vaccines, based on farm-specific PRRSV isolates. Desperation drove farmers and veterinarians in the direction of 'controlled infections' with farm isolates with biosafety problems. Recently, a new methodology has been developed in our laboratory for adaptable inactivated vaccines with success. These vaccines allow to booster existing immunities in sows, giving a strong protection in sows and in the offspring via colostral immunity. This vaccine will not be the wonder product for naïve animals. In this context, adaptable live vaccines are necessary. Because PRRSV starts with a replication in the respiratory tract, the induction of a local immunity would be most suitable. At present, adaptable attenuated and vector vaccines are under development in the author's laboratory, based on new cellular and molecular insights of the PRRSV pathogenesis and immunity. A combination of adaptable attenuated/vector and inactivated vaccines will be the best tools to control and, if efficient enough, to eradicate PRRS.

Between-sow variation in tolerance in response to Porcine Reproductive and Respiratory Syndrome

Rashidi, H.[1], Mulder, H.A.[1], Mathur, P.[2], Van Arendonk, J.A.M.[1] and Knol, E.F.[2], [1]Wageningen University, Animal Breeding and Genomics Centre, De Elst 1, 6700 AH Wageningen, Netherlands, [2]Institute for Pig Genetics (IPG), Schoenaker 6, 6641 SZ Beuningen, Netherlands; hamed.rashidi@wur.nl

Genetic improvement of disease tolerance for PRRS may alleviate the effect of Porcine Reproductive and Respiratory Syndrome (PRRS) on reproductive performance of sows. However, for estimating the sow's merit for tolerance, virus load should be known on an animal basis, which is not known in practice. Therefore, our objective was to study the between-sow variation in response to PRRS in the absence of virus loads using derived phenotypic parameters as proxies. A total of 5,048 sows with 24,897 records from a commercial multiplier farm were analysed. Given the substantial effect of PRRS on reproduction of sows, outbreaks may be detected based on year-week estimates of reproductive performance. Therefore, a mixed model was fitted in ASReml with fixed effects for daylight, parity, year, and year-week and a random effect for sow. The year-weeks were classified as either healthy or diseased by performing a t-test ($\alpha=0.05$) for each estimated value of them. A year-week was specified as diseased (status = 1) if the estimated value of it is significantly less than zero, else it was classified as healthy (status = 0). Subsequently, variation among sows in response to PRRS for total number of piglets born alive (TBA) was estimated with a linear reaction norm model on the studentized year-weak estimate using random regression. Outbreaks of PRRS clearly led to an increased phenotypic variance of TBA. Some variation was found between sows in their response to PRRS. Correlations between sow performance in healthy and diseased periods were on average 0.9 and ranged between 1 and 0.4. These correlations indicate substantial re-ranking among sows in performance during diseased and healthy periods. Part of the between-sow variation may have a genetic background and therefore improvement of tolerance to PRRS by genetic selection may be feasible.

Genomics-enabled approaches to understanding host responses to infection with PRRSV

Archibald, A.L., Ait-Ali, T., Doeschl-Wilson, A. and Bishop, S.C., The Roslin Institute and R(D)SVS, University of Edinburgh, Division of Genetics and Genomics, Easter Bush, Midlothian EH25 9RG, United Kingdom; alan.archibald@roslin.ed.ac.uk

Porcine Reproductive and Respiratory Syndrome (PRRS) is the most costly disease to pig industries of Europe and North America and new PRRSV variants have the potential to be even more devastating. Tools, including microarrays for monitoring transcript levels of thousands of genes in parallel and SNP chips for determining genotypes at tens of thousands of loci have been used to dissect host responses to PRRSV infection. Our analysis of the innate immune response in PRRSV susceptible alveolar macrophages has revealed that PRRSV replication is associated with a sophisticated temporal and differential control of some type-1 interferon mRNA accumulation, or modulation of de-ubiquitination regulatory function. It is unknown whether these observations are attributable in part to the UK H2 PRRSV isolate used as various PRRSV isolates have been shown to modulate the expression of immunologically relevant molecules with different intensity. There is growing evidence from in-vitro and in-vivo challenge experiments and field studies, that there is host genetic variation in responses to and outcomes of PRRSV infection. We have shown that alveolar macrophages of the Landrace breed origin were less susceptible to the replication of the H2 PRRSV isolate than the other breeds tested. Our conclusions overlapped those of another *in vitro* study of monocyte-derived macrophages using similar breeds, but with a North American virus isolate. Genome scans have revealed loci with effects on PRRS in growing and pregnant pigs. Taken together, these studies highlight the potential scope for genetically improving traits related to the capacity of pigs to cope with PRRS infection and disease at the innate immune level.

Product quality from a nutritional and sustainability perspective

Den Hartog, L.A.[1,2] and Sijtsma, R.[1], [1]Nutreco, R&D and Quality Affairs, P.O. Box 220, 5830 AE Boxmeer, Netherlands, [2]Wageningen University, Animal Nutrition Group, Postbus 338, 6700 AH Wageningen, Netherlands; leo.den.hartog@nutreco.com

The quality of animal products such as dairy, eggs and meat is influenced by many factors, including animal feed. The latter has an impact on both 'positive' and 'negative' quality attributes, such as sensory characteristics, nutrient composition and safety. Until recently, the innovation focus was on the conversion of nutrients and the prevention of transfer of undesirable substances from feed into animal products, and the impact of processing on animal products. However, product quality also needs to be seen in relation to the production system. Society and consumers are increasingly concerned where their food comes from and how it is produced. As a consequence, quality attributes related with ecology and ethics are gaining importance in many markets. Such quality attributes may influence the final product quality either directly or indirectly by affecting the consumer's perception. The animal feed industry has a unique position in steering animal product quality with nutrition solutions, also beyond traditional attributes. The choice and origin of raw materials, the final feed composition and feeding programmes all have an impact on environmental footprints, both upstream and downstream in the value chain. The way crops are cultivated, processed and traded may also have a social impact. Moreover, animal welfare and health can be improved with innovative nutritional solutions. In particular the application of functional ingredients seems to be very promising in further reducing the use of antibiotics. In conclusion: animal product quality is highly linked with sustainable development and animal feed plays a crucial role. Nutritional solutions should be designed in such a way that product quality attributes are optimized taking into account ecological and ethical aspects associated with societal and consumer demands.

How do consumers perceive sensorial quality of beef?

Martín-Collado, D.[1], González, M.[2] and Díaz, C.[1], [1]INIA, Animal Breeding, Ctra. Coruña km. 7.5, 28040, Madrid, Spain; [2]Board of I.G.P. Carne de Ávila, Padre Tenaguillo, 8, 05004 Ávila, Spain; martin.daniel@inia.es

There is an increasing interest on considering meat quality in the breeding schemes of beef cattle. Quality is a complex concept that embraces heterogeneous aspects. Sensorial quality is supposed to be one of the most important for consumers. A first step for the inclusion of sensorial quality in the breeding schemes is to establish criteria that allow prioritizing its different components according to the consumer's opinion. Avileña-Negra Ibérica is a Spanish local cattle beef breed that developed the first Spanish label of fresh meat. Thus, it is an interesting case study to analyze what consumers search for when buying high quality products. With that purpose we developed a questionnaire to analyze the opinion of consumers of beef of the breed on the components and factors of the quality of its products and the reason for buying it. It included both open and closed questions. The last were used to quantify the importance of pre-selected component and factors of quality. The former were used to search for unexpected opinions of consumers analyzing the answers that they give spontaneously. Open questions avoid the bias that might result from suggesting the responses to individuals. We interviewed 384 people randomly sampled in the two provinces where the products are distributed. Flavor was valued as the most important component of the sensorial quality followed by tenderness and juiciness. Sensorial quality was only one of the elements behind election of high quality beef being food security another important element for buying labeled products. According to consumers' opinion, the factors that influence the most the quality of the final products were animal diet and sanitary conditions followed by farming system and animal welfare. Factors such as slaughter age, meat maturation or breed were found to be in a third level of importance. These results reflect the distance between the opinion of animal production professionals and consumers.

Eating quality of bull calves fed only grass or herbs match that of concentrate-fed veal calves

Therkildsen, M., Jensen, S.K. and Vestergaard, M., Aarhus University, Department of Food Science and Department of Animal Science, Foulum, 8830 Tjele, Denmark; mogens.vestergaard@agrsci.dk

The experiment aimed at elucidating the effect of purely grass or purely herb feeding of Holstein bull calves for 6 weeks prior to slaughter on the color, fatty acids and vitamin composition and eating quality of the meat in comparison with meat from traditional rosé veal calves fed a concentrate-based diet. Eleven calves were fed a ration of purely grass (Grass, n=6) or purely herb-based green feed (Herb, n=5) for 6 weeks with a daily gain of 987 and 969 g, respectively, before slaughter 10 months old. Meat was also sampled from 9-10 months old rosé veal calves (Con, n=6). The calves had a carcass weight of 178, 185 and 197 kg for Grass, Herb and Con, respectively, and similar pH 2 h (6.62±0.08) and 72 h post mortem (pm) (5.87±0.15). Seventy-two h pm M. longissimus dorsi (LD) and M. semimembranosus (SM) were removed and the color traits L*, a* and b* were measured of each muscle after one h blooming, with no significant difference between the feeding strategies. The muscles were aged for an additional 7 days, stored at -20 °C before a sensory analysis LD were prepared as steaks to an internal temperature of 63 °C and SM were prepared as roasts in an oven (100 °C) to an internal temperature of 63 °C. A 9-membered trained panel evaluated the aroma, flavor and texture traits of the meat. LD from Herb calves had more meat flavor (P<0.05) and was more juicy (P<0.02) compared with the Grass and Con, whereas SM from Herb calves were characterized as having less sweet aroma (P<0.02) compared with Grass and Con, but otherwise there were no significant differences in the sensory profile of the cuts from the three feeding strategies. The meat from Herb calves contained less oleic acid (P<0.001), and more α-linolenic acid (P<0.001), vitamin E (P<0.001) and β-carotene (P<0.05) compared with the Con calves and with the Grass calves in between.

Effect of breed and diet on beef cattle fatty acid profile and SCD mRNA expression
Costa, A.S.H., Bessa, R.J.B., Alfaia, C.P.M., Pires, V.M.R., Fontes, C.M.G.A. and Prates, J.A.M., CIISA, Faculdade de Medicina Veterinária, Avenida da Universidade Técnica, Pólo Universitário da Ajuda, 1300-477 Lisboa, Portugal; adanacosta@fmv.utl.pt

Intramuscular fat composition of ruminants influences the quality of the final product. It is, therefore, important to estimate the fatty acid composition of cattle from a typical production system for each breed and its composition varies within common production systems. On the other hand, differences among cattle breeds in fatty acid composition have been attributed to genetic variations in fatty acid metabolism. In this study, it was hypothesized that there are breed- and diet-induced variations on lipid metabolism in the muscle tissue, which may be modulated through stearoyl-CoA desaturase (SCD) gene expression. To this purpose, forty purebred young bulls from two phylogenetically distant autochthonous cattle breeds, Alentejana and Barrosã (n=20, respectively), were assigned to either low or high silage diets, in a 2×2 factorial arrangement. Meat fatty acid composition, including the detailed conjugated linoleic acid (CLA) isomeric profile, was determined along with quantitative real-time polymerase chain reaction detection of SCD mRNA levels. Meat from Barrosã bulls was richer in MUFA, CLA, trans fatty acids (TFA) and poorer in polyunsaturated fatty acids and plasmalogenic lipids, than that from Alentejana bulls. Feeding low silage diets promoted higher proportions of TFA, lower percentages of branched chain fatty acids, resulting in less beneficial fatty acid ratios. No association was found between the SCD mRNA levels and the desaturase indices, as well as the proportions of some fatty acids classes, except for TFA. In summary, these findings highlight the importance of the genetic background and diet on meat fatty acid composition.

Effect of linseed diet on intramuscular fatty acid profile of bulls slaughtered at different ages
Karolyi, D.[1], Radovčić, A.[2], Salajpal, K.[1], Kljak, K.[1], Čatipović, H.[2], Jakopović, T.[2] and Jurić, I.[1], [1]Faculty of Agriculture, University of Zagreb, Svetošimunska 25, 10000, Zagreb, Croatia, [2]Agrokor d.d., Trg Dražena Petrovića 3, 10000 Zagreb, Croatia; dkarolyi@agr.hr

The study examined the effect of linseed diet and slaughter age on beef carcass and meat quality, and intramuscular fatty acid (FA) profile and oxidative stability. The sample included 80 Simmental bulls reared under intensive production system. The animals (initial age=221±9 d) were equally assigned to 4 groups and fed with linseed (L) or control diet until slaughter at 13 or 17 months. Diets were fed *ad libitum* as total mixture ration (high moisture corn, corn silage, protein supplement and hay), with the addition of whole linseed grain in L diet groups (130 and 160 g per head daily until and after the age of 13 months, respectively). After slaughter, carcass traits (weight, dressing-out %, trimmed fat) were measured and *M. longissimus thoracis* colour (CIE Lab) and pH, and FA composition (of total lipids) and oxidative stability were determined. Data were analysed by GLM procedure, with the model including slaughter age, diet, and their interaction as fixed effects. As regards to slaughter age, all carcass traits were typically higher (P<0.001) in older animals, which also had a darker (P<0.001) and fatter (P<0.01) meat, but with only few differences in FA composition (e.g. more CLA, P<0.001) compared to younger animals. Diet had no influence on carcass and meat quality, except for a higher dressing-out % (P<0.01) in L diets. Intramuscular fat of linseed fed bulls was less saturated (P<0.01), while for several FA a significant (P<0.05) slaughter age × diet interaction existed. In particular, L diet increased (P<0.05) the proportion of most individual as well as total n-3 FA in yearling bulls but without such response in more grown-up animals. Lipid oxidation, measured during the six-day cold storage was generally more developed in younger and linseed-fed animals. Research was supported by EUREKA Project (E! 3983-RFA).

Productive and carcass traits of Alentejana and Barrosã bulls fed high or low silage diets

Costa, A.S.H.[1], Costa, P.[1], Bessa, R.J.B.[1], Lemos, J.P.C.[1], Simões, J.A.[2], Santos-Silva, J.[2], Fontes, C.M.G.A.[1] and Prates, J.A.M.[1], [1]CIISA, Faculdade de Medicina Veterinária, Avenida da Universidade Técnica, Pólo Universitário da Ajuda, 1300-477 Lisboa, Portugal, [2]Unidade de Investigação em Produção Animal, INRB, Fonte Boa, 2005-048 Vale de Santarém, Portugal; adanacosta@fmv.utl.pt

Carcass composition is mainly influenced by the genetic background and feeding regimen. In the present study, the effects of breed (large- vs. small-framed) and diet (high vs. low silage diets) on the productive traits and carcass composition were assessed in a 2×2 factorial arrangement. For this purpose, forty young bulls from two phylogenetically distant Portuguese bovine breeds, Alentejana (n=20) and Barrosã (n=20) were selected and fed high (70% silage/30% concentrate) or low (30% silage/70% concentrate) silage diets. Animals were slaughtered at 18 months-old and their live slaughter and hot carcass weights were determined, as well as the proportions of subcutaneous and visceral fat depots. In addition, intramuscular fat (IMF) levels were determined. Barrosã bulls fed the low silage diet had the highest IMF contents. While diet determined the proportions of total visceral fat and individual fat depots. Barrosã bulls had higher fatness scores relative to Alentejana bulls. To further understand the relationship among different carcass traits, a correlation analysis was performed. A strong association between IMF and most of the fat depots considered was found for Barrosã bulls, although the same pattern was not observed for Alentejana bulls. This appears to indicate that increasing the IMF content in the latter breed, without an increase in visceral fats, would be possible. Under these experimental conditions, genetic background was a major determinant of carcass composition and meat quality, while the impact of diet was limited.

Carcass and colostrum quality of Angus cattle with different myostatin genotypes

Eder, J.[1], Wassmuth, R.[2], Von Borell, E.[1] and Swalve, H.H.[1], [1]Martin-Luther-University Halle-Wittenberg, Institute of Agricultural and Nutritional Sciences, Theodor-Lieser-Str. 11, 06120 Halle, Germany, [2]University of Applied Sciences Osnabrück, Group Animal Breeding and Husbandry, P.O. Box 1940, 49009 Osnabrück, Germany; eder.johannes@t-online.de

The objective was to evaluate carcass and colostrum quality of Angus cattle with different myostatin genotypes. A total of 77 young bulls from one herd and 24 cows from another herd were included in the present study. The genotypes of cows and bulls were analysed by Eurofins Medigenomix GmbH. The typical mutation of the myostatin gene for Belgian Blue and Angus (nt821) was found. The proportions of free cows (no inactive myostatin allele), heterozygous cows (one inactive myostatin allele) were 54.0% and 46.0%, respectively. The frequencies of young bull genotypes were 79.0% and 21.0%, respectively. Heterozygous young bulls were heavier at slaughter (694.3 kg) than free young bulls (680.8 kg) and had a better marketing score in the EUROP-System. Furthermore, a significant difference was found between the two myostatin-genotypes in carcass weight and dressing percentage. Carcasses from free young bulls weighed less (382.2 kg) with a reduced dressing percentage (56.1%) than heterozygous young bulls (410.8 kg and 59.2%, respectively). The colostrum of homozygous free cows contained less magnesium (1.341 mg/kg) than the colostrum of heterozygous cows (1.730 mg/kg). Hence, the heterozygous myostatin genotype of cows led to higher magnesium colostrum concentration and increased carcass weights and dressing percentages in young bulls. In conclusion, unexpected phenotypic differences appeared between homozygous free and heterozygous animals of the autosomal recessive double muscling gene despite that these differences are usually only expected between homozygous genotypes. The proposed dominant-recessive inheritance of the myostatin allele could not be observed.

Short-term supplementation with rice bran in pre-partum primiparous grazing beef cows

Quintans, G., Scarsi, A. and Banchero, G., National Institute for Agricultural Research, Beef and Wool, Ruta 8 Km 281, Treinta y Tres, 33000, Uruguay; gquintans@tyt.inia.org.uy

When native pastures is the primary diet for cattle, pregnant cows in the last trimester of pregnancy are exposed to winter, when the quantity/quality of forage is poor. In extensive production systems, long-term supplementation is not economically profitable. The aim of this study was to evaluate the effect of a short-term supplementation in primiparous cows and its effects on body condition score (BCS), body live weight (BW), milk production (MP) and early pregnancy rate (EP). Twenty five pregnant heifers (AAxHH) that were previously inseminated at 26m, were randomly assigned to two treatments 56 days before the expecting calving date (Day 0): (1) cows grazed native pastures (CON; n=12); (2) cows grazed native pastures and supplemented daily with whole-rice bran at 0.75 kg /100 kg BW during the last 38 days of gestation (SUP; n=13). All animals were managed together. BCS and BW were recorded biweekly. MP was measured at Day 15 and 30 postpartum (pp) and every 30 days thereafter. The mating period started at Day 49 pp and lasted 60 days. EP was performed by ultrasonography. BW, BCS and MP were analyzed by repeated measures using the MIXED procedure with time as the repeated effect and probability of EP was fitted using the GENMOD procedure with the binomial distribution. At the onset of the supplementation cows weighed (mean±sem) 398±8.0 kg and had 3.8±0.08 u of BCS. At calving, there was no difference in BW or BCS between cows from both groups (381±8.0 kg; 3.7±0.08 u). Calves BW at birth was similar between treatments (38.2±1.6kg). There was no effect of treatment on MP but it decreased significantly (P<0.0001) along the pp period, being in average 8.6±0.5 kg/d at Day 15 and 4.3±0.5 kg/d at Day 120. Probability of EP tended (P=0.08) to be greater in SUP than in CON cows (58 vs 23%). Short-term supplementation before calving could be an economical and effective technique to increase reproductive performance but more research is needed.

The relationship between pork quality traits and fatty acids composition

Jukna, V., Meškinytė-Kaušilienė, E. and Klementavičiūtė, J., Lithuanian University of Health Science Veterinary Academy, Laboratory of Meat Characteristics and Quality Assessment, Tilzess 18, 47181 Kaunas, Lithuania; vjukna@lva.lt

Fatty acid composition is a major factor in the nutritional value of meat, with a high polyunsaturated fatty acid to saturated fatty acid ratio of 0.4 or above considered as suitable for human consumption. The objective of performed study was to estimate correlations between fatty acids composition and meat quality traits in M. longissimus dorsi of pork. The experiment was conducted with 25 meat samples. Pigs were grown in the same housing and feeding conditions. The traits included cooking loss, water holding capacity, color intensity, pH, color, dry matter, fat and fatty acids content. All the studies were performed 48 hours after slaughter and were performed according to generally accepted methods. The most statistically significant correlation was between the fatty acid composition of fat and dry matter content in the muscles. Negative and positive correlation coefficients ranged from very small (0.001) to very large (0.92). The mainly fatty acids from all pork technological quality parameters correlated with pH and cooking loss (P<0.01), (P<0.001). Statistically significant positive correlation was between water holding capacity and total monounsaturated fatty acids (P<0.01), the most of them with oleic acid (C18: 1) (P<0.01), cis-11-eicosenoic acid (C20: 1) (P<0.001), and negatively – with arachidonic acid (C20: 4) (P<0.05) and total polyunsaturated fatty acids (P<0.05). Meat color and fatty acids are positively correlated with palmitoleic acid (C16: 1) and total saturated fatty acids (P<0.05) and negatively correlated with docosapentaenoic acid (C22: 5) and homo-gamma-linolenic acid (C20: 3) (P<0.05). In pork dry matter content is negatively correlated with stearic acid (C18: 0) (P<0.05), as well as negatively correlated to the total saturated fatty acids (P<0.10).

Estimation of myostatin gene effect on production traits and fatty acid contents in bovine milk
Vanrobays, M.[1], Bastin, C.[1], Colinet, F.G.[1], Vandenplas, J.[1,2], Troch, T.[1], Soyeurt, H.[1,2] and Gengler, N.[1], [1]Gembloux Agro-Bio Tech, University of Liege, Animal Science Unit, Passage des Déportés 2, 5030 Gembloux, Belgium, [2]National Fund for Scientific Research, rue d'Egmont 5, 1000 Brussels, Belgium; mlvanrobays@ulg.ac.be

The aim of this study was to estimate the genetic parameters of milk, fat, and protein yields, saturated (SFA) and monounsaturated fatty acid (MUFA) contents in bovine milk and to estimate the Myostatin (mh) gene effect on these traits. For this purpose, 51,614 test-day records (24,124, 16,145, and 11,345 for first, second, and third lactation, respectively) of 3,098 dual purpose Belgian Blue cows in 38 herds from the Walloon Region of Belgium were used. Because only 2,301 animals, including 1,082 cows with test-day records, were genotyped for mh, the gene content of non-genotyped animals was predicted from animals with a known genotype using the relationships with these animals. Variance components were estimated using Restricted Maximum Likelihood. A 3-lactations, 5-traits random regression test-day mixed model, based on the official Walloon genetic evaluation model for production traits, was used with an additional fixed regression on mh gene content to estimate allele substitution effects. Daily heritability estimates (average of 3 lactations) were 0.34 for SFA and 0.16 for MUFA and were higher than those for production traits (0.11, 0.10, and 0.09 for milk, fat, and protein yields, respectively). Allele substitution effects (± approximate standard-errors) for mh through the three lactations were -0.628 (±0.343), -0.024 (±0.014) and -0.021 (±0.009) kg per day for milk, fat, and protein yields, respectively. Concerning SFA and MUFA contents in milk, the average allele substitution effects were -0.001 (±0.027) and 0.029 (±0.023) g/dl of milk. To conclude, results from this study showed that milk performance traits and milk fatty acid profile are influenced by mh genotypes.

Effects of clove and cinnamon essential oils on milk yield and milk composition of dairy goats
Rofiq, M.N. and Görgülü, M., Cukurova University, Animal Science, Ziraat Fakultesi, Zootekni Bolumu baskanligi, Balcali Kampusu, 01330 Adana, Turkey; nasir_rofiq@yahoo.com

The aim of the present study was to evaluate the effect of oral infusion of clove (CO, eugenol) and cinnamon essential oils (CIN, Cinnamaldehyde) on milk yield and milk composition of lactating German Fawn goats. Clove and cinnamon essential oils were used for rumen manipulation in ruminant animal production. Their major component, eugenol and cinnamaldehyde were proved to optimize rumen metabolism and rumen microbial composition in in-vitro experiments. Only few studies examined *in vivo* the effects of the essential oils and their combination on performance of dairy goats. Forty lactation German Fawn Goats having 46.2±1.9 kg live weight, 84±9 day in milk and 1.52±0.32 kg milk/day were allocated in 4 treatments. The treatments were 1) control, 2) oral infusion with 1.8 g/d CO, 3) oral infusion with 1.8 g/d CIN, 4) oral infusion with combination of CO (1,8 g/d) and CIN (1.8 g/d) which were assigned and analyzed in two by two factorial arrangement in a completely randomized designed. Goats were fed *ad libitum* with total mix ration (TMR) containing 60% concentrate and 40% alfalfa hay sized 1-2 cm, and had free access to water at all time. Oral infusion of essential oil was made everyday in the morning for 5 weeks experimental period. Daily milk yield was measured twice a day in milking parlor or small ruminant. Milk composition was analyzed by FOSS® Milko scan FT-120 twice a week. Oran infusion of CO, CIN and their combination had no significant effect ($P>0.05$) on dry matter intake (DMI), milk yield, fat corrected milk yield (FCM) and milk composition (TS, SNF, fat, protein, lactose, casein and urea) concentration. The results revealed that essential oil may not have any effect on milk yield and milk composition under the feeding condition of 60/40 concentrate roughage ratio in the relatively lower yielding dairy goats.

Slaughter value of Limousine breed calves slaughtered at different ages and different body weights

Litwinczuk, Z.[1], Stanek, P.[1], Jankowski, P.[1], Domaradzki, P.[2] and Florek, M.[2], [1]University of Life Sciences in Lublin, Departament of Breeding and Protection of Genetic Resources of Cattle, Akademicka 13, 20-950 Lublin, Poland, [2]University of Life Sciences in Lublin, Department of Commodity Science and Processing of Raw Animal Materials, Akademicka 13, 20-950 Lublin, Poland; zygmunt.litwinczuk@up.lublin.pl

The present research involved 29 calves of Limousine breed (19 bulls and 10 heifers) reared with their mothers on the pasture until the moment of slaughter. The slaughter value was analyzed in relation to calves age; i.e. up to 6 months (up to 180 days), approximately 7 months (181-220 days) and approximately 8 months (221-245 days) and their slaughter weight; i.e. up to 250 kg, 251-300 kg and above 300 kg. The calf sex affected the body weight growth rate and the carcass conformation of analyzed animals; however, their carcass fatness was corresponding. The slaughter age of animals had no significant effect on the dressing, which exceeded 60% in all analyzed groups (ranging from 61.0 to 63.3%). Also, the carcass conformation and fatness were not affected significantly by the slaughter age. This indicates on a possibility of fattening calves with their mothers to the age of 8 months. No significant differences were determined in the carcass fatness for animals slaughtered at different body weights. Regardless of the slaughter age, calves carcasses were qualified to fatness class 1. However, the increase of the slaughter body weight had a significant (P<0.01) impact on the better conformation of carcasses, from the average level of 8.7 points (in 15 points scale) in the group of the lightest animals (slaughtered at weight under 250 kg) to 11.1 points in the heaviest animals group (over 300 kg).

Self performance test results of British candidate beef bulls

Cseh, G.[1], Hollo, I.[1] and Marton, J.[2], [1]Kaposvár University, Guba Sándor Str. 40, 7400 Kaposvár, Hungary, [2]Association of Hungarian Hereford, Angus and Galloway Breeders, Dénesmajor 2, 7400 Kaposvár, Hungary; lastrebel20@gmail.com

The authors evaluated the self-performance test results of Angus (n=50), Hereford (n=62) and Galloway (n=12) candidate bulls on the Experimental and Research Farm of the Kaposvár University, between the time line of 2005-2011. The production traits (365 days adjusted weight, average gain during test, back-fat thickness), the reproduction traits (205 days adjusted weight, size points, height at hip, scrotal circum) and the selection index were analysed. Microsoft Excel data management software, and the SAS 9.1 statistical software were used. The appraisal of the impact of age-group and breed was done with a multivariate analysis of variance (MANOVA). The assessment of the differences between the mean values was made with the Tukey's test, and the relationship between attributes was qualified with correlational analysis. Significant differences were found among breeds in all traits, except for the size points and the scrotal circum. Based on relationships among traits, measurement of back fat thickness at the same bodyweight was advise.

Fatty acid profile of various adipose tissue depots in bulls of different breeds
Bures, D., Barton, L. and Rehak, D., Institute of Animal Science, Cattle Breeding, Přátelství 815 Praha Uhříněves, 104 00 Prague, Czech Republic; bures.daniel@vuzv.cz

The objective of the present study was to compare the fatty acid (FA) profile of intramuscular, subcutaneous and cod adipose tissue (AT) of Aberdeen Angus (AA), Gascon (GS), Holstein (H) and Czech Fleckvieh (CF) bulls. The breeds were selected on the basis of their differing propensity to deposit AT. Thirty-six bulls (9 per breed) were reared and finished under identical housing and feeding conditions. Days on feed and slaughter age were similar for breed groups. The AT site was found to be a significant source of variation (P<0.001). Whereas intramuscular AT contained the highest proportion of polyunsaturated FA and polyunsaturated FA/ saturated FA ratio, cod AT had the highest concentration of saturated FA, and subcutaneous AT contained the highest proportion of monounsaturated FA, regardless of breed. The lowest concentrations (P<0.001) of c9t11CLA (conjugated linoleic acid) were observed in intramuscular AT compared to the other AT sites. The AA and H bulls exhibited higher (P<0.001) intramuscular AT (petroleum ether extract) contents (36.2 and 27.7 g/kg, respectively) than the GS and CF bulls (15.0 and 16.9 g/kg, respectively). These differences give plausible explanation of the tendency towards lower PUFA and PUFA/SFA (P<0.1) found in the AA and H bulls. The AA bulls had the highest proportions of C14:1n-5 and C16:1n-7 in most AT. In addition, the AA bulls had the highest concentrations of c9t11CLA in subcutaneous (P<0.001), cod (P<0.01) and intramuscular (P=0.0528) AT. Based on the results it is suggested that the differences may exist between breeds and between adipose depots in the activity of stearoyl-CoA desaturase, which is the enzyme responsible for the conversion of saturated FA into monounsaturated FA and trans-vaccenic acid into c9t11CLA, as well as in the activity of other enzymes involved in de novo FA synthesis. The study was supported by the Ministry of Agriculture of the Czech Republic (project QH 81228).

Effect of markers in FABP4, LEP and RORC genes on taste in Spanish South-western beef cattle
Aviles, C.[1], Peña, F.[1], Barahona, M.[2], Campo, M.M.[2], Sañudo, C.[2] and Molina, A.[1], [1]Grupo Meragem, Universidad de Cordoba, Campus de Rabanales, 14071 Cordoba, Spain, [2]Facultad de Veterinaria, Universidad de Zaragoza, Miguel Servet, 177, 50013 Zaragoza, Spain; v92avrac@uco.es

Fatness is significantly related to the ultimate eating quality of beef. Several traits such as color, texture, odor or taste depend on the fat composition of meat. Nowadays, there is an increasing concern about nutritional and healthy habits among consumers. As a consequence, a trend to decrease fat in meat has taken place in recent times. However, fatness is necessary to maintain the desirable organoleptic characteristics of meat. The fatty acid binding protein 4 (FABP4), the leptin hormone (LEP) and the retinoic acid receptor-related orphan receptor C (RORC) are involved in the synthesis and regulation of the amount of fat deposited at different levels. The aim of this study was to asses de association among markers in these three candidate genes for fat deposits and taste acceptability measured trough a consumer panel. Two hundred eighty-six individuals belonging to a commercial population from the Spanish South-western beef industry were genotyped for the three genes and 166 were analyzed by the consumer panel. The data were tested using a linear mixed model with different fixed and random effects (year, ageing time, consumer and feedlot), and with the slaughter weight as a linear covariate. The three markers were significantly associated (P<0.05) with differences in taste acceptability in the population studied. Once these associations were confirmed in a wider population, these markers could be used by the Spanish South-western beef industry as predictors of the potential meat quality.

Association of CAPN1 and CAST markers with technological and sensory traits in Spanish beef cattle
Aviles, C.[1], Peña, F.[1], Barahona, M.[2], Campo, M.M.[2], Sañudo, C.[2] and Molina, A.[1], [1]Grupo MERAGEM, Universidad de Cordoba, Campus de Rabanales, 14071 Cordoba, Spain, [2]Facultad de Veterinaria, Universidad de Zaragoza, Miguel Servet, 177, 50013 Zaragoza, Spain; v92avrac@uco.es

Single nucleotide polymorphisms (SNPs) in the μ-calpain (CAPN1) and calpastatin (CAST) genes and their association with tenderness and technological traits have been widely studied. However, not many studies have been developed in such candidate genes related to meat quality in the Spanish beef population. The aim of this study was to find out an association between two SNPs in CAPN1 and CAST genes and measurements related to meat tenderness in a commercial beef population from the South-western of the Iberian Peninsula. Two hundred eighty-six animals mainly belonging to Charolais, Limousin and Retinta breeds were genotyped to assess the frequency of the SNPs in the population. Water holding capacity (CRA), pH, colorimeter readings (L*, a* and b* values) and sensory tenderness assessed by trained panelists were determined in 164 individuals from the previous population to carry out the association study. This association was analyzed with a linear mixed model. Year, ageing time, panelist and feedlot were included as fixed and random effects and the slaughter weight as a linear covariate. Least squared means for each of the genotypes were estimated in those traits where a positive association with the marker was found. Results showed a significant association (P<0.01) between both markers and sensory tenderness. SNP in the CAPN1 gene was significantly associated (P<0.05) with CRA, pH, L* and b* values. However, least squared means of pH and L* values were not as they were expected. No significant association was found between the SNP in the CAST gene and any of the technological traits. Therefore, no clear conclusions can be obtained from the results of the technological traits analysis. A study with a larger data set would be necessary to confirm the association between the markers and sensory tenderness and to reject these markers as a way to predict meat colour.

Polymorphisms of single nucleotide (SNP) in genes related to fighting bull breed meat quality
Pelayo, R.[1], Azor, P.J.[1], Avilés, C.[1], Molina, A.[1] and Valera, M.[2], [1]University of Córdoba, Dpt. of Genetics, C.U.Rabanales. Ed. Mendel, 14071 Córdoba, Spain, [2]University of Seville, Dpt. of Agroforestry Science, Crta. Utrera, Km 1, 41013 Seville, Spain; rociopega55@hotmail.com

Most of the efforts made in the breeding of the fighting bull breed are aimed to look for the 'casta' and the bull bravery. In addition, the economic value of meat is essential in the operating account of this type of farms. In order to improve the profitability of a fighting bull farm some studies are being carried out to characterise the meat quality. In this work, the meat quality potential was assessed using genetic markers (CAPN1, CAST and DGAT1) related with the degradation of muscle fibers during the postmortem process and the different fat depots. The enzymatic system of calpains and calpastatin, and the activity of the dyacilglicerol O-acyltransferase are involved in this process. In this study, 109 samples were analysed belonging to 5 different strains of animals (encastes): Domecq, Marqués de Albaserrada, Murube Urquijo, Núñez and Santa Coloma. The aim of this study was to detect the polymorphisms (SNP) in the fighting bull breed and determinate the allelic frequency in each 'encaste'. In order to asses the association between the different polymorphisms and the 'encastes', a Fisher's exact test was performed. In addiction, the variability and population structure was analysed. Statistical analysis showed that the allelic frequency of 8 of the 12 SNP studied were significantly associated with the different 'encastes'. Furthermore, fighting bull breed presented higher allelic frequencies of the favourable allele of the CAPN1 and DGAT1 genes (0.45 and 0.72 respectively, compared with the average of 0.19 and 0.12 in other breeds published). In this sense is of particular interest to assess the relationship between genetic markers (CAPN1, CAST and DGAT1) and tenderness or lipid depots of meat, with the purpose to perform a genetic selection program in the fighting bull breed to meat production.

Non-additive effects on weight traits in South African beef breeds
Theunissen, A., Neser, F.W.C. and Scholtz, M.M., University of the Free State, Animal Wildlife and Grassland Sciences, P.O. Box 339, 9300 Bloemfontein, South Africa; neserfw@ufs.ac.za

Five sire lines; Afrikaner, Brahman, Charolais, Hereford and Simmentaler were evaluated in crosses with the Afrikaner as dam line and their F1 female progeny. The non-additive effects of Brahman genotypes on birth weight were higher than the other genotypes for individual heterosis, but lower for maternal heterosis. Continental × Afrikaner genotypes had a small individual non-additive effect (2.6%), but a large maternal effect (13.6%) on birth weight, while Brahman (-7.0%) and Hereford (-1.0%) sires had a negative maternal non-additive effect, respectively. All genotypes, except the Hereford, had a positive non-additive effect on weaning weight in the Afrikaner; ranging from 4.0-8.0%. The largest individual non-additive effect (16.1%) was found in Brahman × *Bos taurus* genotypes. The maternal non-additive effect for the Brahman × Afrikaner on weaning weight was 3.0%, compared to 10.0 and 12.0%, respectively for Continental × Afrikaner and British × Afrikaner genotypes. The combined (individual plus maternal) non-additive effect for weaning weight was highest in Continental × Afrikaner genotypes (14.0%). Positive individual non-additive effects (17.8 and 12.0%) were found for 19 month heifer- and cow weight in the Brahman *B. taurus* crosses. *B. taurus* × Afrikaner genotypes had positive maternal non-additive effects ranging from 5.1 to 10.0% on post weaning weight traits, while the Brahman × Afrikaner dams had negative maternal non-additive effects (-5.3 and -2.4%) on the same traits, respectively. The results suggest that that the combined non-additive effect of the Brahman on cow weight (1.7%) was smaller than in Continental × Afrikaner (22.5%) and Hereford × Afrikaner (10.3%) genotypes.

Effects of different lairage times after 30 h transportation on meat quality in Simmental bulls
Teke, B.[1], Akdag, F.[1], Ekiz, B.[2] and Ugurlu, M.[1], [1]Ondokuz Mayis University, Faculty of Veterinary Medicine, Department of Animal Breeding and Husbandry, 55200 Atakum, Samsun, Turkey, [2]Istanbul University, Faculty of Veterinary Medicine, Department of Animal Breeding and Husbandry, 34320 Avcilar, Istanbul, Turkey; bulentteke@gmail.com

The objective of this study was to determine the effects of three lairage times (24 h, 48 h, and 72 h) on meat quality of Hungarian Simmental bulls subjected to long commercial transportation. A total of 30 Hungarian Simmental breed male cattle, with an average age of 24 months were used in this study. The cattle were transported from Bugyi, Hungary to a slaughterhouse in Ankara, Turkey. The transportation of approximately 1,800 km took 30 h, including a 2 h rest period with water and feed available after 14 h transportation and continued other 14 h. The cattle were randomly divided into three groups after transportation. The three groups (n=10 in each group) were kept in lairage for 24 h, 48 h and 72 h. The cattle were slaughtered after lairage times. Carcass pH was measured on longissimus thoracis muscle between 12-13th thoracic vertebrae and meat quality characteristics were determined. In order to determine the effect of lairage time on carcass and meat quality characteristics, one-way ANOVA was performed and Pearson correlations were calculated among meat quality characteristics. The effect of lairage time on pHult value of cattle was significant ($P<0.05$) while the effect of lairage time on WHC, cooking loss and Warner Bratzler shear force value were not significant ($P>0.05$). The b* value was considered the best predictor of muscle pHult. In conclusion, 72 h lairage time could be needed after transportation in order to prevent adverse effect of transportation on meat quality.

Transport shrink and mortality rate of beef cattle during long commercial transportations

Teke, B., Ondokuz Mayis University, Faculty of Veterinary Medicine, Department of Animal Breeding and Husbandry, 55200 Atakum, Samsun, Turkey; bulentteke@gmail.com

The aim of this survey was to determine transport shrink by months and mortality rate of beef cattle during long distance transportations and to find out the transport time at which the transport shrink was the least. This survey was conducted on 121 transfers of bulls by road from commercial finishing units (Bugyi, Hungary) to a public slaughterhouse (Ankara, Turkey) in July and December 2010. A total of 3,874 bulls (96% Hungarian Simmental) were transported and the journeys took approximately 30 h, including a 2 h rest period with water and feed available. The maximum and minimum temperature values during transportations were obtained from meteorological units on the route of journey. The average deviation of environmental temperature during six months was determined month by month (d value). Transport shrink and dressing percentage were found by months. In order to determine the effect of month on shrink, ordinary mixed model was performed and Pearson correlation was used. The effect of months on body weight loss during transport was significant (P<0.001) and average transport shrink was 5.57% during six months. Body weight losses were the highest in August (8.39%) and they were the lowest in October (2.99%) and in November (1.77%). The mortality rate was 0.464% during transportations. There was a positive correlation between the transport shrink and dressing percentage (r=0.294; P<0.01), and d value (r=0.488; P<0.01). The body weight losses were lower during transport at comfort zone temperature intervals. In conclusion, transport shrinks under higher and lower ambient temperatures are much more important and expensive to the cattlemen receiving the cattle.

Biochemical parameters of goat meat raised on conventional and organic farms in Latvia

Birģele, E., Keidāne, D. and Ilgaža, A., Latvian University of Agriculture, Faculty of Veterinary Medicine, Helmaņa 8, 3002 Jelgava, Latvia; edite.birgele@llu.lv

Scientists have started to pay attention to the evaluation of nutritional value and quality of the goat produce meat only in the recent years. The authors have reported that the meat quality is affected by the breed of the goat, animal nutrition and keeping conditions as well as the environmental stress during the transportation of the animal and at the slaughter house. In Latvia, there is no information on the meat quality and nutritional value of local goats, and there is no information on some of the meat quality indices. Therefore, our work was to compare organic (n=12) and conventional (n=12) farmed adult goats (male, carcass weight 40±2.3 kg) biochemical parameters. The main objectives of the research were the following: investigate and compare essential and nonessential amino acids composition of meat obtained from animals raised on organic farm or conventional and compare obtained from treated animals in meat protein, crude fat and cholesterol amount. In conclusion, it was established that animal housing conditions of conventional and organic farms significantly affected by essential amino acids – phenylalanine, leucine, isoleucine, methionine, valine, arginine, cystine and lysine levels and meat, but the amount of essential amino acids is not significantly affected. Essential amino acids (conventional farm 61.2% and organic farm 57.3%) in goat meat are significantly (P<0.05) less than the replaceable (respectively 38.8% and 42.7%). Cholesterol levels were statistically significantly (P<0.01) lower in meat obtained from organic farmed animals (81.6±3.2 mg%; 62.5±4.5 mg%). Goat meat is generally low amount of fats, but relatively high protein content, however, significant differences between conventional and organic farms goat meat samples without a finding.

Genomic selection for new traits: optimal prediction and reference population design
Calus, M.P.L.[1], De Haas, Y.[1], Pszczola, M.[1,2] and Veerkamp, R.F.[1], [1]Wageningen UR Livestock Research, Animal Breeding and Genomics Centre, P.O. Box 65, 8200 AB Lelystad, Netherlands, [2]Poznan University of Life Sciences, Department of Genetics and Animal Breeding, Wolynska 33, 60-637 Poznan, Poland; mario.calus@wur.nl

Genomic selection (GS) is the most successful application of –omics technologies in livestock production. An important promise is that it allows selection for difficult or expensive to measure traits. Our objective is to review requirements to start GS for such new traits, and possibilities to optimize use of limited reference populations. In practice, size of reference populations may be dictated by required number of records to estimate genetic parameters, phenotypic data availability, phenotyping and genotyping costs, and population size. When cost of measuring phenotypes is very high (e.g. methane emission), the most optimal strategy may be to obtain single phenotypic measurements of genotyped animals. To maximize average accuracy of direct genomic breeding values (DGV) with a small reference population, the following strategies can be used: (1) optimize reference population design in terms of relationship within the reference population and to evaluated animals (selective genotyping); (2) optimize selected set of included phenotypes (selective phenotyping); (3) use a model that optimally fits the architecture of the trait and used marker density; (4) measure indicator traits with higher heritability or a larger scale in the reference population to cheaply increase its power; (5) measure indicator traits on evaluated individuals to increase DGV accuracy; (6) combine national reference populations through international collaboration. Limited size reference populations with up to a few thousand individuals with single records yield DGV with relatively low accuracies, compared to traditional breeding values. Combined with a reduced generation interval and an increase in selection intensity, using such DGV can still lead to considerable genetic gain, compared to progeny testing schemes for easy to record traits.

Genotyping of cows for genomic EBVs for direct health traits: genetic and economic aspects
Egger-Danner, C.[1], Schwarzenbacher, H.[1] and Willam, A.[2], [1]ZuchtData EDV-Dienstleistungen GmbH, Dresdner Straße 89/19, 1200, Austria, [2]University of Natural Resources and Life Sciences (BOKU Vienna), Division of Livestock Sciences, Gregor-Mendel-Str. 33, 1180, Austria; schwarzenbacher@zuchtdata.at

In Austria a health monitoring system based on veterinarian diagnoses has been implemented recently. Presently estimated breeding values for direct health traits are available for Fleckvieh (Simmental) cattle. Currently the reference population for genomic evaluation includes about 6,000 bulls for the dairy traits whilst for direct health traits just less that 1000 bulls are available with a reliability higher than 50%. Due to the general reduction of the number of young bulls the gap between traits with genomic EBV and the direct health traits without genomic EBV will widen if the reference population relies on progeny tested bulls only. The impact of genotyping cows with reliable phenotypes for direct health traits on the annual monetary genetic gain (AMGG) and discounted profit was analysed. The calculations are based on a deterministic approach using ZPLAN software. An increase of the reliability of the total merit index (TMI) of 4, 15 and 25% are assumed through genotyping 5,000, 25,000 and 50,000 cows respectively. Costs for genomic EBVs are varied between 150 and 20 € per cow. Genotyping cows benefits medium to high heritable traits to a higher extend than low heritable direct health traits. The AMGG is increased by 1% if the reliability of the TMI is 4% higher (i.e. 5,000 cows genotyped) and a 6% higher AMGG can be expected when the reliability of the TMI is 25% increased (i.e. 50,000 cows genotyped). The discounted profit depends on the costs of genotyping but also on the population size. Genotyping cows with reliable phenotypes can speed up the availability of genomic EBVs for direct health traits. But, due to the huge amount of valid phenotypes and genotypes needed to establish an efficient genomic evaluation, financial aspects will be the limiting factors for most of the dairy cattle breeding programs.

Reference population designs affects reliability of selection for (un)genotyped animals

Pszczola, M.[1,2,3], Strabel, T.[1], Van Arendonk, J.A.M.[2] and Calus, M.P.L.[3], [1]Poznan University of Life Sciences, Department of Genetics and Animal Breeding, Wolynska 33, 60-637 Poznan, Poland, [2]Wageningen University, Animal Breeding and Genomics Centre, P.O. Box 338, 6700 AH Wageningen, Netherlands, [3]Wageningen UR Livestock Research, Animal Breeding and Genomics Centre, P.O. Box 65, 8200 AB Lelystad, Netherlands; mbee@jay.up.poznan.pl

Reliability of direct genomic values (DGV) increases when moving from traditional to genomic selection. The objective of this study was to investigate the origin of this increase and explore the reference population design impact on the reliability when genotyped and ungenotyped animals are analyzed jointly. A dairy cattle population was simulated for a trait with heritability of 0.3. Reference populations were small and consisted of cows only, reflecting a difficult or expensive to measure trait, e.g. methane emission. Reference populations had different family structure: highly, moderately, lowly, and randomly related animals. Evaluated animals were chosen from one generation after the reference population. Four scenarios in which reference population and evaluated animals were not genotyped (AA), reference population was genotyped and evaluated animals were not genotyped (GA) reference population was not genotyped and evaluated animals were genotyped (AG), or both groups were genotyped (GG), were considered. DGV reliabilities were predicted deterministically using selection index theory. For GG, reliabilities were considerably higher than in the other cases. In AG, reliabilities were somewhat higher, than in GA. AG achieved substantially higher reliabilities than AA. Reliabilities increased with decreasing average relationship within reference population, and thus, reference population design is important when genotyped and ungenotyped animals are analyzed jointly. The main origin of gain in the reliability is due to genotyping evaluated animals. Genotyping only one group of animals will, however, always yield substantially less reliable estimates than when all animals are genotyped.

Selection for beef meat quality using ultrasound or genomic information

Pimentel, E.C.G. and Koenig, S., University of Kassel, Nordbahnhofstr. 1a, 37213 Witzenhausen, Germany; pimentel@uni-kassel.de

Selection index theory was used to compare different selection strategies to improve meat quality in beef. Alternative strategies were compared to a reference scenario with three basic traits in the selection index: weight at 200 (W200) and 400 days (W400) and muscling score (MUSC). These traits resemble the combination currently used in Germany. Traits in the breeding goal were the three basic traits plus marbling score (MARB). Economic weights were either the same for all traits, or doubled or tripled for MARB. Two additional selection criteria for improving MARB were considered: intramuscular fat content measured by ultrasound (UIMF) as indicator trait and a genomic breeding value (GEBV) for the target trait directly (gMARB). Results were used to estimate the size of a calibration set required for genomic selection on meat quality. Adding UIMF to the basic index increased the genetic gain per generation by 15% when the economic weight on MARB was doubled, and by 44% when tripled. When the accuracy of gMARB was 0.5, adding it to the index provided larger genetic gain than adding UIMF. Greatest genetic gain per generation was obtained with the scenario containing GEBV for four traits (gW200, gW400, gMUSC and gMARB) when their accuracies were 0.7 or greater. Adding UIMF to the index substantially improved response to selection for MARB, which switched from negative to positive when the economic weight on MARB was doubled or tripled. For all scenarios that contained gMARB, the response to selection in MARB was positive for all relative economic weights on MARB, when the accuracy of GEBV was greater than 0.7. Results indicated that setting up a calibration set of ~500 genotyped animals with carcass phenotypes for MARB could suffice to obtain a larger response to selection than measuring UIMF. If the size of the calibration set is ~2,500, adding UIMF to an index containing already the GEBV would hardly bring some benefit, unless the relative economic weight for marbling is much larger than for the other traits.

Heritability estimates for methane emission in Holstein cows using breath measurements
Lassen, J.[1], Madsen, J.[2] and Løvendahl, P.[1], [1]Aarhus University, Department of Molecular Biology and Genetics, Blichers Alle 20, 8830 Tjele, Denmark, [2]Copenhagen University, Department of Large Animal Science, Bülowsvej, 1850 Frederiksberg C, Denmark; jan.lassen@agrsci.dk

Enteric methane emission from ruminants contributes substantially to the greenhouse effect. Few studies have focused on the genetic variation in enteric methane emission from dairy cattle. The objective of this study was to estimate the heritability for enteric methane emission from Danish Holstein cows. On a total of 683 dairy cows a Fourier Transformed Infrared (FTIR) measuring unit was used to make large scale individual methane emission records. The cows were measured in 7 herds during their visits to automatic milking systems (AMS). The FTIR unit air inlet was mounted in the front part of an AMS close to the cows head for 7 days, recording continuously every 5 seconds. The phenotype analysed was the mean methane to carbon dioxide ratio across visits during the measuring period, as this ratio reflects the proportion of the metabolisable energy exhaled as methane. The statistical linear model included fixed effects of herd, month, days in milk, lactation number, and random effects of animal and residual. Variance components were estimated in an animal model design using a pedigree containing 9,661 animals. The heritability of the methane to carbon dioxide ratio was moderate (0.21). The results from this study suggest that individual cow's methane emission can be accurately measured using FTIR equipment and that the trait is heritable. The data can be used for both management and genetic analysis and opens for future studies on the correlation between methane emission and other traits of economic importance in the breeding goal as well as the possibility to use methane as an indicator of feed efficiency. It is concluded that FTIR breath analysis is effective for measuring GHG emissions and may find further applications with a wider panel of gases including acetone and its relation to ketosis.

Methane emissions as a new trait in genetic improvement programmes of beef cattle
Roehe, R., Rooke, J., Duthie, C.-A., Ricci, P., Ross, D., Hyslop, J. and Waterhouse, A., Scottish Agricultural College, Edinburgh, EH9 3JG, United Kingdom; Rainer.Roehe@sac.ac.uk

One objective of this experiment was to obtain fundamental insight into the influence of breed and diet effects and their interactions on methane emissions in order to account for these effects in genetic improvement programmes. Furthermore, proxy methods to predict methane emissions used in this experiment will be discussed and potential associations of methane emissions and fatty acid profiles of meat investigated. The experiment based on a balanced factorial design comprising of 72 beef steers which originated from 2 breed types (rotational crosses between Limousin (L) and Aberdeen Angus (AA)) and fed 2 diets (51:49 or 8:92 forage (F):concentrate (C) ration on dry matter (DM) basis). Methane emissions were individually measured for a 48h period within respiration chambers. The coefficient of variations of daily methane emissions within diet and breed ranged from 14% to 32%. Daily methane emissions were significantly different (P<0.05) for breed and diet with least squares means (LSM) of 184 g/d (AA) vs. 164 g/d (L) and 142 g/d (C) vs. 205 g/d (F), respectively. Based on a comparison of methane emissions per kg DM intake, only the effects between diets were significant with LSM of 13.7 g/kg DM (C) vs. 21.7 g/kg DM (F). Genotype by diet interactions were non-significant. The sire variances associated with methane emissions per g/d and g/kg DM were substantial with 31% and 37% of the residual variance, respectively; but because only a small number of animals could be tested in respiration chambers, these estimates were non-significantly different from zero. Proxy methods such as the Laser Methane Detector (LMD) or ventilated hoods above electronic feeders may provide the necessary information for genetic improvement programmes. The results showed that there is a high potential variation in methane emissions in beef cattle to be used in genetic improvement programmes.

Challenges for closing the phenomic gap in farm animals

Hocquette, J.F.[1], De La Torre, A.[1], Meunier, B.[1], Le Bail, P.Y.[2], Chavatte-Palmer, P.[3], Le Roy, P.[4] and Mormède, P.[5], [1]INRA, UMR 1213, Herbivores, 63122 Theix, France, [2]INRA, PPGP, 35000 Rennes, France, [3]INRA, UMR 1198, BDR, 78352 Jouy-en-Josas, France, [4]INRA, UMR 1348, Pegase, 35590 Rennes, France, [5]INRA, UMR 444, Cellular Genetics, 31326 Castanet-Tolosan, France; hocquet@clermont.inra.fr

Since the advent of genomics, phenotyping has become the limiting factor for the practical implementation of genomic selection to increase the efficiency of animals and to produce high quality products. Animal production must also integrate new challenges related to the impacts of animal farming on the environment, on healthy food and citizen concern on animal welfare. These evolutions converge towards robust animals which maintain a high production level in a wide range of climatic conditions and production systems. The accurate, reproductive and standardized measurements of the corresponding traits (animal efficiency, wellbeing, environmental impacts, product quality), and their integration in a systems approach are the basis for modeling biological functions of animals and to develop predictive approaches of performances as well as to understand the role of epigenetics in the building up of a phenotype. This is crucial for the development of precision livestock farming to enhance the efficiency and sustainability of production and the competitiveness of the whole food-animal supply chain. To achieve this goal, new strategies should include (1) the development and implementation on farms of tools and devices for reliable, automated, standardized, and high-throughput measurements associated with robustness; (2) the determination of quantitative predictors for animal traits (adaptability, precocity, body development, resilience to stressors, product characteristics, etc); and (3) the development of a central computational system between research organizations and private partners for data sharing and modeling purposes. Therefore, the challenge is highly multidisciplinary and involve new investments in both human and financial resources as well as a collaborative network e.g. with SMEs.

Should one aim for genetic improvement of host resistance or tolerance to infectious disease?

Doeschl-Wilson, A.[1], Villanueva, B.[2] and Kyriazakis, I.[3], [1]The Roslin Institute and R(D)SVS, University of Edinburgh, Genetics & Genomics, Easter Bush, EH25 9RG, United Kingdom, [2]INIA, Departamento de Mejora Genética Animal, Carretera de La Coruña km 7,5, 28040 Madrid, Spain, [3]Newcastle University, Agriculture, Food and Rural Development, Newcastle upon Tyne, NE1 7RU, United Kingdom; andrea.wilson@roslin.ed.ac.uk

Recent advances in genomics provide new opportunities for dissecting host genetic response to infectious pathogens and to accelerate genetic improvement. There are two alternative host defence mechanisms to infectious pathogens that could be targeted for genetic improvement – host resistance vs host tolerance. Resistance refers to mechanisms that restrict the within host pathogen reproduction rate, whilst tolerance mechanisms aim at minimising pathogen inflicted damage on the host. Both strategies reduce the devastating impact of infectious disease on animal production, but can have contrasting effects on population performance and epidemiology. Improving host resistance may result in disease eradication, whereas this may be difficult for tolerant hosts who harbour pathogens without showing symptoms. However, evidence suggests that increasing host resistance may drive pathogen evolution towards higher virulence. The aim of this study was to develop a systematic and comprehensive framework for determining under what conditions improving host genetic disease resistance would be favourable over improving tolerance, and vice versa. Influencing factors and evidence for their effects from literature considered in this framework include (1) the breeding goal (e.g. disease eradication or sustainable production), (2) theoretical and empirical evidence for host genetic variation (and co-variation) in resistance or tolerance and underlying genetic architecture, (3) required phenotypic measures and statistical tools to disentangle resistance from tolerance, and (4) undesirable side-effects of improving resistance or tolerance. Theoretical concepts will be illustrated with examples from various livestock diseases.

A genetic epidemiological function for estimating genetic variance in infectivity and susceptibility
Lipschutz-Powell, D.[1], Woolliams, J.A.[1], Bijma, P.[2] and Doeschl-Wilson, A.B.[1], [1]The Roslin Institute, Easter Bush, EH25 9RG, United Kingdom, [2]Wageningen University, Wageningen, 6701, Netherlands; debby.powell@roslin.ed.ac.uk

Selection for improved host response to infectious disease offers a desirable alternative to chemical treatment. However, a major barrier to closing the genotype – phenotype gap is uncovering the genetic variance in important phenotypes. In a previous study we demonstrated that current tools for genetic analysis ignore potential genetic variation in infectivity. The lack of attention to host variation in infectivity in genetic studies stands in stark contrast to the well-recognised important role of host infectivity in epidemiology. The current framework for genetic analysis of disease data relies on the assumption that the genetic component of a disease phenotype is fully determined by an individual's genetic predisposition to becoming infected, whereas the individuals' exposure to infectious pathogens (or disease prevalence) is considered as an environmental constant. Here we combine epidemiological theory with quantitative genetics theory to expand the existing framework to allow for genetic variation in both susceptibility and infectivity. Most readily available disease data is often binary, indicating whether an individual became infected or not following exposure to an infectious disease. Following quantitative genetics theory it is assumed that this binary outcome may be partitioned into an inherent probability of becoming infected and a stochastic deviation. We derived a genetic epidemiological function which describes the time-dependent infection probabilities inherent in an epidemiological SIR model in terms of individuals' infection status and underlying susceptibility and infectivity and thus captures the dynamic aspect of disease progression in the population. We then validated this function with simulated disease data generated by a stochastic genetic epidemiological SIR model and examined implications for implementing this function in genetic analyses.

Metabolites as new molecular traits and their role for genetic evalution of traditional milk traits
Melzer, N., Wittenburg, D. and Repsilber, D., Leibniz Institute for Farm Animal Biology, Wilhelm-Stahl-Allee 2, 18196 Dummerstorf, Germany; melzer@fbn-dummerstorf.de

In dairy cattle science, it is more and more common to consider metabolite profiles to detect new traits or indicators in addition to traditional milk traits. These new traits may help, e.g. to detect diseases like mastitis at early stages or to improve management decisions. In the recent literature, few new traits were presented and proposed to be used as indicators for such specific topics. In this context, we show that it is beneficial to measure metabolite profiles and to use such profiles to predict various milk traits, which offers the possibility to identify new molecular traits. We applied random forest regression and partial least squares regression to predict several milk traits from metabolite profiles. Both prediction methods allow to determine a measure of importance for each metabolite on a milk trait. A specific cutoff value was set to determine the most important metabolites. Afterwards, the overlap of the important metabolites from both methods was further considered. This selection step revealed metabolites, some of which correspond to proposed indicators in the literature. The detected metabolites important for prediction of a milk trait are also useful to improve the model of the classical genotype-phenotype map (genomic selection): In our approach, we used them to find single nucleotide polymorphisms (SNPs) with significant genetic effect on a specific milk trait. To validate this outcome, we compared the precision of genetic value prediction for a milk trait using different SNP subsets: (a) all SNPs; (b) significant SNPs from (a); and (c) significant SNPs from metabolite investigations. We applied a spike and slab stochastic variable selection method to determine significant SNPs for metabolites as well as for milk trait analyses. For two milk traits (fat content and pH value), we observed that precision from (c) was larger than (b) and equal to (a).

Performance testing for boar taint: a pivotal step towards ending surgical castration of pigs
Baes, C.[1], Luther, H.[2], Mattei, S.[3], Ampuero, S.[4], Sidler, X.[3], Bee, G.[4], Spring, P.[1], Weingartner, U.[5] and Hofer, A.[2], [1]HAFL, Langgässe 85, 3052 Zollikofen, Switzerland, [2]SUISAG, Allmend 8, 6204 Sempach, Switzerland, [3]Vetsuisse, Winterthurerstr. 190, 8057 Zürich, Switzerland, [4]ALP, Rte de la Tioleyre 4, 1725 Posieux, Switzerland, [5]Coop, Thiersteinerallee 12, 4052 Basel, Switzerland; Christine_Baes@gmx.de

This study describes the establishment of a performance test for boar taint in the Swiss terminal sire line PREMO®. A biopsy device for tissue sampling selection candidates is introduced, data obtained from biopsy cores is presented and repeatability is validated. Statistical models for estimation of variance components based on biopsy data are tested, and models suitable for routine breeding value estimation are identified. A biopsy device for collecting adipose tissue samples from breeding candidates was developed in accordance with the Swiss Federal Veterinary Office (BVET) and the Swiss Animal Protection (SAP). Phenotypic data (A, S and I) was obtained from adipose tissue sampled by biopsy from the neck of 516 boars (100-130 kg live weight); A (mean=0.578, σ=0.527), S (mean=0.033, σ=0.002) and I (mean=0.032, σ=0.002) were within plausible ranges. A total of 36 different mixed linear models were tested for each boar taint compound using R. Example model variations included consideration of age and / or live weight as covariables or as categorical fixed effects. Models were ranked according to their information content, and best models were used for variance component estimation. Pedigree information on 2,245 ancestors was included. Quantification of A, S and I from small samples was accurate and repeatable; heritability estimates (h^2_{lnA}=0.452±0.108, h^2_{lnS}=0.524±0.041, h^2_{lnI}=0.571±0.099) and other genetic parameters were plausible. With the establishment and validation of a performance test suitable for use in large scale breeding programmes, a pivotal step has been made towards genetically resolving the problem of tainted meat and ending surgical castration.

Differences in variance components of gilts' aggression in genetically highly connected populations?
Appel, A.K.[1], Voß, B.[2], Tönepöhl, B.[1], König V. Borstel, U.[1] and Gauly, M.[1], [1]University of Göttingen, Department of Animal Science, Albrecht-Thaer-Weg 3, 37075 Göttingen, Germany, [2]BHZP GmbH, An der Wassermühle 8, 21368 Dahlenburg, Germany; aappel@gwdg.de

In the present study, a total of 543 purebred Pietrain gilts from two nucleus farms (farm A: n=302; B: n=241) of one breeding company, were tested at the age of 30 weeks for agonistic behaviour. Observations were taken by the same observer in both farms during the years 2010 to 2012 and included the frequency of uni- and bilateral aggression during mixing with unfamiliar gilts of the same weight. In both farms 40% of the gilts were purebred Pietrains, and they were mixed with Landrace and Duroc gilts (farm A) or Large White gilts (farm B). The Pietrain gilts were genetically closely linked and the offspring of 96 sires, 64% of these sires having tested progeny in both farms. Pens in farm A had a solid concrete floor with shavings and were equipped with a dry feeder. Pens in farm B had partly slatted floors and an electronic sow feeder. In both farms, mean space allowance was 2 m²/gilt. Gilts from farm B performed more unilateral aggression (1.12±1.42 vs. 0.71±1.20, mean ± sd.) as well as more bilateral aggression (0.78±0.98 vs. 0.44±0.82) compared to gilts from farm A. The heritabilities for behavioural traits were estimated with a linear model and were on a low level in farm A (h^2=0.17 and h^2=0.09 for uni- and bilateral aggression, respectively) and on a moderate level in farm B (h^2=0.35 and h^2=0.34, respectively). For both uni- and bilateral aggression, genetic correlation of the same trait between farm A and farm B was 1.0. These high genetic correlations give an indication for the expression of the same behaviour pattern. The marked differences in estimated heritabilities were mainly caused by lower variance, especially lower additive genetic variance in farm A.

Variance component and breeding value estimation for environmental sensitivity in dairy cattle
Rönnegård, L.[1,2], Felleki, M.[1,2], Fikse, W.F.[1], Mulder, H.A.[3] and Strandberg, E.[1], [1]Swedish University of Agricultural Sciences, Dep of Animal Breeding and Genetics, P.O. Box 7023, 750 07 Uppsala, Sweden, [2]Dalarna University, Statistics Unit, Rödavägen 3, 781 70 Borlänge, Sweden, [3]Wageningen University, Animal Breeding and Genomics Centre, P.O. Box 9101, 6700 HB Wageningen, Netherlands; Han.Mulder@wur.nl

Animal robustness, or environmental sensitivity, may be studied through individual differences in residual variance. These differences appear to be heritable, and there is therefore a need to fit models to predict breeding values explaining differences in residual variance. The aim of this report is to study whether breeding values for environmental sensitivity (vEBV) can be predicted in a large dairy cattle data set having around 1.6 million records. Two traits recorded between 2002 and 2009 from Swedish Holstein dairy cattle were analyzed separately, somatic cell score and milk yield. Estimation of variance components, ordinary breeding values and vEBVs was performed using ASReml, applying the methodology for double hierarchical generalized linear models. Estimation using ASReml took less than 7 days on a Linux server. The genetic standard deviations for environmental variance were 0.21 and 0.22, for somatic cell score and milk yield, respectively, which indicate a moderate genetic variance for environmental variance. This study shows that estimation of variance components, EBVs and vEBVs, is feasible for large dairy cattle data sets using standard variance component estimation software. Especially for somatic cell score selection for lower vEBV seems economically attractive and therefore vEBV may be considered in a total merit index.

Genomic prediction for new traits combining cow and bull reference populations
Calus, M.P.L., De Haas, Y. and Veerkamp, R.F., Wageningen UR Livestock Research, Animal Breeding and Genomics Centre, P.O. Box 65, 8200 AB Lelystad, Netherlands; mario.calus@wur.nl

Genomic selection enables selection for expensive or difficult to measure traits, like feed efficiency and methane emission in dairy cattle. However, only a few thousand phenotypic records are likely to be collected for these expensive traits. Therefore, we suggest that an optimal strategy may be to combine this genotypic and phenotypic information, with information of predictor traits that are available from national recording schemes. To combine this information optimally we investigated a Bayesian genomic prediction model that allows us to analyse two traits while animals only have a phenotype on either of the traits. The impact of adding the bull information was evaluated for accuracy of prediction, and for posterior QTL probabilities to assess effects on power for a genome-wide association study (GWAS). The model was tested on a data set with 1,609 cows and 296 bulls with phenotypes for fat and protein yield and with genotypes for 36,346 SNPs. All bulls had highly accurate daughter yield deviations (DYD) for fat and protein yield from the Irish national evaluations. Estimated genetic correlations between the cow and bull traits were either low (0.22-0.26) or moderate (0.55-0.56). Prediction accuracies were calculated via cross-validation, while the whole data set, including or excluding one of the bull traits, was analysed to investigate effects on GWAS results. Results indicated that adding information of just a few hundred bulls did not significantly increase prediction accuracy, despite the high accuracy of their DYD. To achieve higher accuracy for the genomic predictions, apparently much larger national bull reference populations need to be added. Adding the bull information did however increase the power for GWAS. So, indicator traits used in a genomic prediction model improve power to identify genomic regions associated with traits of interest, even when the accuracy of genomic prediction with similar models remains unaffected.

Implementing a genetic evaluation for milk fat composition in the Walloon Region of Belgium

Gengler, N.[1], Troch, T.[1], Bastin, C.[1], Vanderick, S.[1], Vandenplas, J.[1,2] and Soyeurt, H.[1,2], [1]ULg – GxABT, Passage des Déportés, 2, 5030 Gembloux, Belgium, [2]FNRS, Rue d'Egmont, 5, 1000 Brussels, Belgium; nicolas.gengler@ulg.ac.be

A genetic evaluation system for milk fat composition is currently under development in the Walloon Region of Belgium. Based on the currently used genetic evaluation model for production traits a multi-trait (milk, fat and protein yields, saturated (SFA) and monounsaturated (MUFA) fatty acid contents in milk) and multi-lactation (first, second, third lactation) random regression test-day model was developed. (Co)variance components were developed from results obtained in previous research. Milk production test-day records are available since 1974 (currently 7,019,000 in first, 5,253,000 in second and 3 757 000 in third lactation) and, fatty acids predicted from MIR spectral data since 2007 (currently 499,000 in first, 392,000 in second and 277,000 in third lactation). Mixed model equations were solved as the current routine evaluations with a preconditioned conjugate gradient implementation with integrate outlier checking and correction. Reliabilities were computed using the procedures developed for production traits. By evaluating SFA and MUFA contents in milk together with correlated traits, milk, fat and protein yields were used as predictors in order to improve estimation of EBV for those novel traits. Therefore a total of 525,048 respectively 249,027 evaluated cows achieved reliabilities equal or greater than 0.35 for SFA and MONO compared to 640,072 for milk yield. Among the sires of cows with fatty acid records 876 respectively 613 had reliabilities equal or greater than 0.75 for SFA and MUFA compared to 1,074 for milk yield. Breeding values will be expressed using two indices that represent the desaturation (or mono-desaturation) of milk fat compared to the expected level. Additional research is ongoing to extend this genetic evaluation system to a genomic single-step GBLUP using INTERBULL breeding values as a priori information for correlated indicator traits.

Genetic evaluation of mastitis liability and recovery using transition probabilities

Franzén, J.[1], Thorburn, D.[1], Urioste, J.I.[2] and Strandberg, E.[3], [1]Stockholm University, Dept of Statistics, 106 91 Stockholm, Sweden, [2]UDELAR, Departamento de Producción Animal y Pasturas, Fac. de Agronomía, Garzón 780, 12900 Montevideo, Uruguay, [3]Swedish University of Agricultural Sciences, Department of Animal Breeding and Genetics, P.O. Box 7023, 75007 Uppsala, Sweden; Erling.Strandberg@slu.se

Many methods for genetic analyses of mastitis are performed with a cross-sectional approach, leaving out information such as repeated mastitis cases during lactation, somatic cell count fluctuations, and the recovery process. Acknowledging the dynamic behavior of mastitis during lactation and viewing it as more than a binary variable, can enhance the genetic evaluation of mastitis. Genetic evaluation of mastitis was carried out by modeling the dynamic nature of somatic cell count (SCC) within lactations. The SCC patterns were captured by modeling transition probabilities between assumed states of mastitis and non-mastitis. A widely dispersed SCC pattern generates high transition probabilities and vice versa. The method simultaneously models transitions to and from states of infection, i.e. both the mastitis liability and the recovery process are considered. A multilevel discrete time survival model was used for estimation of breeding values (BVs) on simulated data of different sizes, mastitis frequencies, and genetic correlations. The correlation between estimated and simulated breeding values (accuracy) for mastitis liability was similar (0.75-0.77 with 150 daughters/sire) to those from previously tested methods, despite that these methods used data of confirmed cases, while the current method is based on SCC as indicator of mastitis. In addition, this method also generated breeding values for the recovery process. The accuracy of these BVs was, however, lower: 0.38-0.47. The model provides an improved tool for genetic evaluation of mastitis, where both the mastitis liability and the ability to recover from mastitis can be included.

Multivariate genomic prediction improves breeding value accuracy for scarcely recorded traits

Pszczola, M.[1,2,3], Veerkamp, R.[3], De Haas, Y.[3], Strabel, T.[1] and Calus, M.P.L.[3], [1]Poznan University of Life Sciences, Department of Genetics and Animal Breeding, Wolynska 33, 60-637 Poznan, Poland, [2]Wageningen University, Animal Breeding and Genomics Centre, P.O. Box 338, 6700 AH Wageningen, Netherlands, [3]Wageningen UR Livestock Research, Animal Breeding and Genomics Centre, P.O. Box 65, 8200 AB Lelystad, Netherlands; mbee@jay.up.poznan.pl

Decreasing genotyping costs enabled setting up reference populations, consisting of animals with known genotypes and phenotypes, required for genomic selection. Yet, for some traits obtaining phenotypic observations may still be problematic if measurements are difficult or expensive (e.g. methane emission or dry matter intake in dairy cattle). Therefore, number of observations is usually limited and selection methods should aim to optimally use those to enable accurate selection. To overcome this problem, a common solution in traditional breeding programs is to use easily recordable predictor traits, inexpensive to measure, and genetically correlated with a trait of interest. Using predictor traits, a trait of interest can be improved at a reduced cost. The impact of using a predictor trait in a genomic selection approach has not been studied yet on real data. Our aim, therefore, was to empirically investigate the effect of using predictor traits on the accuracy of direct genomic values (DGV) of dry matter intake with a small reference population. Multivariate genomic BLUP was used to simultaneously evaluate dry matter intake, fat protein corrected milk yield and live weight. Traits were moderately to highly genetically correlated. When predictor traits were recorded only for the reference population, no clear gain in accuracy was observed despite of number of traits used. When traits were recorded for reference and evaluated animals, using two predictor traits in trivariate analyses yielded somewhat higher accuracy than using bivariate analyses with only one predictor trait. Overall, adding predictor traits recorded on all animals improved the DGV accuracy for dry matter intake.

Genetics and genomics of energy balance measured in milk using mid-infrared spectroscopy

Mcparland, S.[1], Calus, M.P.L.[2], Coffey, M.P.[3], Wall, E.[3], Soyeurt, H.[4], Bastin, C.[4], Veerkamp, R.F.[2], Banos, G.[5], Lewis, E.[1], Bovenhuis, H.[6] and Berry, D.P.[1], [1]Animal and Grassland Research and Innovation Centre, Teagasc, Moorepark, Fermoy, Co. Cork, Ireland, [2]Animal Breeding and Genomics Center, Wageningen UR Livestock Research, P.O. Box 65, 8200 AB Lelystad, Netherlands, [3]Sustainable Livestock Systems Group, Scottish Agricultural College, Midlothian EH25 9RG, United Kingdom, [4]University of Liège, Animal Science Unit, Gembloux Agro-Bio Tech, B-5030, Gembloux, Belgium, [5]Department of Animal Production, Faculty of Veterinary Medicine, Aristotle University of Thessaloniki, GR-54124 Thessaloniki, Greece, [6]Animal Breeding and Genomics Centre, Wageningen University, Wageningen University, Wageningen, Netherlands; donagh.berry@teagasc.ie

A cow's energy balance is the differential between energy intake and energy output. Energy balance early post-calving is recognised to be both phenotypically and genetically related to both subsequent reproductive performance and animal health. Therefore having access to information on phenotypic energy balance and genetic merit for energy balance can be useful in improving cow reproduction and health. The measurement of true energy balance on many cows is not possible although proxies for energy balance (i.e. milk fat to protein ratio) have been suggested. To test if milk mid-infrared (MIR) spectroscopy could be used to predict energy balance across different production systems, 2,992 morning, 2,742 mid-day, and 2,989 evening milk MIR spectral records from 564 lactations on 337 Scottish cows managed in confinement on two diets. An additional 844 morning and 820 evening milk spectral records from 338 lactations on 244 Irish cows managed outdoors on grazed grass were also available. The accuracy of predicting energy balance varied from 0.47 to 0.69 across datasets. Significant genetic variation in predicted energy balance existed and genomic regions associated with differences in energy balance were also identified.

Genome-wide association study for milk fatty acid composition using cow versus bull data
Bastin, C.[1], Gengler, N.[1], Soyeurt, H.[1,2], Mcparland, S.[3], Wall, E.[4] and Calus, M.P.L.[5], [1]University of Liège, Gembloux Agro-Bio Tech, Animal Science Unit, 5030, Gembloux, Belgium, [2]National Fund for Scientific Research (F.R.S.-FNRS), 1000, Brussels, Belgium, [3]Animal & Grassland Research and Innovation Centre, Teagasc, Moorepark, Co. Cork, Ireland, [4]Scottish Agricultural College, Sustainable Livestock Systems Group, EH25 9RG, United Kingdom, [5]Wageningen UR Livestock Research, Animal Breeding and Genomics Centre, P.O. Box 65, 8200 AB Lelystad, Netherlands; catherine.bastin@ulg.ac.be

So far, few genome-wide association studies (GWAS) have been performed on milk fatty acid composition (FAC) in dairy cattle. Those studies used single SNP models applied to cow data. The objective of our study is to perform a GWAS for FAC, using either predicted FAC on a few hundred genotyped cows or 226 genotyped bulls with estimated breeding values (EBV) for predicted FAC or both. FAC in milk (g/dl) for both the genotyped cows and the daughters of the genotyped bulls was predicted from mid-infrared profiles using calibration equations developed notably within the RobustMilk project. The following groups of FA were included: saturated, monounsaturated, polyunsaturated, unsaturated, short chain, medium chain and long chain. EBV for FA for the 226 bulls were obtained from a dataset including 345,723 Walloon Holstein cows using a series of 4-trait (milk, fat, and protein yields plus one of the FA traits) 3-lactation random regression models. The 226 bulls had FA EBV with reliabilities $\geq$0.44. Of the 226 bulls, 44 had no daughters with predicted FA, so their FA EBV were only based on the yield traits. Of the other 182 bulls, 140 had between 10 and 1,691 daughters with predicted FA. A Bayesian model was used to estimate the effects of all SNP simultaneously. Genotypes were available for 36,346 SNPs after usual edits. When both cows and bulls were included, the traits were analysed as separate traits, using a bivariate model. It is expected that combining the cow and bull data will increase the power of the analysis.

On-farm measuring of milk progesterone content as a basis for defining new fertility traits
Martin, G., Rosner, F., Schafberg, R. and Swalve, H.H., Martin-Luther-University Halle-Wittenberg, Institute of Agricultural and Nutritional Sciences, Theodor-Lieser-Str. 11, 06120 Halle, Germany

In dairy cattle breeding, selection for improved fertility has been hampered by very low heritabilities exhibited by classical traits that are based on A.I. records. In the era of genomic selection, accurate phenotypes are more important than ever. Female fertility can be broken down into several complexes. One complex is the ability to re- cycle after calving. A traditional trait attempting to measure this is days to first insemination (DFS). By measuring progesterone content in the milk at regular intervals after calving, traits like commencement of luteal activity (CLA) can be defined. However, due to cost restrictions, intervals as large as one week may be advisable. For the present study, progesterone measurements were done using an ELISA based on-farm test in intervals of one week from week 3 to 9 post partum. Data comprised 2,529 cows from four large dairy farms and a total of 22,129 progesterone measurements. On one of the farms, recording could be continued up to week 14 pp. and was also validated by measuring progesterone in the lab based on RIA. Several definitions of CLA were used, either based on various threshold values separating low values from increased values, or based on comparisons with an animal specific average base line. The estimate of heritability was highest (0.19) for a definition of CLA based on a fixed threshold of 5 ng/ml. Preliminary genetic correlations with DFS based on direct estimation or from correlating estimated breeding values were around 0.50 to 0.67. Other definitions of the ability for re- cycling were also used. Among these, the proportion of values above a threshold of 5 ng/ml relative to all measurements for an individual cow was highly correlated with CLA.

Genetic network of bovine reproductive traits as revealed by QTLs reported in literature

Utsunomiya, Y.T.[1], Perez O'brien, A.M.[2], Do Carmo, A.S.[1], Zavarez, L.B.[1], Sölkner, J.[2] and Garcia, J.F.[1], [1]Sao Paulo State University (UNESP), Rua Clovis Pestana 793, 16050-680 Aracatuba, Brazil, [2]University of Natural Resources and Applied Life Sciences (BOKU), Gregor Mendel Straße 33, 1180 Wien, Austria; anita_op@students.boku.ac.at

Despite the advent of high density SNP chips and methodological improvements of genome wide association studies (GWAS) genetic dissection of quantitative traits in cattle remains a challenge. Reproduction traits (RT) are particularly difficult to explore as most of these exhibit low heritability. QTL information published in the past 20 years can be useful for building lists of genes that may be associated with RT. We report results on the building of a comprehensive genetic network of bovine RT through an enrichment analysis of an in silico-generated gene list based on genes mapped to RT QTLs within Cattle QTLdb. Genes were added to a list exclusively when they were within QTL boundaries and the target QTL presented a P-value ≤0.05 with whichever statistical analysis adopted by the original publication. Hypergeometric tests were performed for enrichment of pathways using KEGG, Reactome and Panther databases. Genetic network nodes were added as either enriched genes or traits, and edges were drawn directed from genes to traits. The network suggested Calving Ease, Dystocia, Stillbirth and Non Return Rate, RT that can be related to embryo and calf development problems, to share a genetic component through genes from the histone family (H4). Variations within histone family member genes may impact on how nucleosomes regulate DNA accessibility for gene expression. Embryo gene expression problems have been largely studied and incriminated as important cause of early embryo losses and disturbances on fetus and calf development, especially those derived from *in vitro* procedures. These findings illustrate how QTL databases and enrichment analysis can be valuable resources for exploring the genetic architecture of complex traits.

Iron metabolism phenotype: the example of SLC11A1 genotype in Italian Friesian calves

Abeni, F.[1], Petrera, F.[1], Dal Prà, A.[1], De Matteis, G.[2], Scatà, M.C.[2], Signorelli, F.[2] and Miarelli, M.[2], [1]Centro di Ricerca per le Produzioni Foraggere e Lattiero-casearie (CRA-FLC), Sede di Cremona, via Porcellasco 7, 26100 Cremona, Italy, [2]Centro di Ricerca per la Produzione delle Carni e il Miglioramento Genetico (CRA-PCM), via Salaria 31, 00015 Monterotondo (RM), Italy; fabiopalmiro.abeni@entecra.it

The solute carrier gene SLC11A1 has been associated with a natural resistance to many animal diseases. The objective of this study was to evaluate the effect of a SLC11A1 genotype for a coding SNP at the exon 11 (C>G) on iron metabolism in young cattle. Forty-two newborn Italian Friesian calves were genotyped sequencing the exon 11 amplifying a region of 578 bp. The distribution was: CC n=31; CG n=10; GG n=1. Blood samples were drawn at 1, 2, 3, 4, 6, and 8 wk of age, to determine haematological profile, plasma Fe, total iron-binding capacity (TIBC), unsaturated iron-binding capacity (UIBC), and TIBC % of saturation (TIBC-sat). Data were analysed by a mixed model, with genotype (CC vs CG, excluding GG), wk of age, and their interaction as main factors, with the animal repeated in time. Age affected all the considered haematological variables (erythrocytes count, haemoglobin, haematocrit, mean erythrocyte volume, mean corpuscular haemoglobin, mean corpuscular haemoglobin concentration, red cells distribution wideness), and plasma Fe, TIBC, UIBC, and TIBC-sat, evidencing a normal post-natal haematopoiesis. The genotype and its interaction with age did not affect those haematological traits and TIBC; at the same time, the CG genotype had higher (P<0.05) plasma Fe, and TIBC-sat. In light of the role of macrophage in the clearance of senescent erythrocytes, we suppose that the studied SNP at exon 11, which affects the protein structure in the transmembrane domain 8, could determine a different capacity in Fe recycling.

Milk metabolites and their genetic variability

Wittenburg, D., Melzer, N., Reinsch, N. and Repsilber, D., Leibniz Institute for Farm Animal Biology, Genetics and Biometry, Wilhelm-Stahl-Allee 2, 18196 Dummerstorf, Germany; wittenburg@fbn-dummerstorf.de

The composition of milk is clue to evaluate milk performance and quality measures. Milk components partly contribute to breeding scores, and they can be assessed to judge metabolic and energy status of the cow as well as to serve as predictive markers for diseases. In addition to the milk composition measures (e.g. fat, protein, lactose) traditionally recorded during milk performance test (MPT) via infrared spectroscopy, novel techniques, such as gas chromatography-mass spectrometry (GC-MS), allow for a further decomposition of milk into its metabolic components. GC-MS is suitable for measuring several hundred metabolites in high throughput, and thus it is applicable to study sources of genetic and non-genetic variation of milk metabolites in dairy cows. We studied heritability and mode of inheritance of metabolite measurements in a linear mixed model approach including expected (pedigree) and realised (genomic) relationship between animals. We analysed the genetic variability of 210 milk metabolite intensities from 1,295 cows held on 18 farms in Mecklenburg-Western Pomerania. Beside extensive pedigree information, genotypic data comprising 37,180 SNP markers were available. Based on the full model, including marker- and pedigree-based genetic effects, we evaluated goodness of model fit and determined significance of genetic variance components based on likelihood ratio tests. Broad-sense heritability varied from zero to 0.699 with median 0.129. Significant additive genetic variance was observed for high-heritable metabolites, but dominance variance was not significantly present. As some metabolites are particularly favourable, e.g. for human nutrition, ongoing work should address the identification of locus-specific genetic effects as well as investigating metabolites as the molecular basis of the traditional MPT traits.

Automated milk-recording systems: an experience in Italian dairy cattle farms

Biscarini, F., Nicolazzi, E.L., Stella, A. and Team, P.R.O.Z.O.O., Parco Tecnologico Padano, Bioinformatics, Via Einstein – Loc. Cascina Codazza, 26900 Lodi, Italy; filippo.biscarini@tecnoparco.org

Within the framework of the Italian national project 'ProZoo', automated milk-recording systems have been installed in three large Holstein-Frisian dairy cattle farms in Northern Italy (region Lombardy, area of the Po valley). Such automated systems record milk yield, milk components and other parameters, on a daily basis thus providing a wealth of information to be analysed for the management of the dairy farm and, on a broader perspective, for the statistical description and genetic improvement of the population. In particular, data on milk, fat and protein yield, and on somatic cell counts (SCC) -as well as other parameters such as lactose and urea content- constitute an excellent complementary information to the existing official test-day milk records, and might help fill in the gap, should the latter become less frequent (e.g. moving from a monthly test-day to a sparser recording scheme with fewer test days along the lactation). Our preliminary results show that lactation curves for milk yield and -especially- for fat and protein yield, calculated by the automated system, tend to be overestimated, and the algorithm behind them currently still needs to be regularly gauged with data from the official test-day records. In the same farms where the automated systems have been installed, also clinical diagnosis on metabolism-related diseases (e.g. ketosis, milk fever, displaced abomasum) are being recorded with the aid of trained veterinarians. The objective is to relate them with routinely collected milk parameters (urea and lactose concentration, fat/protein ratio, the shape of the lactation curve, etc ...) in order to find putative (early alert) indicators for dysmetabolic syndromes in dairy cattle.

PEMD delivers increased carcase lean and redistribution of lean to the saddle region in lambs
Anderson, F., Williams, A., Pannier, L., Pethick, D.W. and Gardner, G., Murdoch University, School of Veterinary and Biomedical Sciences, South Street, Murdoch WA 6019, Australia; F.Anderson@murdoch.edu.au

Increasing lean meat yield % and redistribution of lean tissue to more highly priced parts of the carcase will increase its value. Selection for the Australian Sheep Breeding Value (ASBV) for greater post weaning eye muscle depth (PEMD) increased eye muscle area and weight of the eye of the short loin, although had minimal impact on carcase lean meat yield %. We hypothesised that selection using the PEMD-ASBV would increase saddle lean weight, without altering whole carcase lean weight, when animals were compared at the same carcase weight. Lamb carcases (n=1218) from the Sheep CRC Information Nucleus were scanned in 'quarters' (fore, saddle, and hind) using Computed Tomography (CT) to determine fat, lean and bone weights. Data was analysed using the allometric equation $y=ax^b$, fitted in its log-linearised form $\log y = \log a + b.\log x$. Fixed effects were site-year, sex, sire type, birth-type rear-type and kill group within site-year, with random terms sire and dam by year. At a given carcase weight, the lean tissue was 4.2% heavier (P<0.01), and fat 8.7% lighter in the whole carcase (P<0.05) across the 7 unit PEMD range. When compared at the same lean weight, the lean tissue in the saddle was 4.9% heavier (P<0.01), and lean in the forequarter was 4.8% lighter (P<0.01) across the PEMD range. Aligning with our hypothesis, there was more lean tissue in the saddle, although unexpectedly this was at the expense of the forequarter only. The mechanistic reason for this redistribution is not clear, and will be investigated with more extensive sampling from tissues across the carcase. In contrast to our hypothesis, PEMD was associated with increased total carcase lean, and reduced fat. The leaner and more muscular composition appears to be independent of maturity as there was not a corresponding increase in bone weight. These impacts on lean weight and distribution to the loin will increase carcass value.

Estimating hot carcass weight by body measurements for Karayaka lambs
Onder, H. and Olfaz, M., Ondokuz Mayis University, Animal Science, Ondokuz Mayis University, Agricultural Faculty, Dep. of Animal Science, 55139 Samsun, Turkey; hasanonder@gmail.com

Hot carcass weight (HCW) is most important factor for the sheep breeders because their income depends only on this trait. This study was conducted to determine which body measurements (body length, height at withers and rump, chest width, chest depth, shin girth) on hot carcass weight obtained from 46 male Karayaka lambs at slaughterhouse and HCW was recorded approximately 1 h after slaughter. To evaluate the estimator reliability, mean square error, value of the determination coefficient (R^2) and significance were used. Collinearity diagnostic was evaluated among variables and multicolinearity cannot be observed. Linear regression with stepwise procedure was performed to estimate the best model fit. According to the results, best model was determined as HCW=-21.842+0.642*(body length) with determination coefficient of 0.866, superior on live weight (LW) model that HCW=1.85+0.412LW with determination coefficient of 0.281. These results showed that body length can be used successfully to estimate hot carcass weight than others.

Horse genetic resources in Central Europe

Bodó, I., Debrecen University, Animal Breeding, Debrecen Böszörményi út 138, 4032, Hungary; bodoi@hu.inter.net

Based upon the common history of horse breeding Central Europe includes the following countries: Austria, Bohemia, Slovakia, Slovenia, Hungary, Croatia, Serbia, Romania. For warm blooded stock the oriental / Arabian/ influence, the impact of Spanish-Neapolitan horse and the import of English Thoroughbred was significant, also the racing developed well at the 19th century. The influence of imperial or military studs was decisive, started from the last decades of the 18th century: Mezőhegyes, Radautz, Bábolna, Kisbér, Fogaras, Lipizza /1580/ and Kladrub. Based on Arabian /Shagya/, Spanish-Neapolitanian breeds Lipizzan, the Kladrub Gala carossier and later influenced by Thoroughbred Gidran, Furioso, Kisbér and Nonius (Anglonorman) breeds were established. They are protected in official preservation programs. Original cold blooded horses are represented by the Noric, Murinsulaner and Posavina breeds. The mountain Bosnian type and the Hutzul are characteristic ponies of the region. After World War II the mechanization of agriculture and transport resulted in drastic decrease of horse population through the whole region. New national sport horse breeds are established on international gene basis. The population size of traditional breeds decreased and they became threatened by extinction. The recently established breeders' associations and the survived traditional studs have the task of preservation of endangered breeds. Developing their breeding goal for modern life and market is important and not easy The effect of enthusiastic breeders and subsidies should not be neglected in this respect. The use of traditional pony breeds for children is slowly developing (Hutzul) and the role of cold blooded horse breeds is more and more also in meat production.

Preservation of indigenous horse breeds in Slovenia

Kaić, A.[1] and Potočnik, K.[2], [1]University of Zagreb, Faculty of Agriculture, Department of Animal Science, Svetošimunska 25, 10000 Zagreb, Croatia, [2]University of Ljubljana, Biotechnical Faculty, Department of Animal Science, Groblje 3, 1230 Domžale, Slovenia; klemen.potocnik@bf.uni-lj.si

Breeding of indigenous horses in Slovenia is based on the Lipizzan, Posavje and Slovenian cold blooded horses. In Slovenia breed is declared as indigenous only if the breed is a traditional one and if the population of the breed has been closed at least during the last five years. Since 2002 preservation of locally adapted animal breeds has been based on the legislation of Livestock Law and since 2004 Regulation on conservation of farm animal genetic resources (RCFAGR), which describes in detail preservation ways for locally adapted breeds, has been applied. Indigenous horse populations are very small so the maximum weight of preservation work is to prevent inbreeding. To uphold and preserve effective population size, breeding program was written for each breed which was confirmed and allowed by the Ministry responsible for Agriculture. In addition, RCFAGR also prescribed Register of breeds with a zootechnical assessment (RBZA) which must be filled every year for each local breed. Through this way, changes in breeds could be easily monitored. RBZA was filled for all three indigenous horse breeds from year 2005, so in last seven years number of registered animals increased 37% in case of Lipizzaner and Slovenian cold blooded horses and 89% in Posavje horse. One reason for increasing number of indigenous horses is subsidies for the breeders paid by Ministry responsible for Agriculture. Despite this, during last three years, number of covered breeding mares significantly decreased because of economical horse market situation. This clearly shows that preservation of indigenous breeds cannot be preserved in a long-term without the usability of breed in economic sense. In case of horses, as special brand mark incorporate local breeds, there are some possibilities in tourism and in meat and mares milk production.

Horse genetic resources in Czech republic

Majzlik, I., Hofmanova, B., Čapkova, Z. and Vostry, L., CULS in Prague, Animal Sci and Ethology, Kamycka 129, 165 21 Prague, Czech Republic; majzlik@af.czu.cz

The aims of this paper is to summarize the characteristics of four endangered horse breeds – gene resources bred in CR, namely Old Kladruber Horse (OKH), Czech-Moravian Belgian Horse (CMB), Silesian Noriker (SN) and Hutzul (H). The OKH is a warmblooder created during last 400 years on basis of Old Spanish and Italian horses in type of robust (baroque) horse in a grey and a black variety, the gene ressource OKH population consist of 33 sires in 9 lines and 351 mares in 24 families (N_e=120), the average inbreeding for sires is 5.52%, for mares 4.88% with slowly decrease in the last 10 years. CMB and SN are coldblooded breeds created during last 130 years. CMB enroles 52 sires in 19 lines and 369 mares in 42 familie (N_e=52.2) with F_x for sires is 3.21% and for mares 2.82% resp. SN population consist of 39 sires from 15 lines and 235 mares (N_e=43.1) with F_x for sires 3.55% and 3.38% for mares resp. The F_x is continuously growing during last 15 years. H population was included into gene ressource in 1993 – at present with 14 sires in 5 lines and 146 mares from 18 families, (N_e=51), the population rate of F_x=3.62%. According to N_e the OKH is endangered breed, whereas CMB, SN and H are critically endangered breeds – all these breeds are included in conservation program, but its efficiency is lowed by unfavorable economical condition. This project was supported by grant MSM 604 607 09 01.

Socio-economical aspect of cold blooded horse conservation program in Poland

Polak, G.M. and Krupinski, J., National Research Institute of Animal Production, National Focal Point, Wspolna, 30 Str, 00-930 Warszawa, Poland; grazyna.polak@minrol.gov.pl

Two local types of cold blooded horses in Poland: Sztumski and Sokolski were typical horses for agricultural work between XIX and the second part of XX century. Further development and mechanization in agriculture resulted in disappearing of these types. The process of breed erosion was stopped thank to a genetic resources conservation program, which was carried on from 2008. In 2011 in Poland there are 555 Sztumski and 680 Sokolski mares included in the genetic resources conservation program. The herds belong to 379 breeders. The general criteria for participation in the conservation programme include: a typical morphological traits and a desired well defined pedigree requirements. Every breeder must have at least two mares of a given breed recorded in the Stud Book of cold blooded horses. The important aspect of the conservation programe is the possibility to conserve the predisposition of traditional working use of these horses. The aim of the study was to better understand motivation of breeders to participate in the conservation programme. A questionnaire was prepared for all breeders including nine questions. The analysis based on information obtained from 142 breeders show that more than 60% have a medium size farm (10-50 ha). A high number of breeders keep also cows (61%) and pigs (31%), but only 3% of them received any type of subsidies for these species. Near 70% of breeders breed horses for at least 15 years and for 27% of them horse breeding is a strong family tradition. Most of the breeders keep horses for commercial causes and for the desire to save the breed, but for 63% of breeders a key reason to participle in the conservation programme is the possibility to receive subsidies. Unexpectedly, near 1/3 of breeders declared to use the horses in agriculture for draft purposes. This work was conducted as part of multiannual programme 08-1.31.9., financed by the Ministry of Agriculture and Rural Development.

Assessment of inbreeding parameters in two Estonian local horse breeds
Rooni, K. and Viinalass, H., Estonian University of Life Sciences, Kreutzwaldi 1, 51014, Estonia; krista.rooni@emu.ee

The aim of this analysis was to assess the situation of inbreeding in the endangered Estonian Heavy Draught Horse and Estonian Native Horse breeds, including changes in inbreeding in the breeds' histories, the current situation and the influence on the breeds. The full pedigrees of the Estonian Heavy Draught Horse and the Estonian Native Horse breed were used. The data set consisted of the pedigree data from 5,107 Estonian Native horses and 9,714 Estonian Heavy Draught horses. The data were analysed with the software package EVA Interface 1.3. The analysis showed that, over history, considerable changes have occurred in the population sizes of the breeds. The analysis also indicated that the level of inbreeding, and the number of inbred horses, has increased constantly throughout the breeds' histories. Nevertheless, selective breeding in recent years has proved to be effective in controlling the increase in average inbreeding per year of birth. Since the breeding populations of the Estonian Heavy Draught Horse and Estonian Native Horse are relatively small (119 and 669 horses, respectively), the danger of inbreeding is great. Hence, breeding activities and breeding schemes should be closely monitored, and more attention should be given to minimising inbreeding. It is necessary to increase the number of stallions used in breeding. Otherwise, the increase in inbreeding, probably accompanied by inbreeding depression, could cause the genetic bottleneck effect.

Assessment of inbreeding depression on linear type description by the Czech draft breeds of horses
Vostry, L.[1,2], Hofmanova, B.[2], Capkova, Z.[2], Majzlik, I.[2] and Pribyl, J.[1], [1]Institute of Animal Science, Pratelstvi 815, 10401 Prague – Uhrineves, Czech Republic, [2]Czech University of Life Sciences Prague, Faculty of Agrobiology, Food and Natural Resources, Kamycka 129, 165 21 Prague, Czech Republic; vostry@af.czu.cz

Inbreeding depression for 22 traits of linear type description and of 4 body measurements were evaluated in three original Czech draft breeds. The inbreeding depression was analysed by using the linear model with fixed effect: sex, year of description, breed, classifier and inbreeding depression and random effect of the individual. Inbreeding depression was in the range of -0.0992 to 0.0242 points for the particular traits. The inclusion of inbreeding depression in the model resulted in a moderate change in h^2 in one-third of the traits. In two-thirds of traits, the value of r_G increased or decreased by 0.01. Among breeding values estimated by a model without inbreeding depression and a model with inbreeding depression for the particular traits were 0.916-0.999 (all horses sample), 0.710-0.992 (10% of the best horses) and 0.827-0.998 (10% of the worst horses). If the average value of the inbreeding coefficient is low (0.03), it is not necessary to include the influence of inbreeding depression in the model for the genetic evaluation of individuals of original cold-blooded horses kept in the Czech Republic. Supported by the project No. MSM 6046070901, and by project no. MZE 0002701404.

Historical background of piebald colour in the Hucul horse in the conservation programme in Poland
Tomczyk-Wrona, I., National Research Institute of Animal Production, Department of Animal Genetic Resources Conservation, Krakowska 1, 32-083 Balice, Poland; iwrona@izoo.krakow.pl

Huculs, the small primitive mountain horses, are one of the oldest Polish breeds with a consolidated genotype. In 1979 Polish Ministry of Agriculture acknowledged Hucul horses to be a conservation breed safeguarding valuable genetic traits. The biometric standard of the Hucul horse was approved in 2004 by member countries of the Hucul International Federation, HIF (Austria, the Czech Republic, Slovakia, Hungary, Romania, Poland, and also Germany and Ukraine) and is contained in the Source Herd Book concerning the origins of the Hucul breed. The standard specifies the colour of Hucul horses, which is mostly different shades of bay and mouse-gray, also piebald, less often black or dun. Some member countries of HIF (Hungary, the Czech Republic and Slovakia) negate the occurrence of piebald colour as typical of the Hucul horse populations. Historical sources document piebald colour in Hucul horses kept in Hutsulshchyna. Starzewski (1927) described 20 out of 66 Hucul horses to be piebald. Holländer (1938) described 212 piebald horses out of 701, and Hackl (1938) also described the presence of piebald colour in Huculs. In modern studs of Hucul horses in: Poland, the piebald colour originated from Larynka and Sroczka mares by Zefir and Jaśmin stallions and from Jagódka mare [founder animal Agatka (A)–Pisanka b. 1940 bay-piebald–Larynka b. 1957 mouse gray-piebald–Zefir b. 1967 bay-piebald-Jaśmin b. 1977 black sorrel-piebald]. The leading stallion was Jaśmin, which had excellent disposition and performance, the traits that have been transmitted to his numerous progeny. Today, out of about 1500 mares and 200 stallions, almost 30% are piebald purebred horses of documented Hucul origin. They are included in the Polish Hucul Horses' Stud Book, which has been closed since 1984 and allows no other breeds to be introduced.* Work conducted as part of multiannual programme 08-1.31.9., financed by the Ministry of Agriculture and Rural Development.

Morphological characteristics of crossbred horses in Ulupamir village of Van province
Alarslan, E. and Aygün, T., Yüzüncü Yıl University, Department of Animal Science, Agricultural Faculty, 65080, Van, Turkey; taygun@yyu.edu.tr

This study was planned to determine the morphological characteristics of horses in Ulupamir village of Erciş district of Van province. For this aim, the body measures were taken on horses in Ulupamir village. Investigation was carried out on a total of 37 horses (71%), 20 stallions and 17 mares, belonging to 32 horse breeders (72%). The least-squares means of shoulder height (SH), body length (BL), rump height (RH), chest depth (CD), chest width (CW), chest circumference (CC), and shinbone circumference (SC) of horses were 133.97, 136.91, 133.56, 57.16, 34.91, 151.00 and 17.24 cm, respectively. The effect of age groups on rump height and shinbone circumference was statistically important ($P<0.05$). Correlation coefficients between SH and BL, RH, CW, CC and SC were statistically found important ($P<0.05$; $P<0.01$). It is concluded that the horses raised at village were similar to East Anatolia horse breed for the morphological traits.

Effect of parents age on phenotypical development of Hutsul horse offspring
Tomczyk-Wrona, I., National Research Institute of Animal Production, Department of Animal Genetic Resources Conservation, Krakowska 1, 32-083 Balice, Poland; iwrona@izoo.krakow.pl

330 Hutsul horses were analyzed based on three biometric measurements taken at three years of age: height at withers (cm) as measured with a measuring staff; chest girth (cm) as tape measured; fore cannon circumference (cm) as tape measured. The experimental population was grouped into three age ranges depending on the age of mother at birth of the analyzed animal (cm; SD): (1). offspring born from mothers aged less than 5 years; (136.6 cm ; 0.55); (170.0 cm; 1.32); (18.4 cm ; 0.11); (2). offspring born from mothers aged from 6 to 10 years; (137.9 cm ; 0.50); (173.4 cm ; 1.19); (18.6 cm ; 0.09); (3). offspring born from mothers aged over 10 years (137.4 cm ; 0.45) ; (171.5 cm.; 1.04); (18.5 cm ; 0.09). In the case of stallions, statistical analyses showed that the effect of sire's age on body measurements was not significant. As regard mares, statistical analysis showed differences for all measurements between the first and second group: highly significant for chest girth and significant for height at withers and fore cannon circumference. Work conducted as part of multiannual programme 08-1.31.9., financed by the Ministry of Agriculture and Rural Development.

Monitoring of horse breeds in Lithuania
Anskiene, L.[1], Muzikevicius, A.[1] and Svitojus, A.[2], [1]Lithuanian University of Health Sciences, Veterinary Academy, Animal technology faculty, Tilzes street 18, 47181, Kaunas, Lithuania, [2]Baltic Genofond, Tyzenhauzų strret 39A, 02118 Vilnius, Lithuania; arunas_svitojus@yahoo.com

The aim of the study was to analise a distribution of horse breeds in Lithuania according to the monitoring data, estimate and predict the changes in horse breeds. Data was analysed from Agricultural Information and Rural Business Centre, Departament of Statstics, Foreign Trade Statistics Division. Comparing the horse situation in Lithuania and the European Union, in Lithuania there is no effective operating system comprehensive evaluation of horse breeding and horses for sale system. One of Lithuania's horse breeding objectives is to preserve the local horse population genetic potential, as in Lithuania bred and kept horse native breeds are in dangerous condition. Žemaitukai, Žemaitukai large, Lithuanian heavy are included in the native farm animal genetic resources. It was predicted the Žemaitukai breed number increases 39 offspring in 2012 year (y = 39.321x+105.68; R2=0.9953), the large Žemaitukai breed number increases 45 offspring (y = 45.583x + 33.25; R2=0.70), the Lithuanian heavy breed number increases 115 offspring (y = 115.18x+249.32; R2=0.70) Forecasted that the total number of horses will continue to decline in 2.7964 thousand horses per year (R2=0.92). Action plan to improve the horse breeding should be: (1) prepare for work stallions testing centre regulations and testing of stallions work rules; (2) according to the approved horses breeding selection programs carry out purposeful horse breeding program, organize working capacity and genetic potential estimations; (3) select and form in a stud horse herd's high value breeding horse breeding nucleus; (4) ensure the best breeds stallions semen seizure, storage and distribution, protected breeds the best stallions long term storage in licensed laboratories.

The domestic livestock resources of Turkey: national horse history, breeds and conservation status
Yilmaz, O.[1], Boztepe, S.[2], Ertugrul, M.[3] and Wilson, R.T.[4], [1]Igdir University, Department of Animal Science, Faculty of Agriculture, Igdir, Igdir, Turkey, [2]Selcuk University, Department of Animal Science, Faculty of Agriculture, Selkuk University, 42100 Konya, Turkey, [3]Ankara University, Department of Animal Science, Faculty of Agriculture, Ankara University, 06100 Ankara, Turkey, [4]Bartridge Partners, Bartridge House, EX37 9AS Umberleigh, United Kingdom; trevorbart@aol.com

Horses have been important in Turkey for more than 5,000 years. First used as food they were then used in war as cavalry and draught animals, then in agriculture and transport and now largely for leisure and sport. National horse numbers were about 1.3 million in the 1930s having built up from an earlier population reduced by wars in the nineteenth and early part of the twentieth centuries. By 2009 there were about 180,000 horses in the country. Concomitant to reduction in numbers was a narrowing of the gene pool and the total loss of some breeds or distinct populations. Native breeds had evolved to meet various conditions including environmental and economic ones and concurrent changes in these facets of production were in large part responsible for the changes in horse numbers and genetic resources. Since the founding of the Turkish Republic (following the fall of the Ottoman Empire) in 1923 there has been much modification of the natural gene pool driven largely by public institutions in response to new challenges. At least nine breeds of various production functions have been imported and crossed with indigenous resources. In 2011 it is possible to identify 23 Turkish functional breeds whose description is the main thrust of this paper. In response to the threat of extinction and to impoverishment or loss of this important aspect of biodiversity Government has established programmes for conservation and preservation of five native breeds. Government, research institutions and producers should work together to ensure that the local gene pool is preserved and can thus continue to contribute to biodiversity and sustainable livestock production.

Challenges in horse production
Von Velsen-Zerweck, A. and Burger, D., European State Stud Association ESSA, Gestütshof 1, 72532 Gomadingen, Germany; Astrid.vonVelsen-Zerweck@hul.bwl.de

In this contribution we describe the general context of challenges in horse production as it is perceived by breeders, private stud farms as well as state studs in Europe. Status quo and trends of economical, ecological as well as public and ethical aspects are analyzed and discussed. Potential actual solutions of interdisciplinary networks and research are demonstrated.

Evaluation of allergen specific ige elisa as diagnostic test for equine insect bite hypersensitivity

Peeters, L.M.[1], Janssens, S.[1], Schaffartzik, A.[2,3], Wilson, A.D.[4], Marti, E.[3] and Buys, N.[1], [1]KULeuven, Kasteelpark Arenberg, 3001 Heverlee, Belgium, [2]University of Zürich, Obere Strasse, 7270 Davos, Switzerland, [3]University of Bern, Länggassstrasse, 3001 Bern, Switzerland, [4]University of Bristol, Langford House, BS40 5DU Langford, United Kingdom; lies.peeters@biw.kuleuven.be

Insect bite hypersensitivity (IBH) is a hypersensitivity to bites of Culicoides species and possibly other insects. The diagnosis of IBH is based on the history, a physical examination and ruling out other conditions causing pruritus, but misdiagnosis occurs. IBH has a genetic basis and the chances to find molecular markers associated with IBH are greatly affected by the quality of the case and control groups. A better definition of the phenotypes, i.e. presence or absence of allergen-specific IgE in the IBH cases and controls should allow to improve a genetic analysis. Therefore, the aim of this study was to evaluate the performance of allergen specific ELISAs as diagnostic tests for IBH in Warmblood Horses and the effect of the severity of IBH symptoms on the performance of the ELISAs was also investigated. IgE level in 343 plasma samples against seven C. nubeculosus and two C. obsoletus recombinant proteins were measured using ELISA. The horses were subdived in four classes: class I (unaffected, n=176), class II (affected and no symptoms at sampling, n=66), class III (affected and mild symptoms at sampling, n=50) and class IV (affected and clear symptoms at sampling, n=51). The combination of the IgE level against the three best performing r-allergens (CulSumoff3) resulted in a Area under the Curve (AUC) of 0.80. An effect of the severity of the clincial symptoms was found and this study showed that the IgE ELISA is most suitable to differentiate between IBH unaffected and affected horses if the affected horses showed clear clinical symptoms at sampling. The pairwise accuracy between class I versus class IV was 0.84 for the 'CulSumoff3'-test.

Globalisation of movements and spread of equine infectious diseases

Herholz, C.[1], Leadon, D.[2], Perler, L.[3], Binggeli, M.[3], Füssel, A.-E.[4], Cagienard, A.[5] and Schwermer, H.[3], [1]School of Agricultural, Forest and Food Science, Equine Science, Laenggasse 85, 3052 Zollikofen, Switzerland, [2]Irish Equine Centre, Johnstown, Naas, Kildare, Ireland, [3]Federal Veterinary Office, Schwarzenburgstrasse 155, 3003 Bern, Switzerland, [4]European Commission, 2, rue Van Maerlant, 1049 Brussels, Belgium, [5]Merial SAS, Veterinary Public Health, 29, av. Tony Garnier, 69348 Lyon, France; conny.herholz@bfh.ch

Increased international and intercontinental movement carries inherent risks of transfer of infectious disease. Diseases well known in some countries can cause havoc when they are imported into naïve horse populations elsewhere. Quantifying horse populations, disease incidence and the numbers of horse being transported for trade, competition and breeding are important in any emerging disease risk-assessment. This report will describe OIE reports of major equine infectious disease status, together with updates on horse movements from the EU's TRACES database. Examples of equine disease emergence will be presented in this context. Certain notifiable infectious diseases in equidae have the potential to spread across borders and may therefore have a drastic direct and indirect impact on the entire horse sector, involving horse holdings, competition and race establishments, veterinary clinics, but also state veterinary services and local and national governments. We will discuss emergency responses to disease suspicions and outbreaks. Further strategies for risk minimization and prevention will be presented.

Protection of horses against lineage 2 West Nile virus (WNV) in Europe
Venter, M.[1,2], Williams, J.[2], Mentoor, J.[2], Van Vuren, P.J.[1], Pearce, M.C.[3] and Paweska, J.[1], [1]National Institute for Communicable Diseases, Private Bag 4, 2131 Sandringham, South Africa, [2]University of Pretoria, Lynnwood Rd, 0002 Pretoria, South Africa, [3]Pfizer Animal Health, Hoge Wei 10, 1930 Zaventem, Belgium; michael.c.pearce@pfizer.com

WNV is a mosquito-borne flavivirus and important neuropathogen of horses and humans. Two genetic lineages exist. Lineage 1, linked with encephalitis in humans and horses already occurs in Europe. Lineage 2 is endemic in Southern Africa and has recently been associated with encephalitis in humans and horses in central Europe, Greece, and Italy. Surveillance in South Africa shows that WNV lineage 2 accounts for 15% of equine neurological disease with a case fatality rate of 42%. We conducted two experiments to see if Duvaxyn WNV vaccine confers cross lineage protection against WNV. In the first experiment, 3 groups of mice were vaccinated with 0.1 ml Duvaxyn WNV and 3 groups of control mice were administered 0.1 ml PBS on days 0 and 21, then challenged on day 49 with either NY385 (lineage 1) or SPU93/01 (lineage 2) WNV. Clinical signs in mice were monitored for 21 days after challenge. Neutralising antibodies were measured on days 49 and 70. In the second experiment, 16 horses were vaccinated with Duvaxyn WNV on days 0 and 28, then again on day 399 and 6 horses were enrolled as unvaccinated controls. Plain blood samples were taken on days 0, 7, 14, 28, 49, 140, 231, 322 and 413 and tested for the presence of neutralising antibodies against lineages 1 and 2 WNV. Titres of antibody against lineage 1 and 2 were compared using a repeat measures generalised linear mixed model. Vaccinated mice were completely protected but all unvaccinated mice demonstrated clinical signs of infection after challenge with lineages 1 and 2 WNV. Mice vaccinated with Duvaxyn WNV had detectable neutralising antibodies at challenge, and had higher neutralising antibody titres than unvaccinated mice 21 days after challenge. Antibody titres to lineages 1 and 2 following vaccination were comparable in both mice and horses.

Preventive management program in a large equestrian and breeding farm
Dobretsberger, M., Spanish Riding School Vienna and National Stud Piber, Piber 1, 8580 Köflach, Austria; maximilian.dobretsberger@vetmeduni.ac.at

Preventive management and quality management has repeatedly been proven to be the best tool against emerging diseases. This contribution describes the introduction of a novel and exemplary 'health book' for a very large combined equestrian and breeding farm, which includes a modern vaccination and deworming concept, an optimalized nutrition and reproduction as well as screening program concerning infectious diseases, quarantine regulations and hygienic measures. In addition, the monitoring of genetical fitness of the animals, documentation and administration strategies as well as a collaboration program with scientific partners for anticipation and solution of new problems are part of the interdisciplinary handbook.

Monitoring of biochemical parameters in the aging process in Norik Muráň type

Noskovičová, J., Novotný, F., Pošivák, J., Boldižár, M., Dudríková, K. and Tučková, M., University of Veterinary Medicine and Pharmacy, Equine Clinic, Komenského 73, 040 01 Košice, Slovakia; noskovicova@uvm.sk

The aim of the study was to monitoring changes of selected biochemical parameters in the aging process in 27 cold-blooded mares breed Norik Muráň type, divided into group (A, B, C, D) according to age: A (n=7, 5-8 years), B (n=6, 9-15 years), C (n=7, 16-20 years) and D (n=7, 21-23 years). Blood samples were collected from v. jugularis using vacuum tubes (Serum-SSTTMII Advance, BD Diagnostics, USA). Biochemical parameters of aspartate aminotransferase (AST), alkaline phosphatase (ALP), creatinine (Crea), total protein (TP), albumin (Alb), gammaglutamyl-transferase (GGT) and cholesterol (Chol) were determined by COBAS c111. Results were statistically processed using Student's t-test (P<0.05, Microsoft Excel 2003).The measured values of AST had decreasing character with age (A=5.69±1.13 µkat/l, B 5.36±0.9 µkat/l, C 5.13±1.1µkat/l), but the oldest group was increased (D 5.88±1.37 µkat/l). The concentration of ALP had an upward trend with increasing age (A 2.47±0.3 µkat/l, C 2.66±0.43 µkat/l, D 2.77±1.17 µkat/l), only concentration of the group B was lower (2.44±0.46 µkat/l). Concentration of GMT was maintained at the same level in all age groups (A 0.28±0.05 µkat/l, B 0.29±0.03 µkat/l, C 0.25±0.04 µkat/l, D 0.29±0.15 µkat/l). Cre declined with age (A 139.1±37.49 µmol/l, B 118.45±23.18µmol/l, C 119.31±19.85 µmol/l, D 111.59±21.08 µmol/l), as well as Alb (A 31.34±3.24 g/l, B 30.03±2.41 g/l, C 30±2.77 g/l, D 29.67±2.85 g/l) and TP (A 70.07±5.29 g/l, B 65.87±7, 78 g/l, C 63.96±6.59 g/l), but in the oldest age group increased (D 67.41±8.2 g/l). With aging the concentration of Chol eleveted (A 2.33±0.56 mmol/l, B 2.3±0.25 mmol/l, C 2.36±0.29 mmol/l, D 2.43±0.37 mmol/l). It seems that age affects biochemical parameters in mares Norik Muráň type, which are result of particular breeding and selection of the breed. Study was supported by Mini. of Educ. VEGA 1/0498/12 of the Slovak Rep.

Lamb mortality: current knowledge

Gautier, J.M.[1] and Corbiere, F.[2], [1]Institut de l'Elevage, BP 42118, 31321 Castanet Tolosan Cedex, France, [2]Ecole Nationale Vétérinaire, UMR 1225 INRA ENVT, 23 chemin des Capelles, 31706 Toulouse, France; jean-marc.gautier@idele.fr

Lamb mortality rate before weaning is on average 15 to 20% and impacts the profitability of sheep farms. Causes of lamb mortality are infectious (viral, bacterial, fungal or parasitic, systemic or localized) or non infectious (as a result of dystocia, starvation-exposure syndrome, congenital defects, mineral deficiencies...) and depends on age of lambs. Risk factors are numerous and linked either to the dam (age/parity, litter size, maternal behavior, feeding and body condition of pregnant ewes, transfer of passive immunity, genetics), the lamb (weight of birth, birthcoat type, behavior and temperature of the lamb at birth) or the environment (climatic conditions outside or inside sheepfold, lambing place, health status of flock...). There are strong interactions between various risk factors which makes diagnosis particularly complex in absence of reliable records on lamb mortality. To reduce the lamb mortality rate in France, studies must be carried out to i) estimate importance of different risk factors, ii) improve genetic selection by taking into account lamb survival.

The effect of colostrum intake on lambs: plasma proteomic profile and immunoglobulin concentration

Hernández-Castellano, L.E.[1], Almeida, A.[2,3], Ventosa, M.[3], Sánchez-Mácias, D.[1], Moreno-Indias, I.[1], Torres, A.[4], Coelho, A.[3], Castro, N.[1] and Argüello, A.[1], [1]Universidad de Las Palmas de Gran Canaria, Animal Science, Arucas, 35413 Gran Canaria, Spain, [2]Instituto de Investigação Científica Tropical, Rua da Junqueira, Lisboa, Portugal, [3]Instituto de Tecnología Química e Biologica, Oeiras, Lisboa, Portugal, [4]Instituto Canario de Investigaciones Agrarias, La Laguna, Tenerife, Spain; lhernandezc@becarios.ulpgc.es

Knowledge about what minor proteins from colostrum, despite immunoglobulins (Ig), have an immune function on newborn lambs is necessary. For this reason the objective of this study was to evaluate differences in blood plasma proteomic profiles due to the colostrum intake and consequences of the delayed intake of colostrum on Ig concentration during 5 days after birth. Two groups of 6 lambs each were used in this study. One group received three colostrum meals at 2, 14 and 26 hours after birth. The other group was fed with colostrum at 14 and 26 hours after birth. At the end of the colostral period each animal of both groups took the same amount of Ig per live body weight at birth (4 mg of IgG/kg) from colostrum and then were fed with a commercial milk replacer. Blood plasma samples were collected at 2 and 14 hours after birth for proteomic assay and every 24 hours for the determination of immunoglobulin concentration. Differential in Gel Electrophoresis (2D-DIGE) was performed using Immobiline DryStrips with pH 3-10 and 24 cm length (GE Healthcare). Spots showing differential expression were detected using Progenesis SameSpots software (Nonlinear Dynamics) and identifed using MALDI TOF/TOF. Immunoglobulin concentration was determined with a commercial ELISA kit (Bethyl Laboratories, Montgomery, Texas, USA) and the statistical analysis was performed using a PROC. MIXED procedure (SAS, Version 9.00). Results showed that an early colostrum intake increased the expression of several proteins with immune function like Serum Amyloid A and tended to increase immunoglobulin concentration during 5 days after birth.

The effect of low birth weight on brain DHA status and implications for pre-weaning mortality

Tanghe, S.[1], Millet, S.[2] and De Smet, S.[1], [1]Ghent University, Laboratory for Animal Production and Animal Product Quality, Department of Animal Production, Proefhoevestraat 10, 9090 Melle, Belgium, [2]Institute for Agricultural and Fisheries Research (ILVO), Animal Sciences Unit, Scheldeweg 68, 9090 Melle, Belgium; stanghe.tanghe@ugent.be

Pre-weaning mortality of piglets is a serious loss for the pig industry. Birth weight is an important factor determining piglet survival. Recently, attention is given to the supplementation of gestation feeds with n-3 polyunsaturated fatty acids (PUFA) and especially docosahexaenoic acid (DHA), as an approach to reduce pre-weaning mortality. PUFA are essential for the development of the foetus and are highly present in the brain. It has already been shown that fish oil (source of DHA) in the feed of the sow can improve piglet vitality. This study aimed to investigate the effects and possible interactions of birth weight and PUFA supplementation on the DHA status of the brain. Two groups of sixteen sows were fed either a control diet containing palm oil or a diet containing 1% fish oil from day 73 of gestation and during lactation. At birth, the lightest and heaviest male piglet per litter was sacrificed and brain samples were taken for fatty acid analysis. The effects of birth weight and diet were analysed by GLM using SAS 4.3. Piglets that died pre-weaning had lower birth weights than piglets surviving lactation, while no effect of feed on mortality was found. Piglets from sows fed fish oil had higher brain DHA levels than piglets from the control treatment. A lower brain DHA content was observed for the lighter piglets compared to the heavier piglets in both dietary treatments. Although the birth weight × diet interaction term was not significant, the difference in brain DHA levels between the weight groups was much smaller for the piglets of the maternal fish oil diet. These results suggest that the higher incidence of pre-weaning mortality in low birth weight piglets is related to their lower brain DHA status.

Effect of generation, parity and year on pig prolificacy in small closed population
Razmaitė, V., Lithuanian University of Health Sciences, Lithuanian Endangered FarmAnimal Breeders Association, Institute of Animal Science, R. Žebenkos 12, 82317 Baisogala, Lithuania; razmusv8@gmail.com

Lithuanian White pigs were developed in the 20th century. However, since 2003 drastic decline in the numbers of Lithuanian White has occurred and the solution adopted by the Institute of Animal Science has been to conserve the remaining of the original Lithuanian White pig breed in a closed herd. Nowadays this herd is the single herd of the old genotype Lithuanian White pigs. The objective of the study was to examine the influence of generation, parity number and year on litter traits in the closed herd of old genotype Lithuanian White pigs. The data on farrowing and litter size were available per parities from 2000 to 2011. The piglets came from 395 litters (104 dams and 28 sires) of five generations. The data were processed by the GLM procedure in Minitab. The generation showed effect on the numbers of total born and males and tended to affect the total number of piglets born alive per litters. However, the generation did not appear to affect the numbers of stillborn piglets. The year of farrowing showed the overall effect on the numbers of piglets total born. However, the effect of year on the numbers of piglets born alive and stillborn piglets was observed only in the first generation. The decline in the numbers of piglets born alive was observed from 2008. The parity showed the overall effect on the numbers of total born piglets and tended to show effect on the numbers of piglets born alive and affected the number of stillborn piglets in the first generation. The numbers of stillborn piglets decreased in parity 2, however significant increase was reached in parity 5 and the maximum number of stillborn piglets was reached in parity 10. This study showed that the effect of generation on the litter size was negligible and that breeding of old genotype Lithuanian White pigs in a small closed population over four generations had no clear negative influence on the litter size.

Neonatal lamb behaviour contributes to improved postnatal survival
Dwyer, C.M.[1], Nath, M.[2] and Matheson, S.M.[1], [1]SAC, Animal Behaviour and Welfare, Animal and Veterinary Sciences, King's Buildings, West Mains Road, Edinburgh EH9 3JG, United Kingdom, [2]BioSS, King's Buildings, West Mains Road, Edinburgh EH9 3JG, United Kingdom; cathy.dwyer@sac.ac.uk

High lamb mortality remains a welfare and economic concern for sheep producers. Earlier studies have considered ewe maternal behaviour to be a significant contributor to lamb mortality, but the behaviour of the lamb was not often considered. However, newborn lambs are active soon after birth and play a critical role in reaching the udder and achieving successful sucking, which is fundamental for survival. We have shown that the early behaviours of the neonatal lamb (e.g. speed to stand or suck after birth) are independent of maternal behaviour, and associated with improved survival. Using detailed behavioural and physiological data collected from 582 lambs and analysed by Cox proportional hazard models, we show that lambs that needed assistance with delivery or to suck successfully from their mothers are statistically less likely to survive until weaning in comparison to unassisted lambs ($P=0.007$ and $P=0.02$ respectively). Other significant contributors to lambs survival in the model were litter size (larger litters had higher mortality, $P=0.001$), ewe age (offspring of younger ewes had higher mortality, $P=0.04$), and lamb rectal temperature at 2 h old (lambs with a higher temperature had better survival, $P=0.0003$). Lamb breed, and line and sire within breed also have a significant effect on lamb neonatal behaviour, and survival. Data collected from over 11,000 lambs on commercial farms recording assistance with delivery and sucking were used to estimate genetic parameters for these traits. Heritability for birth assistance and sucking assistance were 0.26 ± 0.03 and 0.32 ± 0.04, suggesting that selection on these traits is possible and would be expected to improve lamb survival. For more extensive systems, where these measures cannot be readily obtained, alternative methods to assess the survivability of lambs are required.

Lambs behaviour during suckling in the period of creep feeding
Margetinova, J., Apolen, D. and Broucek, J.; apolen@cvzv.sk

The purpose of this study was to investigate the effect of age, litter size and weight of lambs on time that lambs need to find mother, select udder on left (L) or right (R) side, and start sucking. The observations included 72 lambs (42 single (S), aged 14.5 days; 30 twins (T), 14.8 days). Average weight at entry into the trial was 5.21 kg. Observations were made in the morning after opening of kindergarten and releasing of suckling lambs to mothers. At the same time the live weight of lambs was recorded at weekly intervals. Lambs were divided into 3 groups according to body weight (GW) (1st = 3.5 to 4.5kg; 2nd = 4.6 to 5.5; 3rd = 5.6 to 7. 5), according to age (GA) at entry into the trial (1st = 11-13 days; 2nd = 14-16; 3rd = 17-18). Data were analyzed using General Linear Model ANOVA by the statistical package STATISTIX, Version 9.0. Significant differences between groups were tested by Comparisons of Mean Ranks. Values are expressed as means ± SE. 24 lambs (33%) always favoured the same side of udder when they found their mother. Out of them 12 lambs favoured L side, 12 lambs R side. This means that each lamb-twin sucks always on the same side of udder. Lambs of 2nd GA needed most time to find the mother (41.17 s), least time (31.52 s) lambs of 3rd GA. The time has decreased in the 1st GA by 24.06 s, in the 2nd GA by 5.38 s, in the 3rd GA by 5.94 s from first to last observation. Significant effect of lambs' age on time to find mother was not found. It was found out that the average time needed to locate the mother by lamb was in favour of single. Significant effect of litter size on time was recorded at the last measurement (S 23.17±2.520 s vs T 52.43±11.337 s; P<0.01). Weight of lambs at last observation was about 16.62 kg in S (average daily gain ADG 0.32kg); 13.91kg in T (ADG 0.26 kg). Comparing the weight categories, most of the time needed lambs in 1st GW (51.73 s), least time lambs in 2nd GW (32.54s). Significant effect of LW of lambs on time necessary to search the mother was not found.

Session 14b   Theatre 1

Effectiveness of a deactivator on mitigating the impact of endophyte alkaloids in Merino ewes
Leury, B.J., Henry, M.L.E., Digiacomo, K., Ng, C., Kemp, S. and Dunshea, F.R., The University of Melbourne, Agriculture and Food Systems, Parkville, 3010, Victoria, Australia; brianjl@unimelb.edu.au

Perennial ryegrass (PRG) shares a symbiotic relationship with an endophyte (*Neotyphodium lolli*) that produces two classes of toxins, ergot alkaloids (e.g. ergovaline) and indole-diterpenes (e.g. lolitrem B). These alkaloids are beneficial for plants but harmful to sheep when ingested, causing a variety of symptoms including hyperthermia. This study investigated a commercial alkaloid deactivator, Elitox® (AD), in alleviating adverse effects associated with alkaloid intake. Twenty-four 6 month old Merino ewes (35 kg) were housed indoors for 3 weeks and fed at 1.2 × maintenance (50%, 09:30 h and 50%, 16:30 h) with access to water. Sheep were allocated to one of six treatments: 0, 2 or 4g AD/d with or without infected PRG seed (PRGs; ergovaline 100 µg/kg and lolitrem B 80 µg/kg LW). Sheep were weighed weekly and feed intake measured daily. Respiration rate (RR) and skin (T_S) and rectal (T_R) temperatures were obtained at 09:00, 12:00 and 16:00 h daily. T_R was increased in sheep given PRGs (39.2 vs. 39.5 °C, P<0.05) and tended to be decreased by AD (39.5, 39.2 and 39.4 °C for 0, 2 and 4 g AD/d, P=0.10). There was a PRGs × AD × Week interaction for T_R (P<0.01) such that T_R increased by around 0.1 °C for all treatments except PRGs + 0 g AD/d which decreased by 0.15 °C. T_S (38 vs. 38.3 °C, P<0.05) was increased in sheep given PRGs. RR was unchanged. All feed was consumed except for PRGs + 0 g AD/d (-9.5% by week 3) and 2 g AD/d (-7% in week 1 which improved by week 3) (PRG × AD × Week interaction, P<0.001). Liveweight was unchanged for nil PRG groups, but decreased in PRGs + 0g AD/d (-0.4kg) and increased in PRGs + AD (+1.1 and +2.6 kg for 2 and 4 g AD/d; PRG × AD interaction, P<0.05). These data suggest that inactivation of the alkaloids mitigates some of the adverse physiological responses associated with alkaloid ingestion in sheep and may improve weight gain, most likely by reducing alkaloid absorption and/or modifying activity in the rumen.

Mediterranean shrub *Pistacia lentiscus* L. as a potential tool in the control of nematodes in sheep

Saric, T.[1], Rogosic, J.[1], Beck, R.[2], Zupan, I.[1], Zjalic, S.[1], Musa, A.[3] and Skobic, D.[3], [1]University of Zadar, Department of Ecology, Agronomy and Aquaculture, Obala Kralja Petra Krešimira IV./ 2, 23000 Zadar, Croatia, [2]Croatian Veterinary Institute, Department for Bacteriology and Parasitology, Savska cesta 143, 10 000 Zagreb, Croatia, [3]University of Mostar, Faculty of natural science and education, Ulica Matice hrvatske b, 88000 Mostar, Bosnia and Herzegowina; tosaric@unizd.hr

Gastrointestinal nematodes are among the most important causes of production loses in sheep and goat farming. The major exposure of sheep to the infections is influenced also by their diet poor in tannins. During the experimental period the treated group of sheep fed with mixture of *Pistacia lentiscus* L. dry foliage, barley and soybean meal, while the control group of sheep fed with mixture of barley and soybean meal. The number of oocists and nematode eggs in gram of feces (EPG) was monitored on days 0, 5 and 7. The blood parameters are measured on days 0 and 7, while the concentration of tannins in kidney and liver were determined on day 7. Animals fed with *P. lentiscus* dry foliage showed a significant decrease in EPG number ($P<0.05$) while the blood parameters and tannin concentration seem not to be impaired by the treatment. According to results obtained the foliage of *P. lentiscus* could be used as healthy and environmental friendly tool in control of nematodes in ruminants, contributing to reduction of pasture contamination with nematode eggs and consequently slow down the dynamics of animal infection.

Valeric and isovaleric acid concentrations: useful biomarkers for subacute ruminal acidosis?

Fanning, J., Cockcroft, P. and Hynd, P., The University of Adelaide, School of Animal and Veterinary Science, School of Animal and Veterinary Sciences, Roseworthy Campus, The University of Adelaide, Roseworthy 5371, Australia; joshua.fanning@adelaide.edu.au

Subacute Ruminal Acidosis (SARA) causes significant economic losses in the dairy industry. Little is known about the production or economic losses in the red meat feedlot industries, but it is believed to be significant. At present the only direct means of detecting SARA is by repeat rumen fluid sampling or telemetric pH logging. We hypothesised that the concentration of longer chain volatile fatty acids, valeric and isovaleric acids, are inversely correlated with ruminal pH. Six fistulated lambs (six month old South African Meat Merinos) with pH loggers inserted intra-ruminally, were transitioned over one week from an *ad libitum* lucerne/oaten chaff diet (8.6 MJ/kg DM) to an *ad libitum* 80:20 rolled barley:lucerne/oaten chaff diet (12.7 MJ/kg DM) and allowed to adapt for ten days. During the transition and adaptation periods, rumen pH was measured every 15 minutes by the loggers and a rumen fluid sample was taken via the rumen fistula three hours post feeding for volatile fatty acid analysis. The rumen fluid samples within sheep were bulked over a seven day period for analysis. Valeric acid concentration was moderately correlated with time spent in the SARA pH range, pH 5.0 to 5.5 ($r=0.27$, $P>0.05$). Isovaleric acid concentration was more highly correlated with time in the SARA range ($r=0.57$, $P<0.05$). Further research is warranted to validate the use of C5 fatty acids as markers of SARA in ruminants and to determine the microbiological and biochemical mechanisms of the relationship between C5 acid concentrations and ruminal pH.

Preliminary assessment of sheep welfare on pasture

Mialon, M.M.[1], Robin, C.[2], Verney, A.[2], Brule, A.[2], Pottier, E.[2], Davoine, J.M.[2], Ribaud, D.[2], Boivin, X.[1] and Boissy, A.[1], [1]Inra, UMR1213 Herbivores, 63122 Saint-Genès-Champanelle, France, [2]Institut de l'Elevage, Monvoisin, BP 85225, 35652 Le Rheu, France; marie-madeleine.richard@clermont.inra.fr

On farm assessment for animal welfare has been more developed for intensively reared farm species than for extensively reared ones. The aim of our study was to test a new approach to assess welfare in sheep on pasture, estimating inter- and intra-observer reliability and variability among French farms. We proposed animal-based measures inspired from the 4 principles of WelfareQuality® method. Feeding (body condition score), housing (animal dirtiness, wool humidity) and health (lameness, lesions, respiratory disorders, hoof overgrowth and udder) were scored at the individual-level on 30 ewes per farm. A human approach test and Qualitative Behaviour Assessment (QBA) were also performed on two or three 30-ewes groups per farm to assess appropriate behaviour. The inter- and intra-observer reliability was assessed on ten experimental farms using intra-class correlation or kappa coefficients. Between-farms variability was estimated in summer time on 53 commercial farms by variance analysis with a mixed model. Inter-observer reliability was good for individual measures as well as for group measures (>0.66). Intra-observer reliability was generally good for individual measures (>0.65) but was poor (<0.2) to moderate (<0.6) for group measures. Most measures presented occurrence rates below 2% except lameness (5.6% of ewes), nasal discharge (3.6%) and hoof overgrowth (5%). The highest occurrence rate was noticed for animals' back dirtiness (26%). For these measures and for the QBA, within-farm variability was higher than between-farms. By contrast, between-farms variability represented more than 60% of the total variability for the human approach test, suggesting a promising measure for discriminating farms. Our approach will be further tested on winter-pasturing sheep where more variability should be expected, particularly in health indicators.

Curved allometries reveal no constraint in horn length in Rasquera White Goat

Parés-Casanova, P.M. and Sabaté, J., University of Lleida, Animal Production, Av. Alcalde Rovira Roure, 191, 25198 Lleida, Spain; peremiquelp@prodan.udl.cat

The size of the sexual elaborate ornaments (defined as exaggerated or novel structures used to visually attract mates) carried by many animal species has been traditionally explained in terms of the trade-off between the benefit, in the form of increased mating success, and the costs that arise from carrying these traits. Many studies have demonstrated the advantage of these elaborate secondary sexual traits (for instance when they are employed in male-male competition), but few have shown compelling evidence for the limits to the elaboration of these traits that must exist. No many insights if any have devoted to study the pattern of ornament expression and other body parts in domestic mammals. In this study we describe such ornament evidence in the exaggerated horns of Rasquera White Goat, an ancient local breed from Catalunya (SE Spain) managed under extensive conditions for meat purposes. Twenty-two male and twenty-seven female well preserved adult skulls were studied. The curvature of the allometric line for each sex was quantified by performing a second-order polynomial ($y = ax^2 + bx + c$, where y = log horn length and x = log skull length). Residual horn length explained 99.4% of the variance in curvature ($F=1.379$, $P=0.261$). The value of a gave indication of the amount of curvature, with positive values in both cases thus indicating an increasing slope, but being most pronounced in females ($a=11.32$) than in males ($a=1.276$). This was consistent with the hypothesis that the increase in size horn did not account the increasing costs, so horns had no limit to their size. This pattern could suggest that in the Rasquera White Goat breed there is no limit in horn exaggeration, which in some cases reaches to more than 60 cm for each horn.

The sheep sector in greenhouse gas inventory in Hungary

Borka, G., Németh, T., Krausz, E. and Kukovics, S., Research Institute for Animal Breeding and Nutrition, Gesztenyés út 1, 2053 Herceghalom, Hungary; sandor.kukovics@atk.hu

Production characteristics of the Hungarian sheep sector relevant to greenhouse gas (GHG) emissions were surveyed. On the basis of the survey, GHG emissions from sheep sector were calculated by using of IPCC Tier 2 method for the period of 1985-2009, and the GHG emission reduction possibilities were examined. Finally the current Hungarian agricultural GHG emission inventory was analysed by gas and source, the nitrous oxide (N_2O) and methane (CH_4) emission trends of the Hungarian agricultural sector and the importance of the sheep sector were reviewed. Main results of the project are the following. (1) As regards the average GHG emissions of the Hungarian economy between 1985 and 2008, agriculture takes 13% of it; so it is the second largest emission sector after energy consumption (78%). The emissions from agriculture decreases in absolute terms, and its share in total emissions also shows a decreasing trend; in 1985 it was 15%, but in 2008 it was slightly below 13%, while energy sector is still responsible for 81% of the emissions in 2008. (2) In Hungarian animal production the cattle sector is the largest GHG emitter with the share alternating between 42% and 47% (average: 45%). Out of the total GHG emissions of animal production, swine, sheep and poultry sectors take in average 30%, 11% and 10%, respectively. All the other animal production sectors (goat, buffalo, equines, rabbit) are responsible for 4-6% of the emissions. (3) The share of sheep sector was 0.9% in average (0.6-1.3%) regarding the examined period compared to the total GHG emissions of the national economy. The weight of sheep sector (even together with goat sector) does not reach 1%. As regards gases, sheep sector takes 2.6% (1.7-3.6%) for CH_4 and 5.1% (3.8-8.3%) for N_2O. (4) The realistic possibilities for reduction of GHG emissions from sheep sector (nutrition: increasing of concentrate and fat ratio and optimization of protein feeding; manure storage and application: adequate storage and application methods) are very limited.

Exploring skull shape and sexual dimorphism in a local goat breed

Parés-Casanova, P.M., Sabaté, J. and Soto, J., University of Lleida, Animal Production, Av. Alcalde Rovira Roure, 191, 25198 Lleida (Catalunya), Spain; peremiquelp@prodan.udl.cat

As it has been traditionally said, goats are sexually dimorphic animals, with a general pattern of males larger than females. Most of research has been done on a pure biometrical basis of linear measurements. This research was intended to assess differences in sex-specific skull shape based on 2D geometric morphometrics. Twenty-four skelotonized heads from adult animals (12 males and 12 females) belonging to the 'Cabra Blanca de Rasquera' meat breed, reared under extensive conditions at Catalunya (SE Spain). Skulls were dorsally photographed and on each picture there were digitized 15 landmarks (SEM=1.60). The analysis identified two components, overall accounting for almost 99% variance. PC1 was consistently related to general head width and muzzle length, while C2 explained the muzzle width. But skull shape was not significantly discriminated between sexes. This similarity between sexes may be interpreted in relation to equal extensive management-styles of the animals that moreover are under a low anthropogenic influence.

Constraints to efficient protein utilization in the dairy cow and on the dairy farm
Broderick, G.A., USDA-ARS, US Dairy Forage Research Center, 1925 Linden Drive, Madison, WI 53706, USA; glen.broderick@ars.usda.gov

Dairy cattle and other ruminants are moderately productive on diets poor in true protein because ruminal microbes synthesize protein with a good essential amino acid (EAA) pattern. Moreover, ammonia utilization by ruminal microbes allows feeding nonprotein N plus utilization of urea N that would otherwise be excreted in the urine. However, microbial protein is diluted by nucleic acid and cell wall N that is no use to the cow. The dairy nutrition literature suggests maximal N-efficiency (milk-N/dietary-N) is in excess of 40%; experimental measurements of 32-35% are common. However, efficiency under normal management is typically 20-25% and often much lower. Low N utilization necessitates feeding supplemental protein, reducing both the economic and environmental sustainability of dairy production. Sophisticated models allow the balancing of EAA needs of the cow with supply from microbial protein and rumen-undegraded protein. However, inadequate quantitative data on factors influencing microbial yields, protein escape, intestinal digestion and tissue EAA metabolism prevent these models from being reliable. When N-efficiency is improved, the improvement is in the form of gram-for-gram reduction in urinary urea-N; this translates into reduced ammonia volatilization from the farm. Although a number of manure handling practices will improve agronomic capture of excreted N, improving protein utilization in the animal is our principal tool for reducing the environmental impact of dairying.

The use of antibodies in order to alter bacterial population in the rumen
Foskolos, A., Cavini, S., Ferret, A. and Calsamiglia, S., Univ. Autónoma Barcelona, Ciéncia Animal i dels Aliments, University campus, 08193 Bellaterra, Spain; sergio.calsamiglia@uab.es

Recent efforts to control N losses in the rumen are focusing on the manipulation of the proteolytic and hyper ammonia producing bacteria (HAP). Our objective was to produce polyclonal antibodies against proteolytic and main HAP in order to improve N utilization and decrease N release to the environment. Bacterial strains obtained from ATCC were: Prevotella ruminicola, Clostridium aminophilum and Peptostreptococcus anaerobius. Bacteria were grown according to recommendations, freeze dried and used to immunize rabbits. Blood samples were collected and analysed by ELISA. A 24 h batch culture was used to test the effects of polyclonal antibodies on ruminal fermentation. Treatments were: control (CTR; serum of non-immunized animals), polyclonal antibodies against Prevotella ruminicola (APr), Clostridium aminophilum (ACl) and Peptostreptococcus anaerobius (APa), tested at 0.005 and 0.05 ml of serum / 30 ml of medium. Selected tubes were withdrawn at 3, 12 and 24 h and sampled for ammonia-N (NH_3) and volatile fatty acids (VFA). In a second study, eight continuous culture fermentors were inoculated with ruminal liquid from a dairy cow fed a 50:50 concentrate:forage diet in 2 replicated periods to test the effect of the same treatments at 3.2 ml of serum/fermentor/day. Each experimental period consisted on 8 days: 5 days of adaptation and 3 days of sampling. Fermentors were sampled 0, 2, 4 and 6 h post feeding. Samples were analysed for NH_3 and VFA. Polyclonal antibodies had no effect on NH_3 and VFA concentrations at 3, 12 and 24 h of batch fermentation. In the fermentors study, NH_3 of the effluents varied from 7.31 (CTR) to 7.91 (APa) mg·/100 ml without significant differences. Similarly, hourly variation of VFA and NH_3 concentrations of the fermentors did not differ significantly. Polyclonal antibodies against deaminating and proteolytic bacteria tested in the current study did not affect ruminal protein degradation either in short or in long term fermentation.

Protein level and forage digestibility interactions on dairy cow production

Alstrup, L. and Weisbjerg, M.R., Aarhus University, Department of Animal Science, Blichers Allé 20, 8830 Tjele, Denmark; lene.alstrup@agrsci.dk

The objective of the study was to test whether requirements for crude protein (CP) supplementation depend on the digestibility of the forage. The hypothesis was that requirements for protein are less when forage digestibility is high, i.e. that reduced CP supply will cause less reduction in feed intake and milk yield if the forage digestibility is high than if it is low. A 4×4 Latin square experiment was conducted with a total of 48 lactating Danish Holstein cows during four 21-day periods and with four treatments. Treatments were arranged 2×2 with 2 levels of forage digestibility (low vs. high) and 2 CP levels (low vs. high) giving the four treatments LL, LH, HL and HH. All rations consisted of 55% forage (62% maize silage and 38% grass clover silage) and 45% concentrate on dry matter (DM) basis. Organic matter digestibilities were 79.8% and 74.7% in high and low digestibility ration, respectively. The protein level was altered by replacing barley and beet pulp with rapeseed- and soybean meal to increase CP concentrations in DM from 13.9-14.0% (low) to 15.7-16.0% (high). DM intake (DMI) was 21.9, 22.5, 23.9 and 24.9 kg/d, and yield in energy corrected milk (ECM) was 31.9, 32.9, 33.5 and 34.8 kg/d and N efficiency (N in milk/N intake) was 0.35, 0.32, 0.35 and 0.30 for LL, LH, HL and HH, respectively. As such, DMI increased by 2.2 kg/d from low to high forage digestibility, whereas the high protein level increased DMI by 0.7 kg/d compared to low protein level. DMI effects were reflected in the milk yield, as the high digestibility increased ECM yield by 1.7 kg/d and high CP increased ECM by 1.2 kg/d. There were no significant interactions between forage digestibility and CP level. Compared to previous observations, low CP level reduced feed intake and milk yield as expected, whereas low forage digestibility reduced feed intake and milk yield slightly less than expected. Requirements for protein supply appeared independent of the digestibility of the forage.

Feed intake and milk yield responses to reduced protein supply

Weisbjerg, M.R., Kristensen, N.B., Hvelplund, T., Lund, P. and Løvendahl, P., AU-Foulum, Aarhus University, Animal Science and Molecular Biology and Genetics, P.O. Box 50, 8830 Tjele, Denmark; martin.weisbjerg@agrsci.dk

Protein supply to dairy cows below requirements might reduce feed intake and production, whereas oversupply normally is economically suboptimal. Furthermore, manure and urine from animal production can lead to N leaching into open waters and ground water, and to ammonia evaporation, which can result in severe environmental problems in countries with high density animal production. Therefore, it is important to have precise knowledge on cow's response to protein concentration in the diet. The sensitivity of feed intake and milk production in dairy cows to reduced protein content in diets was tested. Four protein concentrations, 167, 150, 134 and 121 g crude protein (CP) per kg ration DM, were fed to 64 Danish Holstein cows in 4×4 Latin square designs (in total 256 observations). The three lowest protein concentrations were below Danish recommendations for dairy cows with milk yield potential of approx. 10,000 kg milk per lactation. Compared to the highest protein concentration (167 g CP/kg DM), effect of first reduction step (150 g CP/kg DM) was modest, but further reduction steps (134 and 121 g CP/kg DM) progressively led to lower feed intake, 21.4, 21.4, 21.0 and 20.0 kg DM/d and reduced production, 31.1, 30.6, 29.1 and 27.1 kg energy corrected milk/d, resulting in increased N efficiency (N in milk/N in feed), 0.297, 0.325, 0.344 and 0.357, respectively. Individual cows' responses depended on their daily milk production, so that high yielding cows responded most in energy corrected milk production, whereas responses in feed intake were independent of daily milk yield level. We conclude that dietary protein restrictions to dairy cows are possible but reductions below 150-160 g crude protein per kg ration DM cause reduced milk production especially in high yielding cows.

Dietary energy source modifies the N utilization and the whole-body leucine kinetics in dairy cows
Cantalapiedra-Hijar, G., Savary-Auzeloux, I., Cossoul, C., Durand, D. and Ortigues-Marty, I., INRA, INRA-Theix, 63122 Saint Genès Champanelle, France; gonzalo.cantalapiedra@clermont.inra.fr

Although diets based on starch improve the efficiency of N utilization (ENU) in dairy cows compared to diets based on fibre, little is known on the metabolic adaptations responsible for this improvement. The effect of the dietary energy source (ES) on ENU and whole-body Leu kinetics were studied at 2 different crude protein contents (CP) in 4 catheterized dairy cows in order to test whether glucogenic nutrients from starchy diets could improve ENU, especially at low CP, by decreasing the AA catabolism. Four iso-energetic diets with 2 ES (based on starch [NDF/starch=0.95] vs fibre [NDF/starch=9.8]) and 2 CP (Low [12.0%] vs High [16.5%]) were fed during 28-d experimental periods according to a 4×4 Latin Square design. On days 27 and 28, L[1-^{13}C] Leu and [^{13}C] sodium bicarbonate, respectively, were infused into one jugular vein, and blood samples were taken hourly from the mesenteric artery once the isotopic enrichments (IE) reached the plateau. The IE of plasma Leu and CO_2 released from blood were analyzed. Feed, refusals and milk were sampled and analyzed for N content the days of tracers infusions. In addition, milk urea-N content (MUN) was assayed by infrared spectroscopy. The DM intake was similar across dietary treatments. As planned, N intake and milk N secretion were higher for high compared to low CP diets. Higher Leu irreversible loss rate (ILR) and oxidation (both absolute and fractional) as well as higher MUN found in high vs low CP diets were consistent with the higher ENU found for the latter. Although N intake was similar across ES, milk N, and accordingly the ENU, was higher for diets based on starch vs fibre. This improvement of ENU with diets based on starch was accompanied by a higher Leu ILR, Leu oxidation as well as a higher MUN compared to diets based on fibre. Present data show that whole body Leu oxidation and MUN alone did not explain the higher ENU found for diets based on starch vs fibre.

Effect of a dietary escape microbial protein on production and fertility in Italian dairy cows
Agovino, M.[1], Warren, H.[1] and Segalini, D.[2], [1]Alltech Bioscience Centre, Summerhill Road, Co Meath, Ireland, [2]Ferraroni Mangimi, Bonemerse, (CR), Italy; hwarren@alltech.com

Excess protein represents significant financial and biological costs to dairy enterprises. However, overly high levels of crude protein (CP) are often still fed with little attention to the quality of protein included. This trial investigated the potential for lowering overall ration CP using a dietary escape microbial protein (DEMP®, Alltech Inc., KY). Holstein Friesians (n=138) were allocated to one of two dietary treatments: Control (C; n=69, DIM 151, parity 2.6) – basal diet (maize silage, commercial concentrate, alfalfa hay, sugar beet pulp, straw; CP 163.2 g/kg DM); Treatment (T; n=69, DIM 173, parity 2.6) – basal diet reformulated to contain 0.3 kg DEMP (CP 155.2 g/kg DM). Intakes averaged 25 kg DM/d. Animals remained on trial for five months (May–September 2011). Milk yield, composition and urea, as well as days from calving to conception were measured. Data were analysed using Two-way T-test unless abnormally distributed (Wilcoxon Rank Sum test). Initial fat-corrected milk yields (FCM) were similar at 36.2 for C vs. 34.8 kg/d for T, but diet reformulation with 0.3 kg of DEMP resulted in a significant (P<0.001) increase by the end of the trial (29.2 and 34.5 kg FCM/d for C and T, respectively). Similar trends were noted for milk fat content where non-significant increases were seen (32.6 for C vs. 32.8 g/kg for T). There was no effect of treatment on milk protein. By the end of the trial, milk urea levels were significantly (P<0.001) lower for the animals fed the reformulated diet compared to C (20.7 vs. 28.8 mg/dl, respectively). Days from calving to conception were non-significantly reduced from 143 to 137 for C and T, respectively. Nitrogen use efficiency was non-significantly increased by 0.012 (0.286 for C vs. 0.298 for T). These data demonstrate benefits of improving N use efficiency by reformulating rations to provide a more suitable protein profile.

Milk protein responses to dietary manipulation of amino acids at two levels of protein supplies
Lemosquet, S.[1,2], Haque, M.N.[1,2], Rulquin, H.[1,2], Delaby, L.[1,2], Faverdin, P.[1,2] and Peyraud, J.L.[1,2], [1]Agrocampus Ouest, UMR1348,PEGASE, 35000 Rennes, France, [2]INRA, UMR1348, PEGASE, 35590 Saint-Gilles, France; sophie.lemosquet@rennes.inra.fr

Improving the efficiency of metabolisable protein (PDIE in INRA) utilisation by dairy cows remains a key issue in dairy industry that can be achieved by manipulating the essential amino acids (EAA) profile of dietary proteins. Two experiments were conducted to analyse the effect of balancing the EAA profile for Lys, Met (using MetaSmart® dry) and Leu of diets providing 2 levels of PDIE (94 vs. 112 g/kg DM) in Holstein cows with a 'non-limiting' supply of rumen-degradable protein to avoid any interaction with digestibility. Four corn silage diets (56.3% of DM) were offered in fixed quantities in Exp. 1 and *ad libitum* in Exp. 2 to analyse the effects on dry matter intake (DMI). In Exp. 1, 15 cows were allocated in a 4×4 Youden square design (1 missing cow) with periods of 3 weeks. In Exp 2, 16 cows were allocated in a split-plot design with 2 groups of 8 cows receiving each one level of PDIE (94 or 112 g) during 12 weeks while the EAA profiles were given in each group according to cross over design with periods of 6 weeks. Results of both experiments were analysed by using MIXED procedures of SAS (2004) including the effects of PDIE, EAA profiles and interaction, with a random effect of cow. The levels of Lys, Met, Leu in % of PDIE were at 6.4 vs. 7.3, 1.9 vs. 2.3, 8.6 vs. 8.8, respectively in unbalanced vs. balanced EAA profile diets. At both PDIE levels, balancing the EAA profile increased milk protein content by 0.7 g/kg and protein yield by 21 g/d (Exp. 1: P<0.01). At high PDIE level, the unbalanced EAA profile increased DMI (Exp. 2: P<0.01), but not milk protein yield, decreasing gross efficiency of PDIE utilization from 0.52 to 0.48. Balancing EAA profile through diets significantly improves the protein efficiency irrespective of the PDIE levels. However this improvement remains limited. It mainly occurs through a higher metabolic efficiency of intestinal AA for milk protein.

Effect of a source of sustained-release non-protein nitrogen on beef cattle
Rossi, C.[1], Compiani, R.[1], Baldi, G.[1], Vandoni, S.[2] and Agovino, M.[2], [1]Department of Science and Technology for Animal Nutritional Safety, University of Milan, Milan, Italy, [2]Alltech Italy, 40033, Casalecchio di Reno (BO), Italy; magovino@alltech.com

Improving nitrogen (N) use efficiency in beef cattle diets contributes to cost-effective production. This study investigated the effects a sustained-release (SR) ruminal non-protein N (NPN) source in the diet of young bulls as partial replacement of true protein on beef production parameters. Charolais bulls (n=56) were allocated to 1 of 2 diets: control (n=28; basal diets of corn silage, corn meal, concentrate, brewer's grains, straw and SBP) and treated (n=28; basal diets reformulated to include SR NPN (Optigen®, Alltech Inc.) decreasing CP from 147 to 135, 152 to 136 and 153 to 137 g/kg DM for adaptation, fattening and finishing, respectively). Animals were on treatment for 140 d (40 d adaption, 30 d fattening, 70 d finishing). Dry matter intake (DMI), feed conversion rate (FCR) and daily gain (ADG) were measured at d0, d40 and d100. Chemical (DM, CP, EE, ash, starch and NDF) and visual evaluation of faeces (d45, d75) and analyses of blood urea N (BUN; d40, d100) and rumen N (TVB N at slaughter) were carried out. Data were evaluated by ANOVA using the GLM procedure (SAS 8.02). The treated group had significantly (P<0.05) higher ADG (1.63 kg/d) from d40 to d100 compared with control (1.46 kg/d). DMI was lower (P<0.05) in treated animals indicating that increased ruminal N availability resulted in positive effects (P<0.05) on FCR (7.40 vs. 6.85 for control and treated, respectively). Ruminal [N] and faecal chemical composition were not affected. Visual evaluation showed greater but non-significant presence of mucin in faeces for control (1.54) compared to treated (0.62). BUN was similar indicating SR allowed for complete utilisation for microbial synthesis. Treated animals used less N to produce 1 kg ADG (138 g) vs. control animals (181 g). These data support the optimisation of soluble protein in beef cattle diets using SR NPN to improve ruminal synthesis whilst reducing dietary CP.

Mammary metabolism to the supply of an 'ideal' amino acids profile for dairy cows

Haque, M.N.[1,2], Rulquin, H.[1,2], Guinard-Fament, J.[1,2] and Lemosquet, S.[1,2], [1]Agrocampus Ouest, UMR1348, PEGASE, 35000 Rennes, France, [2]INRA, UMR1348, PEGASE, 35590 Saint-Gilles, France; muhammad.Naveed-Ul-Haque@rennes.inra.fr

Balancing the essential amino acids (EAA) profile of metabolisable protein (PDIE in the INRA system) is relevant to reduce the protein supply for dairy cows by increasing the efficiency of EAA utilization for milk protein. The aim of this study was to analyse mammary metabolism of EAA in response to the 'ideal' EAA profile (EAA+) to understand the increased PDIE efficiency. Four dairy cows received 2 different EAA profiles (Ctrl and EAA+), provided by a corn silage based diet and duodenal infusions of 10 EAA, combined with 2 levels of PDIE·NE$_L^{-1}$ (14.5 vs. 16.8 g/MJ i.e.128 vs. 145 g crude protein/kg DM, respectively), according to a 2×2 factorial design. Blood samples were collected from a carotid artery and a mammary vein. Mammary plasma flow was estimated using the Fick principle applied to Phe+Tyr. Results were analysed using the MIXED procedure of SAS (2004) with a random effect of cow. The intestinal concentrations in EAA+ and Ctrl (in % of PDIE) were at the levels expected for Met (2.4% vs. 1.7%); Leu (9% vs. 7.9%), and His (3% vs. 1.9%), but was lower than expected for Lys (7.0% instead of 7.3% vs. 5.8%). Nevertheless, the EAA+ profile increased milk protein yield from 262 to 290 g /12 h per half udder (P=0.01), milk protein content from 30.5 to 32.0 g/kg (P=0.04) and the efficiency of PDIE utilisation for milk protein from 0.54 to 0.58 (P=0.02) at low and high PDIE supply with no interaction. Mammary plasma flow was not modified by treatments (389±24 l/h). With the EAA+ profile, mammary uptakes of Lys, Leu, Met and Val increased (P<0.05) by 22%, 12%, 9% and 9% respectively and the efficiency of intestinal Lys and Leu utilisation by mammary gland did not decrease (Lys: 0.40 and Leu: 0.34). The results suggest that balancing EAA profile improves the efficiency of metabolisable protein utilisation, at both PDIE levels, through an increased uptake of EAA by the mammary gland.

Effect of normal and high NaCl intake on PDV urea-N flux and renal urea-N kinetics in lactating cows

Røjen, B.A.[1] and Kristensen, N.B.[2], [1]Aarhus University, Department of Animal Science, Blichers Allé 20, 8830 Tjele, Denmark, [2]Agro Food Park, Agro Food Park 15, 8200 Aarhus, Denmark; betina.amdisenrojen@agrsci.dk

The objective of the present study was to investigate if high NaCl intake would induce hypouremia from increased water and urea diuresis. A NaCl induced hypouremia was hypothesized to affect ruminal and portal-drained visceral (PDV) extraction of urea when compared with an equal hypouremia induced by decreasing dietary N intake. Nine Holstein cows (21±2.1 kg DMI·/d, 37±4.0 kg milk·/d) fitted with ruminal cannulas and permanent indwelling catheters in major splanchnic blood vessels were assigned to normal NaCl diet (CON) and high NaCl diet (HIGH; supplemented with 2.5% feed NaCl in DM) in a cross-over design experiment. Cows were fed equally sized portions at 8 h intervals. Eight hourly sets of rumen fluid, urine, and arterial, portal, hepatic, and ruminal blood samples were obtained, starting 30 min before morning feeding on the last day of each 21-day period. Data was analysed using the mixed procedure of SAS. It was not possible to detect a response in blood urea N concentration in HIGH compared with CON. Diuresis, and the amount of blood urea-N cleared by the kidney were increased (P<0.01 to P=0.01) in HIGH compared with CON, while urine urea-N concentration decreased (P<0.01) resulting in unchanged urinary urea-N excretion between treatments. Renal plasma flow and urea-N reabsorption were not affected by treatments. Increasing NaCl intake did not affect the total transfer of urea-N from blood to gut (net portal flux of urea-N). Additionally, there was no change in ruminal and PDV extraction of blood urea-N, and ruminal ammonia concentration, reflecting no changes in epithelial permeability for urea-N. In conclusion, high dietary intake of NaCl did not elicit the expected decrease in blood urea-N concentration, and had no apparent effects on the mechanisms regulating urea-N transfer across gut tissues. The kidneys maintained homeostatic balance of inorganic ions.

Milk iso C17:0 indicates low dietary rumen degradable protein supply in a Cuban monitoring study

Fievez, V.[1], Fuentes, E.[2], Noval, E.[3] and Lima, R.[3], [1]Ghent University, Laboratory of Animal Nutrition and Animal Product Quality, Proefhoevestraat 10, 9090 Melle, Belgium, [2]Universidad Nacional Agraria La Molina, Facultad de Zootecnia, Apartado 12-056, Lima, Peru, [3]Universidad Central 'Marta Abreu'de Las Villas, Department of Veterinary Medicine and Zootechnics, Carretera de Camajuani km. 5.5, 50100 Santa Clara, Cuba; veerle.fievez@ugent.be

Milk iso C17:0 derives from branched-chain volatile fatty acids, originating from rumen fermentation of amino acids (valine, leucine and isoleucine). Whether this fatty acid can be a biomarker of low supply of rumen degradable protein was assessed in a monitoring study performed on dairy farms of the 'Desembarco del Granma' Cooperative (Santa Clara, Cuba) during the Cuban rainy season (July and August 2010). Cows grazed natural pastures, mainly consisting of *Dichantium* spp. (average CP: 80 g/kg DM) or were fed diets based on freshly harvested elephant grass, *Pennisetum purpureum* var. CT-115 (average CP: 93 g/kg DM). Supplementation with concentrates was limited to marginal. The loading plot of a principal component analysis was used to find relations between milk protein and fat content as well as concentrations in milk fat of odd and branched chain fatty acids. A positive association between milk protein, urea and iso C17:0 (ranging between 14.5 and 35.5 g/kg milk, 115 and 200 mg/kg milk, 2.34 and 5.71 g/kg milk fatty acids, respectively) was observed and linked to situations of increased dietary protein supply. Lowest status of milk iso C17:0 were associated with grazing of natural pastures. We suggest that iso C17:0 can be an indicator of limited supply of rumen degradable protein.

Effect of maize silage and Italian ryegrass silage on nitrogen efficiency of organic milk production

Baldinger, L., Zollitsch, W. and Knaus, W., University of Natural Resources and Life Sciences, Department of Sustainable Agricultural Systems, Gregor Mendel Str. 33, 1180 Vienna, Austria; lisa.baldinger@boku.ac.at

In Austrian organic agriculture protein-rich grass and grass-clover silage is often combined with maize silage to balance protein and energy supply and thereby increase nitrogen efficiency. However, reports about Italian ryegrass (*L. multiflorum*) silage providing better feed efficiency than maize silage raised the question whether it might improve nitrogen use efficiency as well. Therefore a feeding trial in a cross-over design (two 6-week periods) with 22 lactating Holstein Friesian cows was conducted to test two diets. Both contained grass silage (38%) and hay (10%), produced from clover grass leys (mainly *L. perenne* and *T. pratense*) and permanent grassland (balanced plant stands with 50% grasses, 30% legumes and 20% herbs). The maize diet (M) contained 40% maize silage and the ryegrass diet (R) 40% Italian ryegrass silage [dry matter (DM) basis]. By adding low amounts of barley and soybean cake partly mixed rations with similar amounts of utilisable crude protein at the duodenum (uCP; R: 133; M: 136 g/kg DM), but different energy contents (R: 6.15; M: 6.33 MJ NEL/kg DM) were produced. Statistical analysis was done using proc MIXED of SAS 9.1, with a model including the random effect of cow(order). Irrespective of treatment energy supply was adequate and milk production level (R: 21.3; M: 23.3 kg ECM/d) satisfactory. However, feeding the maize diet significantly increased total feed intake (R: 17.0; M: 18.4 kg DM/d) as well as energy intake (R: 109; M: 120 MJ NEL/d). Combined with significantly higher milk protein yields (R: 629; M: 737 g/d) feeding the maize diet resulted in an above-average nitrogen use efficiency of 0.304 as compared to an average level of 0.259 when feeding the ryegrass diet. The negative ruminal nitrogen balance (RNB; -29.4 g/d) in treatment M, indicating a scarcer nitrogen supply to ruminal microbes than in treatment R (RNB 9.4 g/d), did not influence performance negatively.

Effect of protein source on feedlot performance of early weaned beef calves
Beretta, V.[1], Simeone, A.[1], Elizalde, J.[2], Gamba, D.[1] and Terzián, A.[1], [1]Facultad de Agronomía-Universidad de la República, Animal Science, Ruta 3 km 363, 60000, Uruguay, [2]Consultant, Rosario, 2000, Argentina; beretta@fagro.edu.uy

An experiment was conducted to evaluate the effect of protein source on feedlot performance of spring-born early-weaned beef calves. Twenty seven Hereford calves (93.7±15.5 kg; 110±18.2 days old) were randomly allocated to 9 pens outdoors and 1 of 3 total mixed rations (26% sorghum silage, 74% concentrate) differing in supplementary protein source (PS): urea (U); soybean meal (SBM); or fish meal (FM), (n=3 pens/treatment). Concentrates were iso-energetic (ME: 12.64 MJ/kg DM) and iso-nitrogenous (CP: 191 g/kg DM) across treatments, varying in rumen undegradable protein (RUP) content (U: 20.8%, SBM: 40.9%, FM: 58.1% CP). Feed was offered at 2.5% of live weight (LW) in 3 meals, during 75 days. Calves were weighed every 14 days, and subcutaneous back fat (SBF) and Longissimus dorsi area (LDA) were determined by ultrasonography on d75. Dry matter intake (DMI) was determined daily and feed: gain ratio (F:G) calculated based on mean adjusted values. Records of LW were analysed as repeated measures using the Mixed Procedure of SAS, while F:G, SBF and LDA were analysed with GLM procedure. Statistical model included treatment effect and the initial LW as covariate. Animal LW increased linearly with time (P<0.01) and was affected by initial LW (P<0.01) and PS: calves fed the U-diet showed lower LWG compared to SBM and FM, which did not differ (U: 0.859[b], SBM: 0.990[a], FM: 0.995[a] kg/d; SE 0.032, P<0.01). DMI was not affected by PS (3.03 kg/d SE: 0.02; P=0.10) while a tendency was observed for differences in F:G ratio (U: 3.7[a], SBM: 3.2[b], FM: 3.1[b]; SE 0.23; P=0.07). No differences were detected in LDA (29.4 cm[2] SE 3.5; P=0.54) or SBF (3.5 mm; SE 0.7; P=0.80). Results suggest that under restricted DMI (2.5% LW) early-weaned beef calves could benefit from including part of the supplemental CP in the ration as RUP. This would allow for higher metabolisable protein intake and possibly a better N utilization.

Assessing urinary metabolites as markers of nitrogen use in ruminants using a LC-MS and NMR approach
Boudra, H., Doreau, M., Nozière, P. and Morgavi, D.P., INRA, 1213 UMRH, Site de Theix, 63122, France; morgavi@clermont.inra.fr

In ruminant production systems it is important to optimise the incorporation of dietary nitrogen (N) in milk and meat while minimising excretion losses contributing to water and air pollution. The excess, deficiencies and utilisation of dietary N by the animal affect its metabolism and, in this context, a metabolomic approach may be a strategy for discovering markers of N use efficiency. Liquid chromatography coupled with mass spectrometry and 1H NMR spectroscopy (LC-MS/NMR) were developed to simultaneously measure metabolic profiles in dairy cows urine. The aim of this study was to compare spot versus total urine collection as the most adequate sampling technique for discriminating the metabolome and, potentially, for finding new biomarkers. Urine samples were collected from dairy cows fed two different levels of N with starch or fibre as the main energy source in a 4×4 Latin square design. Multivariate MS data analysis showed that metabolic profiles were strongly influenced by sampling with only 24% of significant ions common to both sampling techniques. Total collection seemed better than spot to discriminate diets as quantitative criteria such as a higher number of significant ions and the better quality of the predictive parameter Q2 for partial least square-discriminant analysis (PLS-DA). The superiority of total collection needs to be validated based on markers identification. PLS-DA showed a separate grouping for each diet in both sampling techniques. There was a clear distinction between contrasting dietary treatments: high N-Starch vs. low N-Fibre and high N-Fibre vs. low N-Starch but the separation between high and low N diets was less clear. From MS data, a list of 37 and 21 marker candidates was established for total and spot urine samples, respectively. The identification of these potential marker candidates is under progress. For NMR, chemical shifts for lactate, pyruvate, phenylalanine and hippurate – the most discriminating metabolites – were confirmed by NMR-2D.

Effect of rumen-protected methionine on buffalo rumen environment

Chiariotti, A.[1], Huws, S.A.[2], Contò, G.[1] and Pace, V.[1], [1]CRA-PCM Research Centre for Animal Production, Via Salaria, 31, Monterotondo (Roma), 00015, Italy, [2] IGER- Institute of Grassland and Environmental Research, Plas Gogerddan, Aberystwyth, SY23 3EB, United Kingdom; vilma.pace@entecra.it

The aim of this experiment was to investigate microbial diversity of Buffalo rumen resulting from a diet containing rumen protected-methionine (RPM). Four rumen-cannulated Mediterranean buffalo cows were fed *ad libitum*, two iso-energetic diets (0.90 MilkFU/kg DM) containing, on DM basis, 44% corn silage,13% soybean meal, 15% corn meal, 26% lucerne hay (treatment A; CP=15.5%) and 44% corn silage, 9.5% soybean meal, 18.5% corn meal, 26% lucerne hay (CP=14.2%) +12 g/head/d of RPM (MEPRON®) (treatment B) according to a cross-over design with two 3-months periods. Traditional and molecular techniques were applied to assess microbial population diversity (DGGE and Q-PCR) in ruminal samples after 15d of diet adaptation before morning feeding. The data obtained were analysed according to the SAS GLM procedure. No statistical differences were detected in bacterial density both by traditional (cfu/ml) and Q-PCR technique (16S rDNA), but the number of fungi was significantly lower for treatment B compared with A (1.18×10^7 vs. $2.34 \times 10^7 \cdot$/ml; P<0.01) as was the number of protozoa (17×10^7 vs. $7.2 \times 10^7 \cdot$/ml; P<0.01). However, DGGE-ciliate 18S rDNA-based dendograms did not show any clustering by diet when based either on the Dice or on the Pearson algorithm. Plasma urea level was significantly higher for treatment A compared with B (7.4 vs. 6.2 mmoles/l, P<0.01). Urinary urea-nitrogen was also higher for treatment A compared with B (154 vs. 132 g/d; P<0.05). The amount of total organic nitrogen detected in urine was 208 for treatment A vs. 179 g/d for B (P<0.01). The reduction of protein level in the diet supplemented with RPM seems to negatively affect the growth of some rumen microorganisms, but it exerts a positive effect lowering the urinary nitrogen excretion, thereby contributing to reduce the impact of buffalo herds on the environment.

An analysis of eco-efficiency scenarios in dairy farming: simulations of calving interval

Hutu, I. and Chis, C., Banat's University of Agricultural Sciences and Veterinary Medicine Timişoara, 119 Aradului Street, 300645 Timişoara, Romania; codrutachis@gmail.com

The eco-efficiency of dairy farms presumes the approach of economic efficiencies and the impact upon environment. The study started from a real situation of a dairy farm with 38 cows and 50 hectares of farmland with quantified productions, rations, cost and income and calculated nitrogen balance and value of profit. By simulating the calving interval (CI) from 15 to 26 months, two scenarios were performed by using the N-CyCLE software: minimization of the N-balances and the maximization of the profit. The two scenarios significantly influenced the value of profit – the scenario towards maximizing the yearly gross margin generated the highest average profit (€ 798·/cow), and the scenario towards minimum nitrogen-balance generated the lowest average profit (€ 402·/cow). The annual N-balance was 126 kg N/ha in the economic scenario and 70 kg N/ha in the ecological scenario. The loss of profit for the each kg N/ha/yr reduction was € 7.10·/cow. The eco-economic character of the farm was obtained through a mathematical model constructed on barycentric coordinates with equal weights from the two scenarios above. This resulted in a new scenario for the computation of the values of N and the rates of gross margin, computation which was performed using the Maple 13 software. In this scenario the eco-economic character of the farm can be improved at low CI; in this situation the farm acquires an eco-economic character when the annual N-balance is 98 kg N/ha/yr and the profit is € 629·/cow. In this case the loss of profit for the each kg N/ha/yr reduction becomes € 6·/cow. To improve the eco-economic character of dairy farms by CI, the society should support a yearly subsidy of € 169·/cow. This study was co-financed from project POSDRU/89/1.5/S/63258, Postdoctoral school for zootechnical biodiversity and food biotechnology based on the eco-economy and the bio-economy required by eco-san-genesis.

An analysis of eco-efficiency scenarios in dairy farming: simulations of cows number
Huțu, I. and Chiș, C., University of Agricultural Sciences and Veterinary Medicine Timișoara, 119 Aradului Street, 300645 Timișoara, Romania; codrutachis@gmail.com

The eco-efficiency of dairy farms presumes the approach of economic efficiencies and the impact upon environment. The study started from a real situation of a dairy farm with 38 cows and 50 hectares of farmland with quantified productions, rations, cost and income and calculated nitrogen (N) balances and value of profit. By simulating an increase in the number of cows from 39 to 55, two scenarios were performed by using the N-CyCLE software: minimization of the N-balances and the maximization of the profit. The two scenarios significantly influence the value of profit – the scenario towards maximizing the annual gross margin generated the highest average profit (€ 617·/cow), and the scenario towards minimum N balance generated the lowest average profit (€ 193·/cow). The annual N balance was 136 kg N/ha in the economic scenario and 89 kg N/ha in the ecological scenario. The eco-economic character of the farm was obtained through a mathematical model constructed on barycentric coordinates with equal weights from the two scenarios above. This resulted in a new scenario for the computation of the values of N and the rates of gross margin, computation which was performed using the Maple 13 software. In our study, the eco-economic character could not be improved by increasing number of cows without land increase – the farm acquires an eco-economic character when the annual N balance is 112 kg N/ha and the profit is € 351·/cow. Thus, increasing the number of the cows is economically not feasible to increase sustainability of dairy farms; in this case the society will support a yearly subsidy of € 267·/cow which is too much in comparison with costs of other management practices as increasing milk or protein productions or optimizing feeding and manure management. Acknowledgements: This study was co-financed by project POSDRU/89/1.5/S/63258

Rapeseed-cake supplementation as strategy to reduce milk urea nitrogen concentration in dairy sheep
Mandaluniz, N., Arranz, J., Ruiz, R., Ugarte, E. and García-Rodríguez, A., NEIKER, Animal Production, P.O. Box 46, Vitoria-Gasteiz, 01080, Spain; rruiz@neiker.net

As livestock has been identified as one of the sources of nitrogen (N) pollution, ways to reduce its excretion must be studied. Improvement of N utilization efficiency by livestock decreases N losses from farms. The objective of this study was to evaluate locally produced oil-cake as feeding supplementation strategy to reduce N excretion. Two experiments were conducted (2009 and 2010) over 5-6 weeks each with multiparous Latxa dairy ewes. Sheep were blocked into 3 homogeneous groups of 12 and assigned to 3 different concentrates (commercial with soya as protein source, rapeseed oil-cake and sunflower oil-cake). The concentrates were isoenergetic (1.0±0.01 UFL/kg DM) and isonitrogenous (170±10 g CP/kg DM). Indoor feeding was composed by 0.6 kg concentrate DM and 1.0 kg hay DM and were allowed to graze on a polyphite grass pasture for 4 h/d. Milk yield was monitored individually and milk samples were collected weekly to determine protein, fat and milk urea nitrogen (MUN). The concentration of MUN was used as a simple and non-invasive measurement to predict N excretion. Data were analysed considering type of concentrate as fixed effect. According to the results, concentrate type did not affect significantly ($P=0.13$) milk yield (1.15, 1.24 and 1.27 kg/d for rapeseed-cake, commercial and sunflower-cake, respectively) and milk gross composition (protein: 6.10, 6.03 and 5.90%; fat: 5.89, 5.60 and 5.48%, respectively). However, MUN was significantly ($P<0.001$) lower in rapeseed-cake than for commercial or for sunflower (449[a], 518[b] and 495[b] mg/l; respectively). According to these results there are no significant differences between protein sources with respect to production performances (milk yield and milk gross composition) which indicate that commercial protein source can be replaced by locally produced protein sources like rapeseed or sunflower cake. Finally, the supplementation of rapeseed-cake causes lower MUN values.

Effect of feeding rumen protected methionine and choline on milk yield and milk composition in early
Soleimani, A.[1], Nourozi Ebdalabadi, M.[2], Kousary Moghaddam, M.[1] and Ahmadzadeh Bazzaz, B.[1], [1]Islamic Azad University,Kashmar Baranch, Department of Animal Science, Kashmar, Seyyed Morteza Blvd, 91857, Iran, [2]Khorasan Razavi Agricultural and Natural Research Center, Department of Animal Science, Mashhad, Toroq, 91888, Iran; babak.ahmadzadeh@gmail.com

The purpose of the experiment was to determine the effect of supplemental rumen-protected methionine (RPMet) (Methilock™) and rumen-protected choline (RPC) (Vicomb™) on milk yield and milk composition of dairy cows. Thirty multiparous cows were randomly divided into 5 treatments directly after parturition: (1) 0 g of RPMet and RPC per head per day (Control); (2) 15 g of RPMet per head per day; (3) 30 g of RPMet per head per day; (4) 50 g of RPC per head per day; and (5) 100 g of RPC per head per day (as top dressed). All cows were fed 3 times a day with a standard ration containing alfalfa and corn silage as roughage and corn and barley as main ingredients in the concentrates. Cows were milked at 04:00, 13:00 and 21:00 h and milk yields were recorded at each milking until 42 days postpartum. Milk sampling was done weekly. The data were analysed using the PROC MIXED model of SAS for a completely randomized design with repeated measurements. Milk yield was 38.9[a], 44.3[ab], 47.4[b], 42.3[ab], and 45.2[b] kg/d (P<0.05), milk fat yield was 0.91[a], 1.04[a], 1.28[ab], 1.24[ab] and 1.44[b] kg/d (P<0.05), milk protein yield was 1.22[a], 1.39[a], 1.50[b], 1.31[a] and 1.39[a] kg/d and milk lactose yield was 1.83[a], 2.09[a], 2.25[b], 1.96[a] and 2.10[a] (P<0.05) for treatments 1, 2, 3, 4 and 5, respectively. Results of this study showed that RPMet at a level of 30 g/d improved milk lactose yield and milk protein yield, the latter due to a higher milk yield and a higher milk protein content. RPC at a level of 100 g/d improved milk yield and milk fat yield.

Effect of total replacement of soybean meal with a sustained-release non-protein nitrogen source
Nollet, L.[1] and Warren, H.[2], [1]Alltech Belgium, Gentsesteenweg 190/1, Deinze, 9800, Belgium, [2]European Bioscience Centre, Summerhill Road, Sarny, Dunboyne, Co. Meath, Ireland; hwarren@Alltech.com

Protein use efficiency in ruminants is increasingly important as feed costs rise. This, and the increased awareness of raw material availability, has fuelled interest in enhancing microbial protein supply. Hence, this study investigated the complete replacement of soybean meal (SBM) with a sustained-release non-protein nitrogen (NPN) source. Seventy Holstein dairy cows (average 10,300 kg, 4.08% fat, 3.61% protein) were allocated to one of two groups based on DIM, parity and milk production. Two treatments were tested in a cross-over design: Control (grass and maize silage-based, beet pulp, wheat, distillers grains and SBM; CP 167 g/kg DM, 6.95 MJ/kg DM, € 2.20 /h/d); Treatment (Control where 2.5 kg rapeseed meal and 0.13 kg sustained-release NPN replaced 2.5 kg of SBM; CP 167g/kg DM, 6.98 MJ/kg DM; € 2.11 /h/d). The trial consisted of two, 21d experimental periods each preceded by one, 7d adaptation period. Standard commercial individual milk recordings were taken at the end of each experimental period. Group intakes, milk yield, fat and protein levels, milk urea and SCC were measured during the trial. Data were tested for normality and analysed using the Two-sample T Test (Statistix 9.0). Overall milk yields were similar for Treatment animals (30.0) and Control animals (29.4 kg/d). There was no effect on concentrations of milk fat, protein, lactose, urea or SCC. The Treatment group tended to have higher daily intakes (estimated from group intake data) at 24.4 compared with 23.9 kg DM for the Control group. Despite higher daily intakes, the reduced cost of the Treatment ration resulted in additional income of € 0.29 /h/d. N use efficiency was similar for both diets at 0.25. These data support the hypothesis that SBM can be completely replaced in dairy cow rations with a sustained-release NPN source without negatively affecting performance.

Specific components and mechanisms that are unique to the formation of colostrum

Baumrucker, C.R., The Pennsylvania State University, Dairy & Animal Science, 302 Henning Bldg, 16802 University Park, PA, USA; crb@psu.edu

Colostrum of all species is defined as a mammary secretion produced during late pregnancy and typically extracted as the first milk following parturition. The mechanism to produce colostrum is termed colostrogenesis. Bovine colostrum is uniquely different from mature milk with its increased fat, protein, lactoferrin, and specific ion content and less of the typical components found in mature milk. This pattern is generally similar for most species with variations in the immunoglobulin types. Minor components of colostrum are various vitamins, hormones, growth factors, and cytokines. While colostrum components may originate from mammary gland synthesis, others arrive from blood. Interestingly, some specific blood provided hormones appear to be concentrated in colostrum, but not milk, suggesting specific mechanisms that only occur during colostrogenesis. A known process of colostrogenesis is the specific transcytosis transfer of immunoglobulin G1. The capacity to move IgG1 into the colostrum has been reported with an average concentration of ~5-7 mg/ml in blood to ~39 mg/ml in colostrum. Furthermore, the mechanism is restricted to the prepartum phase and has been linked to circulating steroids originating from the placenta. The abrupt termination of the process appears to be either due to the decline in steroids at parturition and/ or the stimulation of mature milk production by lactogeneic hormones (prolactin). Mass transfer analysis of the process shows that the mechanism is extremely variable and not linked to subsequent milk production. The FcRn is a known IgG1 receptor that accounts for IgG1 uptake by the rodent intestine. Using the rodent system as a model, the bovine system will be compared and contrasted utilizing the literature. We hypothesize that the FcRn transcytosis system provides the mechanism for IgG1 transfer, but also includes colostrum hormones.

Sampling procedure during milking and between quarters on the assessment of colostrum IgG content

Le Cozler, Y.[1], Guatteo, R.[2], Le Dréan, E.[3], Turban, H.[3], Leboeuf, F.[4], Pecceu, K.[4] and Guinard-Flament, J.[1], [1]UMR1348 Agrocampus-Ouest INRA, Rennes, 35000, France, [2]UMR1300 ONIRIS-INRA, Nantes, 44000, France, [3]ISAE35, Rennes, 35000, France, [4]3MSD Santé Animale, Beaucouzé, 49000, France; yannick.lecozler@agrocampus-ouest.fr

Adequate passive transfer relies on both quality of colostrums and calf's ability to absorb an adequate volume. A key issue is to assess the quality (ie the amount of IgG) of the colostrums, with a threshold objective of 50g/l to consider a colostrum as good. It is generally recommended to assess the IgG content from a composite sample (of the 4 quarters), taken at the end of the 1st milking. Present study investigated the influence of the sampling procedure on the assessment of colostrums IgG content: quarter (QC) vs composite colostrum (CC) and time of sampling during 1st milking. QC and CC samplings were performed on 79 healthy Holstein cows. Just after calving, colostrum from each quarter and a global sample were collected. To investigate the influence of sampling time, 9 healthy Holstein primiparous cows were milked within 3 hours after calving. A sample of 5 mL was taken every minute during 1st milking leading to 4 to 9 samples per cow. The variation of IgG values was analysed over time and compare to the repeatability of the method (15%). In both studies, the IgG content was measured using radial immunodiffusion. The comparison between QC and CC samples was done using a Student-Newman-Keul test. The concentrations in the left front teats (52.8 g/l) was lower than in the back ones (55.9 to 56.7 g/l), but not different from the right front one (54.6 g/l; $P<0.05$). While the IgG content from individual quarter was slightly higher than the composite one, there was no influence of the type of sample. During 1st milking, slight differences (<10%) were observed between sampling times, leading to consider as stable the IgG content. The present findings provide evidence to facilitate the daily work of the farmer and therefore enhance the transfer of passive immunity.

Secretion of water into milk: specific constituent or unregulated diluent?
Knight, C.H., Nazemi, S. and Klærke, D., University of Copenhagen SUND, Grønnegårdsvej 7, 1870 Frb C, Denmark; chkn@life.ku.dk

The basic mechanism of secretion of lactose and water, together with ions, as the aqueous phase of milk was elucidated by Malcolm Peaker and others some 35 years ago. The synthesis of lactose within an intracellular compartment (the Golgi vesicle) from which it cannot escape provides an osmotic drive for water to enter, and the subsequent exocytosis of the vesicle contents across the apical membrane results in the appearance of that water in milk. At the time, no consideration was given to the means by which the water was drawn into the Golgi vesicle. Subsequent efforts to produce concentrated milks by knock-down of alpha lactalbumin (and hence inhibition of lactose synthesis) imply an assumption of passive diffusion unregulated by any factor other than lactose concentration. Lipid bilayers are not in themselves readily permeable to water, which must inevitably flow through a hydrophilic pore formed by a membrane protein of some sort. In the early 1990s the first identification of an aquaporin was made; a hydrophobic membrane protein consisting of six domains in an hour-glass structure with a central hydrophilic pore. There are at least 13 distinct aquaporins, some of which permeate water alone whilst others also transport small solutes including glycerol, ammonia and urea. Their presence on the plasma membrane has been demonstrated in a variety of cells including mammary epithelial and ductular cells. There has been more debate about the possible role of aquaporins in regulating water movement into intracellular compartments, although there is evidence of aquaporin expression in both mitochondrial and Golgi membranes. Distinct physiological roles linked to cell polarization and secretory processes have been suggested. In this paper we shall review the evidence for the existence of mammary aquaporins and their tissue and subcellular localization, and then consider from a theoretical point of view the physiological and practical implications of variable aquaporin expression on the mammary Golgi membrane and elsewhere.

Serotonin (5-HT) regulates calcium mobilization at the onset of lactation
Laporta, J., Peters, T.L., Merriman, K.E. and Hernandez, L.L., University of Wisconsin, Department of Dairy Science, 1675 Observatory Dr., Madison, WI 53706, USA; llhernandez@wisc.edu

5HT is a known homeostastic regulator of lactation and recently demonstrated to be a regulator of bone turnover. Circulating calcium (Ca^{2+}) decreases at the onset of lactation, and often results in milk fever in dairy cattle. 5HT is synthesized from the amino acid L-tryptophan. The rate-limiting step is catalyzed by tryptophan hydroxylase (TPH1) to form 5-hydroxytryptophan (5HTP). To explore 5HT's role on Ca^{2+} homeostasis in the transition period (10 d pre and post-partum) we fed rats (n=15 per treatment) 2 diets: control (CON) and 5HTP (0.2% total diet). Milk samples were collected on d 1, 5 and 9 of lactation to measure Ca^{2+} concentrations. We collected serum and plasma on d 20 of gestation and d 9 of lactation to measure 5HT, Ca^{2+} and parathyroid hormone-related protein (PTHrP) levels. mRNA was isolated from mammary gland tissue and femurs from d 9 lactating animals and analyzed for PTHrP, TPH1, plasma membrane Ca^{2+} ATPases 1 and 2 (PMCA1, 2), sodium-Ca^{2+} exchanger 1 (NCX1), secretory Ca^{2+} ATPase 1 and 2 (SPCA1, 2), sarcoplasmic reticulum Ca^{2+} ATPase 2 (SERCA2), tartrate-resistant acid phosphatase (TRAP), cathepsin K (CATK), and receptor-activated nuclear factor kappa B ligand (RANKL). Serum 5HT levels increased (P<0.001), as did plasma PTHrP (P<0.05) in the 5HTP group. Additionally, milk Ca^{2+} was increased in 5HTP fed animals (P<0.05), while serum Ca^{2+} remained unchanged. mRNA expression of PTHrP, TPH1, NCX1, PMCA2, SPCA2, and SERCA2 were increased and SPCA1 was decreased in the mammary glands of the 5HTP cohort (P<0.05). mRNA expression of TRAP, CATK, and RANKL were significantly increased in femurs of 5HTP fed animals (P<0.05). In conclusion, feeding 5HTP increased PTHrP production by the mammary gland, resulting in increased Ca^{2+} mobilization from bone to support milk synthesis. Our data suggests feeding 5HTP to transition animals can increase the amount of Ca^{2+} mobilization from bone necessary for milk, preventing severe decreases in circulating Ca^{2+} concentrations.

Milk fat globule membrane proteomics: a 'snapshot' of mammary epithelial cell biology

Cebo, C.[1], Henry, C.[2] and Martin, P.[1], [1]INRA, UMR 1313 GABI, Equipe LGS, bâtiment 221, Domaine de Vilvert, 78350 Jouy en Josas, France, [2]INRA, UMR 1319 MICALIS, Plateforme PAPSSO, Domaine de Vilvert, 78350 Jouy en Josas, France; christelle.cebo@jouy.inra.fr

Lipids are released in milk as fat globules, which are droplets of apolar lipids surrounded by a complex membrane deriving from the mammary epithelial cell (MEC) and called the Milk Fat Globule Membrane (MFGM). Milk lipid synthesis initiates into the endoplasmic reticulum (ER) of the MEC by the budding of cytoplasmic lipid droplets (CLD). CLD then migrate to the apical pole of the MEC where they are progressively wrapped up by the plasma membrane to be released as fat globules into the lumen of mammary acini. The structure of the MFGM is thus highly complex and closely related to the mechanisms of milk fat globule biogenesis and secretion by the MEC. We have recently characterized MFGM proteins in several species, including the goat, the horse, and the camel species. We have highlighted prominent differences with the bovine species, especially regarding lactadherin, a major MFGM protein. Recent technological breakthroughs in proteomics (primarily, the development of one dimensional-gel electrophoresis approach coupled to high resolution tandem mass spectrometry (1D-LC-MS/MS)) led to the identification of hundreds of proteins associated to the MFGM. Newly identified MFGM proteins are not only involved in lipid metabolic or exocytosis-related biological processes, but also in translation, or cytoskeleton organization. Identification of proteins in the MFGM will probably contribute to identify genuine partners of lipid droplets formation, growth and transit in the MEC and ultimately their release as fat globules in milk. In addition, the fact that the MFGM most likely reflects the mammary epithelial cell content and the emergence of quantitative proteomics will help to improve our understanding of biological mechanisms occurring in the mammary gland under different physiological or pathological conditions.

Transfer of blood constituents into milk during mastitis

Wellnitz, O., Lehmann, M. and Bruckmaier, R.M., Vetsuisse Faculty, University of Bern, Veterinary Physiology, Bremgartenstrasse 109a, 3001 Bern, Switzerland; olga.wellnitz@vetsuisse.unibe.ch

Mastitis is accompanied by changes in milk composition. Besides secretory changes of the mammary gland the blood-milk barrier integrity is impaired and molecules can cross the border from blood into milk or vice versa. This study aimed to investigate the transfer of several blood constituents into milk during a mammary immune response due to intramammary endotoxin challenge. Five cows who received a beta hydroxy butyrate (BHB) clamp infusion to increase BHB in blood to a steady concentration (1.5-2.0 mmol/l) were challenged intramammarily into one quarter with 200 µg *E. coli* lipopolysaccharide (LPS). They were immunized against bluetongue virus (BTV) 2 years before. Blood and milk were analyzed hourly for 8h. Changes were considered significant if $P<0.05$. Blood concentrations of LDH, IgG, and specific antibodies against BTV did not change. In milk the somatic cell count increased in LPS challenged quarters within 4h but not in controls. Lactate dehydrogenase (LDH), BHB, lactate, and immunoglobulin (Ig) G was 33±8 U/l, 0.08±0.01 mmol/l, 8±2 mg/l, and 0.1±0.0 mg/l, respectively, and increased within 3h to maximum levels after 8h of 326±146 U/l, 0.56±0.01 mmol/l, 182±26 mg/l, 0.72±0.24 mg/l, respectively, in milk of LPS challenged quarters but not of control quarters. IgG increase in milk was paralleled by an increase of antibodies against BTV. Lactate concentrations in blood and milk of control quarters increased 2 h after challenge and were correlated (r=0.74). In conclusion, the blood constituents LDH, BHB, IgG, and lactate increase in milk after LPS challenge due to impaired blood-milk barrier integrity. This increase depends on the respective blood concentrations. Increase of antibodies against BTV in milk is obviously of no use for mammary immune defense. It shows that transfer of blood components into milk during mammary inflammation is not only specifically targeted to mastitis pathogens and can also be a side effect of the blood-milk barrier opening.

Diurnal and seasonal hormone patterns in blood and milk of dairy cows

Castro, N.[1,2], Wellnitz, O.[2], Lollivier, V.[3] and Bruckmaier, R.M.[2], [1]Universidad de Las Palmas de Gran Canaria, Animal Science, Arucas, 35413, Spain, [2]University of Bern, Veterinary Physiology, Vetsuisse Faculty, Posieux, 1725, Switzerland, [3]Agrocampus Ouest, Production du lait, Rennes, UMR1080, France; ncastro@dpat.ulpgc.es

Hormones such as melatonin (MEL), prolactin (PRL) or cortisol (COR) show diurnal patterns of their blood concentration. Hormones can pass from blood into milk. It is assumed that the hormone patterns are parallel in blood and milk and the hormone concentration in the milk obtained at milking can serve as a mean representing the period between milkings. The aim of this study was to investigate if MEL, PRL and COR plasma concentrations are reflected by their concommitant concentration or mass transfer into milk. Blood and milk samples of 12 dairy cows housed without artificial light in Switzerland (46° 46' N, 7° 6' E) were collected every hour during 24 h in June and December at 16 and 9 h of daylight, respectively. Blood and milk MEL concentration were analyzed by ELISA and PRL and COR by RIA. For statistical analyses ANOVA for repeated measures and Tukey´s test were performed. Pearson´s correlations were calculated between blood concentration and milk mass of the studied hormones. Blood and milk MEL reached higher levels (P<0.05) in both seasons, showing diurnal changes with high values during the night and low values during the photoperiod (18.3 and 7.1 pg/ml vs. 4.0 and 0.8 pg/ml in blood and milk, respectively). However, blood and milk PRL levels were lower (P<0.05) in short photoperiod (11.9 vs 20.9 and 9.8 vs 13.3 ng/ml December and June in blood and milk, resp.). Similarly COR was lower in December (7.6 and 1.7 nmol/l in blood and milk, resp.) than in June (9.8 and 2.2 nmol/l in blood and milk, resp.). Milk MEL, PRL and COR showed positive correlation (P<0.001) with blood MEL concentration (r=0.29), PRL (r=0.32) and COR (r=0.37). In conclusion, all studied hormones showed a similar pattern in blood and milk, however seasonal differences were observed depending on the endocrine function of the hormone.

Changes in intensity of biosynthesis of milk fat fatty acids during lactation in grazing dairy cows

Kirchnerová, K.[1], Foltys, V.[1] and Špička, J.[2], [1]Animal production research centre Nitra, Nutrition, Hlohovecká 2, 951 41 Lužianky, Slovakia, [2]South Bohemian University, Branišovská 31a, 370 05 České Budějovice, Czech Republic; kirchnerova@cvzv.sk

The aim of the paper was to extend the knowledge about correlation of current fatty acids (FAs) profile of cow milk fat at herds of cows held at pasture in mountain dairy farms in Slovakia to milk production and quality parameters. The milk samples were taken in total from 134 cows at summer pasture period. The FAs composition of individual milk was determined by GC-MS, where 54 FAs were identified. Saturated fatty acids (70.48±4.04% in the milk fat), which are considered undesirable constituents of milk fat, show in the first third of lactation highly significant positive correlation coefficients (r>0.45, P<0.01) with all indicators of milk production (days, the total amount of milk fat and protein in kg). Monounsaturated fatty acids MUFA (26.26±3.59% in the milk fat) have to the total amount of milk production significant indirect relationship. Their content in milk fat decreases with the rise of the total amount (kg) of produced fat (r=-0.426), protein (r=-0.494), milk (r=-0.514), and with the increasing number of days of lactation (r=-0.583, P<0.001). Polyunsaturated fatty acids PUFA (3.26±0.069% in the milk fat) identified in this work are in the carbon chain from 18 to 22 carbons. The whole group PUFA has correlation coefficients to total amount of produced milk, fat, protein (kg) as well as to the number of days in lactation in an indirectly oriented relationship at the same level of significance from r=-0.468 to r=-0.485 (P<0.01). Grazing of dairy cows has a better value of the composition of milk fat from a health perspective, but at the account of lower production as seen by correlations at mountain farms with dairy grazing systems. This article was written during realization of the project CEGEZ no. 26220120042, supported by the Operational Programme Research and Development funded from the European Regional Development Fund.

Timing of milk fatty acid profile responses to dietary oil additions: 21 days vs. shorter periods
Martínez Marín, A.L.[1], Carrión, D.[1], Gómez-Cortés, P.[2], Gómez Castro, G.[1], Juárez, M.[2], Pérez Alba, L.M.[1], Pérez Hernández, M.[1] and De La Fuente, M.A.[2], [1]Universidad de Córdoba, Producción Animal, Ctra. Madrid-Cádiz, km 496, 14014, Córdoba, Spain, [2]Instituto de Investigación en Ciencias de la Alimentación (CSIC-UAM), Nicolás Cabrera, 9, 28049, Madrid, Spain; domingo.carrion@genusplc.com

There is a need for a minimum adaptation period to dietary oil addition before determining milk fatty acid (FA) changes. Twelve midlactation multiparous goats were used to compare milk FA composition at 21 d (504 h) with that of shorter adaptation periods (1, 12, 24, 72, 120, 192 and 312 h) after introducing no oil (CON), or 48 g/d of high oleic sunflower (OSO) or regular sunflower (RSO) or linseed (LO) oils into the same basal diet. Basal diet was made of alfalfa hay (0.33) and pelleted concentrate (0.67). Milkings at 0 (covariate), 1 and 12 h were stripped out by hand after an intravenous dose of oxytocin. FA composition was analyzed by gas chromatography. PROC MIXED of SAS was used to analyze data. Seventy two FA were identified and quantified in milk fat. Only 22 FA from CON had differences ($P<0.05$) with their respective 504 h value; differences disappeared by 72 h or shorter times for 15 of those FA. Twenty two FA in milk fat from OSO had differences ($P<0.05$) with their respective 504 h value, 16 disappearing by 72 h. Fifty six FA in milk fat from RSO had differences ($P<0.05$) with their respective 504 h value, 37 of them disappearing by 72 h or before. Fifty three FA in milk fat from LO had differences ($P<0.05$) with their respective 504 h value; the last time at which differences were observed was 72 h (26 FA), 120 h (5 other FA), 192 h (11 other FA), and 312 h (11 other FA). Vaccenic and rumenic acids had their last difference with their 504 h value at 72 h or before in all treatments, while 18:3n-3 in LO kept increasing in milk fat up to 504 h. These results suggest that changes in milk FA profile due to dietary oil addition were less and occurred faster with OSO, were more in number with RSO and LO, and occurred slower with LO.

Primary bovine mammary epithelial cells demonstrate transcytosis of IgG1 *in vitro*
Baumrucker, C.R.[1], Stark, A.M.[2] and Green, M.H.[3], [1]The Pennsylvania State University, Dairy & Animal Science, 302 Henning Bldg, University Park, PA 16802, USA, [2]University of Bern, Veterinary Physiology, Vetsuisse Faculty, Bremgartenstrasse 109a, CH-3012 Bern, Switzerland, [3]The Pennsylvania State University, Nutritional Sciences, 202A Chandlee Laboratory, University Park, PA 16802, USA; crb@psu.edu

IgG1 colostrum transcytosis mechanisms across bovine mammary epithelial cells has not been established, but the Fc Receptor of the neonate (FcRn) is suspected. Primary bovine mammary epithelial cells (BMEus) were plated (1.5×10^5) onto Corning collagen coated transwells in DMEM/F12 with 10% FBS and antibiotics overnight for attachment and media was changed to IMDM plus insulin / transferin / selenium (ITS) daily until a confluent layer of cells was observed by microscopy. To test epithelial barrier (tight Junction [TJ] establishment), Lucifer Yellow (LY) was utilized to determine if LY flux occurred. After TJ establishment, IgG1 (plus labeled material) was applied to test movement of the IgG1 from each compartment established by the transwell. In addition, the cells were treated with 2% FBS in IMDM media (control) or 1 µg/ml each of 17-beta estradiol (E2) and progesterone (P4) in control media. Labeled IgG1 movement was sampled over a 34 hour period that was concluded by LY test to determine the continuation of the TJ barrier in the transwell. Analysis of the flux of labeled IgG1 with Simulation, Analysis and Modeling (SAAM) showed that control flux from the top (transwell) to the bottom (culture plate) was 1.7%/h with a delay compartment of 1.6 h. Flux from the bottom to top was <1%/h without any delay. Treatment with E2/P4 showed a top-to-bottom rate of 1.5%/h with a delay compartment of 1.2 h while bottom to top flux was again <1%/h without delay. These results show IgG1 directional movement across a BMEus epithelial barrier that has directional preference, but also indicates there is a bi-directional flux capacity. This system will provide a means to establish the mechanism of IgG1 transfer in the formation of colostrum.

The effect of different sources of buffers on the performance of lactating dairy cows

Moodi, D.[1,2], [1]Azad University, Varamin, 9197854854, Iran, [2]Azad University, Varamin, 12345, Iran; danielmoodi@gmail.com

There is little agreement among research workers about the influence of buffering materials on the performance of ruminants, but While some studies have shown beneficial responses to buffer additions, others have found either no response or even a reduction in animal performance so this experiment was carried out to determine the effectiveness of different source of buffers on production parameters of lactating Holsteins. Our objectives were to test the efficacy of a commercial buffer and other buffers on feed intake, milk yield and milk composition of cows fed a low fiber diet. In this trial, eight early-lactating Holsteins fed total mixed ration *ad libitum*, Cows were arranged in a 4×4 Latin square. Basal diet was 65% concentrate: 35% forage (which contains 20% corn silage and 15% alfalfa hay). Dietary treatments were control, $NaHCO_3$ at 1% of diet DM, sodium bentonite at 2% of diet DM and commercial buffer at 2% of diet DM. Daily milk production was recorded and daily milk samples were collected for analysis of fat, protein and lactose. Feed intake and total tract apparent digestibility were similar for all treatments ($P \leq 0.05$). Milk yield for the four treatments were 38.9, 39.37, 41.39, 42.5% respectively ($P \leq 0.05$). Milk fat percentage was increased by commercial buffer ($P \leq 0.05$). Milk protein percentage was not affected by treatment ($P \leq 0.05$) and milk Lactose percentage was improved by sodium bentonite and commercial buffer ($P \leq 0.05$). The total tract apparent digestibility of acid detergent fiber (ADF) and neutral detergent fiber (NDF) also were similar for all treatments ($P \leq 0.05$). Ammonia concentration in fluid rumen was decrease by sodium bentonite at two hour after feeding ($P \leq 0.05$). Blood pH was unaffected during the trial by treatments ($P \leq 0.05$).

Across-breed genomic evaluations in cattle

Berry, D.P., Animal and Grassland Research and Innovation Centre, Teagasc, Moorepark, Fermoy, Co. Cork, Ireland; donagh.berry@teagasc.ie

Simulations, undertaken within-breed, suggest that genomic selection can increase genetic gain by >50%. However, interest in alternative dairy cattle breeds and crossbreeding is increasing necessitating an across-breed genomic evaluation for dairying in some countries. Beef cattle breeding and sheep breeding is based on many different breeds and crossbreds, which when coupled with their smaller population size, signify also a requirement for across-breed genomic evaluations. Also genomic information can be exploited to maximise the benefits of crossbreeding (at the genomic level). Here we investigate approaches to genomic selection across breeds using real-life data from Ireland cattle. The accuracy of prediction is discussed.

Genomic prediction within and between dairy cattle breeds with an imputed high density marker panel
Erbe, M.[1], Hayes, B.J.[2,3,4], Bowman, P.J.[3,4], Simianer, H.[1] and Goddard, M.E.[3,4,5], [1]Georg-August-University, Department of Animal Sciences, 37075 Göttingen, Germany, [2]La Trobe University, Bundoora, 3086, Australia, [3]Dairy Futures CRC, Bundoora, 3083, Australia, [4]Department of Primary Industries, Biosciences Research Division, Bundoora, VIC, 3083, Australia, [5]University of Melbourne, LFR, Parkville, 3010, Australia; merbe@gwdg.de

In the last years, genomic breeding value prediction in dairy cattle has become widely used especially for predicting breeding values of young bulls without progeny. Accuracy of prediction is clearly determined by the size of the reference set. Since building a purebred reference set of a useful size is challenging for small breeds, one strategy would be to build a multi-breed reference set by combining the reference animals of different breeds. The benefit of using a multi-breed reference set is expected to be even higher when denser marker sets are used since then phases between markers and QTL should be more stable over breeds. We used a data set of 2,257 Australian Holstein and 540 Australian Jersey bulls to study the accuracy of genomic prediction within and between breeds using 50K and imputed 777K (Illumina HighDensity (HD) SNP chip) SNP data. Phenotypes used were daughter yield deviations for three production traits. For predicting genomic breeding values of the youngest bulls, we used a new Bayesian method (BayesR). BayesR models the variances of the SNP effects as a series of four normal distributions. The proportion of SNPs in the distributions is not fixed, but modeled with a Dirichlet distribution. BayesR worked well in our data set and was comparable or in many scenarios even better than a genomic BLUP model. Using imputed HD data rather than 50K did not lead to a significant increase of accuracy for within breed prediction neither for Holstein nor for Jersey. Comparing pure-bred to multi-breed reference set, the minor breed, Jersey, could benefit from the augmented size of the reference only when using the imputed HD panel.

Genomic evaluation combining different French dairy cattle breeds
Karoui, S.[1], Carabaño, M.J.[1], Diaz, C.[1] and Legarra, A.[2], [1]INIA, Mejora Génetica Animal, Ctra La Coruña km 7.5., 28040, Madrid, Spain, [2]INRA, SAGA, UR 631, F-31326, Castanet-Tolosan, France; sofiene.karoui@inia.es

We investigated the impact of increasing the number of genotyped candidates in the training set on the accuracies of genomic estimated breeding values (GEBV's) using a multi-breed French dairy cattle reference population, in contrast to single breed. Three traits (milk, fat content and female fertility) were analysed using a multi-breed and single breed reference populations. Three breeds of French dairy cattle were used in this study: Holstein (H), Montbéliarde (M) and Normande (N). Training populations included 2,976, 950 and 970 bulls; validation ones, 964, 222 and 248. All animals involved were genotyped with the Illumina Bovine SNP50 array. Two models were applied. The first, a random regression model, was used for estimation of genetic variances and genetic correlations between breeds for each trait and for estimation of GEBV's. The second, a multiple trait model between breeds for one trait was used only for estimation of GEBV's. Accuracy of GEBV's was evaluated under three scenarios for the genetic correlation between breeds (r_g): a) unknown r_g ; b) breeds highly related, r_g=0.95; c) uncorrelated, r_g=0. Accuracies of the GEBV's were assessed by the coefficient of determination (R^2) and the quality of prediction by the regression coefficient (δ) of daughter yield deviations on GEBVs of sires in the testing sets. Posterior means for rg ranged from -0.01 for fertility between M and N to 0.79 for milk yield between M and H. Differences in R^2 for the three scenarios were manifest only in the case of fat content for Montbéliarde (from 0.27 in scenario c) to 0.33 in a) and b)). Accuracies for fertility were lower than for other traits and values of δ showed severe overestimation of GEBVs for this trait in M. The use of the multi-breed reference population only helped to increase accuracy of GEBVs for traits and populations that showed the largest correlation and in the breed with the smallest data set.

Genomic selection in small breeds using multi-breed reference populations

Mészáros, G.[1], Gredler, B.[2], Schwarzenbacher, II.[3], Meuwissen, T.[4] and Sölkner, J.[1], [1]University of Natural Resources and Life Sciences, Vienna, Division of Livestock Sciences, Gregor-Mendel Str. 33, 1180, Vienna, Austria, [2]Qualitas AG, Chamerstrasse 56, 6300 Zug, Switzerland, [3]ZuchtData EDV-Dienstleistungen GmbH, Dresdner Str. 89/19, 1200, Vienna, Austria, [4]Norwegian University of Life Sciences, Institute of Animal Science and Health, Sorasveien 5B, 1432 Aas, Norway; gabor.meszaros@boku.ac.at

Genomic selection became a routine procedure when evaluating animals of major cattle breeds. In these populations a set of reference bulls is selected, providing the basis for prediction of accurate genomic enhanced breeding values (GEBV). In some other breeds however it is not possible to build up a large reference group simply because of the small population size. The objective of our study is to explore possibilities of genomic selection in small populations using Tyrolean Grey cattle and Pinzgau as example cases. We genotyped 100 bulls with Illumina BovineSNP50 BeadChip (50K) and 120 bulls with the Illumina BovineHD BeadChip (700K) in each breed representing all important and accessible bulls in these endangered populations. The quality check for the genotype data was done jointly for all breeds in order to maintain SNPs segregating only in some of the breeds. SNPs located on the X chromosomes were excluded. Estimated breeding values (EBV) and deregressed proofs from 20 traits were used as dependent variables; SNP effects were estimated using BayesB and GBLUP methodologies. The 60 youngest Tyrolean Grey and Pinzgau bulls were selected as the validation populations, the rest of the group as the reference. In a stepwise procedure we also added different genotypes to the reference sets in order to increase the reliability of the estimation. Accuracies of GEBVs were surprisingly high (0.32-0.81 in most cases) using the small single breed reference sets. They did not increase substantially when multi breed reference populations were used. The only exceptions were GEBVs for Pinzgau when Tyrolean Grey was added to the reference.

Genomic selection for cow traits in beef cattle

Todd, D.L.[1,2], Woolliams, J.A.[1] and Roughsedge, T.[2], [1]Roslin Institute, Genetics and Genomics, Rib, Easter Bush, Midlothian, EH25 9RG, United Kingdom, [2]SAC, Animal Breeding, Rib, Easter Bush, Midlothian, EH25 9RG, United Kingdom; Darren.Todd@sac.ac.uk

Beef cattle selection in the UK is dominated by terminal traits with only 15% of beef sired matings in commercial herds resulting in a female retained for breeding. Although cow trait EBV are available in most beef breeds, genetic improvement is difficult, with selection in seedstock herds inevitably concentrated on terminal traits. Nevertheless a commercial demand exists for bulls with improved cow performance traits. This study investigates whether genomic selection can provide a solution to this issue and allow breeders to select effectively and practically for cow traits. A scenario was considered whereby a nucleus of herds breeding for cow traits was established within the main seedstock population. A cow index, consisting of five key cow traits, was used as the breeding goal in a deterministic selection index programme to predict response in nucleus herds. Two selection programs were considered; one adopting a conventional progeny test where selection decisions were made on sires at 6.5 years of age, and the other using young sires selected at 1.5 years of age which is reflective of current breeding practice. It was assumed that a genomic SNP key developed from a training set in the main population, with either 2,000 or 5,000 genotypes and phenotypes, would be available and was incorporated via GEBV in the index simulation. With EBV only, the progeny test was predicted to achieve a response of £1.94/animal/year which was 30% better than the response predicted when using young sires (£1.49). However when evaluating GEBV, young sire response was improved by 32% to £1.96 (2,000 Training set) and 66% to £2.47 (5,000 training set). Genomically selected young sires could therefore produce a similar response in nucleus herds to that when using a progeny test and therefore offer a potential route of improving cow traits in UK beef cattle.

The genomic selection system in Italian Holstein

Finocchiaro, R., Van Kaam, J.B.C.H.M. and Biffani, S., ANAFI-Italian Holstein Association, Via Bergamo 292, 26100 Cremona, Italy; raffaellafinocchiaro@anafi.it

Since December 2011, Holstein genomic evaluation in Italy has become official for bulls. The initial genotyped population consisted of more than 3,000 bulls but after entering a collaboration with USA, Canada and UK the overall population size has increased to over 15,400 bulls. After SNP data editing and imputation, a full pedigree deregression is used to derive EDPs from national or MACE EBVs. A linear SNPblup model with 41,472 autosomal SNPs and a residual polygenic component is used to estimate SNP effects. DGVs are computed for 31 single traits (5 production, 19 conformation and 7 functional traits) and composite traits are derived from single trait proofs. Successively GEBVs of proven bulls are calculated by weighting EBVs and DGVs by their respective EDCs, while GEBVs of young bulls are based on DGVs and a residual polygenic component. Genomic proofs are not yet propagated to non-genotyped relatives. Reliability is based on scaling computed reliabilities to the theoretical level derived with the approach from Goddard. This approach should strike the right balance between the over-estimated reliability from inversion and the under-estimated reliability from validation. For validation purpose a reduced dataset of only Italian proven bulls, with a cutoff birth date on bulls born before February 2005, was created and then deregressed. Exchange of genotypes with the north-american consortium resulted in about 19% and 6% higher reliability for most traits of young animals as compared with pedigree index and previous genomic predictions from only Italian data, respectively. Increase in reliabilities is evident in production and conformation traits, however functional traits need still some further investigation. Final ranking list showed the dominance of certain sire lines.

Application of genomic-assisted selection in swine breeding

Forni, S., Cleveland, M.A. and Deeb, N., Genus Plc, Hendersonville, TN, 37075, USA; selma.forni@genusplc.com

Marker-assisted selection has been used in swine breeding and genetic gain has increased for traits associated with QTL of large-effects. Studies have shown that a small number of markers could improve the predictive ability of breeding values by 40-60% for traits such as scrotal hernia, mortality and litter size. Disease resistance and physiological disorders are important traits that may benefit from genomic selection. Wide-genome scans have reported small genomic regions that may be responsible for up to 15% of the additive variation in phenotypes related to boar taint and respiratory infections. The use of high-density genomic information imposes computational challenges in pig breeding because information is accumulated and selection decisions are made as often as weekly. For this reason, the single-step genomic evaluation has appealing features for the swine industry. Increases in EBV accuracy of 30% for selection candidates have been observed in lowly heritable traits with the single-step evaluation. Accuracy improvement was also obtained when only the parents of selection candidates were genotyped. Nonparametric methods present similar advantages regarding computational efficiency and can be used to account for non-linear additive and non-additive effects. Predictive ability of nonparametric methods was similar or better than GBLUP in crossbred populations. Simulations have indicated that genomic information on parental lines could significantly impact the evaluation of performance at the commercial level, if non-additive effects were contemplated. Genotype imputation has become imperative for the full implementation of genomic selection in pig breeding because the costs of high-density genotyping selection candidates are still prohibitive. High-density genotypes could be imputed from low-density panels with accuracy of 0.90 to 0.99, depending on availability of parental genotypes. The essential tools for genomic evaluation are available and the implementation of genomic-assisted selection in swine breeding is a reality.

Alternative single-step type genomic prediction equations

Gengler, N.[1], Nieuwhof, G.[2], Konstantinov, K.[2] and Goddard, M.[3], [1]ULg – GxABT, Passage des Déportés 2, B-5030 Gembloux, Belgium, [2]ADHIS, 1 Park Drive, Bundoora, Vic 3083, Australia, [3]University of Melbourne, Parkville Campus, Melbourne, Vic 3010, Australia; nicolas.gengler@ulg.ac.be

Current derivations of single-step equations are based on modifed relationships among animals replacing for genotyped animals and on an inverted scale, pedigree based relationshps, by modified ones. These relationships are obtained as linear combination of strictly genomic and pedigree based relationships, therefore implicitly 'weighting' SNP and polygenic effects. Alternative equations were recently proposed de-absorbing the genomic relationships out of the equations. This derivation did not change basic assumptions, but was derived using a matrix of relationship differences. This presentation will show a new and alternative derivation of single-step type genomic prediction equations allowing joint estimation of GEBV and SNP effects based on the partitioning of genetic (co)variances. The method was derived from a random mixed inheritance model where SNP and residual polygenic effects are jointly modelled. The derived equations were modified to allow non-genotyped animals and to estimate directly and jointly GEBV and SNP effects. Equations resemble superficially recently proposed alternative single-step equations but were derived differently and are based on completely different assumptions. They also avoid certain issues in de-absorbing derivation linked to the matrix of relationship differences by using (co)variances. Several other advantages of the new equations are that weighting of SNP and polygenic effects becomes explicitly and that SNP effects are also estimated. This method makes better use of High-Density SNP panels and can be easily modified to accommodate other genetic effects as major gene effects or copy-number variant based effects. Finally these alternative equations combine advantages of single-step and of explicit SNP effect estimation based methods. Additional research is required to test and validate further the proposed method.

Compatibility of pedigree-based and marker-based relationships for single-step genomic prediction

Christensen, O.F., Aarhus University, Blichers Alle 20, 8830 Tjele, Denmark; OleF.Christensen@agrsci.dk

Single-step methods for genomic prediction have recently become popular because they are conceptually simple and in practice such a method can completely replace a pedigree-based method for routine genetic evaluation. An issue with single-step methods is compatibility between the marker-based relationship matrix and the pedigree-based relationship matrix. The compatibility issue involves which allele frequencies to use in the marker-based relationship matrix, and also that adjustments of this matrix to the pedigree-based relationship matrix are needed. In addition, it has been overlooked that it may be important that a single-step method is based on a model conditional on the observed markers. When data are from routine evaluation systems, selection affects the allele frequencies, and therefore both observed markers and observed phenotypes contain information about allele frequencies in the base population. Here, two ideas are explored. The first idea is to instead adjust the pedigree-based relationship matrix to be compatible to the marker-based relationship matrix, whereas the second idea is to include the likelihood for the observed markers. A single-step method is used where the marker-based relationship matrix is constructed assuming all allele frequencies equal to 0.5 and the pedigree-based relationship matrix is constructed using the unusual assumption that animals in the base population are related and inbreed with relationship coefficient alpha and inbreeding coefficient alpha/2. The parameter alpha should be determined from the markers, but since there is selection in routine evaluation systems the phenotypes in principle also provide information about this parameter. The likelihood function used for inference contains two terms. The first term is the REML-likelihood for the phenotypes conditional on the observed markers, whereas the second term is the likelihood for the observed markers. The performance of the proposed method is studied on simulated data examples.

Comparison of genomic evaluation in Lacaune dairy sheep using single or multiple step GBLUP

Baloche, G.[1], Legarra, A.[1], Lagriffoul, G.[2], Larroque, H.[1], Moreno, C.[1], Robert-Granié, C.[1], Giral, B.[3], Panis, P.[4], Astruc, J.M.[2] and Barillet, F.[1], [1]INRA, UR631 SAGA, 31326 Castanet-Tolosan, France, [2]Institut de l'Elevage, 31320 Castanet-Tolosan, France, [3]Ovitest, La Glène, 12780 Saint-Léons, France, [4]Confédération Générale de Roquefort, 13103 Millau, France; guillaume.baloche@toulouse.inra.fr

In Lacaune Dairy sheep, if a genomic breeding scheme is implemented, strong emphasis will be put on selection intensity of the young genotyped rams. Then, in order to keep unbiased GEBV, the pre-selection of candidates based on genomic information would have to be taken into account in genomic evaluation. With recent methodological and software development, the so-called Single-Step (SS) GBLUP gives Genomic Enhanced Breeding Value (GEBV) with such properties. Consequently, an attempt of such an evaluation was carried out in order to test its ability to predict rams genomic breeding value compared to the current multiple-steps (MS) procedure. Available reference population is a subset of 2,614 AI progeny tested rams (belonging to the Lacaune Breeding organisations Confédération Générale de Roquefort and Ovitest), born between 1998 and 2008 which were genotyped with the Illumina Ovine SNP50 BeadChip. After editing 42,039 SNPs were available for genomic evaluation. Depending on the method, phenotypes used were either dairy ewes own performances (SS) or daughter yield deviation weighted by equivalent daughter contribution of rams (MS). Traits studied were milk traits, somatic cell score and udder traits. GEBV computations followed Interbull's recommandations. Finally, predictions were compared to the most recent available EBV. As expected, SS slightly outperforms MS approach across all traits. Moreover its simple implementation and quick computations, except breeding value accuracy, give SS GBLUP better values for routine evaluation. Acknowledgements for French ANR & ApisGene (SheepSNPQTL project), and for FUI, Midi-Pyrénées region, Aveyron & Tarn departements, & Rodez town (Roquefort'in Project).

Genomic predictive ability for growth, carcass and temperament traits in Nelore cattle

Carvalheiro, R.[1], Perez O'brien, A.M.[2], Nevez, H.H.R.[1], Sölkner, J.[2], Utsunomiya, Y.T.[3] and Garcia, J.F.[3], [1]GenSys Associated Consultants, Porto Alegre, 90000 Porto Alegre, Brazil, [2]University of Natural Resources and Life Sciences (BOKU), Gregor Mendel Str. 33, 1180 Vienna, Austria, [3]Universidade Estadual Paulista (UNESP), Aracatuba, 16012 Aracatuba, Brazil; soelkner@boku.ac.at

Nelore bulls were genotyped with the Illumina Bovine HD Beadchip to assess genomic predictive ability on weighing traits, scrotal circumference, temperament and carcass traits evaluated through visual scores. After quality control, 685 samples and 320,238 SNPs remained in the analyses. Most SNPs were discarded due to MAF (<0.02), call rate (<0.98) and high correlation (>0.995) with other SNPs. Sires with proofs in 2007 were considered as training group and sires without proofs in 2007 but with proofs in 2011 were considered as testing group. The proportions of sires were 75% (training) and 25% (testing), approximately. Average accuracy of expected breeding values (EBV) was 0.81-0.88 in the training group and 0.73-0.88 in the testing group. Predictive ability was assessed through the correlation between direct genomic breeding values (DGV, using GBLUP) and 2011 EBV for the testing group. Predictive ability based on empirical formula was also calculated and used for comparison. Observed and expected correlations for pre (wG) and post (yG) weaning weight gain, temperament (yT), scrotal circumference (SC), weaning muscling (wM) and precocity (wP), yearling muscling (yM) and finishing precocity (yP) were: 0.38 and 0.51 (wG), 0.53 and 0.50 (yG), 0.16 and 0.41 (yT), 0.60 and 0.47 (SC), 0.55 and 0.50 (wM), 0.52 and 0.50 (wP), 0.69 and 0.49 (yM), 0.70 and 0.49 (yP), respectively. Further analyses suggested that greater than expected predictive abilities were observed for traits affected by genotype stratification. Two subgroups were observed in principal component analyses based on genomic kinship coefficients. The strategy of assessing predictive ability within subgroups will be tested after genotyping more animals.

Improvement of genomic models for calving ease in the UK

Eaglen, S.A.E.[1,2], Coffey, M.P.[2], Wall, E.[2], Woolliams, J.A.[1] and Mrode, R.[2], [1]The Roslin institute and R(D) SVS, University of Edinburgh, Easter Bush, EH25 9RG, MidLothian, Edinburgh, United Kingdom, [2]SAC, Animal & Veterinary Sciences Group, Easter Bush, EH25 9RG, Midlothian, Edinburgh, United Kingdom; sophie.eaglen@sac.ac.uk

Genomic selection can potentially result in greater accuracy for maternal traits due to an increase in accuracy of selection without increasing the generation interval. Currently, a univariate ridge regression genomic model is used to estimate direct genomic values (DGV) for direct (CEd) and maternal calving ease (CEm) in the UK. This study was conducted to evaluate the accuracy of DGVs from a univariate GBLUP model which can be extended into a bivariate GBLUP model to account for the genetic covariance between direct and maternal effects. Data from bulls with 50K genotypes were available for analyses as a result of cooperation between the Cooperative Dairy DNA Repository (CDDR), ANAFI (Italy), the UK AI industry and SAC for the purpose of genetic evaluations. After validation, 41703 SNPs were selected for genetic evaluations. Reference and validation populations consisted of 4,556 and 1,333 bulls for CEd and 4,553 and 532 bulls for CEm respectively ($\geq$10 effective number of daughters, $\geq$ reliability of 59%). DGVs were estimated following a univariate GBLUP model where a genomic relationship matrix (G) is replacing the conventional pedigree relationship matrix (A). The accuracy of the obtained DGVs is then evaluated in the validation population by correlating the DGVs to the official deregressed proofs (DRP). Results show accuracies of genomic evaluation with the univariate GBLUP model of 0.55 and 0.52 for CEd and CEm respectively. Currently, bivariate GBLUP models are being investigated which include DRPs of CEd and CEm as the separate dependent variables and allows for a genetic covariance between the traits through a fitted covariance matrix, as in the traditional multivariate BLUP model. Variances and covariances, both additive genetic and environmental, are taken from previous conventional variance component estimation (non-genomic).

Preliminary single-step multitrait genomic evaluation of Holstein type traits

Zavadilová, L.[1], Bauer, J.[1], Haman, J.[1], Přibyl, J.[1], Přibylová, J.[1], Šimečková, M.[1], Vostrý, L.[1], Čermák, V.[2], Růžička, Z.[2], Šplíchal, J.[2], Verner, M.[2], Motyčka, J.[3] and Vondrášek, L.[3], [1]Institute of Animal Science, Přátelství 815, 104 00 Prague 10 Uhříněves, Czech Republic, [2]Czech Moravian Breeding Corporation, U Topíren 860, 170 41 Prague, Czech Republic, [3]Holstein Cattle Breeders Association of the Czech Republic, Těšnov 17, 117 05 Prague, Czech Republic; lida.zavadilova@seznam.cz

Single-step procedure was used for genomic breeding value (GEBV) prediction with a multiple-trait animal model for 20 linear type traits. The phenotypic data on linear scoring were available for 143,208 Holstein cows first calved between years 2005 to 2010 and SNP markers from Illumina BovineSNP50 were available for 631 sires. Data subset included data on 66,285 cows calved between years 2005 to 2007. There were 225 sires with SNP available those daughters first calved after year 2007 and 101 sires with SNP available with minimum 50 daughters first calved after year 2007. Breeding values were estimated by multiple-trait animal model either only with the pedigree-based relationship matrix (breeding value, BV) or with the pedigree-based relationship matrix augmented by the genomic relationship matrix (GEBV). Estimations were carried out for the whole dataset or for the data subset. The model equation included fixed effects of herd-date of classification, classifier, and season of calving, quadratic regressions on age at calving and on days in milk and the random effect of animal. DMU5 program was employed for computations. For mentioned 101 sires with daughters with phenotypic data after year 2007, the correlations between GEBVs estimated in data subset and GEBVs estimated in whole dataset were stronger by 0.10 in average than correlations between corresponding BVs. The highest increase occurred for rear udder height (0.20). For udder traits, the increase was 0.13 in average. The lowest increase (0.03) was found for angularity. Supported by the Ministry of Agriculture of the Czech Republic, Project No. QI111A167.

Artificial neural networks: influence of the network topology on genetic breeding value estimation
Ehret, A.[1], Hochstuhl, D.[2] and Thaller, G.[1], [1]Institute of Animal Breeding and Husbandry, Christian-Albrechts-University, Olshausenstr. 40, 24098 Kiel, Germany, [2]Institute for Theoretical Physics and Astrophysics, Christian-Albrechts-University, Leibnizstrasse 15, 24098 Kiel, Germany; gthaller@tierzucht.uni-kiel.de

Artificial neural networks (ANNs) provide a powerful technique for learning about complex traits by predicting future outcomes based on learning data. The efficiency of ANNs on estimating the underlying function in a given dataset strongly depends on the network topology, e.g. number of neurons in hidden layers. For the analysis data from 2,322 German Holstein bulls with both phenotypic and genotypic information were used. Deregressed proofs of the traits milk yield, fat yield and protein yield were analysed as response variables for SNP data of two chromosomes (BTA14, BTA19) as input variables. To train the networks using back propagation and a non-linear transformation function a subset of 2,000 animals was randomly selected. The remaining 322 animals were used for testing the predictive ability of the different network topologies on unknown phenotypes. In general, if the number of neurons in the hidden layer was strongly increased the predictive ability of future outcomes was reduced. In addition, the computation time increased drastically. For milk yield and genotypic data of BTA14 the correlations between predicted and true phenotype ranged between 0.25 (one neuron in hidden layer) and 0.18 (30 neurons in hidden layer). The prediction optimum was given with a network topology of 5 neurons in the hidden layer (0.28). Similar results were achieved if SNP data from BTA19 was used. In the present study simple topologies of ANNs showed the potential to provide a reliable prediction of future phenotypes and might be an appropriate method for further genetic analyses of quantitative traits under complex interactions. However, possible improvements for the predictive ability of future outcomes under different learning and network parameters have to be examined.

Accuracy of genomic prediction for protein yield using different models in Slovenian Brown bulls
Špehar, M.[1], Potočnik, K.[2] and Gorjanc, G.[2], [1]Croatian Agricultural Agency, Ilica 101, 10000 Zagreb, Croatia, [2]University of Ljubljana, Biotechnical Faculty, Department of Animal Science, Groblje 3, 1230 Domžale, Slovenia; mspehar@hpa.hr

Use of genomic (SNP) information enables more accurate inference of breeding values (BV), especially for young animals. The objective of this study was to compare the correlation between progeny based and genomically based evaluation for protein yield of Slovenian Brown bulls based on different sources of information and scenarios. Emphasis was to integrate genomic information into national evaluation. Four data sets were constructed as follows: (1) DS1 – phenotypic data (PD) from national genetic evaluation (1.342.134 test-day records of 57.670 cows recorded between years 1997 and 2011); (2) DS2 – DS1 + 50K Illumina SNP genotypes for 183 bulls; (3) DS3 – DS1 + direct genomic value (DGV) for 183 Slovenian bulls evaluated at InterBull (InterGenomics) treated as correlated trait; (4) DS4 – DS1 + DGV for 399 bulls in the national pedigree. Two scenarios were evaluated. In the first scenario, all PD was used in analysis, while in the second PD was removed for years 2008 to 2011 to exclude daughter information of 35 genotyped bulls. Animal model was used for the analysis of DS1, while the joint pedigree and genomic relationship model was used for the analysis of DS2. For the analysis of DS3 and DS4, bivariate animal model was used. Both theoretical and validation based accuracies were evaluated. Application of joint pedigree and genomic relationship model did not increase validation accuracies due to the limited number of genotyped animals. Theoretical accuracies were 0.58 (DS1), 0.61 (DS2), 0.84 (DS3), and 0.74 (DS4), while validation accuracies were 0.56 (DS1), 0.54 (DS2) 0.86 (DS3), and 0.72 (DS4). For comparison, theoretical accuracy of genomically enhanced breeding value at InterGenomics for validation bulls is 0.90. Results show that integration of genomic information into national evaluation was successful.

Genomic best linear unbiased prediction by simulated annealing

Esquivelzeta, C. and Casellas, J., Universitat Autònoma de Barcelona, Facultat de Veterinària, Departament de Ciència Animal i dels Aliments, Campus Universitat Autònoma S/N, 08029 Cerdanyola del Vallès, Spain; ceciliaer@gmail.com

Genomic best linear unbiased prediction (gBLUP) models predict genomic breeding values by accounting for the information of thousands of single nucleotide polymorphisms (SNP) summarised into the genomic relationship matrix (G). Although current parametrizations use all SNP to compute G, it is highly questionable to believe that all SNP contribute reliable and relevant information when computing genomic relationship coefficients conditional to a given phenotype. Within this context, we simulated genomic data and performed serial gBLUP analyses by simulated annealing in order to identify which SNP contributed relevant information and which SNP must be removed from the calculation of G. We used as reference the mean square error (MSE) between simulated and predicted data, and the analytical process started with all SNP included in the calculation of G. For each new iteration, a SNP was selected at random and the MSE was calculated after changing its state (i.e., removed or included to the list of SNP for the calculation of G). This change was accepted if MSE reduced from previous iteration. The simulated annealing process stopped after 1,000 iterations without changes in the list of used/discarded SNP. The simulation process involved 1,000 preliminary generations ($N_e=100$) and five more generations ($N_e=200$) contributing phenotypic data ($h^2=0.5$). Genomes had a unique 100 cM chromosome with 5,000 SNP and 500 quantitative trait loci with mutation probabilities of 10-3 and 10-5, respectively. Ten independent data sets were generated and analysed. On average, the full model with all SNP reached a MSE of 1.370 ± 0.004, whereas this parameter reduced until 1.293 ± 0.005 when dropping off between 35% and 47% of the SNP. These results involved a ~6% reduction of MSE when using appropriate SNP for G and suggested a very appealing way to improve the statistical performance of gBLUP models when analysing massive genomic data.

Adaptation of BLUPF90 package for genomic computations

Misztal, I.[1], Aguilar, I.[2], Tsuruta, S.[1] and Legarra, A.[3], [1]University of Georgia, 425 River Rd, Athens GA 30602, Georgia, [2]INIA, Las Brujas, 90200 Canelones, Uruguay, [3]INRA, BP 52627, 32326 Castanet-Tolosan, France; ignacy@uga.edu

The original BLUPF90 package contains programs for renumbering, BLUP, parameter estimation, accuracy approximation and for sample visualization. A renumbering program (RENUMF90) supports national data sets. BLUP programs are for equations in memory (BLUPF90) and iteration on data (BLUP90IOD). Parameter estimation is via REML (REMLF90 and AIREMLF90) and Bayesian programs (GIBBS*F90), which uses optimized algorithms able to support large number of traits (20+). Samples from GIBBS* programs can be analyzed by POSTGIBBSF90, and accuracies of predictions can be approximated by ACCF90. Specific programs are available for threshold-linear models. Nearly all programs have been updated to support the genomic information and several new programs were added. Program PreGSF90 analyzes the SNP information, provides basic quality control, creates a genomic relationship matrix using a large variety of options, and combines pedigree and genomic relationship matrices for a single-step methodology. Preparing matrices for 30k animals with 50k SNP takes about 1 h. PostGSF90 converts GEBV to SNP effects, displays Manhattan plots possibly using moving averages, and estimates weights of SNP effects. Program PredF90 predicts GEBV based only on estimates of SNP effects obtained from PostGSF90. Most of the programs are available online at nce.ads.uga.edu. The package can be used for genomic predictions (including national data sets), parameter estimation (including GBLUP and G-REML), and GWAS. Unequal variances for SNP effects similar to those in BayesX and subsequently 'Manhattan' plots can be computed by iterating on postGSF90, preGSF90, and one of BLUP programs. These operations do not require deregression and are fast. Classical GWAS can be carried out with BLUPF90 fitting one SNP at a time as fixed regression and an animal effect with a genomic relationship matrix. The updated package simplifies genomic analyses in breeding applications.

Reliability of genomic prediction using low density chips
Segelke, D.[1,2], Chen, J.[2], Liu, Z.[2], Reinhardt, F.[2], Thaller, G.[1] and Reents, R.[2], [1]Institute of Animal Breeding and Husbandry Christian-Albrechts-University, Olshausenstraße 40, 24098 Kiel, Germany, [2]vit, Heideweg 1, 27283 Verden, Germany; dierck.segelke@vit.de

With the availability of dense SNP marker chips, genomic evaluation was implemented in many countries. But for an average dairy breeder genotyping his whole herd is still too expensive. Therefore there was a demand to develop low density chips to reduce genotyping costs. To study the accuracy of low density chips, Illumina Bovine3K BeadChip (3K) and BovineLD BeadChip (6K), were simulated using Illumina marker maps. Accuracy of imputing was investigated by using the software packages Beagle and Findhap. Three different genotype data sets were used: EuroGenomics data set with 14405 reference bulls, smaller EuroGenomics data set with 11,670 older reference bulls and a data set containing 31,597 genotyped German Holstein animals. For validation, 1374 bulls for the EuroGenomics data set, 534 bulls for the smaller EuroGenomics data set and 2,205 animals of the genotyped German Holstein data set, their imputed genotypes were compared to their original genotypes to calculate allele error rate. Additionally reliability loss by using imputed instead of original genotypes was computed. Furthermore the difference between the original combined genome-enhanced breeding values (GEBV) and imputed GEBV were calculated. Allele error rate for the EuroGenomics data sets was 3.3% for Findhap 3K, 1.6% for Beagle 3K, 1.7% for Findhap 6K and 0.6% for Beagle 6K, respectively. Reliability of the genomic prediction decreased on average about 5.3% for Findhap 3K, 2.6% for Beagle 3K, 1.9% for Findhap 6K and 1.0% for Beagle 6K averaged over 12 selected traits, in comparison to those of using original genotypes. Differences in GEBV of original and imputed genotypes were largest for Findhap 3K, whereas Beagle 6K had the smallest difference. The results demonstrate that Beagle is more accurate than Findhap. The low density chip 6K gave markedly higher imputing accuracy and more accurate genomic prediction than the 3K chip.

Genomic evaluation with SNP chip switched
Alkholder, H.[1], Liu, Z.[2], Reinhardt, F.[2], Swalve, H.H.[1] and Reents, R.[2], [1]Martin-Luther-University Halle-Wittenberg, Theodor-Lieser-Str. 11, 06120 Halle (Saale), Germany, [2]VIT w. V. Verden Germany, Heideweg 1, 27283 Verden/Aller, Germany; alkhoder.hatem@landw.uni-halle.de

For routine genomic evaluation of German Holsteins, genotypes of Illumina BovineSNP50 chip (version 1) have currently been used to estimate effects of the SNP markers on this chip. Since the introduction of a new version of the 50K SNP chip (version 2), animals, in particular young candidates, have genotypes of SNP markers which have no effect estimates available. As the older chip version (v1) has been phased out, we have decided to switch the SNP chip for the routine genomic evaluation of German Holsteins. The objectives of this study were first to impute genotypes of all animals with the old version to the new chip version, and then to estimate SNP effects using genotypes of the new version chip (v2). Of all 50,324 genotyped Holstein animals, 33,315 animals were genotyped with the older chip v1 whereas 17,009 animals, mostly younger candidates, with the newer chip v2. Because the SNP markers on chip v1 but not on v2 (n=1,661) are all covered by the markers of the new chip v2, these missing SNP markers were ignored in our imputation study. As a consequence, a total of 52,340 SNP markers on chip v1 were left, which are all mapped on the chip v2 with 54,609 SNP markers available. Imputation software packages Findhap and Beagle were applied to fill in missing genotypes of v1 animals. Imputed genotypes from both software packages were compared to investigate their accuracy. Effects of SNP markers on chip v2 were estimated a BLUP SNP model using c.a. 23,000 genotyped Holstein reference bulls. As a result of the large number of common SNPs between both chips and the high imputation accuracy, genomic EBV of animals based on the new chip are highly correlated with those of the older chip v1. Therefore, we recommend that the new chip be introduced in the routine genomic evaluation.

Error rate for imputation from BovineSNP50 to BovineHD

Schrooten, C.[1], Dassonneville, R.[2], Brøndum, R.[3], Chen, J.[4], Liu, Z.[4] and Druet, T.[5], [1]CRV BV, P.O. Box 454, 6800 AL Arnhem, Netherlands, [2]INRA, UMR GABI, Domaine de Vilvert, 78352 Jouy-en-Josas, France, [3]Aarhus University, Department of Molecular Biology and Genetics, Blichers Allé 20, 8830 Tjele, Denmark, [4]VIT, Heideweg 1, 27283 Verden, Germany, [5]Unit of Animal Genomics, Faculty of Veterinary Medicine / CBIG, University of Liège, 4000 Liège, Belgium; chris.schrooten@crv4all.com

Starting in 2007, large numbers of cattle have been genotyped with 50k platforms, mainly the BovineSNP50 BeadChip, to obtain more reliable breeding values for selection of animals. Since then, higher density genotyping platforms have been developed, among others the Illumina BovineHD BeadChip with 777k SNP. All animals could be regenotyped with the BovineHD chip, but the expected gain in reliability is expected not to outweigh the additional cost. Therefore, genotyping part of the population with the BovineHD and subsequently imputing the animals genotyped with 50k chips to HD is considered an attractive alternative, provided that imputation can be done at low error rate. The objective of this study was to investigate the error rate for imputation from BovineSNP50 genotypes to BovineHD. BovineHD genotypes were obtained for 548 high impact bulls from the Eurogenomics reference population. Genotypes for all but the BovineSNP50 markers were masked in 60 validation animals, different for each of 4 subsets. These 60 animals did not have descendants with BovineHD genotypes. The BovineHD genotypes of the remaining 488 animals in each subset were used as reference to impute the masked genotypes of the 60 animals. Imputation errors were computed as the number of differences between imputed alleles and observed alleles, divided by the number of compared alleles. When Beagle 3.3 was used, imputation error ranged from 0.55% to 0.76%. Combining Beagle with DAGPHASE did not improve results, indicating no benefit from adding linkage information with DAGPHASE. It was concluded that imputation from 50k platforms to HD can be done at low error rate when using Beagle 3.3.

SNPpit: efficient data management for high density genotyping

Groeneveld, E. and Truong, C.V.C., Institute of Farm Animal Genetics, Mariensee (FLI), Höltystr 10, 31535 Neustadt, Germany; eildert.groeneveld@fli.bund.de

Expanding use of high density genotyping creates substantial data management problems, both in terms of storage requirements as well as computational demands arising from data manipulation. Given the background of ever increasing panel densities across all species and reducing genotyping costs, SNPpit has been developed with the objective to store, manage, and retrieve genotype data of different SNP panels in a uniform way in one database, while leaving the data analysis to other packages like PLINK and GenABEL. Thus, SNPpit can be used as a central repository of all genotyping data accumulating within a lab. Highly compressed internal storage of genotype data allows to efficiently store large volumes of data. As a result, a 500 GB hard disk will be able to hold 100 billion genotypes which is equivalent to 200 panels of 500K genotyped for a total of 200,000 individuals. SNPpit provides for the creation of named subsets of genotypes by viewing them as matrices, with one of their dimensions being defined through a selection list of SNPs and the other a list of individuals. Through their names, SNPpit can export each subset for further analysis by software packages like PLINK. Because of their unique way of definition, derived datasets – as they may arise during editing or by focussing on certain chromosomes – can be stored practically without space requirements as only the SNP and individual selection vectors need to be saved in the database. Any combination of an existing SNP and individual selection vector will then result in a new genotype set, which can be exported for further use. In this way, disk space issues of derived datasets are eliminated. Because of the efficient compressed storage scheme data import and export are fast: importing 20 mio genotypes on 80 individuals took 39 sec while the export finished in 14 sec on an Intel Core i5 laptop. SNPpit is written in Perl using the PostgreSQL database for mass data storage. As all software components are released under the GPL, installations can be made freely.

Estimation of breeding value using one-step approach with random regression test day model

Suchocki, T.[1], Liu, Z.[2], Żarnecki, A.[3] and Szyda, J.[1], [1]University of Environmental and Life Sciences, Department of Genetics, Kożuchowska 7, 51-631 Wrocław, Poland, [2]Vereinigte Informationssysteme Tierhaltung, Heideweg 1, 27283 Verden, Germany, [3]National Research Institute of Animal Production, Krakowska 1, 32-083 Balice, Poland; tomasz.suchocki@up.wroc.pl

Selection based on dense marker panels has become a very important part of dairy cattle breeding programs. In the future, instead of a traditional genetic evaluation model based entirely on cow records, an approach incorporating information on records and marker genotypes in a single step can be used. The most important advantage of such a model is the utilization of polygenic and marker-based genomic relationship data. Such solution allows for using both, genotyped and non-genotyped animals in the prediction procedure. However, it is still a matter of debate how to adjust the two sources of information in a genomic relationship matrix. This study is a pilot towards implementing a one step approach in a random regression test day model context for Holstein Friesian population, considering various ways of a relationship matrix adjustment. Data set and methods Data consists of 890 animals (10 genotyped bulls, 100 cows with phenotypic data and 780 ancestors without neither genotypes nor phenotypes). Random regression test day models with a polygenic effect on milk yield modeled by Legendre polynomials of order 2 were used for the estimation of variance-covariance parameters and for the prediction of genomically enhanced breeding (GEBV). Results and conclusions The highest accuracy of GEBV, both for genotyped bulls and for cows with phenotypes were obtained for weighting parameter w=0 and the lowest for w=1. An implementation of the single-step approach in a random regression test day model framework is very attractive for genomic prediction in dairy cattle since it allows for an incorporation of the new, genomic source of information directly into a conventional genomic evaluation.

Genomic selection for increased farm profit and reduced methane energy production in dairy cattle

Lopez-Villalobos, N.[1] and Davis, S.R.[2], [1]Massey University, Private Bag 11222, 4442 Palmerston North, New Zealand, [2]ViaLactia Biosciences, P.O. Box 109185, 1149 Auckland, New Zealand; N.Lopez-Villalobos@massey.ac.nz

Selection index theory was used to examine the genetic responses for farm profit and methane energy production (CH_4MJ) per cow under a progeny test (PT) selection scheme compared to an alternative genomic selection (GS) scheme. CH_4MJ per cow per year was predicted from total metabolizable energy requirements. The aggregate genotype comprised the following traits and corresponding economic values: lactation yields of milk (-$0.083/l), fat ($1.646/kg) and protein ($7.846/kg), live weight (-$1.171/kg), fertility ($2.790/1%), somatic cell score (SCS; -$28.924/unit) and longevity ($0.042/day). The genomic selection scheme was based on the selection of the 150 best bull calves from 50,000 DNA genotyped bull calves with accuracy of 70% for the aggregate genotype traits and 50% for CH_4MJ. Young bulls were widespreadly used when they become 1 year old. Annual genetic response based on a PT scheme was $15.3 comprised of 46.9 L milk, 2.2 kg fat, 1.9 kg protein, 0.5 kg live weight, 0.2% fertility, -0.01 SCS and 4.6 days of longevity. The correlated response for CH_4MJ was 21.2 MJ. The annual genetic response based on a GS scheme was $21.2 comprised of 105.3 L milk, 3.8 kg fat, 2.04 kg protein, 1.3 kg live weight, 0.6% fertility, -0.02 SCS and 9.9 days of longevity. The correlated response for CH_4MJ was 37.6 MJ. Including CH_4MJ in the aggregate genotype with an economic weight of -$0.20/MJ and including genomic breeding for CH_4MJ in the GS scheme resulted in an economic gain of $15.6 comprised of 83.5 L milk, 2.8 kg fat, 1.05 kg protein, -2.6 kg live weight, 0.8% fertility, -0.03 SCS, 13.5 days of longevity and 31.2 MJ of methane energy. A GS scheme can result in higher genetic gains than those obtained with a PT scheme; changing the breeding goal to reduce CH_4MJ may have unfavourable longer term impact on production and live weight.

SNP identification in Italian pig breeds using next generation sequencing technology

Guiatti, D.[1], Sgorlon, S.[1], Licastro, D.[2] and Stefanon, B.[1], [1]Università degli Studi di Udine, Dipartimento di Scienze Agrarie ed Ambientali, via delle Scienze 206, 33100 Udine, Italy, [2]CBM scrl, Genomics, AREA Science Park, Basovizza, 34149 Trieste, Italy; denis.guiatti@uniud.it

The dissection of complex traits of economic importance to the pig industry requires the availability of a significant number of genetic markers, such as single nucleotide polymorphisms (SNPs). Most of the Italian breeds were selected for heavy pig production to obtain animals with high aptitude for the PDO dry-cured ham production, such as Parma or S. Daniele ham. Identifying genetic variants should become increasingly feasible with improved sequencing methods on a genome-wide scale. Next-generation sequencing (NGS) tools are valuable for the discovery of genetic markers in animal genomes. Pig genomic DNA from a F1 (Italian Duroc × Italian Large White) crossbreed was extracted and normalized at 20 ng/µl and, using a Covaris station, we sheered the DNA to an average of 300 base pair. The purified sample was ends repaired, 3' adenilated and adaptor ligated. The obtained library was amplified to selectively enrich those DNA fragments that have adapter molecules on both ends. The final purified product was quantitated using both qPCR and Agilent 2100 Bioanalyzer (Agilent) and normalized to 10 nM and seeded on two lanes of a Illumina Flow cell to a final concentration of 12 pM using a cBOT system with a TRuSeq PE Kit cluster Kit V3. After cluster generation, sequencing was performed using a 200-Cycle Paired-End Run TruSeq SBS. The output from a sequencing run was mapped against the reference genome Sscrofa 9.2 using the open software BWA. The resulted mapped sequences was quality filtered, duplicate removed and variants called using the software Samtools. Alignments covered 94% of the genome. An high number of SNPs and small indels (insertions/deletions) were found. After a false positives clearing, these efforts can also be directed to the identification of genetic variants in candidate genes for selected traits for future large scale association studies.

Why livestock farming systems are complex objects

Martin, O., INRA, UMR791 MoSAR (Modélisation Systémique Appliquée aux Ruminants), AgroParisTech, 16 rue Claude Bernard, 75005 Paris, France; olivier.martin@agroparistech.fr

Livestock production is faced with many challenges such as climate change, food security, globalization, sustainability, competition for natural resources, animal health and welfare. To address this intricate set of issues, the multiple dimensions of livestock farming systems have to be put in perspective. As a changing paradigm, the scale of approach is enlarged from a local focus to a holistic view on agro-ecosystem functioning and criticality at the territory level, which is intuitively posited as a complex problem. The term complex pertains to the difficulty in predicting properties from the segregation in sub-properties. Complexity thus emerges from the dependencies between the sub-parts of the whole. At first sight, the complexity of livestock farming systems emerges from the multiple relations in time and space between its main components (farmers, farmlands, animals, products, crops, resources, facilities, wastes, etc) and with components of other systems of its environment (landscapes, consumers, stakeholders, institutions, fauna/ flora, pathogens, markets, water, soil, air, etc). Classically, the functioning of livestock farming systems is depicted by way of the cross-linkage of an anthropical (husbandry activities) and a biological (physiological processes) subsystem. An essential feature of the latter one is its organization through hierarchical levels (individuals, batches, herds, breeds, species), which constitutes a structural step in the organization of the whole order of the system and a functional step in the emergence of a global property of the system. Any attempt to disentangle the complexity of livestock farming systems requires considering interactions between components of the system at a given level (horizontal complexity) and integration of the system through higher levels (vertical complexity). The present communication proposes two alternative and complementary views to highlight the horizontal and vertical complexities of livestock farming systems.

Integrated simulation/optimization models to deal with multiple farming objectives

Villalba, D.[1], Díez-Unquera, B.[2], Carrascal, A.[3], Bernués, A.[4] and Ruiz, R.[2], [1]University of Lleida, Animal Production, Avda. Rovira Roure 178, 25198 Lleida, Spain, [2]NEIKER-Tecnalia, Granja Modelo de Arkaute, Carretera N-1, Km 355, 01192 Arkaute- ALAVA, Spain, [3]Tecnalia, Industrial Systems Unit, Paseo Mikeletegi, 2, 20009 Donostia- San Sebastián, Spain, [4]CITA-Aragón, Avda. Montañana 930, 50059 Zaragoza, Spain; dvillalba@prodan.udl.es

The analysis of livestock farming systems must integrate the three pillars of sustainability, i.e. economic, social and environmental. The design of new management strategies requires the valuation of trade-offs within and between these three pillars. Models with different levels of detail have been used as tools to cope with complexity in LFS as they are able to integrate multiple variables and interactions. We present an integrated decision support tool for sheep farming systems that combines simulation and optimization procedures. A stochastic and dynamic Animal sub-model (voluntary intake module and management-nutrition-reproduction interactions module) and a Farm sub-model (feeding resources module and economics module) are able to represent multiple interactions between variables. The optimization model is based on a Genetic Algorithm (GA) (a search of optimal solutions based on natural selection and evolution procedures). The GA searches for optimal solutions within delimited values of simulation parameters (e.g. feeding, reproductive, management decisions) that maximize a 'fitness' function that includes technical, economic and environmental objectives (e.g. income, costs, N and E surplus, GHGs). The GA allows for the evaluation of complex and non-linear problems, specifically in terms of trade-offs between management strategies and objectives. However, this approach is time demanding and it yields no unique solutions.

Management flexibility as a new dimension of the trade-offs occurring in grassland agroecosystems

Sabatier, R.[1,2], Doyen, L.[3] and Tichit, M.[1,2], [1]AgroParisTech, UMR SADAPT, 16 rue Claude Bernard, 75005 Paris, France, [2]INRA, UMR SADAPT, 16 rue Claude Bernard, 75005 Paris, France, [3]MNHN, CERSP, 55 rue Buffon, 75005 Paris, France; rodolphe.sabatier@forst.uni-goettingen.de

Sustainable management of grazed grassland implies a trade-off between ecological and productive performances. Accurate tuning of the grazing strategies is needed to reconcile, to a certain extent, ecological and productive objectives. However, multiplying the objectives of a production system adds constraints that are very likely to limit the farmer's degrees of freedom. The objective of this study was to draw the trade-offs between the management flexibility of a grazed grassland agroecosystem and its ecological performance. We developed a dynamic model to assess the ecological and productive performances as well as the flexibility of a grassland agroecosystem. The model explicitly links grazed grassland dynamics to bird population dynamics. It is applied to Lapwing (Vanellus vanellus) conservation in wet grasslands in France. The model is based on the mathematical framework of the Viability Theory that reveals the set of management strategies (grazing sequences) respecting a given set of constraints (thresholds on cattle densities and grass heights ensuring suitable habitats for birds). Counting these so called viable strategies provides a good index of the flexibility of the system. Beyond the flexibility we also computed an ecological indicator (bird population size) and a productive indicator (overall grazing intensity). We calculated these three indexes for different ecological constraints and showed that a new trade-off occurred between flexibility and ecological performances. Ecological constraints aimed at maximizing ecological performances reduce the flexibility of the system by more than 85%. On the applied point of view, this means that agri-environment schemes based on management constraints may have a strong impact on the ability of a farmer to adapt his management to environmental variations.

Modelling the effect of turnout date to pasture in spring of yearling dairy bred cattle

Ashfield, A.[1,2], Crosson, P.[2] and Wallace, M.[1], [1]University College Dublin, School of Agriculture and Food Science, Belfield, Dublin 4, Ireland, [2]Teagasc, Animal and Grassland Research and Innovation Centre, Grange, Dunsany, Co. Meath, Ireland; austen.ashfield@teagasc.ie

The objective of this study was to determine the effect on profitability of turnout date to pasture in spring of yearling steers which were subsequently slaughtered at 24 months of age. The steers were crossbreds from dairy Holstein-Friesian dams and Belgian Blue sires. The three turnout dates investigated were; 1st March (E), 15th March (M) and 1st April (L). All scenarios were modelled at moderate (170 kg organic nitrogen (ON) /ha) and high (225 kg ON /ha) stocking rates (SR). The analysis was conducted using a bio-economic simulation model of Irish dairy calf-to-beef systems, the Grange Dairy Beef Systems Model (GDBSM). The GDBSM is a whole-farm, static, single-year deterministic model. Dietary components consist of grazed grass, grass silage and concentrates. The model operates with a monthly time-step and consists of four sub models (farm systems, animal nutrition, feed supply and economics). To model the effects of compensatory growth, it was specified that the animals which were turned out on the later dates had greater daily liveweight gain until the end of June than animals turned out on the earlier dates and thus, liveweight per animal at the start of July was the same for all three turnout dates (428 kg). Turnout date had no effect on gross output (GO). Variable costs (VC) were 2 and 3% higher for M and L compared to E due to an increase in grass silage and concentrate costs. This resulted in gross margin (GM) being 3 and 5% lower for M and L compared to E. Fixed costs (FC) were similar for all turnout dates resulting in net margin (NM) being 13 and 27% lower for M and L compared to E. Increasing SR increased GO by 33%, VC and FC increased by 37 and 24%, respectively, which resulted in GM and NM increasing by 24 and 23%, respectively. The most profitable system was E coupled with the higher SR.

Some considerations about multicriteria evaluation of farm sustainabilty

Botreau, R., Inra, UMR1213 Herbivores, 63122 Saint-Genès-Champanelle, France; raphaelle.botreau@clermont.inra.fr

More and more demands arise about the sustainability evaluation of farming systems, and this for very different purposes, from certification scheme to diagnostic tool. Sustainability is a multidimensional concept covering several aspects such as environment protection (global warming, eutrophication, biodiversity...), economic viability and resilience facing hazards, farmers' quality of life, and social aspects like participation to territory development and respect of societal expectations (high quality products, animal welfare...). Even if links may exist between these aspects (eg encouraging grazing of dairy cows on permanent pastures may favour both biodiversity and milk quality), they have different impact on sustainability and must be considered and interpreted independently. Thus multicriteria evaluations of sustainability are to be designed, implying to keep in mind several theoretical and practical considerations. The objective of the evaluation is a key element, influencing the type of evaluation (from a report gathering all sustainability indicators for an advisory purpose, to a standardized normative evaluation to certify a certain level of sustainability to the consumer), and thus the mathematical construction, but also the indicators to be used. They will be different if the whole evaluation has to be done within a 2 hours or a 2 days inspection. Another element to be defined is the evaluation scale: at unit, farm or territory level? Even if the evaluation is made at farm level, some indicators may be required at other scales (eg watershed characteristics must be considered to correctly interpret the amount of PO_4^{3-} emitted by the farm in terms of eutrophication potential). In addition to the necessary interpretation of the indicators, other considerations exist around the aggregation problem (necessary? which weights? compensations?...). This list of considerations is not exhaustive but points out major difficulties to be faced and choices to be made when designing a sustainability evaluation model.

Modeling complexity in livestock farming systems to address trade-offs and synergies

Rossing, W.A.H., Groot, J.C.J., Oomen, G.J.M. and Tittonell, P., Wageningen University, Organic Farming Systems, P.O. Box 563, 6700 AN Wageningen, Netherlands; Walter.Rossing@wur.nl

LFS are complex for various reasons. Firstly, complexity arises from variation among farms even within the same agro-ecological zone. This constitutes a major challenge for policy development, extension and research. Secondly, complexity is due to the many and dynamic biophysical elements that interact on each farm, affected by an ever changing environment. This level of complexity is relevant when aiming at improving farm level performance. Note, however, that the farm may not be the proper unit of analysis! The third element of complexity is that farms are part of social-ecological systems. These systems demonstrate emergent behaviour due to non-linear relations and feedbacks. In concrete cases, development of LFS involves multiple actors who often disagree on norms and values as well as desirable solutions. Within this multi-actor, multi-objective and multi-scale setting, models help by providing insight into key causes of systems behaviour. We present two complementary approaches: functional farm typologies and design-oriented quantitative systems modelling. Farm typologies are typically based on farm structural characteristics. This ignores functional aspects such as choice of markets, role of agriculture in total income, and non-market contributions. We present an example of a functional typology showing three livelihood strategies of African farmers; hanging-in, stepping-up or stepping-out, as well as the hysteresis in development trajectories. Design-oriented quantitative systems models demonstrate the window of opportunity for development of farms or landscapes. Based on quantitative description of the basic components soil, crops, animals, manure, landscape composition and structure, and their interactions, trade-offs between land use objectives are revealed. These provide input for discussion on competing claims on land use, and help to reveal lack of technology. Examples are discussed from an adaptive management perspective.

Planning intensity and its spatial allocation for reconciling livestock production and biodiversity

Teillard, F. and Tichit, M., INRA, UMR 1048 SAD APT, 16 rue Claude Bernard, 75231 Paris Cedex 05, France; teillard@agroparistech.fr

Livestock production is facing a trade-off between food demand and environmental sustainability. Production intensity often drives this trade-off between both performances. Therefore, planning the intensity levels and their spatial allocation is key for the reconciliation. The objective of this study was to develop a model-based scenario approach to evaluate several options for the evolution and reallocation of livestock production intensity. We addressed the scale of the whole France country and assessed scenarios on three criteria: (1) milk and beef production; (2) total income; and (3) the composition of a farmland bird community. The relationships between livestock production intensity (described by input costs per ha) and these criteria were calibrated relying on two data sources: the Farm Accountancy Data Network and the French Breeding Bird Survey. The three relationships were then integrated in a model to evaluate the impact of intensity evolution and reallocation scenarios on the three criteria. The intensity indicator was strongly linked to total income ($r^2=0.63$), production ($r^2=0.55$), and to the farmland bird community ($r^2=0.18$). Moreover, the spatial aggregation of production intensity significantly reinforced the intensity effect on birds ($P<0.001$). Among scenarios, there was a trade-off between economic/production performances, and biodiversity performances. We showed that scenarios spatially targeting intensity changes to specific area (with low and aggregated intensity) were more cost effective: for the same economic and production cost, they yielded higher biodiversity benefits. Reallocating production intensity even enhanced biodiversity at almost no production cost. Our result could provide an opportunity to improve the effectiveness of policies promoting the sustainability of livestock production. We discuss the policy mechanisms that could be used as well as the technical options that would be needed to change intensity at farm scale.

Intensification as a way to reduce cattle greenhouse gas emissions: a question of scale

Puillet, L.[1], Agabriel, J.[2], Peyraud, J.L.[3] and Faverdin, P.[3], [1]INRA UMR791 MoSAR, AgroParisTech 16 rue C. Bernard, 75005 Paris, France, [2]INRA UMRH 1213, Site de Theix, 63122 Saint Genès Champanelle, France, [3]INRA UMR1348 PEGASE, Domaine de la Prise, 35590 Saint Gilles, France; laurence.puillet@agroparistech.fr

Increasing animal production (milk or meat per head) is highlighted as a way to decrease cattle environmental impact. We examined if this option is efficient to decrease greenhouse gas (GHG) emissions, testing whether the interaction between milk and meat production at national level (under a scenario of constant milk and meat outputs) modified the response of cattle GHG emissions to intensification, observed at animal level. We developed a national herd model (dairy and beef cattle), simulating the production of milk and meat (carcass and exports). The cow population (3 dairy and 5 beef breeds) generated calves, which are allocated to replacement or finishing types (slaughter or export). Herd dynamics, calves distribution matrix among finishing types and breed performance are input parameters. The number of cows in each breed is adjusted by optimization to minimize the difference between the calculated meat and milk production and the targeted production (model input). Assuming a steady state herd, demographic categories are calculated and combined with emission factors to calculate GHG emission inventory. The model was used to evaluate contrasted scenarios of dairy intensification (specialized and dual-purpose breed) and beef intensification (carcass weight and finishing cycle). Results (compared to 2010 French situation) showed that both dairy and beef intensification led to small GHG decrease (between -2.0 and -1.4%). If adverse biological effects of increasing animal production (fertility decrease, calf mortality increase) are considered in the simulations, intensification had almost no or negative effects (between -0.5% and +1.8%). In conclusion, multiscale approaches are needed to fully evaluate the response of livestock systems to technical options and the conserved properties of the response at different levels.

Animal integrity: an absolute concept?

Gremmen, B., Wageningen University, Adaptation Physiology, De Elst 1, 6708 WD Wageningen, Netherlands; bart.gremmen@wur.nl

In recent years some violations of the integrity of animals have raised public debates in the Netherlands. For example, the docking of pig tails and the trimming of chicken beaks, have led to certain changes in regulations. Although the concept of animal integrity is part of the Dutch regulatory framework, in the literature it is considered to be a problematic notion. A commonly used definition is: the wholeness and intactness of the animal and its species specific balance, as well as the capacity to sustain itself in an environment suitable to the species. The concept of animal integrity has been developed because we cannot use the notions of interests and rights to accommodate the intuition that we should adjust the farm to the animal and not vice versa. The concept of animal integrity seems helpful because of its objective aspects, implying that the animal is intact or whole. However, often there are good reasons, for example when the health of the animal is involved, to violate the integrity of an animal. A second difficulty mentioned in the literature is the problem of 'gradation': how to distinguish between acceptable and non-acceptable violations of the integrity of an animal? One of the possible answers is to describe different kinds of violations, and thus, for example, to make a difference between cutting part of a tail and cutting a leg. However, integrity seems to be an absolute notion: a body is intact or not. In that case all violations of integrity seem ethically wrong. In this paper two different ways to handle the concept of integrity as an absolute notion will be rejected: to consider the notion as more or less absolute, and to consider the notion of integrity as not absolute at all. We will defend the view that the concept of integrity has to be considered as an absolute right rather than an absolute norm. In this way animal integrity means an absolute right to inviolability of the body. As a result every violation becomes problematic and has to be justified.

Standards of integrity and compromises for the sake economic feasibility
*Jensen, K.K., University of Copenhagen, Institute of Food and Resource Economics, Rolighedsvej 25, 1958
Frederiksberg C., Denmark; kkj@foi.ku.dk*

The notion animal integrity is notoriously contested. In this presentation, I shall not attempt a clear definition. Rather, for the sake of argument, I shall use certain low input animal systems as examples of forms of animal production trying to respect at least in part some aspects of animal integrity. One such example is organic farming; but another is production with local breeds the identity of which needs to be preserved. Low input animal production is typically based on specific values that inform the production. By contrast, conventional production is not committed to special values; in principle, it can produce anywhere and buy input and sell output on the world market. The values underlying low input animal production systems make up their identity, which often find a clear expression in a brand. These production systems are perceived by the consumers as taking ethical responsibility, and in many cases the products can be sold with a price premium. Because of this, they also raise higher expectations among consumers. This again makes low input production more vulnerable in case of problems than high input conventional production systems. This presentation reports from a recent ethics workshop for low input production covering laying hens, dairy cows, sheep and pigs. It was generally observed that these systems have welfare problems of their own, due to compromises to economic feasibility. This involves a risk that consumers may lose confidence to these systems being a risk which is amplified if the problems are concealed from the public. As a tentative conclusion from these observations, I shall argue that low input systems attempting to protect animal will have to address openly which compromises are acceptable and which are not. This may lead to conflicts with the aim of enlarging market shares. But I see no way to avoid addressing this dilemma, if low input production systems want to retain their own integrity.

Do cows (should) have horns?
Marie, M.[1,2], [1]INRA, SAD-ASTER, 662 avenue Louis Buffet, 88500 Mirecourt, France, [2]Université de Lorraine, ENSAIA, 2 avenue de la Forêt de Haye, 54505 Vandoeuvre lès Nancy, France; michel.marie@mirecourt.inra.fr

Keeping hornless cattle is now the preponderant situation in the European context. The reasons for this cover the diminution of risks, the easiness of management, and economic arguments relative to the cost of facilities or the marketing of animals, but, in the other way, dehorning or disbudding is a source of pain for the animals, and the use of genetically hornless cattle may be considered as non-natural. We question here the relevance of this practice, not only from the point of view of animal welfare, but also of animal integrity. A survey conducted in farms from different regions of France explored the reasons why farmers keep horned (or hornless) cattle, the consequences on facilities and practices, and on the frequency of incidents or accidents involving animals or humans. An ethical evaluation, based either on the ethical matrix (after Ben Mepham), taking into account the different stakeholders' positions and referring to well-being, autonomy and justice, or on the reflexive equilibrium method (after van der Burg and van Willigenburg), which enables to introduce other values such as aesthetics, naturalness, intrinsic value or integrity, is then performed. These analyses question the legitimacy of depriving cattle from their natural attributes.

The impact of stockmanship on the welfare of growing pigs
Von Borell, E. and Schaeffer, D., Martin-Luther-University Halle-Wittenberg, Institute of Agricultural and Nutritional Sciences, Theodor-Lieser-Str. 11, 06120 Halle, Germany; eberhard.vonborell@landw.uni-halle.de

Low investment costs and high production efficiency are vital criteria for the competitiveness of pig production. However, with increasing farm and herd sizes, animal welfare and health care monitoring is becoming increasingly important as individual animal control has to be safeguarded by fewer caretakers. On the other hand, automatisation and precision farming management tools may ease working conditions changing from manual work to supervision of daily processes on the farm. Nevertheless, supervision requires the regular presence of the caretaker with daily control of health status, feed and water intake at the animal level and of technical processes and installations such as feeders, ventilation, heating and cooling devices. Indicators for good stockmanship include regular inspection of animal behaviour, provision of functional separate areas and material that allow for rooting and manipulation, immediate repair of damaged installations, avoidance of mixing unfamiliar pigs, group conformity and separation of sick animals. Recently developed check lists for on-farm monitoring of pig welfare have shifted away from strictly resource and management based control points towards animal based welfare indicators such as the reaction of pigs towards humans and novel objects, startling, tail and ear biting, body condition, skin lesions and lameness, aggressive behaviour, exploration and play, vocalization, coughing and sneezing, faeces texture, defecation pattern and cleanliness of body and pen. The integrity of the pig itself is increasingly seen as the main indicator for good stockmanship and housing condition as good health, mental and physical integrity (that includes entire un-castrated boars and tails) are considered as appropriate accumulative outcome-based welfare criteria for the life cycle of a pig.

Assessment of fear and human animal relationship in finishing pigs at the slaughterhouse
Dalmau, A., IRTA, Animal Welfare Subprogram, Finca Camps i Armet s/n, 17121, Monells, Spain; antoni.dalmau@irta.es

The results of three studies on fear assessment and one in human animal relationship in pigs at the slaughterhouse developed into the Welfare Quality® project are presented. The aim of the first study was to validate approach and locomotive behaviour as indicators of fear in pigs. To carry out it, 60 pigs were exposed to visual, auditory and olfactory novel stimuli. The variables studied were feeding behaviour, approach behaviour and locomotive behaviour (general activity, reluctance to move, turning back and retreat attempts). Two groups were studied: saline and midazolam treated group. It was concluded that although reluctant to move was the most common response to the different fear stimuli applied, the combination of reluctant to move and turning back would be the best criterion to assess general fear in domestic pigs. The aim of the second study was to test inter-observer reliability (IOR) when assessing fear during arrival at the slaughterhouse in finishing pigs. Two Belgian and two Spanish slaughterhouses were visited by six to seven observers and reluctance to move, retreat attempts, turning back and vocalizations assessed in the unloading area. Turning back was the measure with the highest IOR and retreat attempts had the lowest. Based on the results, it was suggested to score maximum two measures of fear on the same animals at the unloading area: turning back and reluctance to move. Finally, in the third study, the sensitivity of different measures was assessed in 11 Spanish slaughterhouses. Fear was assessed in a total of 3,472 pigs with clear differences between slaughterhouses in the presence of reluctant to move ($P=0.0002$) and turning back ($P=0.0114$). In addition, the human-animal relationship, measured by means of high pitched vocalizations, showed also clear differences between slaughterhouses ($P<0.001$). In fact, in the slaughterhouses with a restrainer the presence of vocalisations was higher than when automatic doors linked to CO_2 stunning was used.

The impact of stockmanship on the welfare of sows

Spoolder, H.A.M., Hoofs, A.I.J. and Vermeer, H.M., Wageningen UR Livestock Research, Animal Welfare, P.O. Box 65, 8200 AB, Netherlands; hans.spoolder@wur.nl

The welfare of sows is affected to a large degree by the farmer's management skills and his choices in how to operate his farm. At the 2011 EAAP meeting we reported that there is a huge difference in success of first insemination (range: 70-96%), which does not appear to be affected by gestation housing system (stalls, ESF, trough feeding) or timing of introduction to the group (4 days - 4 weeks after service). To a large extent, these and other health and welfare results are affected by management. Good sow management starts when rearing them as gilts: offering environmental enrichment pre-weaning affects play behaviour, which supports social skills in adulthood. In addition, repeated exposure to unfamiliar adults during gilt rearing facilitates mixing in dynamic groups later: gilts that have been regrouped four times only showed 57% of bilateral agonistic actions and 82% of total fighting time, in comparison with control gilts which had not been regrouped. The main success factors identified during a large trial on gestating sow reproduction included professional attitude of the farmer. Finally, management during the farrowing and lactating phases has an impact on sow and piglet welfare. Simple measures such as offering a jute bag pre-farrowing will result in less posture changing and shorter parturition, thus saving piglets and increasing farm profitability.

Challenges and opportunities for low input dairy systems

French, P. and Shalloo, L., Teagasc, Livestock Systems, Moorepark, Fermoy, Co. Cork, Ireland; padraig.french@teagasc.ie

Low input forage based dairy systems have numerous advantages and disadvantages over high input grain based systems. The advantages include a lower cost of production and subsequently a more financially secure business in a volatile milk price environment, lower initial capital costs which is more sustainable for family farm businesses and a lower environmental footprint. However, the major limitation to low input forage based dairy systems is, unlike grain based systems, the technology employed is not generally replicable across regions. The major challenge for low input forage systems is the requirement to develop region specific technology to optimise milk production from the most viable forage available. There are many technologies required to optimise the profitability and sustainability of low input milk production systems however it is important in evaluating any technology that it's impact on the overall performance of the system are understood. The overall objective of low input systems should be to optimise the quantity of forage produced and the efficiency to which it is converted into milk or milk products. Key technologies required include animal breeding, forage production and utilisation and farm infrastructure requirements. Animal breeding should evaluate animals within the system of production that they are required for, and select based on breeding goals that optimise profitability within that system. Forage production should similarly evaluate species and cultivars within the system of production they are required for and select based on profit driven breeding goals. Forage utilisation particularly for grazed pastures has to be evaluated in whole farm system research experiments in order to understand the complex interaction of the animal and the forage over time. Because milk supply from low input dairy systems are usually more seasonal than from high input systems, it is important that milk pricing structures are designed to optimise profitability from both the processor and producer as a single unit.

Role of rumen digestion in low input milk production systems

Shingfield, K.J.[1] and Huhtanen, P.[2], [1]MTT Agrifood Research Finland, Jokioinen, 31600, Finland, [2]Swedish University of Agricultural Sciences, Umeå, 901 83, Sweden; kevin.shingfield@mtt.fi

Efficient utilization of forages is central to meeting the nutrient requirements for milk production in low input and organic systems. Forage legumes are often used as a means to fix atmospheric N to ensure high crop yields and sufficient forage protein without the need for inorganic N fertilizers. The intake potential of legumes, including red clover, white clover and lucerne is also higher than for grasses. Digestibility and intake of forages are determined by intrinsic properties of plant cell walls (potential extent and rate of digestion, rate of particle breakdown) and extrinsic animal and diet factors that influence both digestion and passage rates. The higher intake potential of red clover compared with grasses is related to a lower NDF content and higher rate of NDF digestion in the rumen. Despite a higher content of indigestible NDF, the rate of particle size breakdown and passage through the rumen is higher for lucerne relative to grasses. Decreases in the digestibility of grass silages due to increases in maturity are associated with increases in ruminal outflow rate and NDF pool size. However, rumen fill or digestion and passage rates in the rumen do not explain the differences in intake potential of primary grass growth compared with regrowths. While forage legumes are important to the N economy of low input and organic production systems, the efficiency of N utilization is often lower compared with grasses, due in the most part to a higher protein intake. Red clover lowers ruminal N degradation, improves the energetic efficiency of microbial protein synthesis and increases post ruminal non-ammonia N flow compared with grass silage, but this does not typically stimulate an increase in milk protein output.

Challenges and opportunities for animal welfare in organic and low input dairy farming

Winckler, C., Tremetsberger, L. and Leeb, C., University of Natural Resources and Life Sciences, Department of Sustainable Agricultural Systems, Division of Livestock Sciences, Gregor-Mendel-Strasse 33, 1180 Vienna, Austria; christoph.winckler@boku.ac.at

Achieving and maintaining a high health and welfare status is an important aim in sustainable organic and low-input livestock farming. Animal welfare is often assumed to be high due to the preventive effect of more extensive housing and management conditions as e.g. defined in EC 834/07. However, field studies have shown that although organic standards form a good basis dairy cattle health and welfare may be impaired in these farming systems for example with regard to lameness prevalence or mastitis incidence. Monitoring of health and welfare as well as the development of effective improvement strategies are therefore crucial also for organic and low-input dairy farming systems. Feasible animal-based on-farm welfare assessment schemes have recently been developed which provide knowledge about the status within a given herd and identify welfare problems; outcomes may also be used for benchmarking purposes. Animal health and welfare plans are promising approaches to effectively reduce medicine input and improve health and welfare and may be used in different advisory contexts including common-learning settings such as Stable Schools. They rely on a set of principles such as continuous development and improvement based on evaluation, action and review, a farm specific process and farmer ownership. The challenges and opportunities of these approaches will be discussed based on own experiences in European on-farm studies.

Analyses of different brown cattle breeds and their crosses in Switzerland

Spengler Neff, A.[1], Mahrer, D.D.[1,2,3], Moll, J.[2], Burren, A.[3] and Flury, C.[3], [1]Research Institute of Organic Agriculture (FiBL), Ackerstrasse, 5070 Frick, Switzerland, [2]Qualitas AG, Chamerstrasse 56, 6300 Zug, Switzerland, [3]Bern University of applied Sciences, School of Agricultural, Forest and Food Sciences, Länggasse 85, 3052 Zollikofen, Switzerland; anet.spengler@fibl.org

In Swiss mountain regions it has become common to inseminate dairy cows of the Swiss Brown Cattle breed (BV) with bulls of the Swiss Original Brown Cattle breed (OB). The goal of this practice – which is especially common on organic farms – is the improvement of robustness and animal health. The aim of the study was to derive possible effects of cross breeding (i.e. heterosis) on different phenotypic characteristics. The sample consisted of pure OB-cows, a randomly drawn set of BV-cows, and F1- and F2-crossbred cows born between 2000 and 2010. 163,734 animals were analyzed with general linear models. Dependent variables were milk yield, fat- and protein contents, fertility traits, somatic cell score (SCS), and persistency, all in first lactation. Further life production and lactation number were investigated. Independent variables were: breed, production area, age at first calving, calving month and days open. Interactions between breed and production area were integrated as well. In all models the effect of the breed was significant. Generally, all crosses showed better fertility and SCS but lower levels in production traits than purebred BV-cows. In comparison with OB-cows crosses were inferior in fertility and SCS and at a similar level in production traits (except F2 with 25% OB-blood). It is concluded, that for functional traits crossbreed individuals are inferior in comparison with OB but superior in comparison with BV. Regarding production traits the performance is comparable between the crosses and OB but inferior to BV. Therefore crossbreeding with OB is recommended for the amelioration of functional traits in BV herds, but changing to pure OB individuals is expected to lead to a faster and probably more solid success regarding health and robustness.

Organic farming: an opportunity for dairy farmers in mountain?

Dockès, A.-C., Fourdin, S., Delanoue, E. and Neumeister, D., French Livestock Institute, 149 rue de Bercy, 75012 Paris, France; anne-charlotte.dockes@idele.fr

Implementation of development actions, taking into account barriers and motivations of farmers and dairy products field stakeholders. In mountain regions, organic farming development is a real opportunity in order to offer to farmers a better value-added and to respond to market demand. The near-end of milk quotas, implies to give a special support to farms of these areas and to help them to take place in the market of organic milk production. Research and development MontagneBio project, financed by the French Ministry of Agriculture, aims to elaborate methods and tools to develop organic dairy-farming in piedmont and mountain areas. Within this framework, fourteen pilot French departments have been involved and are drawing schemes of organic milk production development. Preliminary studies on hindrances and motivations towards organic farming have been conducted in these departments in 2010. 264 semi-directive interviews have been conducted on farmers and other agents of dairy products field in order to analyze local contexts and field strategies. Farmers' main hindrances are mostly technical (feeding security, importance of local systems, structural constraints…), but they are very often linked to a lack of knowledge concerning organic farming standards. The first motivation expressed is the economic gain and for some farmers ethics issues. Most of all, the presence or the lack of a local organized collection of organic milk constitutes a major stake. Currently, development actions are held on field to advice and back already converted farmers or those that are willing to. Training courses, visits of organic farms, economic assessments for example are conducted by local coordinators of the project. A national synthesis of these actions and an assessment of their impacts is being done. This synthesis will help to understand organic farming representations among farmers, and it will also provide useful tools and ideas sources to develop organic farming in region with similar issues.

Effect of farming system on productivity, efficiency and sustainability in dairy farming
Guerci, M.[1], Schönbach, P.[2], Kristensen, T.[3] and Trydeman Knudsen, M.[3], [1]Università degli Studi di Milano, Dipartimento di Scienze Animali – Sez. Zootecnica Agraria, Via Celoria, 2, 20133 Milano, Italy, [2]University of Kiel, Institute of Crop Science and Plant Breeding, Hermann-Rodewald-Str. 9, 24118 Kiel, Germany, [3]Aarhus University, Dept. of Agroecology, Blichers Allé 20, Postboks 50, 8830 Tjele, Denmark; matteo.guerci@unimi.it

During the last decade, the dairy farming systems throughout Europe has changed towards more specialization and intensification. Among other aspects, the proposed cancelling of the milk quota system by 2015 will presumably lead to an acceleration of this development. In the present paper we will analyze the productivity and efficiency of a range of dairy farming systems in Italy, Germany and Denmark. The farms represents different production methods (organic vs. conventional) and summer feeding systems (confinement vs. pasture) and different annual production levels (6,000 to 11,000 kg milk per cow). There is large variation in stocking rate (1 to 5.3 LSU/ha) among the farms, which has major impact on production per area of farm land (4 to 30 t milk/ha) and the supply of feed from farm land (50 to 85% of energy). In combination with the geographical differences, this has major impact on type (maize, grass and legumes) and proportion of roughage (45-95%) in the ration and thereby also the land use, with an important variation being permanent grassland vs. grassland in rotation and proportion of land in grass. Utilization of manure is a critical element for a sustainable production and at the farms the amount applied in average per ha annually varies from 125 kg gross N/ha to more than 500 kg/ha. The sustainability will be addressed based on local indicators like nutrient surplus and eutrophication and global indicators like fossil energy use, global warming potential and land use based on a life cycle assessment approach. The effect of methodological choices on the ranking of the systems will be shown, as well as how to estimate new indicators like biodiversity.

Nitrogen use efficiency in grassland based farming systems in Europe and EECCA region
Van der Hoek, K.W., RIVM, P.O. Box 1, 3720 BA Bilthoven, Netherlands; klaas.van.der.hoek@rivm.nl

Grassland based farming systems vary widely over Europe and EECCA region. Nitrogen inputs range from no external inputs as synthetic fertiliser and concentrates, input by biological nitrogen fixation to high external inputs. The duration of the grazing period ranging from full time summer grazing to zero grazing also affects the nitrogen use efficiency and nitrogen losses per hectare. Based on recent detailed national statistics and farm scale balances an overview is presented on nitrogen budgets and balances for grassland based farming systems in different European and EECCA countries. The differences between countries are explained and discussed in terms of nitrogen use efficiency and nitrogen losses per hectare. Finally options for improvement of both nitrogen parameters are explored.

Effect of non-fertilized winter grazing dairy production on soil N balances and soil N dynamics

Necpalova, M.[1,2], Phelan, P.[1,2], Casey, I.[2] and Humphreys, J.[1], [1]Teagasc Moorepark, Fermoy, Ireland, [2]Waterford Institute of Technology, Waterford, Ireland; magdalena.necpalova@teagasc.ie

Soil surface nitrogen (N) balances and soil soluble N pools are often analysed to evaluate nutrient management practises on the soil surface. The objective was to investigate how they are affected by non-fertilized winter grazing dairy production on a clay loam soil in southern Ireland over two years. The systems were a combination of 2 factors: (1) conventional grazing (Feb-Nov) and winter grazing (Apr-Jan); (2) conventional fertilizer N input (100 kg/ha) and no fertilizer input. Each system consisted of 6 paddocks. Balances for each paddock and year were calculated as the difference between N inputs and outputs. Inputs included N entering soil through soil surface as fertilizer N, slurry, excreta, atmospheric deposition and white clover biological N fixation (BNF). Outputs consisted of N leaving the soil as harvested and grazed herbage. Soil samples were taken from 4 paddocks per system 8 times during the study. At each sampling 15 cores per paddock were taken to a 0.9 m depth (0-0.3, 0.3-0.6, 0.6-0.9), bulked to a composite sample per each paddock. Soil N was assessed in KCl extracts. Balances were subjected to ANOVA; soil results were analysed as a repeated measure. In non-fertilized systems, the N input was dominated by BNF (123 kg N/ha/yr) and was only 80% (P<0.05) of that into the fertilized system. Application of fertilizer decreased BNF (P<0.05). There was no difference in N uptake (242 kg N/ha/yr) or N surpluses (144 kg N/ha/yr) between the systems. N uptake by herbage (r^2=0.30, P<0.05) and soil N (r^2=0.43, P<0.05) were positively correlated to N input. Moreover, soil N was correlated to N surplus (r^2=0.36, P<0.05). The non-fertilized winter grazing system did not affect soil N surplus and soil N dynamics compared with the other systems. Seasonal changes in soil N were driven by soil moisture and soil temperature, which are the factors controlling microbial activity and biochemical processes.

Grazing behavior and metabolic profile of 2 Holstein strains in an organic full-time grazing system

Thanner, S.[1,2], Schori, F.[2], Bruckmaier, R.[1] and Dohme-Meier, F.[2], [1]University of Bern, Veterinary Physiology, Bremgartenstrasse 109a, 3001 Bern, Switzerland, [2]Forschungsanstalt Agroscope Liebefeld-Posieux, Route de la Tioleyre 4, 1725 Posieux, Switzerland; sophie.thanner@alp.admin.ch

The challenge for sustainable organic dairy farming is to find cows well-adapted to forage-based production systems with a low concentrate input. Therefore, the aim of the present study was to compare the grazing behavior, physical activity and metabolic profile of 2 different Holstein strains in an organic grazing system. Twelve Swiss Holstein-Friesian (H_{CH}; 587 kg of BW) and 12 New Zealand Holstein-Friesian (H_{NZ}; 546 kg of BW) cows in mid-lactation were paired according to the stage of lactation and kept in a rotational full-time grazing system without concentrate supplementation. After an adaptation to the grazing system, milk yield was recorded daily and grass intake and nutrient digestibility were estimated by the double alkane technique during a 7 d data collection period. Physical activity was recorded with a pedometer and grazing behavior was investigated using a behavior recorder over 3 d. On 3 consecutive d, blood was sampled at 0700, 1200 and 1700 h from each cow by jugular vein puncture. Data were analyzed using linear mixed models. No differences were found in milk yield but milk fat (3.69 vs. 4.05%, P=0.06) and milk protein percentage (2.92 vs. 3.20%, P<0.05) were lower in H_{CH} than in H_{NZ}. Grass intake did not differ between strains but H_{CH} had a greater (P<0.01) digestibility of nutrients compared to H_{NZ}. The H_{CH} cows spent less (P<0.05) time ruminating (439 vs. 469 min/d) and had a lower (P<0.05) number of ruminating boli than H_{NZ}. Time spent eating and physical activity did not differ between strains. The H_{CH} had lower (P<0.05) concentrations of cholesterol and triacylglycerides than H_{NZ} which indicated differences in the fat metabolism of the 2 Holstein cow strains. In conclusion, at similar grass intake heavier H_{CH} cows displayed a greater nutrient digestibility which had no positive effect on milk production.

Development trends of organic livestock farming in Lithuania

Svitojus, A.[1], Skulskis, V.[2], Paulaitiene, J.[1] and Juozaitiene, V.[3], [1]Baltic Foundation HPI, S. Konarskio str. 49, 03123 Vilnius, Lithuania, [2]Lithuanian Institute of Agrarian Economics, V. Kudirkos st. 18-2, 03105 Vilnius, Lithuania, [3]Lithuanian university of health sciences Veterinary academy, Animal Breeding, Tilzes str. 18, 47181 Kaunas, Lithuania; arunas@heifer.lt

Present Lithuanian agricultural policy focuses on organic farming with an environmental, social and economic benefit. This research reviews organic farming over period of 2004-2011. Data from the national certification body 'Ekoagros' was used for the analysis. Very rapid growth in the number of organic farms certified was observed during the period up to 2007. The total number of organic farms in 2007 increased by 2,4 times comparing to 2004, but then slow declined by 9.0% in 2011. It covers about 2.6 thous farms – 158.0 thous of ha or 6.0% of declared agricultural area. We presume that rapid growth before 2007 was due to novelty of process and intense of EU financial support (subsidies per certified ha). Later, the changes in level of support and strengthened requirements for farmers knowledge on organic farming reduced the level of certification. Rather different trend was observed in the sertification of organic livestock farms. The highest number reached in 2007 (817 farms) after slow decrease in 2008-2010 grew by 4.5% in 2011. We suggest that main increase was due the fact that number of farms started to certify livestock as organic. The main strain of organic livestock is cattle and sheep husbandry. Number of milking cows has been growing by 2.9 times in the period showing 8.9 thousand, young cattle – 4.8 times (14.0 thousand). Number of sheep grew from 3.8 thousand in 2004 to 14.3 thousand in 2011 (3.8 times). Our research showed that present number of certified organic farms in Lithuania maintains at stable level, and development of organic farming often depends on changes in the regulation system of the organic farming. The development of organic products market in the country may be next factor for growth of the organic farming.

Determination of fatty acids and conjugated linoleic acids contents in ruminant feces

Cesaro, G., Grigoletto, L., Bittante, G. and Schiavon, S., University of Padova, Department of Agronomy Food Natural resources Animals and Environment (DAFNAE), Viale dell'Università 16, 35020 Legnaro (PD), Italy; giacomo.cesaro@studenti.unipd.it

Three procedures to determine fatty acids (FA) profiles in ruminant feces were compared. Procedures were based on: (1) acid-base esterification of FA performed directly on dry fecal samples (J); (2) acid-base esterification of FA performed on the ether extract (EE) recovered with accelerated solvent extraction (ASE) at 125 °C for 25 min (J_{EE}); (3) acid catalyzed methylation of FA performed on EE extracted from feces with ASE (C_{EE}). Feces were collected from 9 bulls receiving 0, 8 and 80 g/d of rumen protected conjugated linoleic acids (rpCLA; 3 bulls per dose) were analyzed in triplicates by each method using a gas chromatograph, expressing FA contents as mg/g DM. The repeatability of FA and CLA measurements of each method was determined. For CLA, because of variance heteroscedasticity, the procedures were compared by linear regression. Within procedure, fecal contents of CLA were regressed against the rpCLA dose. For many FA measurements J was the most repeatable procedure. Procedures based on EE evidenced for C18:2,c9,t11 and C18:2,t10,c12 regression equations with slopes and intercepts close to 1 and 0, respectively, whereas the relationships of J_{EE} and C_{EE} with J had slopes of 0.264, and 0.284 for C18:2c9,t11, and 0.420 and 0.469 for C18:2t10,c12, respectively. Regressions between C18:2c9,t11 fecal content (mg/g DM) and rpCLA dose (g/d) had slopes of 0.0011, 0.0002 and 0.0003 for J, J_{EE} and C_{EE}, respectively. Regressions between C18:2t10,c12 fecal content and the rpCLA dose had slope of 0.0017, 0.0007 and 0.0008 for J, J_{EE} and C_{EE}, respectively. With increasing rpCLA dosage the EE-based procedures provided lower fecal contents of the two CLA isomers and a higher C18:2t9,t11 content with respect to J. Procedures based on EE likely induced CLA isomerization, and J is recommended for determining the fecal CLA content.

The effect of soybean and fish oil inclusion in goat diet on their milk and plasma fatty acids
Tsiplakou, E., Tsiligkaki, A., Mountzouris, K.C. and Zervas, G., Agricultural University of Athens, Nutritional Physiology and Feeding, Iera Odos 75, 11855, Greece; eltsiplakou@aua.gr

The objective of this study was to determine the effects of soybean and fish oil inclusion in goat diet on their milk and plasma fatty acids (FA) profile. Twelve dairy goats were divided in two sub-groups. The first was fed with the diet without any oil supplementation and the other with a diet containing 5% soybean oil and 1% fish oil. All the goats were fed individually according to their requirements. The results showed that the milk yield and the fat content of the treated group were increased by 23% and 18% respectively, compared to the control, without the results being significant. The concentrations of trans-11 C18:1, trans-10 C18:1, cis-9, trans-11 C18:2, trans-10, cis-12 C18:2, C18:2n6c, polyunsaturated and monounsaturated FA in milk fat were found to be significantly higher, while those of C10:0, C12:0, C14:0, C18:0, short, medium and long chain FA lower in the treated group compared to the control. In addition, in goat blood plasma the soybean and fish oil inclusion caused a significant increase in trans-11 C18:1, trans-10 C18:1, cis-9, trans-11 C18:2, trans-10, cis-12 C18:2 and a decrease in C14:0, C15:0, C16:0, C18:0 and cis-9 C18:1 FA. In conclusion, the soybean and fish oil inclusion in goats diet modified the milk and plasma FA profile.

Effect of dietary oil and seed addition on pork subcutaneous and intramuscular fatty acid profile
Carrion, D.[1], Maeztu, F.[2] and Oficialdegui, M.[3], [1]Pig Improvement Company, Avda Ragull 80 2°, 08173, Sant Cugat del Vales, Barcelona, Spain, [2]INTIA, Avda. Serapio Huici, 22, 31610 Villava, Navarra, Spain, [3]Granja Los Alecos, Ctra Zaragoza Km 24, 31395 Barasoain, Navarra, Spain; domingo.carrion@genusplc.com

There is a need to better understand fatty acid (FA) deposition in the two most important fat compartments in slaughter pigs, especially in Duroc crosses that brings high subcutaneous (SCF) and intramuscular (IMF) fat. A total of 165 (Duroc × (Landrace × Large White)) pigs were randomly allocated in 12 pens for 35 days from 83 to 114 kg live weight to evaluate the effect of three dietary treatments: control (CON), seed (70% extruded linseed+30% wheat middling) blend 4% inclusion rate (SB), and oil (80% linseed+20% fish) blend 1.5% inclusion rate (OB) on the FA profile in the SCF and IMF. Feed and water were provided ad-libitum. Data were evaluated by the GLM procedure of SAS. There was no effect of treatment on growth, carcass and meat quality. Compared to CON, SB and OB increased C18:3 and total n3 FA in both SCF (0.89, 2.08, 2.76 and 0.92, 2.15, 3.63%) and IMF (0.45, 1.22, 1.46 and 0.50, 1.34, 2.00%) compartment respectively ($P<0.001$). Treatment had no effect on total n6 and saturated (SFA) FA in both SCF and IMF content. Compared to CON, SB did not affect, whereas OB increased EPA and DHA content in both SCF (0.35, 0.36, 0.75 and 0.03, 0.03, 0.11%) and IMF (0.35, 0.36, 0.48 and 0.05, 0.05, 0.11%) respectively ($P<0.001$). Compared to CON, SB and OB increased PUFA content in both SCF (11.78, 13.32, 14.76%) and IMF (8.56, 9.66, 9.87%) ($P<0.05$). Compared to CON, SB did not affect, whereas OB increased ($P<0.01$) PUFA/SFA ratio in SCF (0.30, 0.34, 0.37%), whereas this ratio was not affected in the IMF. Compared to CON, SB and OB decreased n6/n3 ratio in both SCF (12.70, 5.20, 3.05%) and IMF (15.80, 6.27, 3.95%) respectively ($P<0.001$). These results suggest that both SB and OB increases n3 and improves n6/n3 ratio. Those positive changes in FA profile takes place earlier in the SCF than IMF compartment.

Effect of a blend of three phytonutrients on performance of broilers compared to a shuttle program
Oguey, C.[1], Sims, M.D.[2] and Bravo, D.[1], [1]Pancosma SA, Voie des Traz 6, 1218 Le Grand Saconnex, Switzerland, [2]Virginia Diversified Research Inc, Virginia, 22801 Harrisonburg, USA; clementine.oguey@pancosma.ch

Previous studies demonstrated that a ME value of 50 kcal/kg was attributed to a blend of cinnamaldehyde, carvacrol and capsicum oleoresin (XT, XTRACT 6930 – Pancosma). XT also led to greater performance than bacitracin. The objective of this trial was to evaluate if XT could lead to similar broiler performance to a shuttle program. A total of 1200 day old broilers were randomly allocated to 40 floor pens of 30 birds each for 42 days. Birds were fed *ad libitum* a corn soybean meal diet in three phases (starter, grower, and finisher) Three dietary treatments were applied: NC: non supplemented negative control; PC: positive control supplemented with 50 ppm bacitracin in starter and grower phases, and with 20 ppm virginiamycin in finisher phase; XT: experimental group supplemented with 125 ppm XT in starter and grower phases, and with 100 ppm XT in finisher phase. BW, FI, FCR, mortality adjusted FCR, feed costs and net benefits were evaluated. Data were analyzed by a 2-tailed distribution basic T-test model with equal variances assumed. Results showed that FI was similar among the treatments. Final BW of birds fed XT and PC were not different but higher than NC (respectively 2.25, 2.27 and 2.19 kg, $P<0.05$). FCR in XT group was intermediate to NC and PC (respectively 1.84, 1.88 and 1.83 g/g) but was different ($P<0.05$) among the 2 control treatments. Mortality adjusted FCR of birds fed XT and PC were not different but significantly lower than NC (respectively 1.82, 1.82 and 1.86 g/g), $P<0.05$). This shows that XT increases performance of broilers at the same level as the tested shuttle program. Feed cost and net income were similar in PC and XT groups and higher ($P<0.05$) than NC. This suggests that XT leads to similar benefits as PC. These results show that birds supplemented with XT have equivalent growth performance and economical results to broilers fed a shuttle program with bacitracin and virginiamycin.

Quality of eggs from Greenleg Partridge hens maintained in organic vs. backyard production systems
Krawczyk, J. and Szefer, M., National Research Institute of Animal Production, Department of Animal Genetic Resources Conservation, Krakowska 1, 32-083, Balice, Poland; mszefer@izoo.krakow.pl

The aim of the study was to evaluate the quality of eggs from 56-week-old hens of the native Greenleg Partridge chicken breed, maintained in three production systems differentiated by feeding – organic (group 1), backyard (group 2) and intensive, kept indoors in a poultry house (group 3 – control). Egg quality and eggshell traits were found to be related to the production groups. The eggs laid by hens from the organic farm (group 1) were characterized by the highest total weight and highest weight of yolk which had the most desirable colour. The yolks of these eggs were characterized by the lowest level of cholesterol and the lowest shell strength, with the highest level of vitamin E. In the backyard production system (group 2), eggs were characterized by the lowest total weight and yolk weight, the highest protein content of albumen, the highest vitamin A and the lowest vitamin E content of yolk as well as lower shell crushing strength compared to the control group. As a result of the study, it was found that the organic production system of Greenleg Partridge hens has a favourable effect on most quality traits of eggs by increasing their nutritive value. Supported by grant no N R12 0083 10 financed by the National Centre for Research and Development.

Changes in net hepatic flux of nutrients by the deacetylation of p-aminohippuric acid in dairy cows

Rodríguez-López, J.M., Cantalapiedra-Hijar, G., Durand, D., Thomas, A. and Ortigues-Marty, I., INRA, INRA-Theix, 63122 Saint Genès Champanelle, France; gonzalo.cantalapiedra@clermont.inra.fr

Recent studies reported that cattle, like sheep, acetylate in the liver the marker para-aminohippuric acid (PAH), frequently used to measure hepatic blood flow, leading to an overestimation of hepatic venous and arterial blood flows. Therefore, it is necessary to assess the quantitative impact of adding a deacetylation step in the PAH analytical process on hepatic metabolism of nutrients. Four lactating Jersey cows equipped with permanent catheters in the mesenteric artery, portal and hepatic veins for blood sampling, and mesenteric and ruminal veins for PAH infusion were used. They were offered four iso-energetic diets with hourly meals. Sixteen series of blood samples were obtained. For each serie, five hourly blood samples were taken over 5 hours starting 45 min after the onset of a continuous PAH infusion (38 mmol/h). Plasma samples were concurrently analyzed with or without the inclusion of a PAH deacetylation step, after method validation according to ISO procedures. Concentrations of volatile fatty acids, glucose, lactate, β-hydroxybutyrate, α-amino-N, urea, ammonia and oxygen were obtained. Plasma and blood flows at portal and hepatic veins and hepatic artery were determined as well as net nutrient fluxes. The paired t-test for means (SAS) was used. The PAH deacetylation did not affect portal plasma flow. However, this process led to a 16 and 66% lower hepatic venous and arterial plasma flows, respectively, as well as a lower arterial contribution to total hepatic plasma flow. The net hepatic flux of nutrients was either moderatly affected (propionic and butyric acid, glucose, lactate, α-amino-N, ammonia and urea) or modified by more than 20% (acetic acid, β-hydroxybutyrate and oxygen). These results confirm the hepatic acetylation of PAH in cattle and highlight the importance of including a deacetylation step when measuring hepatic nutrient fluxes in dairy cows.

A starch binding agent decreases the *in vitro* rate of fermentation of wheat and barley

Dunshea, F.R., Russo, V.M. and Leury, B.J., The University of Melbourne, Department of Agriculture and Food Systems, Royal Parade Parkville, 3030 Victoria, Australia; fdunshea@unimelb.edu.au

The rapid rumen fermentation of the starch in wheat and barley can result in sub acute ruminal acidosis (SARA) with resultant inhibition of rumen function. The aim of this study was to investigate the effect of a starch binding agent (Bioprotect™, RealisticAgri, Ironbridge, UK) on the rate of *in vitro* gas production. The active ingredient in Bioprotect™ is a stable non-volatile organic salt that complexes with the hydroxyl groups of starch at neutral or slightly acidic conditions (pH 6 to 7), as observed in the rumen. These complexes decompose under more acidic (pH 2 to 3) conditions as in the abomasum and duodenum, making the starch available for enzymatic digestion. Wheat and barley were ground and passed through a 1 mm sieve. A sample of the ground grains was treated with Bioprotect™ (8 ml/kg). Samples (1.0 g) of the grain (n=6 for each grain) were added to flasks containing buffered rumen fluid obtained from lactating dairy cows. The flasks were purged with CO_2 and maintained at 39^0C. Gas production was monitored every 5 min for 48 h using the ANKOM™ wireless *in vitro* gas production system. Gas production was modeled using a Gompetz equation to determine maximum amount of gas production (Rmax) and the rate constant (β). The Rmax was higher for the control wheat than the control barley (134 vs 123 ml/g for wheat and barley, P<0.05). The Rmax was decreased (P<0.05) by treatment of barley (123 v 112 ml/g for control and treated barley) and to a lesser extent by treatment of wheat (134 vs 129 ml/g). The β was decreased (P<0.05) by treatment of wheat (0.267 vs 0.207 min^{-1}) and to a lesser by treatment of barley (0.224 v 0.213 min^{-1}). These data demonstrate that wheat is fermented faster than barley and that Bioprotect™ can slow the fermentation of both grains. If more starch can pass through the rumen without being fermented it may reduce the incidence of SARA and allow for greater post-ruminal enzymatic digestion of starch.

Prediction of NDF content and rumen degradability by Fournier ransform infrared spectroscopy

Belanche, A.[1], Allison, G.G.[1], Newbold, C.J.[1], Weisbjerg, M.R.[2] and Moorby, J.M.[1], [1]Institute of Biological, Environmental and Rural Sciences, Aberystwyth University, SY23 3EE, United Kingdom, [2]Animal Science, Aarhus University, P.O. Box 50, 8830 Tjele, Denmark; aib@aber.ac.uk

The concentration of neutral detergent fibre (NDF) and its ruminal degradability are key factors that determine the nutritional value of ruminant feeds. The aim of this study was to examine whether Fournier ransform infrared (FTIR) could be used to develop a universal equation that allows the determination of feed NDF concentration and its rumen degradation parameters. A total of 120 feeds samples were used in the experiment (109 forages, 11 non-forages). In situ NDF degradability was described in terms of rumen degradable but not soluble fraction b, and the rate of degradation of b fraction c. The overall effective degradability (NDFED) was determined assuming a rumen passage rate of 2%/h and the indigestible NDF (INDF) was calculated after 288 h of incubation. For FTIR analysis, samples were dried and grounded at 1mm. Infrared spectra were collected by attenuated total reflectance from 4,000-600 cm^{-1} using an Equinox 55 FTIR spectrometer (Bruker Optik GmbH, Germany) fitted with a Golden Gate ATR accessory (Specac, UK). Duplicate spectra were averaged, derivatized, normalized and mean centred. Data were modelled by partial least squares regression using MatLab. Prior to calibration, spectra were transformed. Models were trained on a randomly selected subset of samples and the precision of the equations was confirmed by external validation (18% of the samples). Samples varied widely in dry matter NDF concentration (17 to 83%), NDFED (23 to 82%), INDF (1 to 59%), b (42 to 100%) and c (1 to 13%/h). FTIR analysis accurately predicted the NDF content (R^2=0.90, SECV=4.8%), NDFED (R^2=0.95, SECV=6.3%) and c (R^2=0.94, SECV=1.1%). Estimations of b (R^2=0.85, SECV=9.5%) and INDF (R^2=0.79, SECV=7.6%) were less accurate. This study shows that the FTIR is an appropriate tool to quickly predict the NDF content and rumen degradability for many feedstuffs used in ruminant diets.

Passage kinetics of fibre in the rumen estimated by external marker excretion curves

Krämer, M., Lund, P. and Weisbjerg, M.R., Institute of Animal Science, Aarhus University, AU-Foulum, 8830 Tjele, Denmark; Monika.Kramer@agrsci.dk

New feed evaluation systems for ruminants depend on estimation of rates of passage of nutrients out of the rumen. The present study aims at describing intrinsic effects of forage type and concentrate level on mean retention time (MRT) of NDF in the rumen. Sixteen lactating Holstein cows were fed total mixed rations based on maize silage (M) or grass-clover silage (GC), each combined with either 25% (25) or 50% (50) of concentrate on DM basis. Cows were fed 95% of their individual *ad libitum* dry matter intake (DMI) divided on three daily feedings. Forage and concentrate NDF were labelled with ytterbium and lanthanum, respectively, and fed as a single pulse dose. Nineteen faecal grab samples were taken 0, 4, 8, 12, 16, 20, 24, 28, 32, 36, 40, 44, 48, 52, 56, 60, 73, 85, 97 h after marker dosage. Rumen passage kinetics were determined by fitting age-dependent one (Gn) and two (GnG1) pool models. Model parameters were statistically analysed using a general linear model including fixed effects of block, forage type, concentrate level, restricted DMI as covariate and interactions between forage type and concentrate level. Among the one pool models tested, the G2 model fitted best, but G4G1 was the overall superior model. MRT of forage derived NDF in the rumen pools varied between 28 h for GC50 and 32 h for M25 and GC25 and for concentrate derived NDF between 18 h (GC50) and 20 h (GC25). MRT of NDF in the age-dependent pool was shorter than in the age-independent pool. MRT of forage NDF in the age-dependent pool differed between forage types (P=0.02). The proportion of MRT spent in the age-dependent pool varied between 0.25 (M50) and 0.38 (GC25) for forage NDF and between 0.29 (M25, GC25, GC50) and 0.33 (M50) for concentrate NDF. The relative time spent in the age-dependent pool of forage NDF differed significantly (P=0.01) between forage type. Based on the present results, a distribution of MRT between the two pools of 0.3:0.7 is recommended for use in feed evaluation systems.

Effects of a fibrolytic enzyme cocktail on *in vitro* and *in sacco* digestibility of forages and feed
Van De Vyver, W.F.J. and Cruywagen, C.W., Stellenbosch University, Animal Sciences, Private bag X1, Matieland, 7602, South Africa; wvdv@sun.ac.za

Exogenous fibrolytic enzymes (EFE) show potential in improving the digestibility of forages, not only *in vitro* and in situ, but also in terms of the production of ruminant animals. In this study, a novel exogenous fibrolytic enzyme cocktail, applied 12 h prior to incubation as a supernatant from the fermentation of fungal strain ABO 374 was evaluated *in vitro* and *in sacco* using six Döhne-Merino sheep. Substrates included lucerne hay (Medicago sativa) and two C4 grasses: weeping love grass (Eragrostis curvula) hay and kikuyu (Pennisetum clandestinum) and a complete sheep feed. Data were fitted to a non-linear model to determine the digestion kinetic values. The gas production fermentation kinetic values showed significant effects on the total gas production (b) as well as the rate of gas production (c) of lucerne and kikuyu. Concurrently the DM, CP and NDF disappearance and the effective degradability of each nutrient of the complete feed was significantly improved due to EFE treatment. Contradictory to expectations, EFE treatment of the complete feed resulted in a significantly lower soluble and rapidly degradable NDF fraction. The *in vitro* digestibility results substantiated the findings that EFE treatment of the complete feed can result in the improved digestibility thereof. Also, in agreement with findings of other groups, it was demonstrated that, although the EFE contained only fibrolytic enzyme activity, its effects were not limited to fibre. Based on these results on nutrient digestibility, it is speculated that EFE partly resulted in subtle changes to the cell wall structure, allowing microorganisms earlier access to the cell contents. In addition, these effects may also be related to the enhanced attachment of microorganisms to the plant cell wall and by the synergistic effect with enzymes produced in the rumen, therefore affecting the *in sacco* disappearance of CP in addition to improving the overall digestibility of the feed.

Starch digestibility in the alimentary tract of dairy cows
Moharrery, A.[1], Larsen, M.[2] and Weisbjerg, M.R.[2], [1]Agricultural College, Shahrekord University, Animal Science, P.O. Box 115, Shahrekord, Iran, [2]AU Foulum, Aarhus University, Animal Science, P.O. Box 50, 8830 Tjele, Denmark; martin.weisbjerg@agrsci.dk

The aim of the present study was to provide mechanistic equations predicting starch digestibilities in the rumen, small intestine, hind gut and the total tract by conducting a metaanalysis of available data on starch digestion. Data for starch digestion was extracted from 64 *in vivo* experiments with 289 dietary treatments involving lactating dairy cows, hereof 159 dietary treatments with rumen (RSD), 45 with small intestinal (SISD) and 47 with hind gut starch digestibility (HGSD). The major starch sources were corn 170, barley 46, wheat 27, sorghum 27 and other starch sources 19 diets. The metaanalysis of starch digestion was conducted by regression analysis. For groups of starch sources, wheat showed highest and corn lowest starch digestibility in total alimentary tract (TTSD). However, TTSD for barley was more variable than for other sources. Lowest RSD was found for corn and highest for wheat. Across all starch sources, without taking starch source into account, the RSD decreased as the amount of starch consumed increased. Generally, the correlation between RSD (g/kg) and starch intake (kg/d) was highly significant and negative (r=-0.402, P<0.0001). Across all diets, the mean RSD was 650 g/kg. Post rumen, SISD was positively correlated with the RSD, however starch type also significantly affected the SISD. Further, HGSD was positively correlated to the SISD (and RSD), and estimation of HG starch digestibility was not improved by including starch source in the model. This indicates generally that highly digestible starch sources have high digestibilities in all digestive compartments, and that there is limited compensatory starch digestion in the lower tract for starch sources with low ruminal starch digestibility, although some starch sources deviate.

Effect of pre-partum beta-carotene supplementation on post-partum status of Holstein cows
Erasmus, L.J.[1], Machpesh, G.[1] and Nel, F.[2], [1]University of Pretoria, Dept Animal and Wildlife Sciences, Lynnwood Rd, 0001 Pretoria, South Africa, [2]DSM Nutritional Products, Brewery Rd, Isando, South Africa; lourens.erasmus@up.ac.za

The greatest benefit of supplemental β-carotene might be on reproductive performance in cows through the conversion of circulating β-carotene to Vitamin A, specifically within the ovaries and uterus. A general recommendation is that cows be supplemented with β-carotene when blood plasma levels are deficient (<1.5 mg/l) or marginal (<3.5 mg/l) especially during the transition period which is characterized by low DMI and significant losses of β-carotene through colostrum. The objective of this trial was to determine to what extent pre-partum β-carotene supplementation could maintain post-partum plasma β-carotene levels above 3.5 mg /l in cows fed a lucerne hay based TMR. Twenty multiparous Holstein cows were blocked on previous milk production and BW into two experimental groups and fed either 8 kg of a control TMR or the control diet supplemented with 1,200 mg of β-carotene. In addition Eragrostis curvula hay was available ad lib. The experimental period was from 60 d pre-partum until 9 weeks post-partum; β-carotene supplementation was terminated at calving. Blood samples were collected once per week and analysed for plasma β-carotene using a handheld spectro-photometer. Average plasma β-carotene levels pre-partum were higher (6.15 mg/l) for supplemented cows compared to control cows (3.10 mg/l (P<0.05). Overall the average post-partum plasma β-carotene values were 1.50 mg/l for the control cows and 2.43 mg/l for the pre-partum supplemented cows (P>0.05). Supplemented cows maintained sufficient β-carotene levels only for the first 2 weeks post-partum. Results suggest a minor carryover effect of β-carotene after pre-partum supplementation and continuous supplementation of β-carotene is recommended until cows are confirmed pregnant when feeding lucerne hay based diets.

Intestinal digestibility of phosphate from ruminal microbes
Sehested, J., Lund, P. and Jørgensen, H., Aarhus University, Department of Animal Science, P.O. Box 50, Foulum, 8830 Tjele, Denmark; jakob.sehested@agrsci.dk

Data on content and intestinal digestibility of P in ruminal microbes are very sparse. The significant microbial synthesis in lactating dairy cows could cause a significant loss of P in feces depending on digestibility of microbial P. The objective of the present study was to measure the intestinal digestibility of microbial P. Ruminal evacuations was performed on 5 ruminally fistulated and lactating Holstein Friesian dairy cows, and microbes were isolated from liquid and particle fractions. Microbial P content was 1.45%±0.25 of DM. Intestinal digestibility of rumen microbial P and N were estimated using the rat model. Five groups of 5 male Wistar rats (65 g LW) were adapted to the 5 experimental diets for 4 d followed by a balance period of 4 d. Diets were based on rumen microbes, adjusted to 16 g N and either 1.5 or 3.0 mg P kg-1 DM with casein and an N-free mixture and fed at 10 g d-1. The true digestibility of microbial P (43%±4.1), N (69%±0.8) and amino acids (88%±0.5) were estimated by linear regression based on increased inclusion of microbial P in the rat diets. Digestibility of microbial amino acids was very close to the default intestinal digestibility of microbial amino acids (85%) used in the Danish protein evaluation system (AAT/PBV), indicating that the rat model is a suitable model for prediction of intestinal digestibility of rumen microbial matter. The relatively very low digestibility of microbial P indicates that this element is included in other parts of the microbial cell wall and content, e.g. as low digestible phospholipids, compared to N. The very low digestibility of microbial P supports that increased rumen microbial synthesis negatively influence intestinal P digestibility due to a reduction in the ratio between P originating from more highly digestible feed and saliva P and low digestible microbial P. In the present experiment the loss of indigestible microbial P with feces in dairy cows can be estimated to approximately 1.5 g P/kg DMI.

Impact of using fibrolytic enzymes on sheep performance in summer season

El-Bordeny, N.E.[1], El-Sayed, H.M.[1], Abedo, A.A.[2], Hamdy, S.M.[1], Soliman, H.S.[1] and Daoud, E.N.[3], [1]Ain Shams Univ., Animal Production Dept., Cairo, 68 Hadayek Shoubra, Egypt, [2]National Research Center, Animal Production Dept., 12311, Dokki, Egypt, [3]Alex. Copenhagen Company, Meat and Milk production, Noubarya, 12311, Egypt; abedoaa@yahoo.com

The present work aimed to evaluate effect of using fibrolytic enzymes (Fibrozyme) to reduce the effect of heat stress on feed intake, nutrients digestibility and rumen parameters of lambs. An ambient temperature was ranged from 33-36 °C and the relative humidity was ranged from 52-56%. Twelve Barki lambs weighted in average 50 ± 1.5 kg were divided into two similar groups; the first fed without Fibrozyme (control) and the second fed diet supplemented with 10 g Fibrozyme /h/d. Animals was fed on concentrate fed mixture at 2% of their body weight and fed sorghum grass ad lib. as a source of roughage, feeding trial extended to 8 weeks, at the end of trial feces and rumen liquor samples were taken to analysis. The results indicated that roughage and total dry matter intakes were significant ($P\leq0.001$) increased (664.0 and 1,536.9 g, respectively) for lambs fed fibrozyme compared with 534.6 and 1,453.4 g/h/day for control group. Digestibility of crude protein, crude fiber, ether extract, nitrogen free extract and feeding value as total digestible nutrients were significant improved (78.09, 66.27, 82.05, 80.46 and 71.89%) compared with 73.96, 55.90, 75.74, 77.29 and 67.12, respectively for control group, while digestibility of dry matter, organic matter and feeding value as digestible crude protein were insignificant increased. Ruminal pH value was insignificant decreased, while ammonia was significant ($P\leq0.001$) decreased, but total volatile fatty acids concentration was ($P\leq0.001$) increased for lambs fed diet supplemented with fibrozyme in comparing with those fed control diet. From the previous results could be concluded that fibrolytic enzymes improved digestibility, feeding value and rumen parameters.

Effect of oral or injectable selenium in serum selenium and vaccine response of heifers Nellore

Zanetti, M.[1], Turic, E.[2], Morgulis, S.[3], Garcia, S.[1], Reolon, E.[2] and Avino, V.[4], [1]University of Sao Paulo, Animal Science, Av. Duque de Caxias, 225, 13.635-900, Brazil, [2]Virbac Mexico S.A., Sucursal Argentina, Juan Jose Diaz, 629, 1642, Argentina, [3]Minerthal, Animal, Sao Paulo, 12324-000, Brazil, [4]Virbac Brasil, Animal, Eusébio Stevaux, 1368, 04696-000, Brazil; mzanetti@usp.br

The goal of this research was to verify the effect of injectable Se or Se in protein supplement on serum Se levels and vaccine (Bovigen Repro Total Se-Virbac®) response of heifers. Thirty two Nellore heifers were divided into two groups of 16, one group received vaccine with Se and other vaccine without Se. Each group of 16 animals was sub-divided into two groups of 8, one received protein supplement with Se (1.5 mg of Se/kg) and another supplement without Se. The experiment lasted 120 days. Treatments: Vaccine with Se and protein supplement with Se; vaccine without Se and supplement with Se; Vaccine with Se and supplement without Se; vaccine and supplement without Se. The animals were vaccinated with 5 ml of vaccine against multiple viral and bacterial strains (IBR, BVD, PI3, leptospira and vibrios), with (2,000 mµ of Se/ml) or without Se, subcutaneously in day 1 and re-vaccinated on day 21. It was collected blood for analysis of selenium. IgG and Igm on 0, 21, 35, 49, 63, 77, 91 and 120 days. The experimental design was in schema 2×2 factorial (vaccine with or without Se and protein supplement with and without Se). The statistical analysis of data was done using SAS PROC MIXED (2009). As the results, on days 49 and 35, the group that received injectable Se had higher serum Se level ($P<0.05$). When animals ingested Se above NRC (1996), the serum Se levels was higher ($P<0.05$). Se in the vaccine keep the IgM serum levels longer than the control ($P<0.05$). It was concluded that injectable Se had effect on serum Se until 30 days after application, while the Se in the supplement only had effect when the intake increased from 0.5 to 1.8 mg/day. Se in the vaccine had effect on IgM serum levels but not on IgG.

Variation of fatty acids profile in the milk of cows raised on improved Romanian sub alpine pastures
Ropota, M., Voicu, I., Blaj, V.A., Voicu, D. and Ghita, E., National Research Development Institute for Animal Biology and Nutrition, Chemical and Physiological Laboratory, Calea Bucuresti nr.1, 077015 Balotesti, Ilfov, Romania; elena.ghita@ibna.ro

The research on the quality of food staples, such as the milk, intensified lately because of the implication of their consumption on human health. Part of the research focused on the quality changes of the milk (fatty acids profile) from the cows raised on improved sub alpine pastures (1,800 m altitude). Six plots of pastures (A, B, C, D, E and T) with an area of 7,500 m^2 each were used. The pastures were located in Blana BUCEGI area (1,800 m altitude) and were treated in different manners: the pastures were amended, fertilized and treated with sheep waste (D, C), or fertilized and treated with sheep waste (B), or just fertilized (A) or not treated at all (E, T). These pastures were grazed during the grazing season in 2009-2011, each plot having a load of 3 UVM (multiparous Brown cows, at the same stage of lactation). Samples of grass and milk were collected at different moments of the vegetation period: at the beginning (month 6), middle (month 7) and end (month 8), with 4 replicates for each sample. The samples were analysed by gas chromatography (GC) at INCDBNA Balotesti. The laboratory analyses revealed variations in the concentration of the conjugated linoleic acid (CLA) from 0.99% to 1.78%; in the proportion of the unsaturated fatty acids from the total fatty acids, from 26.56% to 40.97%; as well as of the ratio of the saturated fatty acids to the unsaturated fatty acids, from 2.67 to 1.36, in the average milk samples obtained from intensely fertilised pastures compared to the pastures less fertilised, which supports the use of natural feed sources for the animals.

Adding phytase to low phosphorus broiler diet and its effect on performance and phosphorus excretion
Hassan, H.M.A.[1], Abd-Elsamee, M.O.[2], El-Sherbiny, A.E.[2], Samy, A.[1] and Mohamed, M.A.[1], [1]National Research Center, Animal Production, 10 El-Bhoss St, 12622 Dokki, Egypt, [2]Faculty of Agriculture, Cairo Univercity, Animal Production, El- Gamaa St, Giza, Egypt; husseinhma@yahoo.com

Broiler growth experiment was conducted using 360 one-day old Ross 308 chicks to study the effect of reducing dietary phosphorus and phytase supplementation on performance and phosphorus excretion. Two diets were formulated in starting period a control diet contained 0.50% AP and a low P diet contained 0.40% AP. Such low P diet was fed without or with phytase supplementation (500 U/kg). At the growing period every group of birds of the first period was divided into two sub-groups. Two grower diets were formulated, a control diet contained 0.40% AP and a low P diet contained 0.30% AP. The low AP diet was offered with phytase supplementation (500 U/kg). The results showed no significant differences on chick performance among dietary treatments during the starting and growing periods. The best FCR value was recorded for birds fed 0.50% AP diet in the first period then grown on 0.30% AP diet + phytase. The results revealed that dietary AP could be reduced to 0.40% in the starting period and to 0.30% supplemented with phytase in the growing period without adverse affect on broiler performance. Phosphorus excretion decreased more than 20% by reducing AP and/or reducing AP with phytase supplementation. The best results were observed with birds fed 0.40% AP in the starting period and 0.30% AP + phytase in the growing period. It could be concluded that reducing dietary P level and using phytase enzyme could be used to limit quantity of P excreted from broilers without adverse effect upon performance. This could reduce P impact on environmental pollution.

Milk yield and N excretion of dairy cows fed diets with carbohydrates of different fermentescibility
Migliorati, L.[1], Boselli, L.[1], Masoero, F.[2], Capelletti, M.[1], Abeni, F.[1] and Pirlo, G.[1], [1]CRA-FLC, Via Porcellasco 7, 26100 Cremona, Italy, [2]UCSC, ISAN, Via Emilia Parmense 84, 29122 Piacenza, Italy; fabiopalmiro.abeni@entecra.it

The aim of this study was to assess the effect of carbohydrate fermentescibility on milk yield and N excretion. Twenty-four multiparous lactating Italian Friesian cows were divided into 3 groups (one pen per group) according to parity, DIM (103±49) and milk yield (33.5±4) and were fed diets with about 14.7% CP content on DM basis. Treatments were: high fermentable starch with corn and barley hull (HF), low fermentable starch with corn seed (LF), high fermentable starch and NDF high degradability with corn and barley hull and sugar beet pulp (HFNDF). Diets were fed *ad libitum* as TMR and were supplemented with rumen protected DL-Met (Smartamine™) and L-Lys (Relys®). Experimental design was a Latin square (3×3) with 3 treatments and 3 periods of four weeks. The first two weeks were for adaptation whereas the last two were sampling periods. Milk yield and composition were recorded three times a week for every cow, while dry matter intake (DMI) was determined daily per group. DMI was measured daily while BCS at ration shift. Nitrogen excretion was estimated through milk urea content, milk yield and protein content. Effects of treatment and group were determined with ANOVA using the MIXED procedure of SAS. The cow was used as a random effect. No difference was found in milk yield (kg and FCM) and milk N fraction (protein, caseins and urea content) among treatments. Milk fat content and yield were higher in HFNDF than in LF (P<0.10), lactose content was higher in HFNDF than in HF and LF (P<0.05). No difference was found for average DMI and BCS. Average DMI of groups were 192±9 kg/d, average BCS were 2.6±0.43. No differences were found in N excretion among treatments. These results were obtained in the context of 'RENAI' Project financed by MiPAF and were applied to demonstrative dairy farms of 'Life+ AQUA' Project financed by European Commission area Environment.

Influence of nutrients level and probiotic supplementation on performance and egg quality in laying
Han, Y.-K.[1], Jeong, E.-Y.[1] and Na, D.-S.[2], [1]Sungkyunkwan University, Food Science & Biotechnology, 300, Chunchun-dong, Jangan-gu, 440-746 Suwon, Korea, South, [2]SynerBig Co., 27-17, Jeunnong-dong, Dondaemun-gu, 130-020 Seoul, Korea, South; swisshan@paran.com

The effects of probiotics on the performance of layers and egg quality were assessed with 96 Lohmann Brown-Lite laying hens from 25 to 35 wk of age. Layers were grouped randomly into 6 treatments, with 8 replicates per treatment and 2 layers per replicate. The 6 diets consisted of a positive control diet with nutrient levels met or exceeded of Lohmann recommendations, and a negative control diet with all nutrients reduced by 3%, and 2 probiotics (probiotic A composed of *Bacillus coagulance*, *Bacillus lichenformis*, and *Bacillus subtilis*, and Probiotic B composed of Probiotic A and *Clostridium butyricum*) at 105 cfu/kg of feed. There were no significant differences (P>0.05) in hen-day egg production, weight of egg shell, shell percentage and yolk colour between treatments. Feed intake was significantly (P<0.01) lower in the positive control compared with either of the negative control or probiotic treatments. Compared with the negative control and supplementation with probiotic A, positive control and supplementation with probiotic B significantly improved the feed conversion (P<0.01). In contrast, yolk percentage and yolk to albumen ratio were higher (P<0.01) for birds supplemented with probiotic A than other treatments. The weight of egg contents, albumen weight and Haugh units of eggs from fed the probiotic B treatment were superior (P<0.01) to the other treatments. The overall results of this study indicate that probiotic B supplementation in laying hen rations has the potential to save the nutrients, and improve the egg quality.

Effect of linseed on lipidic release and on liver phospholipids fatty acids content of overfed ducks

Rondia, P.[1], Sinnaeve, G.[2], Romnee, J.M.[2], Dehareng, F.[2] and Froidmont, E.[1], [1]Walloon Agricultural Research Centre (CRA-W), Animal Nutrition and Sustainability Unit, rue de Liroux 8, 5030 Gembloux, Belgium, [2]CRA-W, Agricultural product Technology Unit, Chée de Namur, 24, 5030 Gembloux, Belgium; rondia@cra.wallonie.be

The experiment was studying the impact of extruded linseed supplementation during the pre-force-feeding period (PFP, 21 days) and/or during the force-feeding period (FP, 13 days) on lipidic release (LR) and fatty acids (FA) content of liver phospholipids of overfed ducks. We supposed that lipidic release, which is prejudicial to the quality of fatty liver, would be reduced through better elasticity of the hepatic cell membranes consecutive to n-3 PUFA enrichment. 572 Mule ducks were arranged in 4 groups. The control group (CG) received corn grain alone (L0/0) and the experimental groups (EGs) corn grain with: 2% linseed added only during PFP (L2/0), 3% linseed added only during FP (L0/3) or linseed added during PFP and FP (L2/3). The analyses were made on liver of 15 individuals per diet. The phospholipids FA pattern was determined by using HPLC coupled to a light-scattering detector. The LR index was measured using the FIL-IDF method for determination of butter non-fat solid content. Analysis of variance (one-way ANOVA) was used to compare diets using MINITAB software. Liver weight was significantly lower with diets including linseed during FP. There was no difference of LR value between CG and EGs but differences emerged within EGs: L2/3 had a twice higher LR value compared with L2/0 and L0/3. Changes occurred in the phospholipids FA pattern, especially between L2/3 and the others groups. No n-3 ALA was observed with L0/3 or L2/3 but its derivatives were increased. The assumption was not verified because no correlations between LR and n-3 PUFA were found. The L2/0 showed a tendency to a lower LR and had no adverse effect on other criteria. The 3% linseed inclusion during FP was not recommended because of its negative effect on liver weight. Moreover, linseed added during PFP and FP tended to increase the LR.

Effects of bacterial inoculants on nutrient composition and fermentation parameters in corn silage

Homolka, P.[1], Jalč, D.[2], Lauková, A.[2], Simonová, M.[2], Váradyová, Z.[2] and Jančík, F.[1], [1]Institute of Animal Science, Prague 10 Uhříněves, 10400, Czech Republic, [2]Institute of Animal Physiology, Slovak Academy of Sciences, Košice, 04001, Slovakia; homolka.petr@vuzv.cz

The effect of three microbial inoculants (*Lactobacillus plantarum* – LP CCM 4000, *L. Fermentum* – LF2, and *Enterococcus faecium* – EF CCM 4231) on the nutritive value and fermentation parameters of corn silage was studied under laboratory conditions. The whole corn plants (288.3 g/kg of dry matter) were cut and ensiled at 21 °C for 105 days. Uninoculated silage served as control. After inoculation, the chopped corn was ensiled in 20 plastic jars (1 litre) and 20 glass jars (0.7 litre) equipped with lid and divided into four groups. All corn silages had a low pH (below 3.55) and 83-85% of total silage acids represented lactic acid after 105 days of ensiling. The microbial inoculants in inoculated corn silages had a certain effect on corn silage characteristics in terms of significantly ($P<0.05$-0.001) higher pH, numerically lower crude protein content and ratio of lactic to acetic acid than in control silage. But, the inoculants did not affect the concentration of total silage acids (acetic, propionic, lactic acids), as well as *in vitro* digestibility of dry matter of corn silages. All three inoculants significantly (EF CCM4231, LP CCM4000) or numerically (LF2) lowered n-6/n-3 ratio of fatty acids in inoculated corn silages compared to control. From the three used inoculants, mainly EF CCM4231 and LP CCM4000 significantly decreased the concentration oleic acid – C18:1, and significantly increased the concentration of C18:2 and C18:3, respectively. (The study was supported by The Ministry of Agriculture of the Czech Republic – MZE 0002701404).

Chicken meat production in Turkey
Sarica, M. and Yamak, U.S., Ondokuz Mayis University Agricultural Faculty, Animal Science, Ondokuz Mayis University Agricultural Faculty Department of Animal Science, 55139 Atakum/Samsun, Turkey; usyamak@omu.edu.tr

Chicken production in the World has increased since 1940s. Meat and egg production has become different production branches in poultry sector. Today, chicken meat has a value of 30% in total meat production in the World. Also, pig meat has an important role in the total meat production in the World. Chicken meat production has become a big sector in Turkey. This rate is over 50% in Turkey because there is no pig meat consumption. There are about 334 breeder and hatchery companies with a capacity of 1,657 houses. Total number of the chicken meat producing farms is 8,908. These farms have 11,623 poultry houses with a total 131,533,719 broiler capacity. In 2010, total chicken meat production of Turkey was about 1.43 million tones. Total production in EU-27 countries was about 11.7 million tones. Top two chicken meat producers of European Union Countries were France and United Kingdom (1,712 and 1,570 tones). Turkey is one of the top three chicken meat producers in Europe. Also, Turkey is in top ten producers of the World. Turkey exported 104.106 tones of chicken meat in 2010. This foreign trade had a value of 167.5 million USD. Export had mostly done to neighbour countries such as Iraq, Georgia, Azerbaijan, Middle East Countries and some Middle Asia Countries. 2011 values are higher than last years. Besides, all of the chickens used in broiler production were imported from international commercial broiler breeding companies. In 2010, 7,500,000 parents were imported and 900,000,000 chicks were hatched from these parents. Also, total chicken meat consumption in a year per capita is 19.13 kg in Turkey. This value is lower than European Union and developed countries. Turkey has to increase the total consumption of chicken meat. Total production and exporting also have to be increased with newly added companies to the sector.

Effect of Jerusalem artichoke dry form on gut morphology and microbiology of broiler chichens
Zitare, I., Valdovska, A., Krastina, V., Pilmane, M., Jemeljanovs, A., Konosonoka, I.H. and Proskina, L., Latvia University of Agriculture, Research Institute of Biotechnology and Veterinary Medicine Sigra, Instituta street 1, 2150 Sigulda, Latvia; biolab.sigra@lis.lv

A study was conducted to test the effect of Jerusalem artichoke (JA) in dry form on digestive tract. The trial involved 300 cross ROSS 308 broiler chickens divided in three groups: 100 birds in each group. To all chickens' groups were fed out commercial basic feed, nutrition value conforms cross ROSS 308 regalement. Feed ration of the first group chickens was supplemented with additive – JA in dry form in 0.5% concentration, the second group – with the same JA dry form in 1% concentration, the third group – without supplement. Two birds from each group were slaughtered at trial beginning in 7th, further 28th and 42nd days of age. Tissue samples from the duodenum and ileum of each chicken were histo-morphologically examined to determine the villi heights, thickness of lamina propria and epithelium, crypt depths. Content of ileum were bacteriologically tested for counts of Lactobacillus spp. and Enterobacteriaceae. Supplement of JA dry form inclusion in basic feed did not change the broiler chicken performance in first and second groups compared with control. An increase ($U=15<U_{0.01;10;10}=19$) in villi height was observed in duodenum, small intestine of birds at 28th day of age to witch was given 1% supplement of JA, in comparison to those given basic feed at the same time, and inflammation was observed in villi mucosa of ileum of these chickens starting from 28th day of age and continued to till 42th day. Compared to the control diet and 1.0% JA concentration, 0.5% JA concentration significantly ($P<0.05$) increased the number of lactic acid bacteria in the ileum. These results confirm that feed supplemented with JA in dry form in 0.5% concentration had not an effect on microstructure and mucosa of small intestine, but had good effect on lactic acid bacteria in comparison with additive in 1% concentration.

The effects of some essential oils on *in vitro* gas production of different feedstuffs
Boga, M.[1], Kilic, U.[2] and Gorgulu, M.[3], [1]University of Nigde, Crop and Animal Production, Bor Vocational School, 51700 Nigde, Turkey, [2]University of Ondokuz Mayis, Department of Animal Science, Faculty of Agriculture, 55139 Samsun, Turkey, [3]University of Cukurova, Department of Animal Science, Faculty of Agriculture, 01130 Adana, Turkey; mboga@nigde.edu.tr

The aim of this study was to determine the effect of essential oils of oleaster; OLE (*Eleagnus angustifolia*), mint;MNT (*Mentha longifolia*), rosemary; ROS (*Rosmarinus officinalis*), coriander; COR (*Coriandrum sativum*), grape seed; GRA (*Vitis vinifera*), orange peel; ORA (*Citrus cinensis*), and fennel; FEN (*Foenicum vulgare*) on *in vitro* gas production and gas production kinetics of barley, wheat straw and soybean meal. *In vitro* gas productions and gas production kinetics were determined in *in vitro* gas production technique by supplying rumen liquor from three infertile Holstein cows. The study was carried out in a completely randomised design in 7 (essential oil) × 4 (dose) factorial arrangement. Barley, soybean meal and wheat straw were incubated for 3, 6, 9, 12, 24, 48, 72 and 96 h. 0, 50, 100, 150 ppm doses were tested for all essential oils. OLE and FEN were lower than control groups in terms of *in vitro* 24h gas production of barley ($P<0.01$). MNT was higher and FEN was lower than control group in terms of *in vitro* 24h gas production of soybean meal ($P<0.01$). While *in vitro* 24 h gas production of wheat straw were found lower for OLE and FEN groups compared to the control group ($P<0.01$). In conclusion, the additions of FEN, OLE-100, OLE-150, COR-150 and GRA-150 ruduced *in vitro* gas production pattern of barley, soybean meal and wheat straw feedstuffs compared to control group for all the incubation periods ($P<0.01$).These EO and theirs different doses or combinations can be used to improve performance and to manipulate rumen fermentation in ruminant diets.

Intake and digestibility of nutrients in steers fed sugarcane ensiled with levels of calcium oxide
Chizzotti, F.H.M.[1], Pereira, O.G.[1], Valadares Filho, S.C.[1], Chizzotti, M.L.[2] and Rodrigues, R.T.S.[3], [1]Federal university of Vicosa, Animal science, UFV, 36571 Vicosa, MG, Brazil, [2]Federal university of Lavras, Animal science, UFLA, 37200 Lavras, MG, Brazil, [3]Federal university of Vale do São Francisco, Animal science, Univasf, 56330 Petrolina, PE, Brazil; fernanda.chizzotti@ufv.br

A trial was conducted to evaluate the effects of calcium oxide levels (CO) as additive for sugarcane silage on intake, nutrient digestibility, and ruminal characteristics. Four Nellore steers (184± 10.2 kg of BW), fitted with ruminal cannulas were used in a 4×4 Latin square design. Diets consisted of 50% roughage and 50% concentrate, formulated to be isonitrogenous (12% CP, DM basis). The four treatments consisted of sugarcane ensiled with four CO levels (0, 0.5, 1.0, and 1.5%, as fed basis). The experiment lasted 72 days (4 periods of 18 d). Each period had 10 days for diet adaptation. DMI was measured daily and individually. Indigestible ADF was used as an internal marker to estimate apparent nutrient digestibility and fecal output. Ruminal contents were obtained at 0, 2, 4, and 6 h after the morning feeding on day 17 of each period. A mixed model was used. The ruminal characteristics data collected over time were analyzed as repeated measures. There were no effects ($P>0.05$) of CO levels on intake of all nutrients. In addition, there were no effects ($P>0.05$) of CO levels on apparent total digestibility of DM, OM, CP, EE, and NDF. Ruminal pH values and ammonia concentration also were not affected ($P>0.05$) by CO levels. The addition of calcium oxide in sugarcane ensilage does not improve sugarcane silage intake and digestibility of nutrients in steers. Sponsored by CNPq and Fapemig, Brazil.

Genetically modified maize as rabbit feed ingredient

Chrenková, M.[1], Chrastinová, Ľ.[1], Formelová, Z.[1], Poláčiková, M.[1], Lauková, A.[2], Ondruška, Ľ.[1] and Szabóová, R.[2], [1]Animal Production Research Centre Nitra, Hlohovecká 2, 951 41 Lužianky, Slovakia, [2]Institute of Animal Physiology SAS, Šoltésovej 4-6, 04 001 Košice, Slovakia; chrenkova@cvzv.sk

Live weight growth, feed conversion and health of rabbits after feeding complete feed mixtures with 12% proportion of Bt (MON 88017 × MON 89034), isogenic maize and reference maize (LG 3475) was tested on 90 broiler rabbits (Hycola). GM maize contains the different transgenic traits: Bt toxins against Lepidoptera (butterflies and moths), Bt toxin against Coleoptera and glyphosate tolerance. The testing period lasted from weaning at day 35 to day 77 of animals' life. The trial ended with slaughter (LW≈2,500 g). The average daily weight gains were 41.9g and 37.6g resp., with a daily feed consumption 129.6 g and 130.6 g, which gives a feed conversion rate 3.43 and 3.49 resp. The results of chemical analysis MLD muscles at the age of 11 weeks showed the high content of total proteins (22.17-22.43 g/100 g) and very favourable content of intramuscular fat that did not exceed 0.93-1.37 g/100 g. Energetic value does not overstep the value of 410.9-422.8 kJ/100 g MLD. We calculated the index of unsaturation of fat (proportion of unsaturated and saturated fatty acids) I_1=1.09 and index of nutritional value of fat (proportion of essential and saturated fatty acids) I_2=0.42 to characterize the intramuscular fat quality. Data were analysed using one-way ANOVA. Differences between the groups were determined by t-test. Obtained results demonstrate minimal, statistically nonsignificant differences in individual nutrient digestibility in tested mixtures and performance, physical and chemical characteristics of meat in MLD muscle substance, and caecal fermentation pattern of rabbits. GM maize deteriorated neither the health in animals nor the production of animal proteins valuable for human nutrition compared with conventional maize. This article was written during realization of the project'BELNUZ č. 26220120052' supported by the OPRaD funded from the ERDF.

The effect of α- and γ-tocopherol on lipid oxidation and lipid stability of meat in broiler chickens

Tomažin, U., Frankič, T., Voljč, M., Levart, A. and Salobir, J., University of Ljubljana, Biotechnical Faculty, Department of Animal Science, Chair of Nutrition, Groblje 3, 1230 Domžale, Slovenia; urska.tomazin@bf.uni-lj.si

The objective of our study was to evaluate the influence of α- and γ-tocopherol alone and in combination on oxidative stress *in vivo* and on stability of meat in broiler chickens. Forty-six one day old broilers were fed five different experimental diets for 30 days. Negative control (Cont-, n=10) received 5% of palm fat and positive control (Cont+, n=10) 5% of linseed oil. Three groups were given 5% of linseed oil plus 67 mg of either RRR-α-tocopherol (α, n=10), RRR-γ-tocopherol (γ, n=8) or their combination: 33.5 mg RRR-α-tocopherol + 33.5 mg RRR-γ-tocopherol (α+γ, n=8). Lymphocyte DNA damage was evaluated by Comet Assay, plasma malondialdehyde (MDA), ferric reducing capacity (FRAP) and antioxidant capacity of lipid soluble compounds (ACL), and concentrations of both tocopherols in plasma, breast and thigh muscles were measured. Lipid stability of meat was evaluated by measuring MDA in both muscles stored under different conditions (fresh, 6 days at 4 °C, 2 and 4 months at -20 °C). Feeding linseed oil increased MDA values in plasma and muscles, but had no effect on lymphocyte DNA damage, ACL and FRAP. All three vitamin E supplemented groups had lower DNA damage than Cont+, while MDA, FRAP and ACL were reduced only in groups α and α+γ. The retention of α-tocopherol in plasma and both muscles reflected the supplementing regimen. Concentrations of γ-tocopherol were much lower and only slightly elevated in groups γ and α+γ. As considers lipid stability of breast muscle, it was improved in groups α and α+γ, while in thigh only α-tocopherol was efficient. These results are indicating that high levels of PUFA in diet are the reason for susceptibility to lipid oxidation, which can be prevented by feeding high levels of α-tocopherol, but not by feeding the same amount of γ-tocopherol. Feeding the combination is as effective as feeding α-tocopherol alone, despite that only half of dose of α-tocopherol was used.

The effect of preparation 'BIOPOLYM' on fermentation processes of red clover silages
Kubát, V.[1,2], Petrášková, E.[2], Jančík, F.[1], Čermák, B.[2], Hnisová, J.[2], Homolka, P.[1], Lád, F.[2] and Kohoutová, H.[3], [1]Institute of Animal Science, Department of Nutrition and Feeding of Farm Animals, Pratelstvi 815, 104 00 Praha Uhrineves, Czech Republic, [2]University of South Bohemia, Faculty of Agriculture, Department of Genetics, Breeding and Nutrition, Studentska 13, 370 05 Ceske Budejovice, Czech Republic, [3]Masaryk University, Faculty of Economics and Administration, Department of Finance, Lipova 507/41a, 602 00 Brno, Czech Republic; kubat.vaclav@vuzv.cz

The aim of this study was to evaluate the effect of seaweed BIOPOLYM (B) and bacterial preparation BIOSTABIL (BIOS) for qualitative parameters of silage from wilted red clover under laboratory conditions. The set of tested red clover silages from the first cut (25-06-2010) was divided into 7 groups; each of them included 4 samples. First group of the samples was untreated (C I). The second group of silage samples was treated using BIOS (BIOS I) preparation. The third group contained the silage samples treated with combination of BIOS and B 1 : 100 (BIOS + B 100 I) preparations. The fourth group was treated using B 1 : 50 (B 50 I), fifth group B 1 : 100 (B 100 I), sixth group B 1 : 200 (B 200 I) and seventh B 1 : 500 (B 500 I). The distribution of samples was similar for the second cut (7-09-2010) as for the first cut: C II, BIOS II, B 100 II, B 200 II and B 500 II. Group of the samples BIOS + B 10 II and B 10 II were treated with B in concentration 1 :10. In the first cut the group C I and B 100 I positively affected the quality of silage. The highest values of lactic acid showed the groups C I and B 100 I (both in average 34.73 g/kg). The lower pH values (P<0.01) was found in group B 100 I compared with groups BIOS I and BIOS + B 100 I. In the second cut the best results were achieved in the group BIOS II. Very high statistical significance (P<0.001) was found for pH in group BIOS II compared with B 10 II, B 100 II, B 200 II and B 500 II. The lowest value of pH showed the group BIOS II (4.26). This study was supported by project NAZV no. QH 92252 and MZE 0002701404.

Recycling of phosphate is not affected by particle size or chewing time in lactating dairy cows
Sehested, J., Storm, A.C. and Kristensen, N.B., Aarhus University, Department of Animal Science, P.O. Box 50, Foulum, 8830 Tjele, Denmark; jakob.sehested@agrsci.dk

Quantitative data on P recycling in lactating dairy cows are very sparse. Recently we showed that recycling of P was not affected by a reduced dietary P intake. The primary route of P recycling is through saliva secretion and increasing chewing time may therefore increase P recycling. The objective of the present study was to quantify the effect of feed particle size on chewing time and recycling of P. Four Danish Holstein cows (121±17 d in milk, 591±24 kg of BW) surgically fitted with a ruminal cannula and permanent indwelling catheters in the major splanchnic blood vessels were used. Recycling of inorganic phosphate Pi (mmol/d) was determined as the net portal flux of Pi – absorbed P, where absorbed P = feed P – feces P. The experimental design was a 4 × 4 Latin square with a 2×2 factorial design of treatments. Treatments differed in forage (grass hay) particle size (FPS; LONG: 30 mm and SHORT: 3.0 mm) and feed dry matter (DM) content of the TMR. Only the effects of FPS are reported here. Feed P concentration was 3.5 g P/kg DM. Cows were restrictively offered 20 kg of DM/d to avoid feed refusal, and feed intake and milk yield were not affected by FPS. Decreasing the FPS decreased the overall chewing time (P<0.01) and rumination time (P=0.02) by 151±55 and 135±29 min/d, respectively, and chewing time was 28.0±2.1 min/kg of DM when fed short particles. No effect of the reduced chewing time was observed on ruminal pH or milk fat percentage. No effect of the FPS (LONG vs SHORT) was observed on P intake (71.4 g/d vs 73.2 g/d), digestibility of NDF (45% vs 41%, SEM=2.9, P=0.17) or P (41% vs 46%, SEM=2.2, P=0.14), on net Pi recycling (42 g/d vs 43 g/d, SEM=12.4, P=0.98) or on P balance (4.8 g/d vs 7.9 g/d, SEM=2.1, P=0.32). Data are not consistent with the concept of salivary Pi secretion being significantly affected by diet particle size or chewing time in lactating dairy cows.

Using probiotic, prebiotic or organic acids on feeding laying hens

Youssef, A.W., Hassan, H.M.A. and Mohamed, M.A., National Research Centre, Animal Production, Tahrir st., 12622 Dokki Cairo, Egypt; amani_wagih@yahoo.com

A layer experiment was conducted to study the effect of supplemented layer diet with commercial probiotics, prebiotic or organic acids on performance, egg quality. Two hundred and forty 27 WK-old HN Brown layers were randomly assigned to 5 groups (6 replicate × 8 layers, each). A control diet was fed without or supplemented with either probiotic 1(Clostat, a strain of Bacillus subtilis); probiotic 2 (Protexin, mixture of microorganisms); prebiotic (Mos, mannan oligosacchride) or organic acid (Galliacid, mixture of coated organic acids). The experiment lasted 11 weeks. The results showed that dietary supplementation improved egg production. Such improvement was not significant with probiotics or prebiotic and significant (P<0.05) with organic acid compared to the control. Egg weight was not affected by dietary treatments. Supplementation of probiotics, prebiotic or organic acids significantly (P<0.05) increased egg mass and feed conversion ratio compared with control. Addition of probiotics, prebiotic or organic acids enhanced egg quality measurements: shell thickness and yolk color (P<0.05) and shape index, yolk%, SWUSA, haugh unit and specific gravity (P>0.05). It could be concluded that, addition of probiotics, prebiotic or organic acids to layer diets improved performance and egg quality.

Effect of feed allocation of daily ration on growth and feed assimilation of rainbow trout O mykiss

Papoutsoglou, E.S., Alexandridou, M., Karakatsouli, N. and Papoutsoglou, S.E., Agricultural University of Athens, Animal Science and Aquaculture, Iera Odos 75 Votanikos, 118 55 Athens, Greece; stratospap@aua.gr

The continuous quest for minimizing cost in aquaculture inevitably leads to issues relating to feed and feeding cost, always aiming to ensure, as animal scientists, the best possible feed quality for farmed animals. In this context, there was an investigation regarding the effect of feed allocation of the same daily ration on growth and feed assimilation of juvenile rainbow trout Oncorhynchus mykiss (initial weight 64.1±0.64 g) using standard commercial rainbow trout feed and recirculating system. Feed was allocated in three daily meals by means of five different duplicated treatments (70-15-15%, 50-25-25%, 33-33-33%, 25-25-50% and 15-15-70% of 2.0-2.5% body weight) for 14 weeks. Digestive enzyme analyses were performed (total protease, total carbohydrase) separately in stomach, pyloric caeca and intestine. Results were expressed as activity per g tissue, capacity (activity × tissue weight) or activity per mg protein. Results indicate that treatment 25-25-50 led to significantly increased growth (242.9±3.67 g) compared to groups 15-15-70 (223.6±3.99 g) and 50-25-25 (227.2±5.00 g), with treatments 33-33-33 and 70-15-15 having intermediate levels. Present results demonstrate that feed allocation of daily ration affects growth significantly in juvenile rainbow trout. Therefore, growth and weight gain may be improved using the same feed energy possibly by means of improved assimilation under favourable and controlled rearing conditions.

Effects of mannan oligosaccharide on performance and carcass characteristics of broiler chickens

Karkoodi, K.[1], Solati, A.A.[2] and Mahmoodi, Z.[1], [1]Department of Animal Science, Saveh Branch, Islamic Azad University, Saveh, Iran, [2]Department of Veterinary Science, Saveh Branch, Islamic Azad University, Saveh, Iran; karkoodi@yahoo.com

In the present experiment, the influence of Mannan Oligosaccharide (Bio-Mos) from the cell wall of Saccharomyces cerevisiae was investigated on performance and carcass characteristics of broiler chickens. Eighty day-old Ross 308 male broiler chicks were randomly allocated in two dietary treatments (control diet and control plus 2 g/kg Bio-Mos diet) in 4 replicates and ten subreplicates. Supplementation of diets with Bio-Mos significantly decreased feed intake (P<0.05). A significant Increase in body weight and improved feed conversion ratio were observed in Bio-Mos treatment (P<0.05). Results of carcass characteristics showed that the breast, jejunum and ileum weights, duodenum and jujenum length and carcass yield were significantly increased in diets containing Bio-Mos (P<0.05). The liver, heart and gizzard weights and also ileum length were significantly lower in Bio-Mos group (P<0.05). Thigh, abdominal fat and duodenum weights were not significantly affected by addition of Bio-Mos to diets (P>0.05). Result showed that supplementation with Bio-Mos had positive effects on growth performance and some carcass characteristics of broiler chickens.

Genomic selection for feed efficiency in dairy cattle: a complex objective

Pryce, J.E.[1], De Haas, Y.[2], Hayes, B.J.[1], Coffey, M.P.[3] and Veerkamp, R.F.[2], [1]Department of Primary Industries, Biosciences Research Division, 1 Park Drive, Bundoora, 3083, Australia, [2]Wageningen UR Livestock Research, Animal Breeding and Genomics Centre, P.O. Box 65, 8200 AB Lelystad, Netherlands, [3]Scottish Agricultural College, Sustainable Livestock Systems Group, Easter Bush, EH25 9RG, United Kingdom; jennie.pryce@dpi.vic.gov.au

Feed efficiency and energy balance complex are attractive traits to genetically improve using genomic selection. This is because the feed intake data required to calculate these traits is expensive and difficult to measure, so can only be recorded on a sample of the population, usually in research conditions. Thus animals with phenotypes and genotypes could become a reference population to generate the equations required for genomic prediction. Feed efficiency can be defined in a variety of ways including gross efficiency, which is the ratio of product (such as yield) per unit of feed eaten and metabolic efficiency or residual feed intake (RFI). RFI is the feed an animal consumes adjusted for the predicted energy requirements for maintenance, production and body condition score change. To be able to select for RFI we need to be able to identify cows that are truly efficient, rather than those that appear to cheat by mobilising body reserves. A relevant question is whether it is possible to make this distinction. One approach is to consider RFI in growing heifers, provided this correlates with lactating RFI. Using data on 2,000 heifers from Australia and New Zealand, accuracies of genomic prediction of around 0.4 are achievable. We have also shown that accuracies can be increased by including phenotypes from other countries. Using data from Australia, the Netherlands and the UK an improvement of up to 5.5% in accuracy of genomic prediction was observed. However, the number of high quality phenotypes still limits achieving acceptable accuracies for bull proofs. To address this, a major international collaboration to assemble dry matter intake data on more than 5,000 cows with high quality phenotypes and genotypes has started.

Methane emissions and rumen fermentation in heifers differing in phenotypic residual feed intake
Fitzsimons, C.[1], Kenny, D.[1], Deighton, M.[2], Fahey, A.[3] and Mcgee, M.[1], [1]Teagasc, Animal and Grassland Research and Innovation Centre, Grange, Dunsany, Co. Meath, Ireland, [2]Teagasc, Moorepark, Fermoy, Co. Cork, Ireland, [3]University College Dublin, School of Agriculture and Food Science, Belfield, Dublin 4, Ireland; claire.fitzsimons@teagasc.ie

Selection of feed efficient animals by way of improved residual feed intake (RFI) has been suggested as a mitigation strategy against enteric methane production. This study aimed to characterise methane emissions, rumen fermentation and total tract digestibility in beef heifers differing in phenotypic RFI. Individual dry matter intake (DMI) and growth were measured in 22 heifers, [(initial live weight 455 kg (SD=17.1)] offered grass silage (dry matter digestibility = 766 g/kg) *ad libitum* for 120 days. Rumen fermentation (transesophageal samples) and total tract digestibility (indigestible marker) were measured. Methane production was estimated using the sulphur hexafluoride tracer gas technique over two 5-day periods during weeks 3 and 11. Residuals of the regression of DMI on ADG and mid-test $BW^{0.75}$, using all animals, were used to compute individual RFI coefficients. Animals were ranked by RFI into high (inefficient), medium and low groups. Statistical analysis was carried out using the MIXED procedure of SAS. Overall ADG and daily DMI were 0.6 kg (SD=0.07) and 7.8 kg (SD=0.24), respectively. High RFI heifers consumed 9 and 14% more (P<0.05) than medium and low RFI heifers, respectively. Rumen propionic acid and acetate:propionate ratio tended (P<0.10) to be higher in high compared to low RFI heifers. Rumen pH, other fermentation variables and total tract digestibility did not differ (P>0.10) between RFI groups. Methane production (g/d) was lower in periods 1 (P>0.10) and 2 (P<0.05) for low than high RFI heifers, with medium RFI animals being intermediate. Methane production per kg DMI was similar (P>0.05) for the RFI groups. Results indicate that selection for RFI will reduce methane emissions without affecting productivity of growing beef heifers.

Effect of phenotypic residual feed intake in beef suckler cows
Lawrence, P.[1], Kenny, D.[1], Earley, B.[1] and Mcgee, M.[2], [1]Teagasc, Animal and Bioscience Research Department, Animal & Grassland Research and Innovation Centre, Teagasc, Grange, Dunsany, Co. Meath, Ireland, [2]Teagasc, Livestock Systems Research Department, Animal & Grassland Research and Innovation Centre, Teagasc, Grange, Dunsany, Co. Meath, Ireland; peter.lawrence@teagasc.ie

Improving feed efficiency is central to improving economically sustainable beef production systems. This study aimed to quantify the phenotypic variation in residual feed intake (RFI) and productivity of beef suckler cows (n=39) offered grass silage during pregnancy (indoors) and, to determine the effect of RFI classification on herbage intake during lactation (grazing). The residuals of the regression of individual silage dry matter intake (DMI) (73 d) on conceptus adjusted, average daily gain (ADG) and mean body weight0.75 (BW), using all cows, were used to compute individual RFI coefficients. Cows were ranked by RFI into high (inefficient), medium and low (efficient) groups. Cow BW, body condition score (BCS), skeletal measurements, ultrasonic fat and muscle depth, muscularity score, rumen fermentation, total tract digestibility, blood metabolites and haematology, feeding behaviour and, grazed herbage intake (n-alkanes) were measured. Data were analysed by ANOVA using the MIXED procedure of SAS. The RFI groups did not differ (P>0.05) in BW or ADG. Residual feed intake was positively correlated with silage DMI (r=0.87; P<0.001) but not with grass DMI (r=-0.03; P>0.05). Low RFI cows consumed 14% (P<0.05) and 22% (P<0.001) less silage than medium and high RFI groups, respectively. High and low RFI groups did not differ (P>0.05) in BCS, ultrasonic fat or muscle depth, muscularity score, total tract digestibility, milk yield, calving difficulty score and, calf birth weight and growth rate. Low RFI cows were more biologically and economically efficient than high RFI cows, as they consumed less feed during the indoor winter period and a similar amount of herbage during the grazing season, without any compromise in economically important traits measured.

Effects of cow temperament and maternal defensiveness on calf vigour and ADG

Turner, S.P., Jack, M.C., Stephens, K. and Lawrence, A.B., Scottish Agricultural College, Edinburgh, EH9 3JG, United Kingdom; simon.turner@sac.ac.uk

Improving responses to handling (temperament) in young cattle may lead to correlated changes in cow temperament. However, little is known about how pre-calving temperament relates to post-calving defensiveness and how these affect maternal behaviour and calf vigour. Temperament of pregnant cows (n=452) was studied on 2 farms during handling (2-4 occasions; 6 point scale of behaviour in the weigh crate, automated measurement of speed of exit from the crate, 6 point scale of response to isolation). Maternal defensiveness during calf handling at 1 (farm 1) or 4 (farm 2) days of age was scored using a 6 point scale of cow position and a 4 point scale of cow behaviour towards the handler. Cows at farm 2 were studied for 2 parities. Cow interactions with the calf and attainment of behavioural milestones indicative of vigour were recorded from videos by continuous observation for 3h after birth (59 calves, farm 1). REML analyses used cow breed and age and calf sex and birth weight as fixed effects and pen, calf sire and year as random effects. Cows were consistent between measurements within a parity in their pre-calving temperament (repeatability 0.33-0.49, $P<0.001$). Repeatability across parities was seen for behaviour in the crate (0.50, $P<0.001$) and post-calving defensiveness (0.31-0.71; $P<0.001$). At farm 2, cows that were restless in the crate showed more vigourous movement during calf handling ($P<0.001$), but otherwise pre-calving temperament and post-calving defensiveness were unrelated at either farm. At farm 1, but not farm 2, cows that exited the crate rapidly had calves with a lower birth weight ($P<0.05$) and those which were restless in the crate and in isolation had calves with a lower ADG to weaning ($P<0.1-<0.05$). Pre-calving temperament and post-calving defensiveness did not affect behaviour towards the calf or calf vigour. Improving cow pre-calving temperament may benefit calf ADG but is unlikely to affect maternal behaviour, calf vigour or safety when handling defensive cows.

Pooling data on energy balance in dairy cows for genetic and genomic analyses

Wall, E.[1], Banos, G.[2], Veerkamp, R.F.[3], Mcparland, S.[4] and Coffey, M.P.[1], [1]SAC, Animal and Veterinary Sciences, Easter Bush, Edinburgh, EH25 9RG, United Kingdom, [2] Aristotle University of Thessaloniki, Department of Animal Production, Faculty of Veterinary Medicine, Thessaloniki, 54124, Greece, [3]Wageningen UR Livestock Research, Animal Breeding & Genomics Centre, P.O. Box 65, 8200 AB Lelystad, Netherlands, [4]Teagasc, Animal & Grassland Research & Innovation Centre, Moorepark, Cork, Ireland; eileen.wall@sac.ac.uk

Genetic (and genomic) studies for difficult to record traits (e.g., feed intake, energy balance) have been hampered by limited data availability. This study merged data from dairy experimental resources and used statistical models to derive harmonised energy balance profiles for further genetic analyses. Data from four experimental herds located in three countries (Scotland, Ireland and the Netherlands) were used. Two datasets (DS) were created – DS1 contained records on 4,708 first lactation Holstein-Friesian cows and DS2 included records on the first 4 lactations of 4,927 cows (12,497 lactations). Weekly records were extracted from the herd databases and included 7 traits: milk, fat and protein yield, milk somatic cell count, live weight, dry matter intake and energy intake. Missing records were predicted with random regression models, so that at the end there were 44 weekly records (per lactation for DS2), corresponding to 305-day lactation, for each cow. Different lactation traits related to milk production, somatic cells, feed intake, and energy balance were derived. Data were merged and analysed with mixed linear models. Genetic variance and heritability estimates were greater ($P<0.05$) than zero. When estimable, the genetic correlation between herds was high suggesting that data from these experimental herds could be merged into a single dataset for genetic and genomic analyses, despite potential differences in management and recording in the four herds. Genomic analysis on the data is currently underway for energy balance traits. Merging experimental data will increase power of genetic analysis and genomic studies, especially of difficult-to-record traits.

Monitoring the transition cow: influence of body condition before calving
Schafberg, R. and Swalve, H.H., Martin-Luther-University Halle-Wittenberg, Institute of Agricultural and Nutritional Sciences, Theodor-Lieser-Str. 11, 06120 Halle, Germany; renate.schafberg@landw.uni-halle.de

The transition period is a critical time for the well-being and survival of a dairy cow. It is well known that involuntary culling mostly occurs in early lactation. Aim of the present study was to analyze the influence of body condition scores (BCS) before calving on the culling rate in early lactation, i.e. within the first 50 or 100 days in milk. Data used for this study was collected in 2010/2011 on seven large dairy farms in Eastern Germany and included 5,853 calvings of Holstein heifers (n=1,826) and cows (n=4,027). Body condition before calving was scored by experienced classifiers around day 42 before calving. Test day milk yields were merged with calving and BCS data and fat-to-protein-ratio (FPR) was calculated as the ratio of fat and protein content of the first regular test day, i.e. on average at 23 days in milk. Data was analyzed using a threshold model for survival up to day 50 or day 100 as the dependent binary trait and employing a probit link function. As main fixed effects, herd and an effect that was a combination of BCS (three classes; <3, 3-4, >4) and FPR (three classes; <1.1., 1.1-1.5, >1.5) were included in the model. Sire was fitted as random effect. Heifer data and cow data was analyzed separately. The results firstly show that the influence of FPR and BCS is much more drastic in cows than it is in heifers. Secondly, heifers or cows seem to be able to cope with metabolic imbalances as long as they were not over-conditioned before calving. This finding can also be interpreted such that an over-conditioning before calving leads to metabolic imbalances and hence to increases in culling rates. For cows, culling rates up to day 50 of the lactation within the FPR<1.1 class were as high as 35.1%, 28.4%, and 54.2% for BCS scores <3, 3-4 and >4. Within the FPR>1.5 group, also a slight increase of culling rates can be observed for over-conditioned cows.

Between and within-individual variation in methane output measurements in dairy cows
Negussie, E., Liinamo, A.-E., Mäntysaari, P., Mäntysaari, E.A. and Lidauer, M., MTT Agrifood Research Finland, Biotechnology and Food Research, Myllytie 1, 31600, Finland; enyew.negussie@mtt.fi

Methane output in dairy cows is an emerging phenotype which could be used as an indicator of feed utilization in ruminants. Furthermore it has also a significant implication on the environmental impact of milk production in dairy cows. Enteric methane from ruminants is thus an important but often difficult source to quantify on an individual basis which is essential for genetic studies. This study evaluated a non-invasive Photoacoustic Infrared Spectroscopy (PAS) technique for quantifying enteric methane output from the breath of individual dairy cows. The study was conducted at MTT experimental dairy herd and included about 40 first-lactation Finnish Ayrshire cows. Feed intake, milk yield, body weight were recorded daily and simultaneously individual cow methane (CH_4), carbon dioxide (CO_2), acetone outputs were measured continuously for over 2 weeks using a multi-point PAS gas analyzer fitted to two feeding kiosks (sampling point 1 & 2). In total, there were 4800 measurements with several records per cow for each of the different gasses. The ratio of CH_4:CO_2 is concentration independent and can be used to quantify methane output in dairy cows and for each cow CH_4:CO_2 ratios were calculated. The data was analysed using the General Linear Models (GLM) procedure and between and within-individual variability were quantified. The fixed effects of milk yield, body weight, and daily intake of concentrates were also estimated. The overall mean of CH_4 and CH_4:CO_2 ratios were 411 ppm, 0.114 and 283 ppm, 0.110 from point1 and 2, respectively. The repeatability of CH_4 and CH_4:CO_2 ratios were 0.56, 0.58 and 0.57, 0.58 for point1 and 2, respectively. Repeatability sets the upper limit to heritability and repeatabilities from this study are in general higher than reports from previous studies. This suggests the suitability of PAS technique for a large scale individual cow CH_4 output measurements that is a prerequisite in genetic studies.

A genome-wide search for harmful recessive haplotypes in Brown Swiss and Fleckvieh cattle

Schwarzenbacher, H.[1], Fuerst, C.[1], Fuerst-Waltl, B.[2] and Dolezal, M.[3], [1]ZuchtData EDV-Dienstleistungen GmbH, Dresdner Str. 89/19, 1200 Vienna, Austria, [2]University of Natural Resources and Life Sciences, Department of Sustainable Agricultural Systems, Gregor-Mendel-Str. 33, 1180 Vienna, Austria, [3]University of Milan, Department of Veterinary Science and Technology for Food Safety, Via Celoria 10, 20133 Milan, Italy; schwarzenbacher@zuchtdata.at

Beagle v3.3.2 was used to phase llumina 50K genotypes from the routine genomic breeding value estimation after comprehensive data edits. Phased haplotypes were corrected for switch errors within half-sib families in a post processing step. Then in a first analysis step employing a sliding window approach we searched for haplotypes that are absent in homozygous status although frequencies and mating scheme suggest that this was highly unlikely. While in Fleckvieh no haplotype met our significance threshold ($P<0.001$ and no homozygous diplotype), we identified a total of four significant haplotypes located on BTA 5, 7 – termed 'BH1' by VanRaden *et al.*, and two on 19 in Brown Swiss. In a second step we analysed the phenotypic effects of these chromosomal segments on the traits non return rate after 56 days, days from first to last insemination and stillbirth rate in heifers and cows, respectively, and survival rate in calves. Effects were estimated using genetic evaluation models including a linear regression effect on the probability that the embryo or calf was homozygous for the haplotype of interest using pedigree information and identified carriers. While we could not confirm phenotypic effects of BH1 on fertility traits, we found highly significant results for a haplotype located on BTA 19 on stillbirth and calf survival rate.

Selection for longevity in Dutch dairy cattle

Van Pelt, M.L., Koenen, E.P.C., Van Der Linde, C., De Jong, G. and Geertsema, H.G., CRV BV, P.O. Box 454, 6800 AL Wageningen, Netherlands; mathijs.van.pelt@crv4all.com

Longevity of dairy cattle is an important trait from an economic and welfare perspective. Farmers want trouble-free cows that can stay in the herd for a long time. Longevity, days between first calving and the last milk recording date, is a complex trait to analyse as farmers cull cows for a variety of reasons. To improve longevity by breeding, a genetic evaluation was introduced in the Netherlands in 1999. This genetic evaluation used to focus on functional longevity and therefore accounted for voluntary culling of cows due to low milk production. From 2008 onwards focus has been on productive lifetime without adjustment for cow's milk production level. As culling data generally become available relatively late in life, accuracy of selection is enhanced by using observations on correlated traits also and, more recently, by using genomics. Since the inclusion of longevity in the Dutch total merit index in 1999, both genetic and phenotypic response increased by 200 days. In the breeding programme of CRV longevity is an important trait. Young waiting bulls show a genetic progress of 300 days in the last four years. The current model (survival analysis) assumes that longevity is the same trait during the entire productive lifetime of a cow. The genetic evaluation of longevity may be improved by using models that take into account that longevity is not the same trait during the lifetime of a cow. Improvement of longevity has been further supported by genetic evaluations for functional traits such as fertility, mastitis and hoof health that have been introduced and/or improved over the recent years. Management information tools that have been developed to monitor longevity at farm level include information on culling reasons, net revenues of lifetime milk production of the cow and comparison with other herds. These tools enable farmers to improve their insight and management decisions on culling cows. In the end the farmer decides which cows are culled.

Genetic variation in milking efficiency: a novel trait for milkability in automatic milking systems
Løvendahl, P.¹, Lassen, J.¹ and Chagunda, M.G.G.², ¹Aarhus University, Dept. Molecular Biology and Genetics, AU-Foulum, 8830 Tjele, Denmark, ²SAC Research, Future Farming Group, King's Buildings, West Mains Road, EH9 3JG, Edinburgh, Scotland, United Kingdom; Peter.Lovendahl@agrsci.dk

Genetic parameters for milking efficiency, milk flow rate, yield, cell count and milking frequency were estimated for cows intensively recorded in a research station. Milking efficiency was calculated at every milking as kg ECM produced per minute a cow occupy the automatic milking station. The study utilised data from Holstein, Red Dane and Jersey cows (n=556) in first parity over a 7 year period. Milk records were filtered to avoid data from disturbed milkings, (i.e. incomplete milkings, and milking following other incomplete milking). After filtering 280,510 undisturbed records remained of which 146,133 also had composition data. The lactation trajectory was divided into 10 segments of 30 days to allow for co-variance components to change during lactation. Covariance components were estimated for genetic, permanent environmental and residual effects using AI-REML and used to calculate heritability and genetic correlations. Heritability of milking efficiency ranged from 0.34 to 0.50 during lactation. Milking efficiency was positively correlated to higher yield per day but either weakly or negatively correlated to flow rate and somatic cell count indicating that higher efficiency could be obtained without concurrent increase in somatic cell count. This study has introduced a new functional trait that could replace flow rate and milkability in describing the efficiency of cows in utilising the milking station.

WebLOAD: a web frontend to create a consistent dataset from multiple text files in animal breeding
Müller, U.¹, Fischer, R.¹, Van Chi Truong, C.², Groeneveld, E.² and Bergfeld, U.¹, ¹Saxon State Office for Environment, Agriculture and Geology (LfULG), Am Park 3, 04886 Köllitsch, Germany, ²Institute of Farm Animal Genetics (FLI), Höltystr. 10, 31535 Neustadt, Germany; ulf.mueller@smul.sachsen.de

Parameter estimates, genetic analyses and genomic selection – all these methods require consistent data and logically correct pedigrees. If datasets from various sources (farms, teststations or breed societies) are merged, different keys or codes for the same indication or different identifications (IDs) for the same animal are very problematic. Purchases and sales of animals resulting in two IDs for the same animal aggregate the problem. Therefore, to obtain one consistent datapool its initial contributing files need to get harmonized. WebLOAD implements this in a generic manner. WebLOAD is a web frontend for the parameterization of the LOAD software providing a set of rules and filters to clean up heterogeneous data from different sources. This includes the proper treatment of multiple animal ID systems as well as harmonization of codes. The pedigree is checked for consistency and loops. Additionally, allels of offspring are checked against their parents for consistency. Apart from the data checking there are various possibilities to recode keys or animal IDs or to mark them as invalid. After completion of the analysis the results can be exported to text files or as a normalized database. Being a web application problem definition and configuration is done via a browser otherwise an editor. The WebLOAD is written in Perl with LOAD using the logic of the APIIS-framework. WebLOAD is freely released under the GPL license. The project is supported by funds of the Federal Ministry of Food, Agriculture and Consumer Protection (BMELV) based on a decision of the Parliament of the Federal Republic of Germany via the Federal Office for Agriculture and Food (BLE) under the innovation support programme.

Use of high density SNP in genomic evaluation in Holstein-Friesians

Schopen, G.C.B. and Schrooten, C., CRV BV, Research, P.O. Box 454, 6800 AL Arnhem, Netherlands; Ghyslaine.Schopen@crv4all.com

The objectives of this study were to reduce the number of HD SNP data to reduce computer requirements and to see the effect on the reliability of genomic breeding values using the reduced HD dataset. DNA was isolated from semen samples of 548 bulls of the Eurogenomics consortium. These bulls were genotyped with the BovineHD Beadchip (777k SNP) and used to impute all genotyped Holstein-Friesian animals from 50k to HD, using the Beagle software package. The final dataset consisted of 30,483 animals and 603,145 SNP genotypes. For each locus, a haplotype score was obtained from Beagle. Three subsets of variable sizes (38,355, 115,690 and 322,360 loci) were made based on deleting obsolete loci (loci that do not give extra information compared to the neighbouring locus). Error of imputation from 50k to HD was assessed by masking genotypes for SNP that are not on the BovineSNP50 in a subset of 60 animals, and impute those based on the remaining HD genotypes. Subsequently, a validation study using the haplotypes was performed to see the effect on reliability of genomic breeding values for nine traits (production, conformation and functional traits). Average imputation error was 0.784%. The dataset of 115,690 loci gave the highest increase (0.10% to 0.60% compared to other subsets) in reliability based on all validation bulls, averaged across all traits. Using only old validation bulls, the increase was 0.90% to 1.70% compared to other subsets, on average. Eliminating obsolete loci will enormously decrease computation time to estimate genomic breeding values based on HD data. The more HD loci used the higher the increase in reliability of genomic breeding value. The reduced dataset resulted in an increase in reliability of genomic EBV of 1 to 2% compared to the routinely used SNP dataset (~38k SNP).

Mid-infrared predictions of fatty acids in bovine milk: final results of the RobustMilk project

Soyeurt, H.[1,2], Mcparland, S.[3], Donagh, D.[3], Wall, E.[4], Coffey, M.[4], Gengler, N.[2], Dehareng, F.[5] and Dardenne, P.[5], [1]National Fund for Scientific Research, Rue d'Egmont 5, 1000 Brussels, Belgium, [2]University of Liège, Gembloux Agro-Bio Tech, Animal Science Unit, Passage des déportés 2, 5030 Gembloux, Belgium, [3]Animal and Grassland Research & Innovation Centre, Fermoy, Cork, Ireland, [4]Scottish Agricultural College, Sustainable Livestock Systems Group, Bush Estate, Penicuik, Midlothian, United Kingdom, [5]Agricultural Walloon Research Centre, Valorisation of Agricultural Products Department, Chaussée de Namur, 24, 5030 Gembloux, Belgium; hsoyeurt@ulg.ac.be

The development of mid-infrared equations to predict the milk fatty acid (FA) content of milk allows prompt analysis of large numbers of samples and was one of the aims of the RobustMilk project. Data on MIR spectra and FA from multiple countries, production systems, and breeds were used to develop equations to predict milk FA. The calibration set contained 1,776 spectrally different English, Irish, and Belgian milk samples collected for over 6 years. FA were quantified by gas chromatography (GC). Equations were built using partial least squares regression after a first derivative pretreatment applied to the spectral data. The robustness of the developed equations was assessed by cross-validation (CV) using 50 groups from the calibration set. The coefficient of determination (R^2) obtained after CV ranged between 0.7101 for the total content of C18:2 and 0.9993 for the saturated FA group. The standard error of CV ranged between 0.0028 and 0.0998 g/dl of milk. Generally, the group or individual FA having the highest content in milk had the highest R^2cv. The results obtained in this study confirmed the usefulness of MIR spectra to robustly quantify the FA content of milk permitting the use of these equations by milk laboratories in UK, Belgium or Ireland. Therefore, these equations could be used to develop selection or management tools for dairy farmers in order to improve the nutritional and environmental quality of milk based on the knowledge of the FA composition of their milk.

Pedigree analysis of the Portuguese Holstein Cattle: inbreeding and genetic diversity trends
Costa, C.N.[1,2], Thompson, G.[1,3] and Carvalheira, J.[1,3], [1]ICETA/CIBIO, Universidade do Porto, Rua Padre Armando Quintas, Crasto, 4485-661 Vairão, Portugal, [2]Embrapa Gado de Leite, Rua Eugênio do Nascimento, 610 – Dom Bosco, 36038-330 Juiz de Fora, MG, Brazil, [3]ICBAS, Universidade do Porto, Largo Abel Salazar, 4099-003 Porto, Portugal; jgc3@mail.icav.up.pt

The objective of this study was to analyze the pedigree information of the Portuguese Holstein cattle to quantify the inbreeding level and the genetic diversity in the population. Pedigree information including 743,107 registered animals born from 1941 to 2009 obtained from ANABLE data base (BOVINFOR), was analyzed using the software Pedig. The current reference population (RP) was defined as the females born from 2004 to 2009 and comprised 123,039 animals with known parents. The equivalent number of known generations was 3.2 and the average number of known ancestors was 42.7 for the RP. Average inbreeding of 110.718 inbred animals was 0.014 and decreased by 0.0011 /yr from 1986 to 2009. The average inbreeding in the all population was 0.0022 and the percentage of inbred animals in the population increased by 2.29%/yr from 1987 to 2009. The average inbreeding in the RP was 0.004. The average generation interval weighted across pathways was 6.05 yr and was higher for sires (8.05) than for dams (4.75) pathway. The number of founders was 75,180 and the effective numbers of founders, ancestors and founder genomes were 359, 199 and 120 respectively. The calculated effective population size was 250. The ten most influential ancestors (all sires) accounted for 17.6% of the RP gene pool. Five out of 24 subpopulations (defined by country of origin of sires) accounted for 87.65% of the RP gene pool. The use of imported germoplasm in Portugal contributed to acceptable levels of inbreeding and genetic diversity in the population. Despite low, increasing rate of inbreeding suggests attention on selection and mating decisions to avoid the known inbreeding depression effects on overall performance of the Holstein dairy herds.

Comparison of estimation models with data of limited number of animals but intensive recording
Oikawa, T.[1], Koga, Y.[1], Hirayama, T.[1], Munim, T.[2] and Ibi, T.[2], [1]Faculty of Agriculture, University of the Ryukyus, Animal Genetics and Breeding, Senbaru, Nishihara-cho, Nakagami-gun, Okinawa, 903-0213, Japan, [2]Graduate School of Natural Science and Technology, Okayama University, Animal Genetics and Breeding, 1-1-1 Tsushima-Naka, Okayama, 700-8530, Japan; tkroikawa@gmail.com

Variance components were estimated for a data set of body weight of Japanese Black calves having small number of animals but intensive recording; monthly from ages of 0 month (mo) to 16 mo. The data set included 10,295 records of 887 calves measured from 1975 to 2008 at an experimental station of Okayama University. Calves were weaned at 2 or 3 mo of age. Selected heifers were dehorned soon after weaning. Culled heifers and all steers of 8 to 10 mo of age were shipped to a local calf market. Variance components of body weight were estimated by REML using VCE602. Statistical model included fixed effects; contemporary group (year-season-sex), age of calves as a covariate and random effects; animal's direct genetic effect, maternal genetic effect, permanent environmental effect, random residuals. Because of very low permanent environmental effect, it was excluded from the analysis. By multiple trait analysis, low to moderate heritabilities (0.13 to 0.36) were estimated, showing no explicit chronological trend. Heritabilities of maternal genetic effect were in a range of low to high heritability (0.04 to 0.66), however, showing clear downward tendency. Genetic correlations between direct and maternal genetic effect (r_{am}) fluctuated around zero. By random regression analysis with 3-order Legendre polynomials for both direct genetic effect and maternal genetic effect, heritability of direct genetic effect showed upward trend (from 0.18 to 0.87) whereas heritability of maternal genetic effect showed downward trend (from 0.65 to 0.097) showing consistent result with the multiple trait model. The r_{am} consistently showed negative correlations (average: -0.47) throughout the ages.

Intra-observer reliability of different methods for recording temperament in beef and dairy calves
Theis, S., Aditia, E., Hille, K., König, U. and Gauly, M., University of Göttingen, Animal Science, ATW 3, 37075, Germany; koenigvb@gwdg.de

The aim of the study was to validate assessment methods for temperamental traits of cattle for their potential use in breeding programmes. Calves (n=113) of different beef and dairy cattle breeds were video-recorded during a 2 min tethering test at the age of 38±25 days after birth. Subsequently, each video clip was analyzed with the aid of the software Interact in intervals of 1-2 days between subsequent viewings three times by the same person blinded to the identity of individual records. Frequency and duration of behaviour patterns such as head movement, tail-movement, locomotion, lying, defecation and urination were recorded. In addition, each behaviour pattern was recorded on a 10 cm visual analogue scale (VAS), and an overall behaviour score on a scale from 1-5 was assigned to each animal. Using variance components from mixed model analysis, results revealed that generally, the VAS yielded similar or superior intra-observer reliabilities compared to the more time-consuming assessment of exact frequencies and duration (e.g. head movement: VAS: r=0.86; frequency: r=0.38; duration: r=0.27; tail-movement: VAS: r=0.99; frequency: r=0.90 and duration: r=0.93). However, for infrequent behaviour patterns such as urination the assessment of frequencies yielded higher reliabilities (r=0.85) compared to the assessment of duration (r=0.53) or via the VAS (r=0.52). The overall behaviour score likewise resulted in an acceptable intra-observer reliability (r=0.91). Thus, results show that the use of subjective scoring systems or the visual analogue scale might be an appropriate tool to assess some aspects of behaviour. This seems to be particularly true for commonly occurring behaviour patterns. However, there is no ideal assessment method that suits all types of behaviour patterns, and optimal recording methods should be determined prior to use in experimental settings or breeding programmes.

Influence of pre-mating diets on oocyte and embyro development
Ashworth, C.J., University of Edinburgh, The Roslin Institute and R(D)SVS, Easter Bush, EH25 9RG, United Kingdom; cheryl.ashworth@roslin.ed.ac.uk

The quantity of feed consumed and the composition of the diet affect how oocytes develop within the ovarian follicle and have major effects on the number and maturity of oocytes shed at ovulation and the viability of embryos formed following fertilisation. Nutritional effects on oocyte quality can originate when ovarian follicles emerge from the primordial pool and become committed to growth (at approximately 6 months before ovulation in ewes and 3 to 4 months in cows). Studies of the effects of altered nutrient supply preceding mating have shed light on key mechanisms by which nutrient supply affects oocyte and embryo development. Changes in the amount and/or composition of feed consumed alter concentrations of intermediary metabolites, reproductive hormones and the composition of the follicular fluid in which oocytes develop. In turn, these alterations can affect the rate of oocyte development, the distribution and function of organelles, oocyte metabolism and the profile of genes expressed in the oocyte. Much of our understanding of the relationship between nutrition, oocyte quality and blastocyst yield comes from studies on the causes of reduced fertility in the high-yielding dairy cow. In such animals, the hormone and metabolic changes associated with the negative energy balance of early lactation may be sub-optimal for oocyte development. In sheep, oocytes obtained from ewes fed 0.5 maintenance rations for 2 weeks had altered expression of genes involved in glucose transport and energy balance. Oocyte and embryo development is also sensitive to pre-mating nutrition in polyovular species, such as the pig. For example, consumption of increased amounts of food, or increased dietary fibre during the oestrous cycle before mating alters female reproductive endocrinology and follicular fluid composition, enabling oocytes to mature longer before ovulation, thereby increasing oocyte quality and embryo survival. Nutritional management of livestock species before mating provides opportunities to improve pregnancy outcome.

Nutritional and metabolic mechanisms in the ovarian follicle

Dupont, J., INRA, UMR 85, Unité Physiologie de la Reproduction et des Comportements, 37380 Nouzilly, France; jdupont@tours.inra.fr

Nutrition is known to have a profound influence on reproductive performance of female cattle. Cattle in poor condition or losing too much body weight generally have poor reproductive performance. The reason for this relationship is often explained by using the argument of nutrient prioritization. Even if the mechanisms through which nutrition controls reproduction are of great interest, they remain poorly understood. Folliculogenesis is a nutritionally sensitive process. The physiological link between folliculogenesis and energy intake involves several metabolic hormones and growth factors acting within the follicular environment. However, the follicle can directly sense nutritional and metabolic inputs and modify its responses accordingly. Indeed, it appears to have a number of 'nutrient sensing' mechanism that may form the link between nutrient status and folliculogenesis. This review describes the presence of some pathways that may sense nutrient flux from within the follicle including the insulin signaling pathway, adenosine monophosphate activated kinase (AMPK), the peroxisome proliferator activated receptors (PPARs) and some adipokines including leptin and adiponectin. It will also speculate on how these 'nutrient sensing' pathways interact with FSH signalling pathways to adjust gonadotrophin-stimulated follicular function. In summary, nutrition influences ovarian follicle development in cattle through changes in metabolic hormones and/or by acting directly on the ovarian cells. These interactions could be manipulated to improve reproductive performance.

Piglet birth weight and uniformity: importance of the pre-mating period

Wientjes, J.G.M.[1], Soede, N.M.[1], Knol, E.F.[2], Van Der Peet-Schwering, C.M.C.[3], Van Den Brand, H.[1] and Kemp, B.[1], [1]Wageningen University, Department of Animal Sciences, Adaptation Physiology Group, De Elst 1, 6708 WD Wageningen, Netherlands, [2]IPG, Institute for Pig Genetics BV, Schoenaker 6, 6641 SZ Beuningen, Netherlands, [3]Wageningen UR, Livestock Research, Postbus 65, 8200 AB Lelystad, Netherlands; anne.wientjes@wur.nl

Piglet birth weight and piglet uniformity affect piglet survival. Insulin-stimulating sow diets before mating can improve piglet birth weight and uniformity. We hypothesized that nutritionally enhanced insulin levels result in more developed and uniform follicles, resulting in improved and uniform embryo development and luteal development, finally ending in more uniform birth weights. In a first experiment, we indeed found positive relationships between pre-mating insulin levels and follicle development, and subsequent progesterone levels and embryo development (but not uniformity) at d10 of pregnancy. Whether and how the improved progesterone levels and embryo development lead to a more uniform development of foetuses at later stages of pregnancy, is currently studied. In organic sows, however, pre-mating insulin-stimulating diets did not improve piglet birth weight or uniformity. Organic sows have longer lactations (41±4 d compared to 24-28 d in conventional sows) and may switch to an anabolic state during the last weeks of lactation. Therefore, insulin-stimulating diets before mating seem beneficial for follicle development and subsequent piglet uniformity, but only in sows with a compromised follicle development at weaning, due to e.g. a severe catabolic state until weaning. Pre-mating conditions that affect sow metabolic state, such as lactation length, lactational backfat loss and length of the weaning-to-estrus interval (WEI), may thus affect piglet uniformity. Based on data of 2,128 conventional litters, it was confirmed that piglet uniformity was decreased in sows with severe lactational backfat losses and improved in sows with a prolonged WEI (>21 d, incl. repeat breeders) compared with sows having a WEI≤7 d.

Two new mutations affecting ovulation rate in sheep

Demars, J.[1], Fabre, S.[1,2], Sarry, J.[1], Rosetti, R.[3], Persani, L.[3], Tosser-Klopp, G.[1], Mulsant, P.[1], Nowak, Z.[4], Drobik, W.[4], Martyniuk, E.[4] and Bodin, L.[5], [1]INRA, Génétique Cellulaire, 31326 Castanet-Tolosan, France, [2]INRA, Physiologie de la Reproduction et des Comportements, 37380 Nouzilly, France, [3]Istituto Auxologico Italiano, Laboratorio di Recerche Endocino-Metaboliche, University of Milan, 20149 Milano, Italy, [4]Warsaw University of Life Sciences, Department of Genetics and Animal Breeding, 02-786 Warsawa, Poland, [5]INRA, SAGA, BP 31627, 31326 Castanet-Tolosan, France; Loys.Bodin@toulouse.inra.fr

A case-control study aiming at identify putative major genes affecting ovulation rate (OR) and litter size (LS) was undertaken in two prolific sheep breeds, the French Grivette and the Polish Olkuska. A genome-wide association study using a 50K SNP chip revealed in both populations a significant association between phenotypes and SNPs close to the BMP15 gene. Sequencing this gene in case and control ewes identified a new mutation in each population, both inducing non-conservative amino-acid substitution, T317I in Grivette and N337H in Olkuska. In Grivette, 151 additional ewes chosen at random in the population among ewes having at least 4 lambing were genotyped for the new mutation. Thirty-three (22%) were wild type, while 67 (44%) and 51 (34%) were respectively heterozygous and homozygous carriers. Mean prolificacy of wild type and heterozygous ewes were 1.86 and 1.97, and it was significantly higher 2.62 for the homozygous carriers (P<0.001). The functional effect of this mutation was tested *in vitro* after directed mutagenesis. Using a BMP-responsive luciferase test in COV434 granulosa cells, the T317I mutation impaired drastically the BMP15 signaling activity. In Olkuska, the new polymorphic locus (1009A>C) significantly associated with the phenotype was found homozygous mutated type (C/C) in none of the 32 control animals, and homozygous wild type in only one of the 22 cases. Mean OR were 1.63, 1.93 and 4.38 for A/A, A/C and C/C ewes respectively. In contrast with the 6 other BMP15 known mutations, these 2 novel ones induce new phenotypes without streak ovaries at the homozygous state.

Expression of chemerin and its receptor, CMKLR1, in bovine ovary: role in granulosa cells?

Reverchon, M., Rame, C. and Dupont, J., INRA, UMR 85, Unité Physiologie de la Reproduction et des Comportements, 37380 Nouzilly, France; jdupont@tours.inra.fr

Chemerin (Chem) is a novel adipokine produced by white adipose tissue in mammals. It acts through its receptor, CMKLR1 (Chemokine like receptor 1) by activating various signaling pathways. Chem is involved in adipogenesis, inflammation, angiogenesis and metabolic functions. However, until now no function of this adipokine has been described in ovarian functions. The aim of the study was to characterize chem and CMKLR1 expression in various bovine tissues and more particularly in different ovarian cells. Then, we investigated the effect of recombinant human chem (rhChem) on steroid production, cell proliferation and on the activation of various signaling pathways in cultured bovine granulosa cells. By RT-PCR we showed that chem and CMKLR1 were expressed in peripheral and reproductive bovin tissues. We also observed that they were present in different ovarian structures (cortex, corpus luteum, large and small follicles) and more precisely in granulosa cells. By immunohistochemistry we confirmed these results in oocyte, cumulus, theca and granulosa cells. By real time RT-PCR we showed that CMKLR1 is more expressed in corpus luteum than in other ovarian structures. *In vitro* in primary bovine granulosa cells from small follicles, we have shown that rhChem (200 or 20 ng/ml) significantly decreased basal and IGF-1 (10-8M) and/or FSH (10-8M) -induced progesterone as determined by radioimmunoassay. Furthermore, rhChem significantly decreased granulosa cell proliferation in response to IGF-1. In bovine granulosa cells, rhChem has also inhibited MAPK-ERK1/2 phosphorylation but not MAPK P38, Akt and AMPK. One-way analysis of variance (ANOVA) was used to test differences with P<0.05 considered significant. Taken together, Chemerin and CMKLR1 are expressed in bovine ovary. In cultured granulosa cells, rhChem decreases progesterone secretion and cell proliferation. These results are associated with a decrease in MAPK ERK1/2 phosphorylation.

Melatonin-dependent timing of seasonal reproduction: molecular basis and practical applications
Dardente, H., Fréret, S., Fatet, A., Collet, A., Chesneau, D. and Pellicer-Rubio, M.-T., INRA UMR085, Physiologie de la Reproduction et des Comportements, 37380 Nouzilly, France; hdardente@tours.inra.fr

Most mammals living at temperate latitudes exhibit marked seasonal variations in reproduction. In long-lived species such as sheep and goat, it is assumed that timely physiological alternations between breeding and sexual rest depend upon the ability of day length (photoperiod) to synchronise an endogenous clock. Melatonin, secreted only during the night, acts as the endocrine transducer of the photoperiodic message. More specifically, there is unambiguous evidence that long day lengths of spring and summer constitute the most potent synchronising cue for seasonal reproduction. Also, it has been known for decades that the thyroid hormone T3 is crucial to terminate the breeding season in sheep. Our recent data demonstrating that a circadian-based molecular mechanism within the pars tuberalis of the pituitary ties the short duration melatonin signal – reflecting long day length – to the hypothalamic increase of T3 in sheep will be presented. We'll also present results in goats showing that the concept of photoperiodic history, which postulates that the direction of the change in photoperiod is key to the reproductive response, might actually hold true only when animals are transferred to intermediate photoperiods. Such data are fully consistent with the existence of a critical photoperiod (i.e. a threshold in the duration of day-length) for the reproductive response in small ruminants. Follow-up experiments are now underway, in sheep and goats, to define precisely when a day is actually interpreted as being a 'long day'. This is crucial as the application of long days is not only the most widely used method to manipulate reproduction in sheep and goats; it is also the only natural (i.e. without exogenous hormones) procedure.

Artificial insemination and reproduction management in small ruminants with special regard to future
Dattena, M., Maura, L., Falchi, L., Meloni, G., Facchin, F. and Gallus, M., AGRIS-DIRPA, Animal Reproduction Department, Olmedo 07040, Italy; mdattena@agriscerca.it

The aim of this work is to present future perspectives in AI and reproduction management in small ruminants looking at more sustainable conditions. In the last a few years the need of more attention in artificial insemination (AI) and reproduction management programs, with special regard to future perspectives, is come from the new demand for animal products in the Mediterranean regions determined by the population growth, urbanization, increasing purchasing power and special requirements of consumers. However research for the future small ruminant production in the Mediterranean system should also focus on an efficient utilization of local genetic resources, innovation and improvement of traditional processing technologies. Until now application of reproductive technologies has been mostly applied in traditional advanced farm conditions. However, in our day the application of these tools, might be taken into consideration for a more wide kind of situations : (1) difficult farm condition from geographical and economical point of view; (2) organic farm; (3) preservation of genetic resources; (4) endangered livestock. These new approaches might need: (a) avoid the use of hormones; (b) innovation on semen preservation technology; (c) reduction of cost. Indeed in the last few years new perspectives have been claimed for all these approaches in many European countries. It is with this new situation in mind that at the Sardinia Research Center (Agris) we started to look up to different perspectives, particularly to set up: (1) an Artificial Insemination protocol without the use of hormones to be applied in an organic farm using long preservation chilled semen; (2) a 7 days synchronization protocol (short protocol) to synchronize dairy sheep to be fertilized with fresh semen; (3) a mating controlled protocol using entire ram with females synchronized by ram effect.

Sheep insemination: current situation in France and work on data of semen production
David, I.[1], Bonnot, A.[2], Raoul, J.[3,4] and Lagriffoul, G.[3,4], [1]INRA SAGA, BP52627, 31326 Castanet-Tolosan, France, [2]Institut de l'Elevage, 149 rue de Bercy, 75595 Paris 12, France, [3]ANIO-Association Nationale de l'Insémination Ovine, BP 42118, 31321 Castanet-Tolosan Cedex, France, [4]Institut de l'Elevage, BP 42118, 31321 Castanet-Tolosan Cedex, France; ingrid.david@toulouse.inra.fr

In France, 812,000 animal inseminations (AI) have been carried out on sheep in 2010, representing 16% of the national herd, but with a strong imbalance between dairy sheep (653,000 inseminations, 50% of the dairy herd) and meat sheep (159,000 or 4% of meat herd). About 95% of AI is carried out with fresh semen. 497 000 AI were performed with dairy rams, mainly for the production of breeding stock. 315,000 AI were performed with meat rams including 260,000 for meat production and 55,000 for the production of breeding stock. AI fertility in sheep varied greatly despite the homogenization of techniques from semen collection to insemination. The aim of the BELIA study (1997/2005) was to explain the variability of success in AI taking into account the information gathered by AI centers. Following this work, the data on ram semen production (motility, concentration, volume…) were gathered into a database. After data validation and linkage with genealogy, new statistics were produced and sent to INRA, which calculated EBVs (genetic motility, volume, concentration, sperm count) and provided a report. This process tracks trends and allows comparisons with other breed. These data open new perspectives for sheep in a context where genomics, particularly in dairy sheep, is ramping up. First, it should allow AI centers to have new traits for the selection of rams, and might provide news decision rules allowing an improvement of semen production.

Realities of sheep artificial insemination on farm level: farm and breed differences
Kukovics, S.[1], Németh, T.[1], Gyökér, E.[2] and Gergátz, E.[2], [1]Research Institute for Animal Breeding and Nutrition, Gesztenyés út 1., 2053 Herceghalom, Hungary, [2]Pharmagene-Farm Ltd. Biotechnical Research Station, Mosonszentjánosi u. 4., 9200 Mosonmagyaróvár, Hungary; sandor.kukovics@atk.hu

Since the mid 1970's the artificial centers for sheep were ceased in Hungary the artificial insemination (AI) of ewes has been gradually reduced. In mid sixties more than 63% of the total number of ewes was inseminated (in some part of the country this number was above the 85%) and nowadays it is only 2-3%. In order to examine the present practice and effectiveness of the AI a survey was conducted between 2003 and 2010 covering the dominant part of the sheep farms using these techniques. The inseminated ewes were belonging to various breeds: purebred and crossbred Awassi and British Milksheep, Bábolna Tetra, Charollaise, German Mutton Merino, German Blackheaded Mutton Sheep, Ile de France, Hungarian Merino, Lacaune, and Suffolk. The number of ewes, details of the techniques used and the results were evaluated concerning eleven sheep farms inseminating more than ten thousands ewes in the first and about 3500 heads in the last year. Every detail of the AI techniques from the selection of ewes up to the weaning rate of the lambs born from AI was evaluated. Descriptive statistics and chi-square test was applied for processing of data. The main conclusions of the study were as follows: well skilled shepherds could apply the AI with very good results on farm level using dominantly fresh semen collected locally; the conception rate (75-95%) was affected by breed, age of ewes, and year. The cost of AI varied from € 0.35 to € 8.5.

Innovative itineraries minimizing hormones to synchronize and induce oestrus and ovulation in goats
Rekik, M.[1], Ben Othman, H.[1] and Lassoued, N.[2], [1]ENMV, S. Thabet, 2020, Tunisia, [2]INRAT, Ariana, 2049, Tunisia; rekik.mourad@iresa.agrinet.tn

Standard protocol (S) to synchronise oestrus and ovulation in goats uses progestagens in association with prostaglandins and gonadotropin. Such protocol has major drawbacks altering ovulation, mobility and viability of sperm and leaves residues in milk. This study aims to develop, for local Tunisian goats, alternative synchronizing protocols during anoestrus and transition to breeding season. In late May, 40 goats were assigned to either the S protocol or to a protocol where oestrus and ovulation were induced by the buck effect in a single injection-progesterone treated-goats and provoking early luteolysis using a prostaglandin 9 days after exposure to bucks (B). Continuous data were analysed using linear models of the SAS software while discrete variables were analysed using $\chi 2$ test. During the 72 hours after the treatments ended, 15 and 5 goats expressed oestrus in the S and B protocols ($P<0.001$). Mean time to oestrus was 24.6 ± 2.54 h for S goats, much shorter ($P<0.05$) than 40.4 ± 4.39 h for B goats. Ovulation rate averaged 2.07 ± 0.22 and 1.60 ± 0.35 for respectively S and B goats ($P>0.05$). During transition to breeding season (mid September), 60 goats of which 46% were cycling based on plasma progesterone, were assigned to either S treatment, PG treatment where estrus and ovulation were synchronized using two injections of prostaglandins 11 days apart or to GnRH treatment where the goats had their estrus and ovulation synchronized with a GnRH (day 0)-prostaglandin (day 6)-GnRH (day 9) sequence. Significantly more S goats were detected in oestrus over the 84 h-period after the end of the treatments (89, 74 and 55% in S, PG and GnRH treatments respectively; $P<0.05$). One goat in the PG and 4 in the GnRH groups had silent ovulations. Mean ovulation rates were 2.3 ± 0.27, 1.33 ± 0.27 and 1.33 ± 0.27 for respectively S, PG and GnRH goats ($P<0.001$). Prostaglandin and GnRH based treatments resulted in 'clean estrus' and their efficiency should be tested in mid breeding season.

Interactions between milk production and reproduction in the Sicilo-Sarde ewe
Rekik, M.[1], Meraï, A.[1] and Aloulou, R.[2], [1]ENMV Sidi Thabet, ENMV, 2020, Tunisia, [2]ISA Chott Meriem, ISA Chott Meriem, 4042, Tunisia; rekik.mourad@iresa.agrinet.tn

This work aimed to investigate interactions that may exist between milk production and reproduction in the milking Sicilo-Sarde ewe. The study of these interactions were assessed using a database consisting of 6866 observations recorded in seven dairy flocks of the studied breed belonging to three large farms in Northern Tunisia during five consecutive productive years. The parameters studied were total milk production (TMP), daily milk production (DMP) and duration of milking period for milk traits and fertility, litter size and interval between the start of mating period and conception for reproductive traits. The selected data were analyzed by the procedure 'Proc Logistic' of the SAS software. The study revealed a highly significant effect of total milk production during the productive year i-1 on fertility during the productive year i with a difference that reached + 19.08 liters of milk during the year 2008/2009 in favor of barren ewes compared to ewes that have lambed ($Pr<0.001$). However, the intensity of this antagonism varied from one year to another and was not verified between milk production and interval between the start of mating period and conception. In a second study, the effect of the genetic merit for milk production on conception rate following artificial insemination revealed that high-merit ewes had a lower ($Pr <0.05$) conception rate (43%) than lower-merit counterparts (62%). It is concluded that a high level of milk production hampers the ability of females to conceive.

Farmers' decision making with regard to animal welfare

De Lauwere, C. and Van Asseldonk, M., LEI Wageningen UR, P.O. Box 35, 6700 AA, Netherlands; carolien.delauwere@wur.nl

Farmers play an important role in the improvement of animal welfare. It is often assumed that their decision making is purely rational and depends on economic considerations. Social psychological factors however are important as well. The theory of planned behaviour (TPB) is a widely applied model to understand conscious choices. It has been proven useful in understanding farmers' choices. TPB however does not cover all aspects of decision making of farmers. Therefore the model is often extended with other constructs. In a study amongst 105 pig farmers about changing to group housing for pregnant sows, TPB for example showed that farmers who did not change to group housing yet were more negative than farmers who completely or partly changed. It appeared that the farmers who did not change yet had more concerns about animal welfare and the occurrence of tail biting. Moreover they appeared to feel less confident about their skills to keep sows in groups and to organize a building process to realize group housing. Other studies showed that other constructs than those related to TPB also can be useful to understand farmers decisions. In a study amongst 127 dairy farmers about the application of pasturing, it appeared that farmers who applied pasturing were more positive about it than farmers who did not apply pasturing. Farmers who applied pasturing were more convinced that this is good for animal welfare and health and leads to less claw problems. Habits – not an element of TPB – however played a role as well as farmers who applied pasturing day and night had higher scores for habits than farmers who applied pasturing only during the day. Besides this, farmers who applied pasturing seemed to be more sensitive for groups norms (an element of social identity which is not a TPB construct), assuming that other farmers perceive pasturing as beneficial or desirable as well. These kind of findings gives indications for possible interventions which may help farmers to change their behaviour with respect to animal welfare.

Dynamic monitoring of litter size at herd and sow level

Bono, C., Cornou, C. and Kristensen, A.R., University of Copenhagen, HERD – Centre for Herd-oriented Education, Research and Development, Dpt of Large Animal Sciences, Grønnegårdsvej 2, 1870 Frederiksberg, Denmark; clbo@life.ku.dk

Monitoring animal production results in real time is challenging. Existing monitoring information systems (MIS) in pig production are based on static statement of selected key figures typically computed every quarter or year. The idea of the paper is to develop a dynamic monitoring system for litter size at herd and sow level, with weekly update. For this purpose, a modified litter size model, based on an existing model found in the literature, is implemented using dynamic linear models (DLMs). The variance components are pre-estimated from the individual herd database using a maximum likelihood technique in combination with an EM algorithm applied on a larger dataset with observations from several herds. The model includes a set of parameters describing the parity specific mean litter sizes (herd level), a time trend describing the genetic progress, and the individual sow effects (sow level). It provides reliable forecasting, on a weekly basis, for future production. Individual sow values, useful for the culling strategy, are also computed. In a second step, statistical control tools are applied. Shewhart control charts and V-masks are used to give warnings in case of impaired litter size results. The model is applied on data from 13 herds, each of them including a period ranging from 400 to 800 weeks. For each herd, the litter size profile, the litter size over time, the sow individual effect and sow economic value, are computed. Perspectives for further development of the model can take into account indices as, for instance, conception rate, fertility rate, pregnancy rate, service rate, mortality rate etc. Such a model can be used as a basis for developing a new, dynamic, management tool.

Sustainable swine production from the point of view of medicine, hyotechnology and economics
Sviben, M., Freelance consultant, Siget 22b, 10020 Zagreb, Croatia; marijan.sviben@zg.t-com.hr

In December 1990 the scientific project 'Sustainable swine production from the points of view of medicine, hyotechnology and economics' was proposed and it was supported from July 1991 till December 1997 by the Ministry of Science, Technology and Informatics of Republic of Croatia. Numerous colleagues and companies made possible the elaboration of the project after 1997. The health status of the population in an area is assumed to be favoured when the quantity of 85 kg of meat is available and at least 45.17 kg of pig meat per inhabitant a year should be consumed. The application of hyotechnology (the science of methods of producing goods using swine), expressed as the design of the production process, was recognized to be an autonomous fifth factor of production in October 1992. After that sustainable swine production was defined as the system in which the factors of production (land, labour, capital, production design and entrepreneurship) considering animal welfare, health of human beings and the protection of nature are kept in balance with required productivity, profit and needs for the porcine market place. The income required to be sufficient for paying the production costs, corporation tax, income taxes, net profit after tax and post-tax earning became the measure of the labour productivity. The equations of the labour productivity required were published. The entrepreneurship can be attracted to organise other factors of production for acquiring greater profit than normal at the no food money rate of 0.35 and L.P.R. coefficient 43.956. Considering the profit and L.P.R. it was justified to invest in large and larger pig production units. It was concluded that the enlargement of production improving the production methods and the category of goods for selling improved the economy of the swine business. Application of more economical methods of production could be combined with better protection of nature.

The effect of grass white clover and grass only swards on milk production and grazing behaviour
Enríquez-Hidalgo, D.[1,2], Lewis, E.[2], Gilliland, T.J.[3], O'donovan, M.[2], Elliott, C.[1] and Hennessy, D.[2], [1]Queens University Belfast, Belfast, BT97BL, United Kingdom, [2]TEAGASC, Animal & Grassland Research and Innovation Centre, Moorepark, Fermoy, Ireland, [3]Agri-Food and Biosciences Institute, Plant Testing Station, Crossnacreevy, Belfast, United Kingdom; daniel.enriquez@teagasc.ie

The objective of the experiment was to compare milk production and grazing behaviour from lactating dairy cows grazing grass only (GR) or grass white clover (GC) swards. Thirty cows were randomly allocated to each treatment from 17 Apr. to 30 Oct. Swards were rotationally grazed. Cows received 17 kg herbage DM/cow/day. Swards received 150 kg N/ha during April to October. Herbage mass and sward clover content were recorded twice weekly. Milk production was recorded daily and milk composition weekly. Eight lactating rumen-fistulated dairy cows were arranged into four 2×2 Latin squares and allocated to each treatment for one period of two weeks. This occurred on 16 May (TS1) and 11 July (TS2). Grazing behaviour data were collected by fitting the rumen-fistulated cows with IGER behaviour recorders. The proportion of time spent grazing, ruminating and idling was measured. The data were analysed using PROC MIXED in SAS. There was no effect of treatment (P>0.05) on herbage mass (1720 kg DM/ha), milk yield (19.4 kg/d), milk protein content (36.4 g/kg) or milk lactose content (46.0 g/kg). Milk fat content tended to be higher on GR than on GC (43.7 vs. 42.5 g/kg; P=0.08). The average clover proportion of GC was 0.13, and was 0.08 in TS1 and 0.11 in TS2. Grazing behaviour was similar between treatments (P>0.05) in TS1 (proportion of time spent grazing 0.41 and ruminating 0.30). In TS2 GC cows spent a lower proportion of time grazing (0.42 vs. 0.46; P=0.04) than GR cows. In conclusion, GC swards had similar herbage and milk production to GR swards. Grazing behaviour was only affected when higher clover proportions were present. Clearer differences between treatments may be observed if sward clover content was greater through the year.

Allocative efficiency of dairy cattle grazing systems in La Pampa (Argentina) applying DEA approach

Angón, E.[1], García, A.[1], Perea, J.[1], De Pablos, C.[2], Acero, R.[1] and Toro-Mújica, P.[1], [1]University of Córdoba, Animal Production, Campus de Rabanales, 14071 Córdoba, Spain, [2]University Rey Juan Carlos, Economía de la Empresa, Paseo de los Artilleros s/n, 28032 Madrid., Spain; z82anpee@uco.es

The objective of this study is to assess the allocative efficiency of dairy cattle grazing systems in La Pampa (Argentina), using a data envelopment analysis (DEAP Version 2.1, by Coelli T.J.). This technique creates efficiency indexes by comparing the performance of each farm with the best production practices observed, which define the efficiency. The DEA analysis calculates a frontier of efficient farms by assuming inputs to be optimally minimized for a given output level. The study was realized in the dairy areas of the province of La Pampa, with a population of 172 dairy farms and it included a sample for 47 grazing dairy farms (27% of the census). The collection of data was carried out by applying direct producer surveys. In this study, we have considered an output orientated model, which seeks to maximize production. The model was constructed by taking into account 1 output and 5 inputs. The milk production (kg/year) has been chosen as the output of the model and the model inputs include: pasture surface (ha), annual work unit, feed supplementary (kg/day), size (total cows) and feed cost ($/year). The allocative efficiency was of 79% at constant returns to scale and 84% at variable returns to scale. The slacks analysis showed that less efficient farms could develop two improvement strategies a) Firstly, would increase production at 20% with the same inputs. b) A 16.4% cost reduction with the following inputs decrease: 47% of supplementary feeding, 6.2% in the herd size, 13.4% in the use of labor, and lastly 8.8% of the pastures surface.

The profitability of seasonal mountain dairy farming in Norway

Asheim, L.J.[1], Lunnan, T.[2] and Sickel, H.[2], [1]Norwegian Agricultural Economics Research Institute, Research, Storgata 2-4-6, 0031 Oslo, Norway, [2]Norwegian Institute of Agricultural and Environmental Research, Arable Crop, Løken, 2900 Heggenes, Norway; leif-jarle.asheim@nilf.no

Abstract Seasonal mountain dairy farming in Norway developed as a strategy for using large mountainous pasture areas while the agricultural area was limited. However, in recent years dairy cows are kept and fed intensively at the farm site bringing about an on-going encroachment process threatening plant biodiversity in many traditional mountain pasture areas. The pastures primarily consist in native grasses, sedges, willow thickets, and various herbs. Particularly the contents of polyunsaturated fatty acids, CLA and various antioxidants (i.e. carotenoids and alpha-tocopherol) in the milk increase when cows graze species rich pastures affecting its processing properties as well as the flavor and chemical content of dairy products. According to Euromontana consumers expect mountain products to be produced from raw mountain materials, have a link to the cultural identity of local communities, connected to specific cultural areas, and produced with traditional methods by small-scale producers. On Norwegian dairy farms located at 400-700 m altitude and operating 10 to 20 cows the mountain production would involve free ranging for 60-70 days at 800-1,000 m from the end of June with grazing in the valley before and after. In the paper calving time, introducing fertilized pastures or night pens, and supplementary feeding to extend the mountain period and sustain milk yields, are compared with keeping the cows at the farm or investing in a common pasture or co-operative farming. The comparisons are based on calculations in a linear programming (LP) farm model taking into account governmental support and the extra work of processing and costs of keeping the mountain hut and production facilities. The size of a premium needed for 'mountain products' and farmer co-operation on marketing is discussed.

Glucocorticoids as biomarkers for feed efficiency in cattle

Montanholi, Y.R.[1], Swanson, K.C.[2], Palme, R.[3], Vander Voort, G.[1], Haas, L.S.[1] and Miller, S.P.[1], [1]University of Guelph, Animal and Poultry Science, 70-50, Stone Road East, N1G 2W1 Guelph, ON, Canada, [2]North Dakota State University, Department of Animal Sciences, P.O. Box 6050, 58108-6050 Fargo, ND, USA, [3]University of Veterinary Medicine, Department of Biomedical Sciences, Veterinärplatz 1, 1210 Vienna, Austria; ymontanh@uoguelph.ca

A better understanding of the association between feed efficiency and glucocorticoid levels is needed. Therefore, plasma cortisol (PC; ng/ml) and fecal cortisol metabolites (FCM; ng/ml) levels as predictors of feed efficiency were evaluated in feedlot steers. Individual daily feed intake of 112 steers fed a high-moisture corn-based and corn silage diet was measured using an automated feeding system during 140 days. Body weight, blood and fecal samples were collected every 14 days and ultrasound measures for body composition were taken every 28 days. Four productive performance traits were calculated, namely, daily dry matter intake (DMI), average daily gain (ADG), feed to gain ratio (F:G) and residual feed intake (RFI). At the end of the feedlot phase, steers were ranked according to RFI. Samples from the feedlot phase were analyzed for PC and FCM from the 32 steers with greatest and the 32 steers with lowest feed efficiency. In addition, 12 steers of each of these two groups with divergent feed efficiency were hourly blood sampled for 24 h and had their breath continuously analyzed for CO_2, O_2 and CH_4 using a head chamber indirect calorimeter. Results indicated no association between productive performance traits and PC and moderate to high correlations between FCM and these traits over the feedlot phase. More efficient steers had higher levels of FCM over the entire feedlot phase. In addition, more efficient steers had numerically lower values for heat and CH_4 production, which support the classification for feed efficiency by using RFI as selection criteria. These finding might have application in selection programs and in the better understanding of the biological basis associated with productive performance.

SNPs explaining genotype by environment interaction in German Holstein dairy cattle

Streit, M.[1], Reinhardt, F.[2], Thaller, G.[3] and Bennewitz, J.[1], [1]Institute of Animal Husbandry and Animal Breeding, University of Hohenheim, Garbenstraße 17, 70599 Stuttgart, Germany, [2]Vereinigte Informationssysteme Tierhaltung w.V. (VIT), Heideweg 1, 27283 Verden, Germany, [3]Institute of Animal Breeding and Husbandry, Christian-Albrechts-University, Olshausenstraße 40, 24098 Kiel, Germany; melanie.streit@uni-hohenheim.de

Reaction norm random regression sire models were used to study genotype by environment interactions (GxE) in the German Holstein dairy cattle population. About 2300 progeny tested sires were included. Corrected test day records of daughters for milk traits and somatic cell score were used. Herd test day solutions obtained from routine sire evaluation for milk traits, milk energy yield or somatic cell score were used as environmental descriptors. Breeding values for intercept and slope of the reaction norms were calculated using a random regression sire model. Heterogeneous error variances were considered. The results revealed a substantial slope variance, indicating the presence of GxE effects. The sires were genotyped with the bovine 54K SNP-chip. The EBVs for slope were used as observations in genomewide association analysis in order to find SNPs for GxE effects. A random sample of 1,800 sires were used for the genomewide association analysis. The model included a fixed mean, a random sire effect and a fixed SNP effect. Correction for multiple testing was done using the false discovery rate. Suggestive SNPs from the reference population were validated using the remaining 500 genotyped sires. This strategy enabled us to detect and validate SNPs that account for the genetic variation of the sire's response to a changing environmental conditions, and hence, for the GxE effects.

Factors affecting plasma progesterone concentration in cows divergent in genetic merit for fertility

Moore, S.G.[1], Scully, S.[2], Crowe, M.A.[3], Evans, A.C.O.[2], Lonergan, P.[2], Fair, T.[2] and Butler, S.T.[1], [1]AGRIC, Teagasc Moorepark, Fermoy, Co. Cork, Ireland, [2]School of Agriculture and Food Science, UCD, Dublin 4, Ireland, [3]School of Veterinary Medicine, UCD, Dublin 4, Ireland; stephen.moore@teagasc.ie

This study investigated the mechanisms responsible for different circulating progesterone (P4) concentrations in cows with similar genetic merit for milk production traits, but with extremes of good (Fert+) or poor genetic merit for fertility traits (Fert-). Dairy cows (Fert+ = 13, Fert- = 9) were enrolled in an ovulation synchronisation protocol at 61 ± 13 days postpartum. P4 concentrations were measured twice daily. On d 7 (d 0 = ovulation), corpus luteum (CL) blood flow intensity (BFI) and area (BFA) were measured by Doppler ultrasonography. Cows received prostaglandin F2α on d 7 pm and d 8 am to regress the CL, and 2 CIDRs were inserted per vaginum on d 8 am. Liver biopsies were collected on d 9 and hepatic mRNA abundance of genes involved in P4 catabolism was determined. On d 10, CIDRs were removed and frequent blood samples were collected to measure P4 metabolic clearance rate (MCR). Data were analysed using PROC MIXED in SAS with repeated measures used where appropriate. PROC NLIN was used to estimate the decay rate coefficient of P4 for calculating the half-life and MCR. Fert+ cows had greater dry matter intake compared with Fert- cows (24.9 vs. 23.4 kg/day, P=0.03), but similar milk production (30.1 vs. 29.9 kg/day). From d 4.5 to 7.5, mean circulating P4 concentrations were greater in Fert+ cows (2.39 vs. 1.97 ng/ml, P=0.02). Fert+ cows had greater CL BFA (2.24 vs. 1.45 cm^2, P=0.03), but CL volume and BFI on d 7 were not different. Fert- cows had greater (P=0.03) abundance of AKr1C1 mRNA, but AKr1C3, AKr1C4, CYP2C and CYP3A were similar. Genotype had no effect on P4 half-life (37.7 vs. 31.8 min) or MCR (1.83 vs. 2.34%/min). The results indicate that greater circulating P4 concentrations were primarily due to greater CL P4 synthetic capacity rather than differences in MCR in this lactating cow genetic model of fertility.

Milk fatty acids as influenced by energy metabolism and diseases in early lactating dairy cows

Knapp, E., Dotreppe, O., Hornick, J.L., Istasse, L. and Dufrasne, I., University of Liege, Veterinary Faculty, Nutrition Unit, bld de colonster, 20, 4000 Liège, Belgium; eknapp@ulg.ac.be

The monitoring of energy metabolism in the early part of lactation of high producing cows is essential for the management of the herd. Blood and milk samples along with a gynecological examination were obtained on 61 cows from 7 private farms on 4 occasions on a monthly basis starting from calving. The fatty acids (FA) data sets were divided in categories based on milk β-hydroxibutyrate (BHB), plasma non esterified fatty acids (NEFA) and on the ratio (C18:0 + C18:2)/C18:1 in the NEFA fraction as energy metabolism indicators. In the fat mobilizing cows, both the amounts and the proportions of the milk C6-C14 FA – neosynthesis in the mammary gland – were lower than in the non mobilizing cows. For example, 14.8 vs 17.6% P<0.01 in the FA ratio comparison, 15.3 vs 17.7% P<0.01 in the BHB classes and 15.1 vs 19.5% P<0.001 in the NEFA classes. On the other hand, the milk C18:1 was higher (P<0.001) in the mobilizing cows (14.4 vs 11.5 g/l in the FA ratio comparison, 14.2 vs 11.0 g/l in the BHB classes and 15.1 vs 10.1 g/l in the NEFA classes). The FA data sets were also divided according to infection diseases. When metritis was observed, the FA from neosynthesis decreased largely (15.5% vs 19.8% P<0.001) and the long chain fatty acids increased (28.3 vs 22.6% P<0.01 of C18:1, 2.8 vs 2.3% P<0.05 of the polyunsaturated FA). Less clear cut observations were found between the cell counts and the milk FA content. For example it was with the cows with cell counts below 400,000 cells/ml that the long chain FA proportion was the highest (28.3 vs 23.4% of C18:1 P<0.05). It is concluded that milk fatty acids profile changed in proportions and in amounts when energy status is modified or when infection diseases were observed in early lactating cows. So milk fatty acids profile during early lactation could be an accurate indicator to manage the waiting period.

Relation between protein source in diet and some reproductive performance of growing Frisian bulls
Elganiny, S.M.[1], Khattab, H.M.[2], Tharwat, E.E.[2] and Zeidan, A.E.[1], [1]Animal Production Research Institute, Cattle Breeding, Nady Elseid st., 56562 Dokki, Giza, Egypt, [2]Faculty of Agriculture, Shupra st., 20182, Egypt; shoshohd4@yahoo.com

The effect of rumen undegradable protein (RUP) of different protein sources on the reproductive efficiency was studied in growing Friesian male calves from March to September 2010. Fifteen healthy male calves, 9 ± 1 months of age, with average live body weight of 180 ± 40 kg were used and allocated to three groups of five animals each. Each group received a basal diet of 60% of concentrate feed mixture and 40% of forage (silage:berseem hay; 1:1). Dietary treatments were 3 different protein sources: 'CSM' (cotton seed meal and maize gluten feed; 132 g CP·/kg DM, RUP: 33% of CP, 660 g TDN·/kg DM), 'SBM' (soybean meal; 167 g CP·/kg DM, RUP: 34% of CP, 670 g TDN·/kg DM) and 'MGM'(maize gluten meal (158 g CP·/kg DM, RUP: 39% of CP, 660 g TDN·/kg DM). Feed intake, body weight and testicular size were measured monthly. Blood samples were taken monthly. Semen was collected by artificial vagina every week to analyse semen volume, sperm concentration and viability. Dry matter intake was 4.57, 4.50 and 4.57 kg/d for CSM, SBM and MGM, respectively. No significant differences were observed in final body weight (253.7, 256.0 and 261.7 kg), daily gain (1.09, 1.15 and 1.12 kg), scrotal circumference (28.0, 29.4 and 29.0 cm^2), semen volume (4.14, 4.25 and 4.10 ml), sperm concentration (1.52, 1.49 and $1.53 \cdot 10^9$·/ml) and sperm viability (80, 77 and 83% for CSM, SBM and MGM, respectively). Scrotal circumference was significantly correlated to body weight (P<0.01, r=0.76). No differences in blood measurements (total protein, albumen, ALT, AST, ACP, ALP and creatinine, testosterone) were observed among treatments. However, T3 hormone concentration was lower (P<0.05) for CSM than for SBM and MGM (110, 130 and 136 ng/ml, respectively). It was concluded that protein source had no effect on reproductive performance of growing male calves.

Linking genomics to efficiency and environmental traits in dairy cattle
De Haas, Y.[1], Dijkstra, J.[2], Ogink, N.[3], Calus, M.P.L.[1] and Veerkamp, R.F.[1], [1]Wageningen UR Livestock Research, Animal Breeding and Genomics Centre, P.O. Box 65, 8200 AB Lelystad, Netherlands, [2]Wageningen University, Animal Nutrition Group, P.O. Box 338, 6700 AH Wageningen, Netherlands, [3]Wageningen UR Livestock Research, Environment Dept., P.O. Box 135, 6700 AC Wageningen, Netherlands; Yvette.deHaas@wur.nl

Measuring CH_4 production directly from animals is still difficult and hinders both management practices and direct selection to reduce CH_4 emissions. However, developments are under way to develop phenotypic CH_4 measurements; e.g., using (1) laser guns; (2) Fourier Transformed Infrared (FTIR) measuring units; or (3) large scale respiration chamber experiments. Next to these phenotypes, indicators of CH_4 emissions can also be defined with either mid infrared profiles in the milk or feed intake records (e.g. residual feed intake (RFI), or CH_4 predicted from feed intake and diet composition (i.e. the International Panel on Climate Change Tier-2)). RFI (MJ/d) is calculated as the difference between net energy intake and calculated energy requirements for maintenance (based on metabolic live weight) and milk production. In an experimental dataset of 548 heifers, we showed that it is possible to decrease predicted CH_4 emission by selecting more efficient cows (genetic correlation of 0.6 with RFI). However, both the direct phenotypes and most of the indicator traits are difficult and expensive to measure on a large scale, and therefore genomic selection is a promising tool to make progress in breeding environment-friendly cows, since it relaxes the need for information on performance of all animals or their close relatives. In our studies, we have shown that with current genetic parameters a reduction in predicted CH_4 in the order of 11 to 26% in 10 years is theoretically possible, using genomic selection. To double this genetic gain a large reference population (>5,000 animals is required). Therefore, a combined approach, including feeding, management and genetic selection, is likely the best approach to successfully reduce CH_4 emission.

Parentage assignment with molecular markers in sheep: first results of an experiment and prospects

Raoul, J.[1], Chantry-Darmon, C.[2], Barbotte, L.[2], Babilliot, J.M.[2], Bosher, M.Y.[2] and Bodin, L.[3], [1]Institut de l'Elevage, GIPSIE, BP 42118, 31321 Castanet-Tolosan Cedex, France, [2]LABOGENA, Domaine de Vilbert, 78352 Jouy-en-Josas, France, [3]INRA SAGA, BP52627, 31326 Castanet-Tolosan, France;
jerome.raoul@idele.fr

Parentage control of animals which is compulsory for genetic evaluation is based at the present time on single mating either by mono-spermic artificial insemination (AI) or single sire natural mating. However, it becomes more and more difficult for two main reasons: (1) the increasing of flock size; (2) the increasing concern about working time. A study has been implemented to test parentage assignment by molecular markers in conditions of natural mating and AI. A first analysis was designed to establish the conditions of correct parentage assignment. Samples were specifically designed from a DNA sheep bank and the actual relationships were compared with that assigned through analysis of DNA microsatellites. Two experiments were then conducted in life size -A natural mating experiment in Noire du Velay breed: In each flock, single and multi-sire mating groups of ewes were set up. -And a AI experiment performed in the frame of the progeny test led by Insem'Ovin: In each flock mono and hetero-spermic AI group of ewes were designed. Fertility got in single-vs multi-sire mating groups and got after mono-vs hetero-spermic AI were compared. DNA of lamb samples from each group were analyzed with a panel of 16 microsatellites markers. The paternity assignment based on molecular markers will be presented. The change of microsatellites by SNPs would improve the technique. Conditions of its use on a large scale will be discussed. They are strongly linked to the delay between sampling of biological material and results of parentage and of course the balance between the costs of this technique and the benefits it provides. At the present time a SNPs panel is considered, it will gather informative markers for parentage assignment but also mutations of several genes of interest such as PrP, myostatin gene or ovulation genes.

Logistic regression and ROC-surfaces on a Lidia breed allocation problem

Martínez-Camblor, P.[1], Carleos, C.[2], Baro, J.A.[3] and Cañón, F.J.[4], [1]FICYT, Oficina de Investigación Biosanitaria de Asturies, c/. Rosal 7, 33009 Oviedo, Spain, [2]Universidad de Oviedo, Departamento Estadística e I.O. y D.M., Avda. de Calvo Sotelo s/n, 33007 Oviedo, Spain, [3]ETSIIAA, Universidad de Valladolid, CC. Agroforestales, Campus de la Yutera, 34004 Palencia, Spain, [4]Facultad de Veterinaria, Universidad Complutense, Genetica, Avda. de Puerta de Hierro s/n, 28040 Madrid, Spain;
baro@agro.uva.es

The potential of a new method based on ROC curves for individual identification and breed assignment of individuals on marker genotype data was evaluated in seventy Lidia cattle breed lines. Binomial logistic regression (LR) applied on each particular line plus the respective area under ROC curve (AUC) were used as criteria in order to assign one individual to a particular line. The dataset consisted of 1,811 animals from 70 different lines. A set of 24 microsatellite loci were considered for the classification. The particular discrimination capacity was high for most of the lines. Following the leave-one-out method, a minimum AUC of 0.895 was obtained. Two lines attained absolutely correct separation (AUC=1). Volume under ROC-surface was 0.991. When we consider all models and assign each subject to the most probable line, the global percentage of true classification was 84.6%. It is worth mentioning that the percentages of true classification vary greatly between lines. The method was unable to classify the lines labeled as '15' and '46' on which we have little information (9 and 7 subjects, respectively). Although the true-classification-rate of the considered method is not bad (0.846), it is still lower than that obtained by the reference method based on maximum-likelihoods (0.91), and similar to other data-mining/matching learning-based methods. We conclude that this method is advantageous as it permits to incorporate classical LR-based tools. Additionally, each model allows to discriminate each line and to identify the role and relevance of each microsatellite locus.

Testing of Inter-chromosomal LD in a half-sib family of cattle
Gomez-Raya, L. and Rauw, W.M., INIA (Instituto Nacional de Investigación y Tecnología Agraria y Alimentaria), Departamento de Mejora Genética Animal, Carretera de La Coruña km 7,5, 28040 Madrid, Spain; rauw.wendy@inia.es

Inter-chromosomal linkage disequilibrium (LD) was investigated using the Illumina 50K beadChip with a half-sib family with 36 calves. This disequilibrium may arise under selection of loci located at different chromosomes. Investigation of inter-chromosomal LD presents two challenges: (a) extremely large number of LD tests; and (b) misplaced SNPs. We propose: (1) to scan the genome performing all possible LD test between SNPs in different chromosomes simultaneous to linkage analyses when the sire is double heterozygote; (2) to analyze quantiles at the tail of the distribution of likelihood ratio tests (LRT) for all LD estimates for each pair of chromosomes; (3) to compute an inter-chromosomal LRT constructed by adding up all LRT from estimates from all pairs of SNPs between two chromosomes (overall LRT test). This statistics is distributed as a normal (k,2k), with k being the number of SNP pairs. Preliminary analyses have been performed using BTA1 to BTA6 with all 29 autosomal chromosomes totaling over 81 millions LD estimates. The highest overall LRT was for LD between chromosomes 5 and 27 (0.21) and had a tail quantile at 99% of 8.50. Three-dimensional graphs (axes are SNP locations at the two chromosomes and LRT or recombination fraction) are proposed to identify inter-chromosomal regions with LD and misplaced SNPs.

Are genomic evaluations free of bias due to preferential treatment?
Dassonneville, R.[1,2], Baur, A.[3], Fritz, S.[3], Boichard, D.[2] and Ducrocq, V.[2], [1]Institut de l'Elevage, MNE, 149 rue de Bercy, 75595 Paris, France, [2]INRA, GABI, 78350 Jouy en Josas, France, [3]UNCEIA, MNE, Paris, France; romain.dassonneville@jouy.inra.fr

Genomic evaluations now are an essential tool for dairy cattle breeding. While young bulls were the initial target, an increasing number of females (both heifers and cows) are being genotyped. Classical genomic evaluations are based on reliable averaged performances of bulls' daughters. An incoming issue in the 'genomic world' is whether own performances of females genotyped or present in the pedigree of genotyped animals should be explicitly included in the evaluations. Whereas performances of a bull dam has limited impact on the index of her progeny-tested son, the impact of an own performance on the index of a genotyped cow is much more important. Several countries using genomic selection chose to discard or to correct yield deviations (YD) – i.e., the own performances of cows – from genomic evaluations. The purpose of the study is to assess the impact of including YD in the genomic evaluation model. The 3 main French dairy breeds (Holstein, Montbéliarde and Normande) were considered. Two traits were studied: milk production (kg), which is the trait most susceptible to preferential treatment and somatic cell count which is among the less susceptible ones to such a bias. Data consisted of 2 different groups: 29,701 elite females genotyped by breeding companies and 7,314 cows genotyped in a research project (and considered as randomly selected among the national population). 2 different genomic evaluations were performed, one only included DYD (daughter yield deviations) of proven bulls, and another model including both YD for females, and corrected DYD for males. Correlations between breeding values from evaluations with or without YD were lower for elite females compared to random cows. For elite dams, the average difference between breeding values (including YD or not) was significantly different from 0. To conclude, genomic evaluations may be biased when explicitly including own performances of elite females.

Applications of haplotypes in dairy farm management

Cole, J.B., Agricultural Research Service, USDA, Animal Improvement Programs Laboratory, 10300 Baltimore Avenue, Beltsville, MD 20705-2350, USA; john.cole@ars.usda.gov

Haplotypes are now available for almost 100,000 dairy cows and heifers in the US. Genomic EBV values are accelerating the rate of genetic improvement in dairy cattle, but genomic information also is useful for making improved decisions on the farm. Mate selection strategies have usually been based on maximization of genetic progress subject to restrictions on inbreeding, and assuming the transmission of average rather than actual chromosomes, limiting selection gains. Genetic progress can be improved by simulating matings of all cows to a portfolio of potential mates and those which provide the desired outcomes selected. In the case of commercial cows, matings which minimize the variance of outcomes while conditioned on some average desired EBV should be selected. In the case of germplasm producers, the skewness parameter of distributions can be compared to identify matings with the greatest likelihood pf producing offspring with superior EBV. This will increase the rate of genetic progress and reduce the number of animals culled for poor performance resulting from inferior genetics. Low-density SNP tests on dairy calves can be used to increase profitability by increasing the genetic value of the calves raised and used as replacements in the herd, increasing gains through the dams-of-cows pathway. Genotyping calves also can result in increased lifetime profitability. In a simulation study in which the top 90% of calves were retained based on parent average selected calves had EBV $110 greater compared to all calves. When a low-density genomic test was used, $14 per kept calf was gained, but the value of testing decreases as the proportion of calves kept increases. Haplotypes also have been used to identify novel recessives in the Brown Swiss, Holstein, and Jersey breeds, as well as to successfully fine-map the Weaver locus. Future uses of haplotypes include identification of animals resistant to common diseases and identifying those most likely to respond to nutritional and reproductive technologies.

Candidate genes affecting twinning rate in Maremmana cattle

Catillo, G., Marchitelli, C., Terzano, G.M. and Buttazzoni, L., Agricultural Research Council, Research Centre for the Meat Production and Genetic Improvement, Animal Production, Via Salaria 31, 00015 Monterotondo (Rome), Italy; gennaro.catillo@entecra.it

Twinning in cattle is a complex trait with multiple environmental and genetic influences. Ovulation rate has been worldwide recognized as the trait with the highest genetic correlation with twinning in mammals, and the genes controlling the ovulation rate have been the object of a number of studies. The transforming growth factor signaling pathway within the ovary is critical for the regulation of ovarian function, ovulation rate and fertility. In different species (sheep, human and mouse) mutation in three genes GDF9 (Growth Differentiation Factor 9), BMP15 (Bone Morphogenetic Protein 15), and BMPR1B (Bone Morphogenetic Protein Receptor-1B), involved in this pathway, have been associated with either poly-ovulation or infertility. In a herd of 92 Maremmana cows, owned by the experimental farm of CRA-PCM, and deriving from a nucleus of 60 cows and 3 bulls, originally established in 1923, high frequency of twinning was registered along the years; the availability of calving records and genealogy registrations from 1923 allowed the reconstruction of the families where the twinning was more frequent. In a previous study Marchitelli *et al.* identified nine SNP in the GDF9, BMP15 and BMPR1B genes. In the present study, all the cows of the experimental farm were genotyped at the detected SNP; allele and genotype frequencies, Hardy-Weinberg equilibrium were calculated, and haplotypes were inferred. We found that the cows with twin calvings had a different allele frequency, at the non-synonymous mutation (rs110553528) of the GDF9, than the cows with no twins. This missense mutation is located in the GDF9 pro-region, therefore if could affect the structure and the function of the GDF9 protein, that is implicated in a correct foliculogenesis. The allele differences in the cows of this trial allow to consider the rs110553528 mutation as a potential indicator of the ovulation rate.

Variations in composition and nutritional value for ruminants of by-products from bio-based energy
Sauvant, D.[1,2], Chapoutot, P.[2], Heuze, V.[1] and Tran, G.[1], [1]Association Française de Zootechnie, 16 rue Claude Bernard, 75005 Paris, France, [2]AgroParisTech-INRA, 6 rue Claude Bernard, 75005 Paris, France; daniel.sauvant@agroparistech.fr

The global production of maize and wheat by products from bio-based energy has been rapidly increasing since 1995. For maize, the 10 major byproducts of maize distillery are spent grains, wet grains, wet distillers grains (WDG), wet distillers grains with soluble (WDGS), dried distillers grains (DDG), dried distillers grains with soluble (DDGS), maize gluten meal (MGM), maize gluten feed (MGF), condenser distillers soluble (CDS) and dried distillers soluble (DDS).These products vary largely in terms of chemical composition and nutritive value. Some of them are particularly rich in crude protein (CP) such as MGM and some DDGS, others can be quite rich in fat (DDGS), others are rich in NDF (MGF). Composition evolved according to the processes, thus during the last 5 years 'high protein DDGS' (40 to 50% DM) appeared. Moreover, composition also varies within a same category. There is still a debate on the energy content of some of these products for ruminants. Concerning wheat-based ethanol production, three different processes exist in European countries: two dry-milling processes with (1) a first step of bran separating before the saccharification-distillation step, brans are re-incorporated ultimately; or (2) without bran separating; and (3) a wet-milling process, similar to starch extraction with gluten separating at the beginning of the process. The chemical composition of resulting by-products is quite different according to the applied process: more (1) or less (2) starch, and less crude protein (3) respectively for the three methods. A great variability of chemical composition can appear between or within processes and factories. High temperature treatments can decrease the lysine intestinal availability while the ADF content increases because of higher nitrogen-linked fractions. In conclusion, variations of chemical composition and nutritive value of products from bio-based industry are wide but are largely explained.

Energy and nutrient content of European DDGS for pigs
Teirlynck, E.[1], De Boever, J.L.[1], De Brabander, D.[1], Fiems, L.O.[1], De Campeneere, S.[1], Blok, M.C.[2] and Millet, S.[1], [1]Institute for Agriculture and Fisheries Research (ILVO), Animal Science, Scheldeweg 68, 9090 Gontrode, Belgium, [2]Product Board Animal Feed, Stadhoudersplantsoen 12, 2517 Den Haag, Netherlands; emma.teirlynck@ilvo.vlaanderen.be

During the production process of bio-ethanol from grains mainly starch is fermented, leaving by-products, amongst which 'distillers dried grains and solubles' (DDGS). Due to its high protein content, DDGS is a potential alternative for the traditional protein sources in livestock feeds such as soybean meal. In Europe, bio-ethanol production mostly relies on wheat, but also maize, barley and triticale may be used. Due to differences in starting material and production processes the quality of DDGS may vary considerably. The present experiment studied the variability in nutrient digestibility, energy value and amino acid availability of 10 European DDGS batches. The fecal digestibility of energy and nutrients and the apparent ileal digestibility (AID) of protein and amino acids of 8 wheat-based and 2 maize-based DDGS batches were determined in 6 pigs, using indicator method. Fecal protein digestibility averaged 76.0% (67.5 to 80.4%). Fat digestibility averaged 83.9% (74.4 to 90.6%) and was higher for the maize-based DDGS than for the wheat-based DDGS. The net energy value averaged 9.4 MJ/kg DM (8.3 to 10.8 MJ/kg DM). The maize-based DDGS contained more energy than the wheat-based DDGS mainly because of the higher fat content. The AID of protein averaged 64.8% (62.0-79.6%), with the extreme values for the 2 maize DDGS. The AID of lysine was generally low, averaging 53.0%, and was highly variable ranging from 38.9 to 67.1%. The AID was higher for the other amino acids, with for methionine: 82.6% (73.7-91.0%), cysteine: 63.4% (53.2-74.8%), threonine: 72.6% (61.1-82.3%), valine:77.0% (69.0-85.7%), leucine: 85.3% (76.5-90.9%), isoleucine: 79.3% (72.4-86.0%) and tryptophan: 72.0% (56.2-80.9%). Because of the large variation in quality there is a need for reliable methods to predict energy and nutrient content of DDGS.

Prediction of the energy and protein value of DDGS for cattle

De Boever, J.L.[1], Teirlynck, E.[1], Blok, M.C.[2] and De Brabander, D.[1], [1]Institute for Agriculture and Fisheries Research (ILVO), Animal Sciences Unit, Scheldeweg 68, 9090 Melle, Belgium, [2]Product Board Animal Feed, Stadhoudersplantsoen 12, 2517 Den Haag, Netherlands; johan.deboever@ilvo.vlaanderen.be

Dried distillers grains and solubles (DDGS) is a by-product of bio-ethanol production based on grains, which is valorized in animal nutrition. DDGS has a high protein content and may be an interesting alternative for soybean meal. DDGS is also rich in fat and digestible fibre, which makes it particularly appropriate for inclusion in cattle diets. However, because of the use of different grains and grain mixtures and of differences in production processes, the nutritive value of DDGS may vary considerably. It would be interesting to have convenient methods to predict energy and protein value of DDGS in a fast, cheap and accurate way. The nutritive value of 8 batches of wheat-based DDGS and 2 batches of maize DDGS from different European plants was evaluated according to the Dutch energy and protein systems. The net energy value for dairy cattle (NEL) was derived from digestion trials with sheep and varied from 7.16 to 8.73 MJ/kg DM. The protein value was based on rumen degradability and intestinal digestibility determined with lactating cows. The content of protein digestible in the intestines (DVE) varied from 178 to 231 g/kg DM and the degraded protein balance (OEB) from -3 to 95 g/kg DM. The contents of intestinal digestible lysine and methionine ranged from 4.9 to 6.8 g/kg and from 2.8 to 4.8 g/kg DM, respectively. The potential of chemical parameters, *in vitro* digestibility and in situ rumen degradability to predict the energy and protein value was investigated by means of multiple linear regression analysis. For the NEL-value a combination of fat content, starch and a fibre parameter could explain 90% of the variation. For DVE and OEB about 80% of the variation could be explained by protein degradability after 48h rumen incubation combined with respectively starch and crude protein content. The content of available amino acids was difficult to predict accurately.

Rapeseed-cake: an interesting row material for livestock feeding

Mandaluniz, N., Arranz, J., Ruiz, R., Landeras, G., Ortiz, A., Ugarte, E. and García-Rodríguez, A., NEIKER, Animal Production, P.O. Box 46, 01080, Spain; eugarte@neiker.net

The evolution of the prices of cereals in the global markets, linked to the price of oil, has led to the need to review certain feeding strategies. Energetic crops are a possibility to reduce the dependence on inputs. Locally produced oleaginous seeds (i.e. rapeseed) permit farmers to diversify their production, but also to increase the energetic and feeding self-sufficiency, mainly of protein, of the farm. The cold pressed rapeseed yields 1/3 oil and 2/3 oil-cake. Oil can be used as fuel in the farm (after decantation) and the oil-cake can be a raw material for livestock feeding. The objective of the current paper was to analyze the nutritional value of rapeseed-cake to fit feeding formulation to livestock requirements. A survey was conducted with 33 rapeseed-cakes samples collected in 2009 and 2010, harvested in two agro-climatic regions (flat and mountain areas) from 16 different varieties. Samples were analyzed to determine crude fat (CF), crude protein (CP), neutral detergent fiber (NDF) and acid detergent fiber (ADF). The average contents in CP and CF of the rapeseed-cakes samples were 27.5±2.7% and 23.1±4.4%, respectively. The average fibers content was 21.7±2.7% and 18.8±2.7% for NDF and ADF, respectively. It can be concluded, therefore, that cold pressed rapeseed cake obtained at farm level is a suitable raw material for livestock feeding in terms of energy and protein content. Nevertheless, the observed variability in protein and fat contents, around 10% and 20 respectively, suggests the need to monitor these raw materials periodically for the suitable formulation of feedstuffs for livestock.

In vitro degradability and energy value of rapeseed cake produced on farm by cold extraction press
Guadagnin, M.[1], Cattani, M.[2], Tagliapietra, F.[2], Schiavon, S.[2] and Bailoni, L.[1], [1]University of Padova, Department of Comparative Biomedicine and Food Science, Viale dell'Università, 16, 35020 Legnaro, Padova, Italy, [2]University of Padova, Department of Agronomy Food Natural resources Animals and Environment, Viale dell'Università, 16, 35020 Legnaro, Padova, Italy; matteo.guadagnin@studenti.unipd.it

This experiment aimed to evaluate *in vitro* NDF (NDFd) and true DM (TDMd) digestibility, gas production (GP), and metabolizable energy (ME) content of rapeseed cake produced on farm by cold extraction press (RSC) in comparison with soybean seed (SBS) and soybean meal (SBM). A fully automated GP system equipped with 64 bottles was used. Each feed (0.5 g) was incubated in 4 replicates with 10 ml of rumen fluid and 65 ml of 2 different media: Menke medium (deficient of energy) or ammonia-free medium (deficient of nitrogen). Two sets of 24 bottles were incubated to evaluate NDFd and TDMd values at 48 h, and GP kinetics for 72 h. Eight blanks for each medium were also included. The ME content of feeds was estimated from NDFd and chemical composition. All data were submitted to ANOVA using feed, medium and their interaction as sources of variation. Compared to SBS and SBM, RSC showed lower (P<0.001) NDFd (68.1, 91.6, 91.0%, for RSC, SBS, and SBM, resp.), TDMd (82.9, 98.1, 98.1%, in the same order), total GP (156, 180, 214 ml/g DM, in the same order), and an intermediate ME content (15.2, 17.0, 13.9 MJ/kg DM, in the same order). Medium composition did not influence total GP but affected GP rate of feeds. With the Menke medium, RSC provided higher (P<0.05) GP rate (14 ml/h/g DM) compared to SBS and SBM (10.5 and 12.1 ml/h/g DM, resp.), but only at 3 h of incubation. With ammonia-free medium, RSC showed the highest (P<0.05) GP rate from 9 to 15 h of incubation (on average 17 ml/h/g CP) compared to SBS and SBM (on average 12.7 and 12.0 ml/h/g CP, resp.). In conclusion, RSC produced on farm by cold extraction press could be a good source of energy and protein for ruminants due to the high fermentability within the first hours of incubation.

Eggs quality from conservation flocks fed a diet with maize distillers dried grains with solubles
Krawczyk, J., National Research Institute of Animal Production, Department of Animal Genetic Resources Conservation, Krakowska 1, 32-083, Balice, Poland; jkrawczy@izoo.krakow.pl

The objective of the study was quality of consumption eggs in hens from conservation flocks fed a diet containing maize distillers dried grains with solubles (DDGS). A total of 360 Greenleg Partridge (Z-11) and Rhode Island Red (R-11) hens, included in the genetic resources conservation programme in Poland, were investigated. Birds of each breed/line were assigned to 2 groups of 180 hens per group. The control group (C) received a complete standard diet and the experimental group (E) was fed a diet containing 10,4% DDGS. The experiment was conducted from 20 to 56 weeks of age. At 56 weeks of age, 50 eggs were randomly collected from each group, 20 of which were analysed for physical characteristics using the EQM (Egg Quality Management) system. Shell strength [N] was determined with an Egg Crusher. In eggs from 56-week-old hens, greater differences in quality traits were found between the breeds than between the experimental and control groups. Eggs from hens fed the DDGS diet for 36 weeks were characterized by significantly lower weight, higher total protein content of egg albumen, and more intense yolk colour (P≤0.01 or P≤0.05). Decreased yolk weight and lower vitamin A and E content were also found in hens fed the DDGS diet, (P≥0.05). Hen's genotype and the diet used had no effect on egg shell quality. The experimental DDGS diet fed to laying hens did not adversely affect egg aroma, egg flavour, and yolk colour. There were no differences in this regard between the lines.n conclusion, the results of the current study demonstrated that maize DDGS can serve as a useful source of protein in the nutrition of hens from conservation flocks, partly replacing imported soybean meal. The use of DDGS in the diets of hens from conservation flocks improved albumen quality and increased yolk colour intensity, and had without affecting sensory parameters of cooked eggs. This work was conducted as part of grant no. N R12 0083 10 financed by National Centre for Research and Development.

Achieving optimal cow performance with the aid of information systems
Nir (markusfeld), O., afimilk, applied research, Kibbutz Afikim, 15148, Israel; oded.nir@afimilk.co.il

Information systems in the dairy herd are used for planning, management, follow-up & control. Herd data analysis, is a continuously evolving process, in which we address the questions: (a) diagnosis and alert (what happens?); (b) retrospective monitoring (what happened?); (c) retrospective evaluation of causality and economical losses (why did it happen? what were the losses in production and fertility? what were the economical losses?); (d) the quality of data; (e) from manual observation to automation; and (f) prediction abilities (what will happen?). There is no 'universal truth', but each herd has its own 'local truth'. Most production and infectious diseases are multifactorial and call for a 'multifactorial approach'. We apply routine causal analysis based on regression models on data collected from individual herds in order to expose their 'local truth' and to evaluate the contributions of various factors to metabolic diseases, lower fertility and milk yield in the individual herds. More automation will lead to better data, both in quantity and in quality. Afimilk© system has already many automated components that replaced, partly or completely the need for manual observations. New ones are now being incorporated: (a) AfilabTM, an in line – on line milk analyzer that performs real-time analysis of individual cow milk solids; and (b) Pedometer+TM, a new leg tag that continuously records activity (number of steps), lying in every milking). Whole herd models, based on among herds' differences and talking into account production, fertility, health, nutrition, and economics are called for. We routinely estimate the independent effects of stocking density, mean DIM and SCC on yield from monthly data of actual marketing Routine health reports based on epidemiological models are today a common tool. The speaker believes that future progress will be in three main fields (a) improvement of data through automation; (b) development of multidisciplinary models including economical evaluations; and (c) improvement of methods applied to small herds.

Accuracy and potential of in-line NIR milk composition analysis
Melfsen, A., Haeussermann, A. and Hartung, E., Christian-Albrechts-University Kiel, Institute of Agricultural Engineering, Max-Eyth-Str. 6, 24098 Kiel, Germany; amelfsen@ilv.uni-kiel.de

The knowledge of daily milk composition changes can assist monitoring of dairy cow health and helps to detect nutritional imbalances. An analytical tool with the possibility to analyze milk during daily milking routine would provide such information. Near infrared spectroscopy (NIRS) can analyze multiple constituents in a given substrate at the same time. In this study, a special near infrared (NIR) in-line milk analyzing device was designed, and its ability to predict the contents of fat, protein, lactose, urea, and somatic cell count in milk during the milking process was evaluated. Near-infrared (NIR) spectra of raw milk (n=3,119) were acquired on three different farms during the milking process of 354 milkings over a period of six months. Corresponding subsamples were taking during the milking process and used for reference analysis. Chemometric tools were applied for preprocessing of spectra, calibration of different milk constituents, and validation of results. The potential and accuracy of milk composition analysis was tested with randomly created datasets in comparison with farm internal and external datasets. The accuracy of prediction was compared to international recommendations for reproducibility (R) for in-line analyzing devices. In addition, the robustness of NIRS milk composition analysis was evaluated and compared by means of different multivariate models comprising internal and external effects. The results in this study underlined that an excellent to good accuracy can be achieved for prediction of fat, protein and lactose content in milk in fully randomized calibration and validation sets. In most cases, the variability of future milk samples is missing in randomized test sets, which causes an insufficient robustness towards spectra of future measurements. A robust calibration that covers up most of the variability of raw milk samples would be more suitable and beneficial.

The modern cow bell: activity and rumination sensing collars
Bar, D., SCR Engineers, 6 Haomanut St., 42504 Netanya, Israel; doronb@scr.co.il

Modern farming doesn't provide the same intimate proximity of a farmer spending many hours with just few cows. Identifying cows in heat or even recognizing sick cows become major challenges in larger herds. Within the last 25 years various technological solutions for heat identification have been developed, but none of them gained general popularity. This has rapidly changed with the development of the H Tag neck collars. The H Tag combines a unique motion sensor, a detailed activity analysis within the tag, and storage of this information until the tag is read. This activity information is further analyzed in the control box (either connected to a PC or as standalone box) to make the system much more robust to management changes than previous technologies. This has lead to a very rapid adoption of the H Tag in many European countries. Just between 2007 and 2011, more than a million such tags were being sold. Soon, various studies confirmed the benefits of using this system on the fertility results. But using such a system has also other 'side effects', such as peace of mind, more valuable time for other chores or for more free time, and the reduction of hormone usage just to mention a few. In this respect, the new addition of rumination monitoring in the recently widely available HR Tag is opening many new possibilities. Rumination is the result of what the cow has eaten and how well she could rest. Recent studies demonstrated that it can serve as a very early indicator of future calving diseases, and as an early indicator for diseases like mastitis. As about the third of cows in a modern dairy suffer from one of the calving diseases and about a third of cows will suffer from mastitis (many times of recurring nature), the importance of such a sensor is obvious. Rumination has been shown to change by ration components, ration physical characteristics and general management. Constant rumination monitoring and management can therefore optimize not only the individual lactation curve but also the whole herd production curve.

Detection of early lactation ketosis by rumination and other sensors
Steensels, M.[1,2], Bahr, C.[1], Berckmans, D.[1], Antler, A.[2], Maltz, E.[2] and Halachmi, I.[2], [1]Katholieke Universiteit Leuven, Division Measure, Model & Manage Bioresponses (M3-BIORES), Kasteelpark Arenberg 30, 3001 Heverlee, Belgium, [2]Agricultural Research Organization (ARO), Institute of Agricultural Engineering, P.O. Box 6, 50250 Bet-Dagan, Israel; machteld@volcani.agri.gov.il

Early lactation ketosis is a common health problem in dairy cows. Sensors can assist stockpersons in monitoring the herd for health problems. However, data signals have to be modeled to interpret the data into a clarified to-do-list. The aim of this study was to evaluate whether rumination can be used for early detection and modeling of early lactation ketosis. Data were collected from November 2010 to February 2012 in an Israeli dairy farm with 1,100 milking cows. Ruminating time, activity by head movements and milk yield were measured online and monitored from calving till 5 days after the clinical diagnosis of ketosis. A dataset of 40 healthy and 32 ketotic cows was built for calibrating logistic regression coefficients. The coefficients were used to validate another dataset of 15 healthy and 22 ketotic cows. Ruminating time was 6-8 min/2h higher in healthy than in ketotic cows. The highest correlations between ruminating time and ketosis diagnosis were 2 to 5 days before clinical diagnosis. A logistic regression model was built by a stepwise addition of these days. The correct classification rate improved in each step. The independent contribution of other variables in the farm were analysed. Correlations between activity and ketosis diagnosis were higher before clinical diagnosis, while correlations between milk yield and ketosis diagnosis were higher after clinical diagnosis. In this study it was found that ketosis alters the behaviour of milking cows. Rumination sensors are useful to detect early lactation ketosis. In the future, the model will be extended with other variables (activity, milk yield) and validated on a larger dataset. The study was funded by the Israeli Agricultural Ministery Chief Scientist Fund, project 459-4426-10 and 459-4369-10.

Comparison between direct and video observation for locomotion assessment in dairy cow

Schlagater-Tello, A.[1], Lokhorst, C.[1], Bokkers, E.A.M.[1], Koerkamp, P.W.G.[1], Van Hertem, T.[2,3], Steensels, M.[2,3], Halachmi, I.[3], Maltz, E.[3], Viazzi, S.[2], Romanini, C.E.B.[2], Bahr, C.[2] and Berckmans, D.[2], [1]Wageningen UR, Edelhertweg 15, 8219 PH Lelystad, Netherlands, [2]Katholieke Universiteit Leuven, Kasteelpark Arenberg 30, Heverlee 3001, Belgium, [3]Agricultural Research Organization, The Volcani Center, 50250 Bet-Dagan, Israel; andres.schlagetertello@wur.nl

The objective of this research was to test the variation of locomotion scoring when assessed directly or from video, in order to obtain the best Gold Standard as possible for the development of an automatic lameness detection system. The locomotion of 299 cows was assessed with a 5 categories score by two trained observers. Every cow was assessed two times in direct scoring sessions and two times in video scoring sessions. The within and inter observer reliability (WOR and IOR), were calculated as Kappa coefficient. Lameness prevalence for direct and video scoring for both observers were analysed. The significance level was established by calculating the 95% confidence interval (95% CI). The average WOR for direct/direct, direct/video and video/video comparison was 0.43 (95% CI=0.35-0.52), 0.36 (95% CI=0.27-0.44) and 0.61 (95% CI=0.54-0.69), respectively. The IOR values do not differ significantly in relation to direct or video observation. The frequencies for healthy/lame cows for direct and video observations was 77.0/22.9% (95% CI=0.75-0.79) and 73.4/26.6% (95% CI=0.71-0.76), respectively. Observer two recorded a significant lower lameness prevalence in direct than in video observation. There was individual variation that was reflected in the values obtained for WOR, IOR and lameness prevalence. The higher values for WOR and the similar distribution of healthy/lame suggest that, the video observation, would facilitate to the observers to focus their attention in the same part of a scene. In conclusion, the individual variation associated to subjective scoring might be diminished when locomotion scoring is performed by video observation.

Evaluation of potential variables for sensor-based detection of lameness in dairy cattle

Van Hertem, T.[1,2], Maltz, E.[2], Antler, A.[2], Schlageter Tello, A.[3], Lokhorst, C.[3], Viazzi, S.[1], Romanini, E.[1], Bahr, C.[1], Berckmans, D.[1] and Halachmi, I.[2], [1]Katholieke Universiteit Leuven, Division Measure, Model & Manage Bioresponses (M3-BIORES), Kasteelpark Arenberg 30 – bus 2456, 3001 Heverlee, Belgium, [2]Agricultural Research Organization (ARO) – the Volcani Center, Institute of Agricultural Engineering, P.O. Box 6, 50250 Bet Dagan, Israel, [3]WageningenUR, Livestock Research, P.O. Box 65, 8200AB Lelystad, Netherlands; tomv@volcani.agri.gov.il

Lameness is an underestimated problem in intensive dairy farming concerning health and welfare of animals due to low awareness, difficult recognition and poor registration. The aim of this study was to quantify the effect of lameness on dairy cow behaviour and performance variables measured by online sensors, in order to evaluate its potential availability for early detection of lameness. Ruminating time, activity by head movements (HM) and milk yield were measured online by sensors. In a first analysis, 660 treated lame cows were identified from the herd health reports. In addition, 18 lame and 85 non-lame animals were identified in a three week locomotion scoring period. Milk yield starts dropping on average 40 days prior to treatment day, and daily activity starts dropping 20 days prior to treatment, which emphasises that cows are treated a substantial period after the health problem started. Ruminating time is less affected by lameness. A group level (non-lame vs. lame) difference was found for milk yield (36 vs. 30 kg/day), daily activity (389 vs. 327 HM/day) and maximal activity (51 vs. 42 HM/2h). This suggests that hoof and leg problems are manifested by changes in behaviour as well as in performance. Activity and milk yield seem to be potential variables to develop a statistical model to quantify behavioural changes by lameness. This study is part of the Marie Curie Initial Training Network BioBusiness (FP7-PEOPLE-ITN-2008), and contribution number 459-4398-951 funded by the Israeli Agricultural Research Organization (ARO).

Changing conditions require higher level of entrepreneurship for farmers
Beldman, A.C.G., Lakner, D. and Smit, A.B., LEI- Wageningen UR, P.O. Box 65, 8200AB Lelystad, Netherlands; alfons.beldman@wur.nl

External conditions for dairy farmers are changing continuously. We can distinguish management decisions in long term en short term decisions. Daily decisions, e.g. how much concentrates will this cow get, are part of operational management. Tactical management is about choices for one season or one year. Strategic management is about long term decisions for the next 3-5 years. Obviously farmers spend a lot of time on operational management decisions. But the changing circumstances require more attention for strategic decision making. Based on the experience in several projects. LEI has developed a method to strengthen the strategic skills of farmers and to support farmers to develop a strategy. The method is called Interactive Strategic Management (ISM). The entrepreneur is in the center of this process. The ISM training is mostly implemented with groups of 8-10 farmers. The focus in the training is on the individual farmers in the group. Interaction within the group and homework assignments are used to stimulate creative thinking and reflection. In the first step the farmer analyses his own situation by looking in detail at (1) himself as entrepreneur (ambitions and skills); (2) at his current enterprise (structure and performance); and (3) at the environment (market and society). A webbased tool (Strategic Management Tool) is used to support the farmer in this process. This tools results in scores and graphs for the mentioned elements. In the following steps the farmer is facilitated to define his own goals and strategy. The webbased tool calculates a score for a number of possible strategies. The final step is translate the choice of strategy to a good action plan. The application of the method is presently studied in three Central and Eastern European countries in the frame of a Leonardo da Vinci program in combination with Wageningen UR Eastern Europe entrepreneurship study. The process will be explained more in detail and illustrated with examples of farmers strategies under various circumstances.

Interactive strategic management methodology for improvement of entrepreneurship: case of a farmer
Prezelj, K.[1], Klopcic, M.[2] and Beldman, A.[3], [1]Farm Pr Kendu, Idrijske Krnice 1, Spodnja Idrija, Slovenia, [2]Biotechnical Faculty, Dept. of Animal Science, Groblje 3, 1230 Domžale, Slovenia, [3]LEI-Wageningen UR, P.O. Box 65, 8200 AB Lelystad, Netherlands; katica.prezelj@gmail.com

In the frame of Leonardo da Vinci program 'Transfer of innovation' in combination with Wageningen UR Eastern Europe entrepreneurship study, a project in three new EU countries is performed trying to improve farmers' entrepreneurial skills and the ability to make long term decisions. As part of activities about 50 dairy farmers in as well Slovenia, as Lithuania and Poland, formulated their future goals and strategies with help of the so called ISM-tool. We present the case of a mountain dairy farm in Slovenia. Family farm 'Pr Kendu' is located 900 m above sea level. Three generations live and work together on a farm with 40 ha of agricultural land and 25 ha of forest. Main activities are milk production and processing and forestry. 40 Brown cows produce 276,000 kg of milk. Nearly 100,000 kg milk is processed on farm into cheeses, yogurts, butter and other. Future goals are to improve fat and protein content and longevity of Brown cows, to enlarge milk processing capacity on the farm and to offer high quality milk products on the local market. These goals are planned to be achieved by improving the quality of feed such as high-quality forage, use of selected bulls and cows with BB kappa casein, construction of a new cheese factory, the acquisition of additional skills about processing, marketing and consumer demands, search for new marketing channels and hiring additional labour force. Critical success factors can be health problems with animals or family members, market situation with milk and milk products, and lack of investment money. Better quality of milk and milk products, higher family income, and new working places are opportunities on one side, but resulting higher costs, needed investment, greater labour intensity and the responsibility towards consumers/clients give some mental pressure (can be seen as treats) of this strategy.

Shortening the dry period for dairy cows: effects on energy balance, health and fertility
Van Knegsel, A.T.M. and Kemp, B., Wageningen University, Adaptation Physiology Group, Wageningen Institute of Animal Sciences, De Elst 1, 6708 AW Wageningen, Netherlands; ariette.vanknegsel@wur.nl

Recently, a short list of studies has shed a different light on the long tradition of a dry period for dairy cows illustrating how health and fertility of dairy cows can be improved after shortening or complete omission of the dry period. The objective of this paper is to review the current knowledge on dry period length in relation to energy balance (EB), fertility, and health of cows. Review is based on experimental studies only. Energy balance and body condition score (BCS) in early lactation were improved in all available studies after omission of the dry period. These improvements in EB were realized by both a reduction in milk yield and an improvement of feed intake around calving. Results on incidence of disease (mastitis, displaced abomasum, retained placenta) are ambiguous, except for ketosis where all studies reported a lower incidence after a short or no dry period. Results on reproductive performance (days until first ovulation, conception rate) show an improvement after a short or no dry period in all studies for older (parity ≥3), but not for young cows (parity 2). Available studies are limited in number and monitor cows only during part of the successive lactation. Currently, we study conventional (60 d), short (30 d) and no dry period in a herd of 168 cows during 2 successive lactations. Preliminary data show an improvement in BCS (2.9 vs. 2.6 vs. 2.2±0.1; P<0.05) and greater body weights (686 vs. 661 vs. 664±1 kg; P<0.05) in early lactation after a dry period of 0 or 30 days compared with 60 days. In conclusion, shortening or omitting the dry period has potential to improve EB, health and fertility of dairy cows in early lactation. Research is ongoing to study long-term effects of dry period length reduction during successive lactations on cow health and fertility.

Technical and economical consequences of extended (18 m) calving intervals for dairy cows
Brocard, V.[1], Portier, B.[2], Francois, J.[2] and Tranvoiz, E.[1], [1]Institut de l'élevage, monvoisin BP 85225, 35652 Le Rheu cdx, France, [2]Chambres d'Agriculture de Bretgane, 5 allée Sully, 29322 Quimper Cdx, France; valerie.brocard@idele.fr

Voluntary extending calving interval may improve fertility, delay dry-off of high yielding cows and reduce lifetime exposure to the sanitary risks associated with calving and early lactation. During a 6-years trial (2005-2011) two groups of 24 high producing Holstein cows (12 vs 18 months calving intervals) within a compact calving pattern were compared in Trévarez experimental farm (Brittany, Western France). The 12 m control group calved 6 times in autumn, the 18 m group calved twice alternately in autumn and spring. Cows were fed with maize silage, grazed grass (0.15 ha per cow) and 1 t of concentrates per cow per year. All the cows were managed as one herd. Over the 6 years period 216 lactations of 95 different cows were recorded. Yearly feed intake was similar in the two groups. Production results were assessed per lactation and for a 365 d production period (lactation and dry off) to compare both groups. At the lactation level, the 18 m group produced significantly more milk (+3,180 kg) than the 12 m group. But on a yearly basis, it significantly produced 510 kg of milk, 30 kg of fat and 14 kg of protein per 365 d period less than the 12 m control group (P<0.05 for all criterion). The extended primiparous cows produced milk lactation curves with a very high persistency. Extending calving interval had no significant consequence on reproductive performances. The number of health problems was not decreased by extended calving intervals (5.6 trouble per cow over the 6 yrs period), but more lameness were assessed in 18 m group than in 12 m group. The rate of sane udders was lower at dry off in the 18m group but this did not lead to a significant penalty level on milk price. Culling rate was slightly reduced by the extended calving interval. Finally, extending lactation of the whole herd increased the farm gross margin by some 3 000 € per year.

The effect of dry period management and nutrition on milk production
Blazkova, K.[1], Cermakova, J.[1,2], Dolezal, P.[2] and Kudrna, V.[1], [1]Institute of Animal Science, Department of Nutrition and Feeding of Farm Animals, Pratelstvi 815, 104 00 Praha Uhrineves, Czech Republic; [2]Mendel University in Brno, Department of Animal Nutrition and Forage Production, Zemedelska 1/1665, 613 00 Brno, Czech Republic; cermakova.jana@vuzv.cz

The objective of this study was to evaluate the effect of shortened dry period length (35±6.3 d) and type of a diet on dry matter intake and milk production in comparison with widely adopted traditional dry period from six to eight weeks. The experiment included 34 multiparous dairy cows, divided into two balanced groups. The control group assigned to traditional dry period of approximately 60 days (57±5.9 d) was fed a diet corresponding with nutrition requirements of dried cows. Approximately three weeks before expected calving the concentrate mixture was added into their diet. The cows of experimental group assigned to shortened dry period were fed a late lactation diet until calving. Shortening the dry period increased average dry matter intake before calving about 4.11 kg per day compared with traditional dry period. Cows with shortened dry period however, produced in the average about 3.23 kg/d less milk during the first 100 DIM (P<0.05) and their lactation curve was flatter. The average daily production of FCM and milk components was higher for cows with traditional dry period when these cows produced about 3.09 kg more FCM than the cows assigned to shortened dry period. There were no differences in body weight of the calves between the control and experimental group. Shortening the dry period length (35±6.3 d) increased dry matter intake prior calving, but decreased dry matter intake and reduced milk production after calving.

Effect of trace mineral supplementation on the reproductive performance of pasture based dairy cows
Watson, H.[1,2], Evans, A.C.O.[1] and Butler, S.T.[2], [1]School of Agriculture and Food Science, UCD, Belfield, Dublin 4, Ireland; [2]AGRIC, Teagasc, Moorepark, Fermoy, Co.Cork, Ireland; hazel.watson@teagasc.ie

The aim of this study was to examine the effects of slow release trace mineral (TM) bolus supplementation on blood indicators of TM status and reproductive performance in primarily pasture-based dairy cows. In total, 1,381 animals on 5 farms were enrolled in the trial, of which 1,311 were retained for data analysis. Within each farm, cows were randomly assigned to one of 4 treatments: CTRL cows received no TM bolus; DRY cows received TM bolus at dry-off; BREED cows received TM bolus 6 weeks before mating start date (MSD); DRY_BREED cows received TM boluses at dry-off and 6 weeks before mating start date. Each TM bolus contained 30 g Cu oxide, 3400 mg I, 500 mg Se, and 525 mg Co. Blood samples were collected from 10 cows per treatment on each farm at 6 time points between dry-off and 6 weeks after MSD. These blood samples were analysed for plasma concentrations of Cu, Se and inorganic I. Herd breeding records and ultrasound scanning results at the end of the breeding season were collated. Treatment effects were analysed using mixed model repeated measures procedures. Treatment, time, treatment × time, parity and farm were included as fixed effects and cow was included as a random effect. Binary reproductive variables were analysed using the Chi-square test. Treatment with TM boluses increased plasma inorganic I (P<0.001) and Se (P=0.03), but Cu was not affected by treatment (P>0.5). For most of the duration of the study, plasma concentrations of Cu, Se, and I were within the normal range for all treatments, including CONTROL. Fertility records indicated that 21 day submission rate, pregnancy rate after first AI, 42 day pregnancy rate and overall pregnancy rate were not affected by TM bolus treatment (all P>0.1). These results show that TM supplementation does not affect herd reproductive performance in well managed pasture-based animals where plasma TM concentrations are in the normal range.

Productivity and health of dairy cows differing in milk yield and milk protein concentration

Wiedemann, S.[1], Sigl, T.[2], Gellrich, K.[2], Kaske, M.[3] and Meyer, H.H.D.[2], [1]Institute of Animal Breeding and Husbandry, CAU, Olshausenstr. 40, 24098 Kiel, Germany, [2]Physiology Weihenstephan, TU Muenchen, Weihenstephaner Berg 3, 85354 Freising, Germany, [3]Clinic for Cattle, TiHo Hannover, Bischofsholer Damm 15, 30173 Hannover, Germany; swiedemann@tierzucht.uni-kiel.de

Effects of a short-term feed restriction during early lactation on metabolism and productivity of Holstein-Frisian dairy cows differing in milk yield and milk protein concentration were studied. From d 23-25 pp (P1) cows were fed a partly mixed ration based on corn and grass silage *ad libitum* and additional concentrates. From d 26-28 pp (P2) cows received 70% of DMI in P1. From d 29-31 pp (P3) cows received P1-diet again. Depending on milk yield and mean milk protein concentration in P1 cows were categorized in 4 groups: high milk yield and high protein concentration (n=6; MP), low milk yield and low protein concentration (n=5; mp), high milk yield and low protein concentration (n=7; Mp), low milk yield and high protein concentration (n=5; mP). All obtained records were averaged for each feeding period. REML in the MIXED procedure was performed for repeated measures using day of lactation, group and day×group interaction as fixed effects (SAS). DMI during P1 was highest in MP-cows and lowest in mp-cows. Feed restriction induced reduction in milk and protein yields as well as in milk protein concentration in MP-cows. Also, milk protein concentration lowered continuously throughout P2 in high protein cows. Serum glucose levels decreased in low protein-cows during P2. NEFA levels rose above critical values of 1,000µmol/l, but not in mp-cows and BHBA levels increased drastically above critical values for subclinical ketosis in all groups. In sum, despite large animal-to animal variation high milk protein concentration cows tended to show better metabolic adaptability to an increase in extent of negative energy balance compared to low milk protein concentration cows.

Comparative analyses of health traits from regional projects for genetic improvement of dairy health

Stock, K.F.[1], Agena, D.[1], Spittel, S.[2], Schafberg, R.[3], Hoedemaker, M.[2] and Reinhardt, F.[1], [1]Vereinigte Informationssysteme Tierhaltung w.V., Heideweg 1, 27283 Verden / Aller, Germany, [2]University of Veterinary Medicine Hannover, Clinic for Cattle, Bischofsholer Damm 15, 30173 Hannover, Germany, [3]Martin-Luther-University Halle-Wittenberg, Institute of Agricultural and Nutritional Sciences, Theodor-Lieser-Strasse 11, 06120 Halle / Saale, Germany; friederike.katharina.stock@vit.de

In the context of increasing importance of functional traits in dairy, a comprehensive key for health data recording has been implemented in different herd management software in Germany, facilitating standardized analyses. However, recording conditions differ considerably between farms, with yet unknown influences on genetic evaluations for health traits. Comparative analyses were therefore performed using two sources of data: In the project GKuh dairy farmers started health data recording with intense expert support. In large dairy farms in Thuringia (THU) routine recording of health events has been practiced for years without specific expert support. Considering health data from 2010-2011 in GKuh (49 farms; 14,307 females) and 2009-2011 in THU (18 farms; 43,668 females), patterns of lactation incidences were similar. Genetic analyses were performed separately in repeatability linear animal models, revealing univariate heritabilities (REML) of 0.02-0.06 in GKuh and 0.03-0.08 in THU. Correlation analyses were based on estimated breeding values (EBV) for health traits and EBV from routine genetic evaluation. Of the 1,709 German Holstein bulls with daughters in the health data only 213 were represented in both datasets. Moderately positive correlations were found between corresponding bull EBV from GKuh and THU (r≤0.4; closer for metabolic and claw diseases than for mastitis) and between EBV for health traits and EBV for longevity and specific functional traits (r=0.4-0.8). Results indicate that combined use of owner-recorded data from different sources will facilitate improvement of functionality of dairy cattle via genetic evaluation for health traits.

Consequences of selecting for feed conversion efficiency traits on fertility and health

Pryce, J.E.[1], Marett, L.[1,2], Bell, M.J.[3], Wales, W.J.[1,2] and Hayes, B.J.[1], [1]Department of Primary Industries, Biosciences Research Division, 1 Park Drive, Bundoora, 3083, Australia, [2]Department of Primary Industries, Future Farming Systems, 1301 Hazeldean Rd, Ellinbank, 3820, Australia, [3]Melbourne University, Melbourne School of Land and Environment, Building 142, Parkdale, Melbourne, 3010, Australia; jennie.pryce@dpi.vic.gov.au

Recently there has been renewed interest in selecting for residual feed intake, which is the feed an animal consumes adjusted for the predicted energy requirements for maintenance, production and body condition score change. Failure to properly account for changes in body composition would result in a trait mathematically equivalent to energy balance. Negative energy balance in early lactation has been the subject of intense phenotypic (and genetic) investigation and is genetically correlated to poor fertility. Conversely, the genetic correlations between RFI and health and fertility are less well understood. Most evidence is encouraging with insignificant associations of RFI with fertility traits being reported. However, more data is required to be able to understand the genetic relationship between RFI and other important traits. Approximations of these correlations can be obtained using regressions of RFI on sire breeding values. The correlation between RFI and fertility estimated in this way was unfavourable (0.10), but was also not significant (P=0.19). Using this genetic correlation estimate, RFI was tested as an additional trait in the Australian Profit Ranking index (APR) that currently includes milk, fat and protein yield, survival, fertility, somatic cell count, liveweight, temperament and likeability. The economic value for RFI included greenhouse gas emissions in addition to feed costs ($62.10 /kg DM). Selection on APR with RFI as an extra trait would still result in a favourable response in fertility, having a very small reduction in response compared to APR excluding RFI. This demonstrates that selection for RFI and improved fertility is feasible using a multi-trait selection index.

Longevity and culling reasons of Polish Red-and-White cattle included in the conservation programme

Sosin-Bzducha, E.M., National Research Institute of Animal Production, Department of Animal Genetic Resources Conservation, Krakowska 1, 32-083 Balice, Poland; sosine@izoo.krakow.pl

Among the attributes of the indigenous populations are longevity, high resistance and good health. The Polish Red-and-White breed (ZR) has the largest population of cattle included in the genetic resources conservation programme (GRCP) in Poland. At the beginning of the programme in 2008 there were 1,715 cows in 276 herds and in 2010 the number increased to 3,258 cows in 445 herds. In 2011 the population declined slightly to 3013 cows kept in 441 herds. The aim of this preliminary study was to describe longevity and reasons for culling ZR cattle from herds enrolled in the GRCP between 1 March 2008 and 1 March 2011. Data on 734 cows were collected from the SYMLEK system and from farmer documentation. All cows sold to other farms or without the exact date of birth were excluded. Data on 318 cows from herds were the basis for further analysis. The average and maximum lifespan for all culled cows was 2,756 and 5,573 days, respectively. About 40% of the cows lived longer than 3,000 days. The main culling reason was infertility and reproduction problems (37.74%) and the average and maximum lifespan of cows was 2,670 and 5,573 days. Only 4.09% of the cows were culled due to age and the maximum lifespan in that category was about 5,521 days. The frequencies of other reasons were (%): 11.94 udder health, 10.69 foot and leg problems, 5.03 metabolic and digestive diseases, 2.83 low production. Over 25% of all cases were classified as 'injury/accident' (17.3%) and 'others' (8.18%). There have been sporadic cases of respiratory and infectious diseases. The results of the present study show that reproduction problems are the main reason for culling, but the max. lifespan for that category is longer even than for the 'age' category, which suggests that 'infertility' is overused or used improperly. Reasons of culling need further detailed analysis. This work was conducted as part of multiannual programme 08-1.31.9., financed by the Ministry of Agriculture and Rural Development.

Real-time individual dairy cow energy balance estimated from body reserve changes
Thorup, V.M.[1], Højsgaard, S.[2], Weisbjerg, M.R.[1] and Friggens, N.C.[3], [1]Aarhus University, Animal Science, Blichers Alle 30, 8830 Tjele, Denmark, [2]Aarhus University, Genetics and Biotechnology, Blichers Alle 20, 8830 Tjele, Denmark, [3]INRA AgroParisTech, UMR 791 Modélisation Systémique Appliquée aux Ruminants, 16 rue Claude Bernard, 75231 Paris, France; vivim.thorup@agrsci.dk

In dairy cows extended periods of negative energy balance have been linked with reduced reproduction, digestive, locomotive, and metabolic diseases. Being able to assess energy balance, and thus detect excessive negative energy balance, in real-time for individual cows on-farm would be an advantageous management tool. Traditional methods to estimate energy balance require measurements of milk yield and feed intake, but the latter is hard to measure on commercial farms. Recently, we presented a simple method that estimates energy balance from changes in body reserves (EB_{body}) using only historic data of frequent body weight (BW) and body condition score (CS) measurements. Here, we adapt that method to estimate individual EB_{body} in real-time, we also evaluate the consequence of omitting CS data when estimating EB_{body}. For this we use lactation profiles of BW (measured at each milking in an automated milking system) and visually assessed CS from 31 Holsteins, 17 Jerseys and 29 Danish Red cows. BW is corrected for milk yield, foetus, and gutfill. BW and CS are real-time smoothed, and from changes in BW and CS real-time individual EB_{body} profiles are calculated and smoothed. Real-time smoothed EB_{body} performs well when compared to traditional measures of EB, although as expected some lag exists at the beginning of profiles. In order to calculate EB_{body} in the absence of CS a standard body protein change profile is needed because this contributes to BW change, particularly in early lactation. Given this, acceptable EB_{body} profiles can be obtained without needing CS measurements. We believe that EB_{body} can become an important practical tool to manage excessive negative energy balance with the advantage of not needing data on feed intake or feed composition.

Genetic analysis of the fat:protein ratio using multiple-lactation random regression test-day models
Nishiura, A.[1], Sasaki, O.[1], Aihara, M.[2], Saburi, J.[3], Takeda, H.[1] and Satoh, M.[1], [1]NARO Institute of Livestock and Grassland Science, 2 Ikenodai Tsukuba, Ibaraki 3050901, Japan, [2]Livestock Improvement Association of Japan, 11-17 Fuyuki Koutou-ku, Tokyo 1350041, Japan, [3]National Livestock Breeding Center, 1 Odakurahara Nishigo, Fukushima 9618511, Japan; akinishi@affrc.go.jp

During the early stages of lactation, the energy input by feed intake cannot compensate for the energy output by increasingly high milk production. This causes a postpartum energy deficit. The fat:protein ratio of milk (FPR) could be used as an indicator of the status of energy balance. We estimated the genetic parameters of FPR in the first 3 parities and the genetic correlations between FPR and milk production in each parity using random regression test-day models. Data included test-day records of Holstein cattle from 2000 to 2008 – 604,147 records of 48,360 cows in Japan. The model included the fixed effect of herd-test-day and the fixed regression coefficients for region-age-season. The additive genetic effect and the permanent environmental effect were modeled using Legendre polynomials of 2nd, 3rd and 4th degree. The phenotypic values of FPR increased soon after parturition and peaked at days in milk (DIM) 20. These values then decreased slowly and remained constant at approximately 1.2 after DIM 100. Heritabilities of FPR were estimated at 0.14-0.33. The trends of heritability with DIM of the 1st parity were different from those of the other parities, especially during the early stages of lactation. During the early stages of lactation, the genetic correlations of FPR among the parities were estimated at 0.43-0.95. These genetic correlations increased up to DIM 70, and thereafter, were maintained at >0.9. The genetic correlations between FPR and milk yield were negative, approximately -0.5, except for the positive genetic correlation estimated in the 1st parity during the early stages of lactation. Our results indicate the possibility of simultaneously improving milk production and energy balance.

Epidemiological reaction norms for mastitis
Windig, J.J.[1], Urioste, J.I.[2,3] and Strandberg, E.[2], [1]Livestock Research, Wageningen UR, Animal Breeding and Genomics Centre, P.O. Box 65, 8200 AB Lelystad, Netherlands, [2]Swedish University of Agricultural Sciences, Dept. Animal Breeding and Genetics, P.O. Box 7023, 75007 Uppsala, Sweden, [3]Fac. de Agronomía, UDELAR, Depto. Prod. Animal y Pasturas, Garzón 780, 12900 Montevideo, Uruguay; jack.windig@wur.nl

Heritabilities for mastitis, and diseases in general, are low. One reason is that generally no distinction can be made in field data between healthy animals that are resistant and healthy animals that have not been exposed to pathogens. To take exposure into account we quantified the prevalence of mastitis in Swedish dairy herds. Herd prevalence of clinical mastitis (CM) was defined as the percentage of cows in a herd that received a recorded veterinary treatment for mastitis in a three month period. Herd prevalence of subclinical mastitis (SCM) was defined as the percentage of cows in the herd with a somatic cell count above 150,000 at a single test day. In a reaction norm analysis individual (S)CM for Holstein heifers was defined as the occurrence of at least one case of (S)CM in a lactation. In a reaction norm analysis the heritability of (S) CM was estimated as a function of herd prevalence for (S)CM. The best fitting model was for CM as a linear function of average herd prevalence of CM during the whole lactation, and for SCM as a 2nd order polynomial of average herd prevalence of SCM during the whole lactation. Range of estimated heritabilities was with 0.02-0.03 for CM and with 0.07-0.11 for SCM similar to estimates in literature that did not take herd prevalence into account. Explanations may be that the relation between exposure risk and herd prevalence, or between actual infection and (S)CM as defined in this study, are not as expected.

Effect of dry period absence on milk cholesterol metabolism and gene expression in dairy cows
Viturro, E., Hüttinger, K., Schlamberger, G., Wiedemann, S., Altenhofer, C., Kaske, M. and Meyer, H.H.D., Physiology Weihenstephan, Technische Universität München, Weihenstephaner Berg 3, 85354 Freising, Germany; viturro@wzw.tum.de

Cholesterol is a crucial molecule for life with a fine level regulation in the mammalian organism. Milk and milk products are the second principal source of cholesterol in human modern diet. The objective of these experiments was to test the effects of the absence of dry period on bovine cholesterol metabolism. Twenty-four multiparous Brown Swiss cows were assigned randomly to a continuous milking group (CM), milked twice daily until calving and during the subsequent lactation, and to a control group (C), dried off 56 days before calving and milked twice daily post partum. Blood and milk cholesterol levels as well as candidate gene expression in liver were obtained for both groups and compared. CM animals presented significantly higher (P<0.01) blood cholesterol concentrations, as a marker of a reduced incidence of fatty liver and negative energy balance, characteristic of the first weeks of lactation in animals on a classic regime. Interestingly, the milk cholesterol amount was dramatically reduced (P<0.01) in the CM group (3,475 mg milk cholesterol per day) compared to the C group (6,148 mg cholesterol per day). The liver mRNA levels of the sterol regulatory element binding proteins 1 and 2 (SREPB1, SREBP2), key regulators of lipid metabolism, as well as of HMGCoA reductase and Farnesyl Diphosphate Farnesyl Transferase (FDFT), two rate-limiting enzymes in cholesterol biosynthesis, were significantly (P<0.01) regulated during the periparturient period and on a different manner in both experimental groups. Our results reinforce the idea of an extremely regulated homeostasis of bovine cholesterol and a huge potential for altering its levels in milk by animal management or feeding strategies.

Genetic parameters for fertility, production and longevity of Holstein cow in the Czech Republic

Zavadilová, L. and Štípková, M., Institute of Animal Science, Přátelstvi 815, 10401 Praha – Uhříněves, Czech Republic; lida.zavadilova@seznam.cz

Heritabilities and genetic correlations among longevity, production and female fertility traits in the first lactation were estimated using 364,705 Holstein cows. Fertility traits were days open (DO), interval from parturition to first service (iPS) and interval from first service to conception (iFC). Production traits were first lactation milk (MY), fat (FY) and protein (PY) yield. Functional longevity was defined as the number of days between the first calving and culling; that is, length of productive life. Because of computations limitations, four subsets were randomly made based on herd identification. The linear animal model accounted for fixed effects of year-month, regressions on age at first calving, regressions on milk yield (only longevity) and for random effects year-herd, animal and residual. Heritability estimates for fertility traits ranged from 0.03 (iFC) to 0.05 (DO, iPS). Heritability of longevity was 0.08. Heritability estimates for production traits ranged between 0.27 (FY) and 0.34 (MY). Genetic correlations between fertility traits and longevity were moderate and favorable, ranging from -0.41 to -0.49. By contrast, genetic correlations between fertility and production traits were moderate and unfavorable, ranging from 0.48 to 0.65. Genetic correlation between iPS and iFC was moderate (0.52). Supported by the Ministry of Agriculture of the Czech Republic, Project No. MZE0002701404.

Trends in reproductive performance of dairy cow breeds in Lithuania

Nainiene, R., Siukscius, A. and Urbsys, A., Institute of Animal Science, Lithuanian University of Health Sciences, R. Zebenkos 12, Baisogala, 82317, Lithuania; arturas@lgi.lt

Most countries, where genetic selection for milk production has been conducted, have faced a continuous degradation of reproductive performance in dairy cattle. The analysis of the changes in the reproductive performance of dairy cow breeds in Lithuania in relation to their milk production in the years from 2001 to 2010 indicated that higher milk production of cows leads to their lover reproduction performance. The following reproduction traits were analyzed: insemination index of cows and heifers, first service conception rate, calving interval and average age of cows in lactations. The data from the black-and white dairy cows breeds indicated that there was high positive correlation between the milk production increase and calving interval ($r>0.7$) which was longer on the average by 19 days and lasted for 419 days in total. However, there was a negative correlation between the milk production and age of cows in lactations ($r<-0.7$), that decreased from 3.8 to 3.4 lactations. The correlation between the production and insemination index ($0.3<r<0.7$) showed that the insemination index for cows and heifers increased, respectively, from 1.7 to 1.8 and from 1.4 to1.5, while the first service conception rate decreased from 70.4 to 65.1%. In the red and red-and- white cow breeds higher productivity of cows in 2001-2010 resulted in 16 days longer calving interval which all lasted for 417 days. The age of cows in lactations was shorter from 3.3 to 2.9 lactations. The average correlation ($0.3<r<0.7$) between the milk production and cow insemination index showed the increase of the index from 2.1 to 2.2, while the first service conception rate decreased from 62.1 to 56,8%.

Investigations of locomotion score in dairy cows

Weber, A.[1], Junge, W.[1], Stamer, E.[2] and Thaller, G.[1], [1]Institute of Animal Breeding and Husbandry, Christian-Albrechts-University, Olshausenstr. 40, 24098 Kiel, Germany, [2]TiDa GmbH, Bosseer Str. 4c, 24259 Westensee, Germany; aweber@tierzucht.uni-kiel.de

Lameness in dairy cows is one of the major animal welfare issues in dairy production. It is a painful condition and causes economic losses through decreased milk production, impaired reproductive performance and involuntary culling. In Germany, lameness is the third main reason for early culling after mastitis and infertility problems. A useful tool for examination herd and individual lameness in dairy cows is locomotion scoring. The first aim of this study was to investigate the locomotion score as a binary trait. Data were recorded on the dairy research farm Karkendamm between September 2010 and February 2012. During the observation period animals were examined weekly for locomotion score and hoof lesions. The locomotion score was rated on a scale ranging from 1 to 5 (from normal to severe) and lameness (disturbed locomotion) was defined as score ≥ 3. In total, data of 324 animals were analyzed (8,306 observations for 10 to 350 days in milk). The prevalence of lameness was 16%. A threshold model was used to analyze the fixed effects using lameness (yes/no) as dependent variable and random effect of cow. The analysis of the fixed effects test-day, lactation number and lactation period was done with the GLIMMIX procedure of the SAS package. All fixed effects showed a significant impact on the probability for the incidence of lameness. The LS estimates indicated that the probability of lameness is higher in cows with lactation number higher then two. The LS means for the lactation periods showed that the probability of lameness increased until the 200 day in milk and decreased shortly after. The repeatability of the binary trait was 72%. The majority of all lameness cases are caused from claw disorders. As a consequence, the next step of this study is the investigation of the interrelationship between locomotion score and claw and foot disorders.

Use of progesterone profiles to define objective traits of physiological dairy fertility

Von Lesen, R., Tetens, J., Junge, W. and Thaller, G., Institute of Animal Breeding and Husbandry, Christian-Albrechts-University, Olshausenstr. 40, 24098 Kiel, Germany; rvleesen@tierzucht.uni-kiel.de

The fertility of dairy cows decreased in the last decades due to the selection of high milk yield. Further, low fertility influences the profitability of dairy enterprise. The aim of this study is to characterize features of progesterone profiles which are able to describe the fertility performance objectively. Furthermore, progesterone profiles are better in reflecting the physiological function than the traditional fertility traits. The progesterone data originate from first and multiparous dairy cows from the bull dam performance test of a German Breeding company on the research farm Karkendamm. Milk samples were collected twice a week. The progesterone profiles were generated by fitting a linear interpolation curve and using the 1st and 2nd derivative to identify extremes and points of inflections. To remove higher frequency noise a linear low pass filter was applied. Interpolating and smoothing were performed using the approaches of Stineman, and Yang and Zurbenko, implemented in R. A threshold was defined by 0.8 ng/ml to identify the zero points of the 1st derivative which classify the meaningful minima in the progesterone profiles. Possible features of progesterone profiles are given: intervals between the extremes, the concentrations of the progesterone at the extremes and at the points of inflection. Another way to identify features was to calculate a step function based on the threshold of 0.8 ng/ml. With a coding of run length, the period of luteal and interluteal phases (values above and below 0.8 ng/ml) was counted. Additionally, the CLA – commencement of luteal activity- was calculated with the run length coding. The CLA was identified by the day at which the first 10 subsequent interpolated values exceeded the threshold of 0.8 ng/ml. Further it will be tested if there are different characteristics of the features after inseminations.

Animal welfare in an enlarged europe: state of play and perspectives

Cavitte, J.-C.[1], Tristante, F.[2] and Krommer, J.[1], [1]European Commission, DG RTD, SDME 8/17, 1049 Brussels, Belgium, [2]European Commission, DG SANCO, F101 06/168, 1049 Brussels, Belgium; jean-charles.cavitte@ec.europa.eu

EU research, particularly since the Seventh Framework Programme (FP7), makes substantial efforts to facilitate the creation of research networks, tackle critical areas, and develop science-based animal welfare indicators. Additionally, several FP7 projects not specific on animal welfare have covered this area. In parallel, the European Commission has adopted, following an extensive consultation process, a new EU Strategy for Protection and Welfare of animals 2012-2015. This strategy intends to address EU wide problems while still respecting the EU diversity. Some of the key concepts in the new strategy will be simplification of the rules and cooperation at all levels. This presentation intends to give an overview of the actions ongoing or planned for animal welfare in an enlarged Europe.

AWARE project: background, objectives and expected outcomes

Spinka, M., Institute of Animal Science, Department of Ethology, Pratelstvi 815, 104 00 Prague – Uhrineves, Czech Republic; spinka.marek@vuzv.cz

The progress of European welfare strategy depends on broad base of support for it across Europe. However, currently there are many gaps in farm animal welfare awareness and implementation that go hand in hand with gaps in farm animal welfare research and university education. Geograhically, the gaps more or less follow a northwest-southeast division. The goal of the 7th FP project AWARE (2011-2014) is to bridge these gaps and promote integration of farm animal welfare research, university education and awareness/implementation across Europe, with special focus of the new, candidate and associated EU countries in four regions: Eastern Balkan, Western Balkan, East Central Europe and Baltic countries. The project achieves this goal in four steps: first, by mapping the situation; second, through the development of Europe-wide networks of scientists, lecturers and students, and by establishing a network of stakeholders active in FAW knowledge transfer and implementation; third, by enhancing the skills of researchers, lecturers and stakeholders; fourth, by working out strategies for durable outcomes of the project. AWARE also creates a Mobility Desk that facilitates match-finding between young researchers/students and job/research stay offers by universities and research institutions. The expected AWARE outcomes include: a wider participation of researchers and institutions from the four Eastern regions in farm animal welfare project submissions and EU research funding; strengthened and mutually enriching farm animal welfare courses across Europe; and harmonized and cross-communicating approaches to farm animal welfare awareness and implementation. These achievements will contribute to a broader support base for the new European welfare strategy; they will also facilitate a better utilization of the human potential of students, researchers and lectures motivated to work for an integrated knowledge-based farm animal welfare across Europe.

The AWARE project: mapping the farm animal welfare research in Europe
Košťál, Ľ.[1], Bilčík, B.[1], Kirchner, M.K.[2] and Winckler, C.[2], [1]Institute of Animal Biochemistry and Genetics, Slovak Academy of Sciences, Ivanka pri Dunaji, 900 28, Slovakia, [2]University of Natural Resources and Applied Life Sciences, Vienna, 1180, Austria; Lubor.Kostal@savba.sk

One of the goals of the FP7 project AWARE (www.aware-welfare.eu) which started in 2011 is to promote integration and increase the impact of European research on farm animal welfare (FAW). In order to facilitate the exchange of knowledge and researchers between already established research groups and research institutions who wish to develop FAW research activities, mapping of FAW research was performed as a first step. Invitation to fill in Web based questionnaires was distributed via regional representatives of the AWARE project (the hub structure) to all research bodies. The responses provide information about the size of the institution (e.g.number of researchers involved in FAW), facilities and expertise available, collaboration, current and future research areas and priorities, interest in participation in various AWARE activities, etc. Results were analyzed using statistical methods and semantic analysis. Response from 131 individuals (each representing an institution or working group) from 29 countries was received. For the analysis, the macroregions East and West, further subdivided into the hubs (Baltic, East Central, Eastern Balkan, Western Balkan, West Central, Mediterranean, North West and Nordic) were taken into account. In several aspects, significant differences between the Eastern and Western parts of Europe were revealed. E.g. respondents from the Eastern part participated significantly less in EU funded projects (FP7) and had less developed international collaboration. Up to 38% of respondents from the Eastern part stated that they have currently no project running on farm animal welfare, compared to 4% in the Western part. The Western part seems to be more involved in applied research, whereas a large proportion of respondents from Eastern part considers their research to be fundamental. More detailed results will be presented during the workshop.

Cell and tissue biology of laminae of the bovine claw in health and disease
La Manna, V.[1,2], Di Luca, A.[2] and Galbraith, H.[1,2], [1]University of Camerino, Environmental and Natural Sciences, Via Pontoni 5, 62032 Camerino, Italy, [2]University of Aberdeen, School of Biological Sciences, 23 St Machar Drive, AB24 3RY, Aberdeen, United Kingdom; h.galbraith@abdn.ac.uk

Endemic diseases affecting the welfare of cattle include non-infectious disorders of claw horn which include bruising and haemorrhage. Such lesions cause lameness. They are associated with impaired function of the laminar body-weight-bearing suspensory system which protects sole tissue from impact injury and which synthesises white line horn. The study aim was to investigate post mortem laminar tissues of adult female cattle by histological techniques and immunohistochemically. In the results, cells and extracellular molecules were identified as components of suspensory dermal connective tissues, and cellular precursors of cornified epidermal horn. On comparison with those undamaged, damaged laminae showed distortion of anatomical structure. Hyperplasia of epidermal keratinocytes was detected. These maintained nuclei suprabasally with delayed differentiation. Such epidermal cells typically express keratin molecules as Types I and II intermediate filaments (IF) of the cytoskeleton. Vimentin, the Type III IF, indicative of mesenchymal fibroblasts, was detected centrally within the dermis of both undamaged and damaged laminae. Its expression was also observed, only in damaged laminae, within the population of basal and suprabasal keratinocytes of the epidermis. A possible explanation may be transition of epithelial stem cell keratinocytes to mesenchymal phenotype in response to injury, a phenomenon previously observed in integumental tissues such as injured human skin. The results illustrate how the structural integrity of laminae in the bovine claw may be impaired by damage to dermal and epidermal tissues and by alterations in the behaviour of constituent epidermal and mesenchymal cells. Husbandry research and education should be focused on preventing such damage to better meet the welfare needs of susceptible animals.

A network of founding bodies to support farm animal health and welfare research in Europe

*Boissy, A.[1] and Benmansour, A.[2], [1]INRA, UMR1213 Herbivores, 63122 Saint-Gènes-Champanelle, France,
[2]INRA, UMR VIM, 78352 Jouy-en-Josas Cedex, France; Alain.Boissy@clermont.inra.fr*

The activities of consumer groups and animal protectionists and the effects of large-scale sanitary crises have increased the awareness that animal production is more than just an industry. Farm animal welfare is now clearly an important issue for people across Europe and there is clear demand for higher animal welfare standards. Building on the experience and achievements of the previous ERA-Net EMIDA, the new ERA-Net Animal Health and Welfare (ANIHWA) started on January 1st 2012. ANIHWA aims to increase cooperation and coordination of national research programmes on animal health and welfare of farm animals, including fish and bees. ANIHWA is a four-year project which gathers 30 partners and 19 countries from all over Europe, including Israël. The project is coordinated by INRA (A. Benmansour, coord.) and is organized in five mutually supportive Work Packages, including coordination (WP1), mapping and analysis of existing national research activities and facilities (WP2), gap analysis and research framework development (WP3), preparation of a dedicated strategic research agenda (WP4), and strategic activities aimed at a sustainable development (WP5). More particularly, the WP2 aims: (1) to identify common research requirements, emerging and perceived needs, in animal health and welfare; (2) to update and enhance the usability of the EMIDA publication database; (3) to evaluate and analyse current research in animal welfare and to increase exchange of information; and (4) to identify new research requirements for the new Animal Health and Welfare policy.

Comparison of two methods for developing a multicriteria evaluation system to assess animal welfare

Martín, P.[1], Buxadé, C.[2] and Krieter, J.[1], [1]Institut of Animal Breeding and Husbandry, Christian-Albrechts-University, Olshausenstr. 40, 24098 Kiel, Germany, [2]ETSIA, Polytechnic University, 28040 Madrid, Spain; pfernandez@tierzucht.uni-kiel.de

The aim of this paper is to develop a multicriteria evaluation system to assess animal welfare. Welfare Quality® distinguishes four welfare criteria (Good feeding, Good housing, Good health and Appropriate behaviour). An animal unit receives one score for each welfare criteria (expressed on 0-1 value scale). Nine animal units (farms) were used as an example illustrated on fattening pigs. Multi-Attribute Utility theory, whose aggregation process is based on the Choquet integral, was used to aggregate together the four welfare criteria scores to form an overall assessment for each farm. Utility functions for each criteria were determined in two different ways, by using the standard sequences method and by using the MACBETH software. Least squares based approaches with Shapley value and interaction indices constraints were used for Choquet integral capacity identification. A comparison of the Choquet integral results obtained through each method was carried out by ranking the nine farms. Results showed that the ranking derived from the utility functions calculated by the MACBETH approach is more consistent with the preferences of the decision maker than the ranking derived from the standard sequences method. This higher consistency lies in the fact that MACBETH allows less compensation between high importance criteria with very low welfare and low importance criteria with very high welfare. It can be concluded that by using together the MACBETH methodology and the Choquet integral, three specific features of animal welfare criteria aggregation are secured: criteria may not fully compensate for each other, they may not have the same importance and interaction may exist between them. These features should be considered in a further study which aggregates not only the different welfare criteria but the different subcriteria and measures described by Welfare Quality®.

A mobility desk for exchanging researchers on animal welfare within the EU and associated countries
Meunier-Salaün, M.C.[1], Tallet, C.[1], Kostal, L.[2], Mihina, S.[3] and Spinka, M.[4], [1]INRA, UMR1348 PEGASE, 35590 Saint-Gilles, France, [2]Institute of animal Biochemistry and Genetics, Ivanka pri Dunaji, Bratislava, Slovakia, [3]Slovak University of agriculture, Tr A Hlinku 2, Nitra 94976, Slovakia, [4]Institute of Animal Science, Pratelstvi 815, 104 00 Prague, Czech Republic; Marie-Christine.Salaun@rennes.inra.fr

A mobility desk has been created within the European AWARE project. The main goal of the AWARE project is to develop sustainable and actively expanding Europe-wide networks of farm animal welfare issues within the EU and associated countries. To achieve an extended network on animal welfare researches and reinforce the links between the European countries, one effective way is to favour mobility between the countries. This relies on strengthening the exchanges of researchers and encourages students to bring new research topics into the networks trough PhD and postdoc applications within the European Research Area. The objective of the Mobility Desk of the AWARE project is to increase this effectiveness and added value of the research on animal welfare in new and candidate EU countries, by facilitating and supporting these exchanges. It works as a support desk, where individual researchers are helped to construct their professional plan and applications to job offers. They can get information on possible host institutions, on existing education courses, on possibilities for financial support and on other topics related to working abroad. The mobility desk also support host institutions that wish to offer opportunities for PhD research or post-doctoral positions to refine their expectation and/or to find an applicant who matches them. After the recruitment, the mission is unfolded by the mobility desk and help will be providing to young researchers about their professional prospects and career expectations at the end of the mission. The access and information related to the mobility desk can be found at the website of AWARE (http://www.aware-welfare.eu/aware).

Animal welfare education in Europe
Illmann, G.[1], Melišová, M.[1], Keeling, L.[2], Košťál, L.[3] and Špinka, M.[1], [1]Institute of Animal Science, Department of Ethology, Přatestvi 815, 104 00 Prague 10, Czech Republic, [2]Swedish University of Agricultural Sciences, Department of Animal Environment and Health, Upsala, Sweden, [3]Institute of Animal Biochemistry and Genetics, Slovak Academy of Sciences, Ivanka pri Dunaji, Slovakia; Illmannova@vuzv.cz

The aim of this study was to map farm animal welfare university education and training throughout Europe with a view to identifying existing differences and gaps. Questions were asked related to Bachelor, Masters and PhD level courses dealing fully, or at least partly, with farm animal welfare. The mapping of courses was carried out using a web-based questionnaire, distributed through a hub-network structure of contact people in EU and candidate countries. So far 89 replies, covering a total of 184 farm animal welfare courses have been received from 25 countries within 8 regions. The number of replies from each region are: North West Europe, NL, UK, IE (n=28); West Central Europe, AT, CH, DE (n=28); Nordic, SE DK FI NO (n=29); Baltic, LT EW LT (n=2); East Central Europe, HU, CZ, SK, PL (n=42); Eastern Balkan, BG RU TK (n=10); Western Balkan, CR SLO MK (n=8); Mediterranean, FR ES IT (n=38). During the workshop the most important data from the ongoing analysis will be presented, for example; entrance requirements, the overall programme in which the course is given, the number of students attending, the main focus of the course, the hours of farm animal welfare teaching, whether it is optional or compulsory, the main teaching methods used, etc. It is anticipated that existing differences and gaps regarding education in animal welfare within Europe will potentially continue to influence future developments and need to be taken into consideration in future educational strategies. Questions were also asked about the availability and willing to share teaching materials from these courses with a view to exploring the potential for developing graduate level resources in the area of farm animal welfare education.

Dutch animal welfare education: collective understanding and individual application

Ruis, M.A.W.[1] and Hopster, H.[1,2], [1]Wageningen UR Livestock Research, Animal Welfare, P.O. Box 65, 8200 AB Lelystad, Netherlands, [2]Van Hall Larenstein, University for Applied Sciences, Animal Welfare Group, Agora 1, 8934 CJ, Leeuwarden, Netherlands; marko.ruis@wur.nl

In the Netherlands, the Animal Welfare Web (AWW), is a national website on animal welfare, supporting the dissemination of scientific and professional knowledge. The website thus bridges welfare science, agricultural education, and societal issues. AWW supports the activities of a national animal welfare education program (AWEP). The program, implemented by representatives from various education and research institutes, generates projects in which state-of-the-art animal welfare knowledge is scrutinized, selected and applied and distributed for use in teaching and in practice. The activities are especially connected with the main educational programmes on livestock, animal and horse husbandry and on veterinary nursing. The program involves three main themes: knowledge infrastructure, public awareness and application in practice. In the longer term the AWEP activities contribute to a society in which animals receive knowledgeable, professional and respectful treatment. The AWEP closely cooperates with research institutions, NGOs and potential partners to intensify cooperation and to implement animal welfare in the national educational landscape in an effective long-term manner. Recently, the AWEP developed a webbased animal welfare course for both vocational education and the staff of the Dutch Society for the Protection of Animals. The driving force behind the AWW and AWEP activities is the Animal Welfare Group (AWG) at Van Hall Larenstein, University for Applied Sciences and part of Wageningen University and Research Centre. The AWG plays a coordinating role in improving the quality of animal welfare education at various levels. Since this animal welfare infrastructure works well at a national level, it may serve as a blueprint for other countries to improve the quality of animal welfare education. The activities of AWW, AWEP and AWG are mainly funded by the Dutch government.

Food industry observations toward farm animal welfare standards and legislatives in Macedonia

Ilieski, V. and Radeski, M., Faculty of Veterinary Medicine – Skopje, st. Lazar Pop Trajkov 5-7, 1000, Macedonia; vilieski@fvm.ukim.edu.mk

Food industry is one of the main actors in implementation and promotion of farm animal welfare in the food supply chain. Farm animal welfare legislation is groundwork for developing, implementation and improving appropriate standards in the country. The main objective of this paper is to present current attitudes and perception of the food industry from Macedonia regarding farm animal welfare and the existing legislation. Representatives (n=21) from the food industry (slaughterhouse, dairy and meat) participated in answering the developed questioner and discussing the given statements related to farm animal welfare. The questioner and statements were consisted of two sections: welfare standards and society and legislation development. According the representatives, society attitude to farm animal welfare is very low rated, average grade 2.25 of 5. The media attention and current trend in the society toward farm animal welfare has lowest rating (with grade 2). All representatives consider that there is need to introduce labeling of animal products produced by animal welfare standards. The involvement rates of food industry representatives in the legislation creating process and developing standards in Macedonia were: 57.9% – no or little involvement, while only 15.8% have very strong involvement. All participants have agreed that the legal requirements are set in a way that many farmers can easily comply, but farmers do not comply with the requirements. The success of the legislation in terms of improving animal welfare, creating awareness, generating a demand among consumers received an average 3.26 rating of 5. Although there is a basic animal welfare legislative in Macedonia, the food industry considers that there is need for its greater involvement in the legislation creating process and development of specific standards. Increased education of food industry employees, farmers and greater consumer awareness, are prerequisites for implementation of farm animal welfare.

OIE's work on animal welfare

Leboucq, N., OIE Sub-Regional Representative in Brussels, Rue Breydel 40, 1040 Bruxelles, Belgium; rsr.bruxelles@oie.int

The OIE has recently launched a number of standards on animal welfare. These are the result of discussions between stakeholder groups across the world. In her presentation, Dr. Leboucq will address these standards and focus on the process of collaboration across countries, to achieve consensus on animal welfare.

Effect of dietary vitamin A restriction on growing performance and intramuscular fat content of pigs

Tous, N.[1], Lizardo, R.[1], Vilà, B.[1], Gispert, M.[2], Font-I-Furnols, M.[2] and Esteve-Garcia, E.[1], [1]IRTA, Monogastric Nutrition, Ctra. Reus-El Morell Km 3.8, Constantí, 43120, Spain, [2]IRTA, Carcass Quality, Finca Camps i Armet, Monells, 17121, Spain; rosil.lizardo@irta.cat

Selection against fatness of pigs has negatively affected the intramuscular fat (IMF) and consequently the organoleptic characteristics of pork. Therefore, increasing its content became one of the main challenges for pig breeders as well as pig nutritionists. It has been hypothesized that a reduction of dietary vitamin A could increase the IMF and so meat quality. The purpose of this study was to evaluate how dietary vitamin A content could affect productive parameters, carcass quality, IMF and liver retinol content. Forty eight barrows were divided in 3 groups corresponded to the experimental diets, which differed in their vitamin A content: without supplemental vitamin A, the required amount for growing pigs (1,250 IU Vitamin A/kg) or the usual amount used in commercial formulations (5,000 IU Vitamin A/kg). Growth performance, carcass and meat quality, and liver retinol content were evaluated after slaughter at around 115 kg liveweight. No statistical differences were observed for any of the productive or carcass quality parameters evaluated. Despite IMF was not statistically affected, a numerical reduction was observed in animals fed diets without vitamin A supplementation and, that observation seems to be contrary to the initial hypothesis. Liver retinol content was increased when the animals were fed higher levels of dietary vitamin A. Results suggest that levels of vitamin A currently used in the diets can be reduced without affecting pig performance or backfat deposition while the IMF could decrease when the dietary level of vitamin A is reduced. It appears that beta-carotene supplied by feedstuffs may be sufficient to synthesize the vitamin A necessary for growth. It can be concluded that reduction of dietary vitamin A did not result in an increase of the IMF content.

Dietary lecithin improves feed conversion rate and dressing percentage in finisher gilts

Akit, H.[1], Collins, C.L.[2], Fahri, F.T.[1], D'souza, D.N.[3], Leury, B.J.[1] and Dunshea, F.R.[1], [1]University of Melbourne, Parkville, VIC 3010, Australia, [2]Rivalea (Australia) Pty Ltd, Corowa, NSW 2646, Australia, [3]Australian Pork Limited, Deakin West, ACT 2600, Australia; hakit@student.unimelb.edu.au

Crossbred (Large White × Landrace) finisher pigs were used to determine the impact of dietary lecithin on growth performance, carcass and meat quality. The 256 pigs (8 pigs of the same sex/pen) were randomly allocated to a 2×2 factorial experiment with the respective factors being dietary lecithin (0 or 5 g/kg) and sex (female and entire males immunized against GnRF (Improvac)). Pen weights were recorded and prior to slaughter individual live weights were obtained. Pen feed intake was measured by feed disappearance and feed conversion ratio (FCR) subsequently calculated. After six weeks of dietary treatment, pigs were slaughtered and individual carcass weight and P2 back fat depth measured. A pre-determined subset of 64 carcasses (2 pigs per pen) was followed through the boning room for collection of L. dorsi. Colour (L*,a*,b*) and pH (45 mins and 24 hrs) were measured. Additional muscle was removed and frozen prior to hydroxyproline, Warner-Bratzler shear force and compression analyses. Data were analysed by ANOVA suitable for a factorial design. Pigs fed dietary lecithin had better FCR than pigs fed the control diet (FCR 2.96 and 2.79 kg/kg respectively for the control and treatment groups, P=0.006), due to a combination of a small reduction in feed intake (2.91 and 2.81 kg/d respectively) coupled with a similar rate of gain (0.99 and 1.01 kg/d respectively). There was no impact of dietary lecithin on final live weight, carcass weight or P2 back fat depth. Dietary lecithin increased (P<0.05) the dressing percentage in gilts but not in immunocastrated males. There was no impact of dietary lecithin on meat quality. The results indicate that dietary lecithin can significantly improve feed conversion rate of both sexes and dressing percentage of finisher gilts.

Modeling the effect of feeding strategy and feed prices on a population of pigs

Quiniou, N.[1], Vautier, B.[1,2], Salaün, Y.[1], Van Milgen, J.[2] and Brossard, L.[2], [1]IFIP-Institut du Porc, BP 35105, 35650 Le Rheu, France, [2]INRA, UMR1348 Pegase, 35590 Saint-Gilles, France; nathalie.quiniou@ifip.asso.fr

Within a population of pigs, variation in performance and nutritional requirements can be important. Nutritionists elaborate feeding strategies that maximize economic return at the batch scale while (in some countries) accounting for N and P excretion. To evaluate the effect of feeding strategy and feed price on performance, its variability, economic return and environmental impact, a simulation study was carried out using both a virtual population of 2,000 gilts and barrows and the InraPorc® growth model. The 2,000 virtual animal profiles were generated randomly from the average of five parameters that describe growth and feed intake, while using the covariance structure of parameters obtained in a preliminary study. The effects of factors such as feed restriction, number of diets used and amino acids level in each diet were simulated with diets formulated using the least-cost principle. Five different contexts of feedstuff prices, corresponding to average prices recorded for five harvesting campaigns (2005/06 to 2009/10) were used. Calculations were performed for the 2,000 individual animals to obtain performance at the average final BW of 115 kg for the group. With reference to the French payment grid, carcass value was calculated using the final individual carcass weights and lean meat contents. Economic return was calculated as the difference between carcass value and the costs of feed and labor. Results suggest that the variability of final BW was reduced with an increased supply of amino acids. Feed restriction had a greater impact on variability. The price of feed ingredients influenced the optimum level of amino acids for economic return and corresponding N output. These calculations should be performed monthly to adapt nutrient supplies to the economic context.

Lipogenic enzyme mRNA expression and fatty acid composition in relation to nutritional value of pork

Marriott, D.T.[1], Estany, J.[2] and Doran, O.[1], [1]University of the West of England, Bristol, BS16 1QY, United Kingdom, [2]Universitat de Leida, Lleida, 25198, Spain; duncan.marriott@uwe.ac.uk

We have previously demonstrated that expression of the lipogenic enzyme stearoyl-CoA desaturase (SCD) plays a key role in intramuscular fat (IMF) but not subcutaneous fat (SF) formation in commercial pig breeds. Our previous studies have also found that protein expression of the lipogenic enzymes delta 6 desaturase (Δ6D) and fatty acid synthase (FAS) have a breed specific correlation with their fatty acid products. The aim of this investigation was to further investigate the molecular mechanisms regulating fat deposition in genetically diverse breeds and the role of lipogenic enzyme mRNA expression in this process. The study was conducted on 4 genetically diverse breeds: Duroc 1 (D1), Duroc 1 × Duroc 2 (D1xD2), Duroc 1 × Iberian (D1xIb) and Large White × Landrace (LWxLr). Duroc 2 is a standard Duroc line whereas Duroc 1 is a line with enhanced IMF. Expression of SCD, Δ6D and FAS mRNA was analysed by qPCR in muscle and adipose tissue. Fatty acid composition was determined by gas chromatography. ANOVA was performed with Mann-Whitney. SCD mRNA expression in muscle was found to be 3-fold higher in the D1xIb breed than in the LWxLr breed (P=0.012). This was accompanied by 8% higher muscle MUFA content in the D1xIb breed compared to LWxLr (P=0.004). Expression of muscle Δ6D and FAS mRNA did not differ between breeds. In the case of SF no breed-differences were observed in the expression of FAS and SCD. Δ6D mRNA expression was 2-fold higher in SF of LWxLr pigs when compared to D1xIb (P=0.028). This was accompanied by a trend towards increased PUFA content (by 45%) in SF of the LWxLr breed. Results of this study (i) support the hypothesis of breed-specific mechanisms regulating fat partitioning in pigs; (ii) suggest that increased levels of MUFA in the muscle of the D1xIb breed is related to enhanced biosynthesis of MUFA mediated by enhanced expression of SCD, the gene encoding for the enzyme catalysing MUFA biosynthesis.

Different management systems in early life have impact on intestinal immune development in pigs

Schokker, D.[1], Smits, M.A.[1,2] and Rebel, J.M.J.[2], [1]Wageningen UR Livestock Research, Edelhertweg 15, 8200 AB Lelystad, Netherlands, [2]Wageningen UR Central Veterinary Institute, Edelhertweg 15, 8200 AB Lelystad, Netherlands; dirkjan.schokker@wur.nl

The objective of this study was to compare the effect of management treatments, applied at 4 days after birth, on the development of intestinal immunity in growing pigs. Three different farrowing management system were evaluated: (1) no treatment; (2) only antibiotic injection; and (3) antibiotic injection combined with husbandry system induced stress factors. The antibiotic used in this study is commonly applied in high intensive production systems. The stress factors included weighing, ear-tagging, and tail docking, all commonly used in pig husbandry systems. To investigate the effect of the two treatments on intestinal immunity, gene expression in both jejunum and ileum was measured at day 8. Microarrays were used to measure gene expression differences in both tissues for all three groups. Subsequently, up- and down-regulated genes were identified by pair-wise comparisons between the experimental groups for each tissue. The resulting up- and down-regulated genes were uploaded to DAVID for functional clustering analysis in order to identify which processes were affected by the two treatments. Compared to the untreated animals, a decrease in activity of immune related processes was observed in piglets due to the antibiotic treatment combined with stress factors. Similar observations were made for piglets that received only the antibiotic injection. Comparing the data of the two treatment groups indicated a slight increase in activity of immune related processes in piglets that were exposed to the stress factors. We conclude that antibiotic treatment and exposure to stress factors at day 4 after birth have an effect on immunological processes in the intestine of growing pigs independently of the sow background. We hypothesize that part of these differences are related to management-induced variations in early colonization of the gut by microflora.

Evaluation of moist barley preserved with a biocontrol yeast and starter culture fed to growing pigs

Borling, J. and Lyberg, K., Swedish University of Agricultural Sciences, Department of Animal Nutrition and Management, HUV/HVC Box 7024, 75007 Uppsala, Sweden; jenny.borling@slu.se

Microbiological properties and ileal and total tract digestibility of nutrients were evaluated in moist crimped barley inoculated at harvest with the biocontrol yeast Wickerhamomyces anomalus and a commercial starter culture with lactic acid bacteria (Fermentationexperts, Denmark). The four different treatments were; W.anomalus inoculated cereal grains (W-diet), W.anomalus and starter culture inoculated cereal grains (WS-diet) non-inoculated cereal grain containing only natural microbial flora (C-diet), and dry crimped cereal grain (D-diet). The grain was stored in airtight plastic barrels until feeding commenced after six weeks. The experimental diets were fed mixed with water to a dry matter content of 30% and fermented an initial 7 days prior to feeding and then with a daily 50% back-slop. All diets were supplemented with vitamins and minerals. The diets were fed to 6 growing pigs with an average initial body weight of 30 kg fitted with post valve t-caecum (PVTC) cannulas, in a change-over design. Yeast, lactic acid bacteria, moulds and Enterobacteriaceae were quantified, and pH as well as organic acids was measured on fresh samples throughout the study. Results show that; Acetic and lactic acid levels of the fermented diets were 2.1 g/l and 16.3 g/l respectively. The average pH level was 4.1 in all fermented diets except for D-diet which had a level of 3.8. Mould cfu (colony forming units) were below detection limit (10 cfu/g) in all W. anomalus inoculated treatments after six weeks of storage; two weeks later Enterobacteriaceae counts had decreased below detection limit as well. After the initial growth phase lactic acid bacteria and yeast cfu stabilised at 109 and 107 cfu/g respectively in the fermented diets; this together with the small fluctuations in pH indicates a stable fermentation process. Digestibility data will be added.

The influence of dietary live yeast on piglet growth performance and nutrient utilization

Lizardo, R.[1], Perez-Vendrell, A.[1], D'inca, R.[2], Auclair, E.[2] and Brufau, J.[1], [1]IRTA, Monogastric Nutrition, Ctra Reus – El Morell, Km 3.8, 43120 Constantí, Spain, [2]LFA – Lesaffre Feed Additives, 137, rue Gabriel Péri, 59703 Marcq-en-Baroeul, France; rosil.lizardo@irta.es

Weaning is a critical period on piglets' life. In the past, risks were reduced by the systematic use of in-feed antibiotic growth promoters. However, the development of bacterial resistance led the EU to ban their use in 2006. Since then, all the animal feed industries have been looking for alternatives and live Saccharomyces cerevisiae Sc47 yeasts (Actisaf®, LFA, France), might be a potential one's. So, the aim of the present study is to evaluate the efficacy of live yeast in diets for weanling piglets. One-hundred twenty-eight Landrace × Duroc male piglets of 6.8 kg liveweight were distributed among 4 groups corresponding to the control diet and 3 supplemented diets with either 0.1, 0.5 and 1 g/kg of live yeast (5×10^9 cfu/kg). They were distributed by weight in 8 blocks and allocated at 4 per pen in 32 pens. The experiment lasted 6 weeks and productive parameters and nutrient utilisation were evaluated. At the end, an indigestible marker was included in the feeds and fresh faeces samples were collected for digestibility measurements. In general, piglets of treatment with the highest dose of live yeast grew more than the others and they achieved a higher final bodyweight. Overall, feed conversion ratio and feed efficiency were significantly improved with the highest dose of live yeast ($P<0.01$). Dry matter, energy and protein digestibility were slightly increased in treatments containing 0.5 and 1g/kg of live yeast relatively to control and statistically improved relatively to the lowest dose of live yeast. Moreover, neutral detergent fibre and hemicelluloses digestibility significantly improved with live yeast ($P<0.001$). From results, it can be concluded that inclusion of live yeast on piglet diets has positive effects on productive performance and nutrient utilisation after weaning and its utilisation could be recommended.

Split nursing as a management routine to ensure colostrum to newborn piglets
Mattsson, P. and Mattsson, B., Svenska Pig, P.O Box 974, 391 29 Kalmar, Sweden;
petra.mattsson@svenskapig.se

Number piglets born alive (NBA) have increased in Swedish commercial piglet production. At the same time has also the piglet mortality increased. Mean piglet mortality was 17% in 2010. A study in 5 commercial herds, all with loose housed farrowing sows, has been preformed to evaluate split nursing as a management routine to ensure colostrum intake in large litters. The statistical analyses were preformed on 10 farrowing batches, 336 litters and a total number of 3,681 weighed piglets at weaning. Each farrowing batch was divided by the herd staff into one control group and one trial group. Piglets in trial group litters were after ended farrowing divided into two subgroups. These subgroups were altered at the sow's udder two times with at least 60 minutes interval. Cross-fostering was done after split nursing. In the statistical analyses were 205 litters with 2,214 piglets in control group and 131 litters with 1,467 piglets in trial group. The analyses were done using SAS PROC GLM. Data on litter number, number born alive, number of piglets in subgroups in trial group, cross-fostering, number of weaned piglets, individual weaning weight and health were collected. Piglet mortality was 19% in litters in trial group with $\geq$14 NBA, compared to control were the mortality was 28% (P<0.01). There was a tendency towards heavier litter weights at weaning in trial litters compared to control litters with $\geq$14 NBA, 104.2 kg in trial group and 100.3 kg in control group. The piglet mortality is possible to decrease by using split nursing in litters with $\geq$14 number piglets born alive.

Meat technological quality traits from Slovenian indigenous Krškopolje pigs in enriched environment
Žemva, M., Kovač, M. and Malovrh, Š., Biotechnical Faculty, University of Ljubljana, Department for
animal science, Groblje 3, 1230 Domžale, Slovenia; marjeta.zemva@bf.uni-lj.si

Technological quality traits and intramuscular fat (IMF) content were investigated for Krškopolje pig (KP), the only indigenous pig breed in Slovenia. A total of 42 KP (19 barrows and 23 gilts) were kept indoors in one group in a straw-bedded pen. Cereal-based diet (ground wheat, barley and maize) were fed restrictively. Good quality hay was available *ad libitum*, being a good source of proteins. The pigs were fattened for four months, when the first group was slaughtered. Ten to twelve animals were slaughtered in successive months at age between 236 and 364 d. At each slaughter, fatteners were arranged into lighter group (between 120 kg and 130 kg) and heavier group (above 150 kg). The pH value and conductivity of in m. longissimus dorsi (LD) and m. semimembranosus (SM) were measured 45 min (45) and 24 h (24) after slaughter. Instrumental and visual colour were measured for LD, 24 h after slaughter. The drip loss of LD was calculated after 24 (24) and 48 (48) h. The IMF of LD was determined in laboratory. Analyses were carried out using the GLM procedure in the statistical package SAS/STAT. The pH24 value in SM decreased with age at the slaughter for $(-2.2\pm0.4)\times10^{-3}$ /day, and trends for decreasing pH45 in SM and pH24 in LD were also observed. Additionally, conductivity45 $(14.9\pm4.3)\times(10^{-3}$ /day) and conductivity24 $(39.3\pm12.6)\times(10^{-3}$ /day) in LD, drip loss48 $(19.2\pm8.0)\times(10^{-3}$ /day), value a* $(14.8\pm7.1)\times(10^{-3}$ /day) and visual colour $(6.8\pm2.8)\times(10^{-3}$ /day) were increasing with age. Changes in drip loss24 (P=0.0556) and value c* (P=0.0519) were not significant. The age did not influence IMF content in LD. The gender and weight did not effect technological quality, except that heavier KP had higher pH45 (6.09) value in LD than lighter (5.91) KPs. Barrows contained more IMF (4.98%) than gilts (4.04%).

Effect of increasing 1,25-dihydroxyvitamin D3 level on growth performance and mineral status in wean

Schlegel, P.[1], Gutzwiller, A.[1] and Bachmann, H.[2], [1]Agroscope, Tioleyre 4, 1725 Posieux, Switzerland, [2]Herbonis Animal Health GmbH, Grellingerstrasse 33, 4001 Basel, Switzerland; patrick.schlegel@alp.admin.ch

Vitamin D3 is involved in calcium (Ca) and phosphorus (P) absorption and homeostatic regulation. This tolerance experiment was conducted to evaluate increasing dietary levels of 1,25-Dihydroxyvitamin D3 glycosides in seventy 28-day old weaned piglets during 6 weeks. Two basal basal diets (9.7 and 5.5 g Ca and P / kg diet and 500 U exogenous phytase) were formulated to contain 1000 (A) or 2,000 (B) IU vitamin D3. Diet A was added with 0 and 0.5 g PAN / kg diet (PAN, Panbonis HVD, Herbonis, Switzerland). Diet B was added with 0, 0.25, 0.5, 1.0 and 2.0 g PAN / kg diet. PAN contained 10 μ 1,25-dihydroxyvitamin D3 glycosides / g from Solanum glaucophyllum. After 6 weeks, all animals were in good health. Neither body weight nor daily weight gain were affected (P>0.05) by diet. Blood samples were collected weekly. Plasma Ca increased with 2.0 g PAN / kg diet versus control A (P<0.05), but not control B (P>0.05). On week 2, all doses of PAN increased plasma P compared to control A (P<0.05). At the end, plasma P was lower (P<0.05) with 2.0 g PAN / kg diet versus control A. Bones (metacarpus) were collected from animals fed control A and B and 1.0 and 2.0 g PAN/ kg diet. Bone breaking strength (P<0.05), ash (P<0.10) and P (P<0.10) contents were reduced when 2.0 g PAN/ kg diet were fed. The present data indicate that 1) 1000 IU vitamin D3 and 0.5 g PAN / kg diet was equivalent in growth and plasma values than 2000 IU vitamin D3; 2) 2.0 g PAN / kg diet resulted in 20% increased plasma Ca, 13% lower plasma P, 17% lower bone breaking strength and 4% lower bone P which indicates first signs of possible adverse effects (e.g. soft tissue calcification); 3) no adverse effect is to be expected in piglets fed levels up to 1 g PAN / kg diet with recommended Ca, P and vitamin D contents.

Histological profile of the LD muscle in pigs and its effect on loin parameters and IMF content

Bereta, A.[1], Tyra, M.[1], Różycki, M.[1], Ropka-Molik, K.[1] and Wojtysiak, D.[2], [1]National Research Institute of Animal Production, Department of Animal Genetics and Breeding, ul. Krakowska 1, 32-083 Balice, Poland, [2]Agricultural University of Krakow, Department of Reproduction and Animal Anatomy, Al. Mickiewicza 24/28, 30-059 Krakow, Poland; abereta@izoo.krakow.pl

Fibre types and their proportions in muscle determine muscle motility but may also indirectly affect physicochemical properties of muscle tissue. From the consumer's perspective, the proportions of individual fibre types may be manifested as sensory (taste) differences in meat. It is also known that this group of sensory traits is determined by intramuscular fat (IMF) content in meat. The objective of this study was to analyse differences in individual fibre types in the histological profile of the longissimus dorsi (LD) muscle and their effect on pork carcass lean content and level of IMF content, which determines palatability of meat and meat products. Analysis showed that the amount of type IIB fibres had a statistically significant (P<0.05) effect on the IMF content of the longissimus dorsi muscle. Animals with more than 70% of type IIB fibres in this muscle were also characterized by larger loin eye area (P<0.01) and loin eye height (P<0.05). Analogous relationships were noted when the analysed group of animals was divided according to the diameter of type IIA fibres. IMF was negatively correlated to the percentage of type IIB fibres (r_p=-0.162). Analogous relationships with the other two fibre types were positive (IIA – r_p = 0.097; I – r_p = 0.187). It was found that increased percentage of type IIB fibres resulted in a slightly greater loin weight (r_p=0.176), higher loin eye height (r_p=0.136), larger loin eye area (r_p=0.265) and higher carcass lean content (r_p=0.204). Likewise, the increase in the number of type IIA and type I fibres decreased these parameters. The study was supported by the Polish Ministry of Science and Higher Education (project no.NN311349139).

Relationships between fattening and slaughter performance and meat quality of Polish Large White pig

Żak, G. and Tyra, M., National Research Institute of Animal Production, Department of Animal Genetics and Breeding, ul. Krakowska 1, 32-083 Balice, Poland; z1zak@cyf-kr.edu.pl

For many years, per capita consumption of pig meat in Poland has fluctuated around 40 kg and continues to grow despite the expansion of poultry meat. Due to the high share of pork in total meat consumption, pig breeders are challenged to find ways of producing pork characterized by good taste, culinary, technological and dietary qualities while maintaining high carcass meat content. The set goals can be accomplished mainly through proper breeding work. The aim of the study was to analyse the relationships between fattening and slaughter performance traits and selected parameters of pig meat quality. The study was conducted with Polish L.W. pigs. The meat of 125 animals was analysed (testing stations). Pigs were sired by 12 boars. Selected slaughter traits were measured and fattening parameters (daily gain and feed conversion) were calculated. The following meat quality traits were determined: pH of loin muscle (m.l.d.) 45 min and 24 h postmortem, pH of m. semimembranosus (ham) 45 min and 24 h postmortem, colour (L* a* b*) of m.l.d., water holding capacity, and intramuscular fat content (IMF) of m.l.d. Analysis of the results showed a very weak correlation between meat quality traits and fattening performance. Most statistically significant relationships were noted between slaughter traits and pH of meat. It is notable that fattening and slaughter performance was not correlated to IMF level. Considering that IMF content of the analysed meat was very low (only 1.3%, compared to the optimum level of 2.5-3%) and breeding efforts are mainly oriented towards improving fattening and slaughter performance, no improvements in this parameter of meat quality are to be expected, especially with regard to taste properties, when using current models for estimating genetic value. In light of the present results, it seems necessary to extend the models for estimation of pig breeding value with IMF content in the first place.

Influence of feeding ω-fatty acids containing products on the boars breeding ability

Liepa, L., Mangale, M. and Sematovica, I., Faculty of Veterinary Medicine, LUA, Helmana 8, Jelgava, 3004, Latvia; laima.liepa@llu.lv

The aim of the study was to investigate the influence of food components containing ω-fatty acids on the boars breeding ability characterizing indices in the sexual hormone activity reduced period (April- June). The investigation was carried out in the company of artificial insemination. The 10 boars venous blood samples for biochemical analyses and ejaculates were obtained and evaluated in identical periods of the years 2010 and 2011: on 12-04., 29-04., 25-05. (one more time on 30-06-2010). Data were statistically analyzed by SPSS 11,5. Until 25.05.2010, boars received 3 kg barley meal, 220 g food additives (50% protein), 100 g minerals and vitamins. At this time, a reduction tendency of boars ejaculate volume (EV), concentration of spermatozoa (CS), bold testosterone (T) and oestradiol (E) level (P>0.05) were detected. Since 25-05-2010. 20 ml linseed oil and 250 g linseed cake had been added to the food ration. In 2011, the boars ration all the time contained: 1.9 kg barley meal, 0.6 kg wheat meal, 0.1 kg corn meal, 0.3 kg soy cake, 20 ml fish oil, 60 g fish meal, 100 g minerals and vitamins. Adding linseed oil and cake for 35 days increased EV, high density lipoproteins (HDL) (P>0.05), T, E concentration (P=0.07), and lowered CS, total cholesterol (H), and low density lipoproteins (LDL) level in the blood. In 2011, ω-fatty acids containing food components (fish oil and meal, corn meal and soy cake) caused similar and more notable differences in previously listed measurements of boars breeding ability compared to the year 2010, especially in 25-05-2011., when differences were significant (P<0.05). In this study, we detected HDL correlation with E (r=0.7; P<0.01) and T (r=0.5; P<0.05); EV correlation with CS (r=-0.7; P<0.01) and HDL (r=0.4; P<0.01). In conclusion, feeding ω-fatty acids containing products improved the boars breeding ability in April-June period: significantly increased EV, E and T, but CS decreased; H and LDL showed a tendency to diminish, but HDL- to grow up.

Chemical composition and quality characteristics of pork in selected muscles

Bučko, O., Petrák, J., Vavrišínová, K., Haščík, P. and Debrecéni, O., Slovak University of Agriculture in Nitra, Tr.A.Hlinku 2 Nitra, 94976, Slovakia; Ondrej.Debreceni@uniag.sk

The experiment included 28 fattening pigs of Large White breed (14 boars, 14 gilts). We analyzed chemical composition and physical and technological parameters of meat quality in the following 3 muscles: Mussculus aductor (MA), Mussculus longissimus thoracis (MLT) and Musculus semimembranosus (MSM). Chemical composition of pig meat was determined by FT IR Nicolet 6700 analyzer in g/100g samples, meat color values in L *, a *, b * by spectrophotometer CM 2600 D. Based on the analysis of total protein in g/100g muscle samples, we can conclude that the highest value of this parameter was recorded in MLT and the lowest value in MSM, the significance of differences between all 3 monitored muscles was at P <0.001. The value of total fat in g/100g muscle samples with the highest in MSM and lowest in MA, while we found a high statistically significant differences between MSM and MA, at the same time also between MSM and MLT at P <0.001. In assessing the total water content ratio in g/100g muscle samples the highest value was measured in MA and the lowest in the MSM, with statistical significance among all the monitored muscles at P <0.001. Our findings in the parameters of the chemical composition of individual muscles shows that in the MSM is the biggest fat content and lowest water content compared with MA, which has the highest water content and the lowest proportion of fat. The indicators of the actual acidity surveyed 45 min. and 24 hours post mortem, we found no statistically significant differences between the muscles acidity. We found the highest value in electrical conductivity (EC) measured 45 min. post mortem in MSM and the lowest in MA, what was similar to EC 24 hours post mortem. In the indicator color of meat L *, we measured the highest value in MSM, pointing to his pallor and vice versa in MA that has the lowest value, indicating his darkness.

Investigation of the relationship between pH and temperature as PSE indicators of pork

Vermeulen, L.[1], Van De Perre, V.[1], Permentier, L.[1], De Bie, S.[2] and Geers, R.[1], [1]Catholic University Leuven, Biosystems, Bijzondere weg 12, 3360 Lovenjoel, Belgium, [2]VLAM-BELPORK, Leuvensplein 4, 1000 Brussel, Belgium; liesbeth.vermeulen@biw.kuleuven.be

Consumers and industry request carcasses with high meat quality in combination with good welfare and minor economic losses. Meat pH is known to be a good predictor of meat quality. However, several difficulties are related to pH measurements, such as the need of cleaning the electrode and the continuing decline of the pH during slaughter. A possible alternative is to measure muscle temperature, since muscle temperature is related to muscle activity and the biochemical reaction producing lactic acid is temperature dependent. The objective of this study was to examine the relationship between pH and temperature. The pH was measured 30 minutes after slaughter in two different 'white' muscles – M. longissimus dorsi (M.L.D.) and M. semimembranosus (M.S.) and in M. adductor (M.A.), a red typed muscle. Moreover internal temperature of the ham and the M.L.D. was measured. In total, 824 carcasses were measured in two different slaughterhouses, one using electrical and one using carbon dioxide (CO_2) stunning. The results show remarkable high temperatures of M.L.D. and M.S., respectively a mean (standard deviation) temperature of 39.42 °C (1.04 °C) and 40.41 °C (0.67 °C). Probably, a high stress level, which is accompanied with heath production due to an increased energy metabolism, will raise muscle temperature. Furthermore, pH and temperature were correlated ($r_{M.L.D.}$=-0.35 and $r_{M.S.}$=-0.29), proving temperature being an indicator for PSE meat. Results show also a wider range of pH values for M.S. than for M.A., which could make the white muscle of the ham more appropriate for detecting PSE. Further research will focus on measuring rectal temperature during lairage and immediately after stunning in order to relate rectal with muscle temperature and thus detect, already in lairage, a risk for PSE meat. This study will be performed and reported more extensively in the near future.

Differences in quantity and composition of intramuscular fat in selected muscles of the pigs
Bučko, O., Vavrišínová, K., Petrák, J., Mlynek, J. and Margetín, M., Slovak University of Agriculture in Nitra, Tr.A.Hlinku 2, 94976 Nitra, Slovakia; klara.vavrisinova@uniag.sk

In the experiment, we analyzed the amount and composition of intramuscular fat (IF) in muscle mussculus aductor (MA), longissimus thoracis mussculus (MLT) and mussculus semimembranosus (MSM) with 10 Large White breed (5 boars and 5 gilts). Determining of meat chemical composition was performed by FT IR Nicolet 6700 analyzer, determination of fatty acids was performed on the gas chromatograph. Infrared spectrum of the muscle was analyzed by the amount of IF in g/100 g sample. We observed a high statistically significant difference between MSM and MA and also between MSM and MLT at $P<0.001$ with the highest proportion of IF in MSM. Based on our findings we can conclude that in the indicator medium-chain fatty acid were high significant differences between MA and MSM at $P<0.001$, and between the MA and MLT at $P<0.01$ with the largest share in MLT and the lowest in MSM. High significant differences are also confirmed in long-chain fatty acid trait and between MA and MSM at $P<0.01$ and between MSM and MLT at $P<0.05$ with the highest proportion in MSM the lowest in MA. In the trait monounsaturated fatty acid, we found significant differences between MA and MSM at $P<0.05$ with the largest share in MSM and the lowest in MA. In the trait polyunsaturated fatty acid was found highly significant difference at $P<0.01$ between MA and MSM while the highest proportion was found in MA and the lowest in MSM. In the analysis of 6 fatty acids, we found a highly significant difference at $P<0.001$ level between MA and MSM, while the largest number were measured in MA and lowest in MSM. Between MA and MSM, we also found a statistically significant difference at $P<0.05$ in quantities of trans-fatty acids and cis-fatty acids. Finally, we conclude that the quantitative content of IF of muscles is different and qualitative structure varies in its composition and content of matte acids either in terms of chain length or number of double bonds and their isomers.

Ultrasound method for intramuscular fat prediction in live pigs using two different procedures
Bahelka, I., Tomka, J., Demo, P. and Oravcová, M., Animal Production Research Centre, Animal Breeding and Products Quality, Hlohovecka 2, 951 41 Luzianky, Slovakia; bahelka@cvzv.sk

In the first period, fifty-two hybrid pigs (gilts and castrates) were included in the experiment. All pigs were measured using ultrasonograph ALOKA SSD 500 with probe UST-5044-3.5 (3.5 MHz/172 mm) at five different ultrasound intensities for one-three days before slaughter. Cross-sectional images of longissimus dorsi muscle (LD) at right last rib area from hybrid pigs were taken. The most favourable result to laboratory analysed intramuscular fat (IMF) content ($r=0.530$, $R^2=0.28$) was found for intensity of 80%. Therefore, in the second period only this intensity was used. Besides the sonograph ALOKA, in this period also the new equipment SONOVET 2000 was used. In contradiction to the first, longitudinal images of LD muscle at right side of pig's back were taken. Using SONOVET and ALOKA equipments, thirty-three and sixty-six pigs, respectively, were scanned. All images were digitalized and consequently analysed using video image analysis to predict IMF content. Day after slaughter, dissection of right half carcass was done and samples of LD (150-200 g) were taken (at the same place as ultrasonic images were made) for laboratory analysis of IMF. Correlations between ultrasound and laboratory analysed IMF at longitudinal images using ALOKA and SONOVET were $r=0.494$ and 0.441, resp. Coeficients of determinations were 0.24 and 0.20, respectively. The results confirmed the suitability of ultrasound method for intramuscular fat prediction and it seems that both cross-sectional and longitudinal images are effective equally.

Pig carcass classification in Slovakia

Bahelka, I., Demo, P., Hanusová, E., Tomka, J. and Peškovičová, D., Animal Production Research Centre, Animal Breeding and Products Quality, Hlohovecka 2, 951 41 Luzianky, Slovakia; bahelka@cvzv.sk

Since the introduction of grading methods in the Slovak Republic the average of lean meat content in pig carcasses has increased from 47.59% to 55.8% in 1995 and 2011, respectively. Number of pigs classified in the S, E and U classes was increasing from 29.7% in 1995 to 92.6% in 2011. This trend proves the justification of using SEUROP system and it corresponds to actual consumer demands for higher percentage of lean meat and lower content of fat in pig carcasses. More than 40% animals within each of the studied years were slaughtered at the weight from 90.1 to 100.0 kg. Almost 100% of the pig production in the Slovak Republic has been covered by cross-bred combinations (3-4 breeds) whereby sire line boars belong to foreign breeds, mainly Pietrain, Yorkshire, Hampshire, Duroc and their crosses. Sows of domestic breeds Large White and Landrase are used in dam positions. Regarding sex of animals for slaughter, the ratio of gilts and castrates is approximately 1:1. New prediction formulas for two instrumental methods and one manual (TP) were derived from the data collected in the dissection trials performed in accordance with the EU reference. The statistical calculations of lean meat percentage were performed in accordance with the Commission Regulation (EC) 1249/2008, laying down detailed rules for the application of the community scale for grading of pig carcasses. The results suggest quite high accuracy of pig carcass grading by these methods (RMSEP = 1.5603 – 1.7031, R^2=0.7276 – 0.7539).

Effect of season on growth and feed intake in piglets from birth to weaning

Čechová, M., Geršiová, J., Hadaš, Z. and Nevrkla, P., Mendel university in Brno, Department of animal breeding, Zemědělská 1, 61300 Brno, Czech Republic; cechova@mendelu.cz

In this experiment the effect of all seasons (i.e. spring, summer, autumn and winter)on growth and feed intake in piglets within the period from birth to weaning was studied. Always 10 sows were evaluated under identical environmental conditions of the farrowing house. The animal house was not air-conditioned so that the temperature and humidity were not constant. The weight of piglets was estimated with the accuracy to 100g at the moment of birth and in age of 10; 20 and 28 days (i.e. at weaning). Average weight gains and feed consumption (in grams) were calculated also with the accuracy to 100g for each of the aforementioned time intervals. Differences between winter and summer periods (215 vs. 321g, respectively) were significant (P≤0,05). Other differences were found between autumn and winter periods were significant (P≤0,01). In spring period feed consumption was recorder 6,17 g/head/day, in summer period 4,41 g/head/day, in autumn period 4,30 g/head/day and in winter period 3,67 g/head/day. Experiments were performed within the framework of the research project QI111A166.

Fatty acid profile of m. longissimus dorsi in indigenous Krškopolje pigs and hybrids 12

Žemva, M., Kovač, M., Levart, A. and Malovrh, Š., Biotechnical Faculty, University of Ljubljana, Department for Animal Science, Groblje 3, 1230 Domžale, Slovenia; marjeta.zemva@bf.uni-lj.si

The aim of the study was to investigate the fatty acids composition of m. longissimus dorsi (LD) in Krškopolje pigs (KP) and maternal hybrids (H12; dam – Slovenian Landrace, sire – Large White). A total of 48 fatteners, 24 of each genotype, were haused in eight pens with six animals in each. They were penned according to genotype and sex. Rearing and feeding conditions were the same for all animals. To enrich the environment, hay and straw were added to each pen on a daily basis. Pigs were slaughtered in two age groups (G1 and G2) with a two month difference. Twenty-four hours after slaughter, LD samples were taken behind the last rib. Intramuscular fat (IMF) content and fatty acid composition were determined. Statistical analysis was carried out using the GLM procedure in the statistical package SAS/STAT. With IMF content increased SFA, C14:0, C16:0, MUFA, C16:1n-6 and C18:1n-9 increased with greater IMF content while PUFA, n-6 PUFA, C18:2n-6, C20:4n-6, n-3 PUFA, ratio PUFA/SFA, and n-6/n-3 decreased with IMF. Gilt meat had lower n-6/n-3 ratio than barrow meat. In comparison with H12, KP contained more n-3 polyunsaturated fatty acids (PUFA) and C18:3n-3 and less C20:4n-6. Additionally, the n-6/n-3 ratio was more desirable in KP meat, indicating a better nutritional meat quality of KP than H12. Muscle of fatteners from the first age group (G1) contained more SFA and C16:0 than from the second age group (G2). Less C20:4n-6 and n-3 PUFA were found in G1 than in G2 which is not desirable from a nutritional point of view. Moreover, trends for better PUFA/SFA and n-3/n-6 were obtained in G1 in comparison with G2. Interaction between the genotype and the slaughter group showed the highest content of C18:3n-3 and the lower ratio n-3/n-6 in KP from G1.

Comparison of growing-finishing performance pigs when diets containing cull chickpeas and DL- Met

Uriarte, J.M., Obregon, J.F., Güemez, H., Rios, F. and Acuña, O., Universidad Autonoma de Sinaloa, FMVZ, Carr. Int. km. 3.5 Sur, 80246, Culiacan, Sinaloa, Mexico; jumanul@uas.uasnet.mx

To determinate the effect of substitution of soybean meal and corn for cull chickpeas on growth performance, 48 hybrid pigs (BW=27.47±0.89 kg) in groups of four were placed in 12 concrete floor pens (2.5×2.5 m). In a complete randomized experimental design, pens were fed one of three diets: 1) Diet with 17.21% CP and 3.35 Mcal ME/kg, containing corn 71.0%, soy bean meal 25%, and premix 4% (CONT); 2) Diet with 17.1% CP and 3.35 Mcal ME/kg with corn 37%, cull chickpeas 50%, soybean meal 9%, and premix 4% (CHP50) and 3) Diet similar to CHP50 with 0.2% of DL-methionine additioanated (CHP50M). Pigs were weighed at days 0, 49 and 97 of experiments and feed intake was recorded daily. ADG and feed intake/gain ratio were calculated from these data; so the same at the end of experiment 12 pigs by treatment were slaughtered and carcass traits measured. ADG at day 49 (0.584, 0.429 and 0.525 kg) was not similar (P=0.05) between CONT and CHP50. Body weight at day 97 (95.175, 96.650 and 96.175 kg) were not affected (P=0.91) by CONT, CHP50 and CHP50M, respectively. ADG (0.701, 0.704 and 0.712 kg) were similar across treatment (P=0.90). Feed intake (1.968, 2.008 and 2.038 kg) was not affected (P=0.70) by treatments. Feed/gain ratio (2.775, 2.865, and 2.855) was similar (P=0.22) by treatments. Hot carcass weight (78.83, 76.13 and 77.45 kg) was similar between treatments (P=0.66), and carcass yield (78.43, 75.87 and 77.82%) was not affected by treatments (P=0.15). Backfat (1.87, 1.90 and 1.78 cm) was no affected; rib eye area (35.80, 37.25 and 35.70 cm^2) was similar between treatments. It is concluded, that cull chickpeas at 50% in diets for growing pigs (27 to 55 kg) affect growth performance; and that cull chickpeas at 50% added with 0.2% of DL-methionine can be used up 50% in diets for growing-finishing pigs without affecting growth performance and carcass traits.

Evaluation of piglet heating system by water heated panels in farrowing house
Botto, L'[1] and Lendelová, J.[2], [1]Animal Production Research Centre Nitra, Hlohovecká 2, 951 41 Lužianky, Slovakia, [2]Slovak University of Agriculture in Nitra, Trieda A. Hlinku 2, 949 76 Nitra, Slovakia; botto@cvzv.sk

The objective of this work was to evaluate piglet heating system in strawless farrowing house on the basis of surface temperature of water heated panels. Farrowing pens with crate for sow were situated across the alley in sections. Two water heated panels with total area 0.48 m^2 were installed for piglet heating in each pen. Panels in first pens in all rows were situated at the enclosure wall adjacent to marginal lengthwise passage in stable and panels of last (8th) pens at external enclosure wall. Surface temperature of 72 panels in 3 sections was measured by infrared thermometer (AMIR 7811-50) in February 2011. At data evaluation we analyzed average surface temperature of panels from first to eighth pens in rows, including frequency distribution of temperatures (in the range from 34 °C to 42 °C with step of 1 °C). The data were analyzed using the Descriptive Statistics procedure and a General Linear Model ANOVA by the statistical package STATISTIX 9.0. Significance of differences between heated panels in pens was determined by Tukey HSD test (at $\alpha \leq 0.05$). Water heated panels of 5th pens had the highest average surface temperature 40.62±0.70 °C; the panels of 8th pens had the lowest temperature 37.16±2.15 °C (P<0.001). As far as the average surface temperature is concerned, almost 90.0% water heated panels had suitable temperature (37-43 °C) and 58.4% optimum temperature (39-41 °C). The panels of 8th pens had the lowest temperature because they were installed in the pens which were last in the rows and they were situated at external enclosure wall. In these pens, in addition, internal wall was not thermal insulated sidelong the heated panels, as it was in the first pens It is possible to get improvement of temperature conditions by additional insulation of the external enclosure wall in the area of pens as well as by optimal heat regulation of panels according to the position of lying piglets.

Expected and observed ages of maternal and the faster growing boars
Sviben, M., Freelance consultant, Siget 22b, Hr-10020 Zagreb, Croatia; marijan.sviben@zg.t-com.hr

At the beginning of the 21st century in USA the value of boars with regard to the growth was estimated by the age at particular live weight. With the data published by Swine Genetics International Ltd., Cambridge, Iowa, in Boar Catalog 2004 it was established that at the live weight of 114 kg 22 Yorkshire boars were old 155.4±2.941 days. According to the equation of regression of the age (Y – days) on the live weight (X – kg) the expected average was 156.1 days. Observed mean age at 114 kg of 29 faster growing (Landrace and Duroc) boars was 150.6±2.847 days. In accordance with equation of regression expected average age of faster growing boars was 150 days. The applicability of equations was tested with the data published by Swine Genetics International, Cambridge, Iowa, in SGI 2011 Boar Catalog for 22 maternal Yorkshire boars and 16 faster growing boars (Landrace and Duroc with normally distributed data on their ages). It had been expected that at the live weight of 250 lbs (113.4 kg) Yorkshire boars were old 155.4 days but they had 145.7±3.447. Expected age of faster growing boars was 149.4 days but it was observed 137.9±2.845. After seven years the intensity of growth was improved but maternal and faster growing boars still differed. Nowadays Yorkshire boars can be evaluated using the equation for faster growing boars in 2004 and the faster growing boars using new formula. The regression curves make breeders able to evaluate animals by their age at any live weight.

BDporc: a pig information system for technical economic management and research

Noguera, J.L.[1], López, P.[1], Alòs, N.[1], Ibañez-Escriche, N.[1] and Pomar, J.[2], [1]IRTA, Genètica i Millora Animal, Alcalde Rovira Roure 177, 25198 Lleida, Spain, [2]UdL, Enginyeria Agroforestal, Alcalde Rovira Roure 177, 25198 Lleida, Spain; Noelia.Ibanez@irta.cat

The Pig Management Information System BDporc is the reference databank of Spanish pig sector at a national level. Its mains purposes are to provide the farm managers the information for the optimal management of their companies, and from a collective point of view, to create a databank to obtain reference standards and to research. BDporc has allowed the establishment of an information model, to build a large databank of individual data farm, and to produce different reference standards, that are used in the farm decision process. Moreover, a decision support system e-BDporc, which uses the World Wide Web, was developed to supply managers, advisers and farmers with just-in-time information and decision support for pig production. A subscription system enables confidential and personalized information. Background data are available from BDporc Spanish pig management information system that collected periodically show productive and reproductive data from different farms integrated in BDporc system. The system support different facilities, as farm and enterprise results monitoring and interpretative analysis multicriterion for individual or grouped farms results. On the other hand, the database creation allows the achievement of research about the pig populations controlled. The system design and implementation was based in real time dynamic pages using Java server page technology. Since BDporc system started, more than 1,100 pig farms have followed it. The sow's total number in the BDporc system is now higher than 650,000. The sample of farms is distributed in a 10% as familiar companies, a 25% as cooperatives, and a 65% as integrated companies. The BDporc system is a useful pig farm management tool for the Spanish pig sector, as indicated by the number of farms included in.

Effect of hot weather on microclimate in tunnel-ventilated pig fattening house

Botto, L.[1] and Lendelová, J.[2], [1]Animal Production Research Centre Nitra, Hlohovecká 2, 951 41 Lužianky, Slovakia, [2]Slovak University of Agriculture in Nitra, Trieda A. Hlinku 2, 949 76 Nitra, Slovakia; botto@cvzv.sk

The objective of this work was to evaluate effect of hot summer weather on microclimate in tunnel-ventilated pig fattening house. The stable was longitudinal divided into two sections with total capacity 500 housed pigs in 6 pens from 30 to 100 kg. In the summer period air exchange in each section was provided by 5 fans situated at backside of stable (2 front, 2 side fans and 1 fan for under-slatted exhaust). Experiment was carried out when both front fans were operating. Basic microclimatic parameters were noticed in each section in pens in animal zone and in zone of stock-keeper (0.5 m and 1.8 m above the floor). Temperature, relative humidity and air flow were measured by ALMEMO 2290-4 device during the day with hot weather. External air parameters were registered in time measurement, too. At evaluation of measured data the average values in pens were expressed for both sections. The data were analyzed using the Descriptive Statistics procedure by the statistical package STATISTIX 9.0. In fattening house was registered lower average temperature than external air temperature in zones 0.5 and 1.8 m above the floor (36.6±0.45 °C and 36.9±0.60 °C vs. 37.5±0.15 °C). Average internal air relative humidity in both zones was higher than external air humidity (28.8±2.25% and 27.4±1.59% vs. 24.4±0.44%). Air humidification or another system of adiabatic cooling was not used in the stable. Average internal velocity of air was 1.20±0.73 m/s in animal zone and 2.56±1.49 m/s in zone of stock-keeper. Enhanced air flow velocity in animal zone, but mainly in zone of swineherd, ensured evaporative cooling of housed pigs; consequently the ambient temperature sensationally decreased. Pigs tolerated better higher temperatures under interoperation with low relative humidity of air. Behaviour of pigs did not indicate deflection in standard manifestations at lying, movement, urination and appetite.

Actions to mitigate greenhouse gas emissions from milk production
Flysjö, A., Arla Foods, Sønderhøj 14, 8260 Viby J, Denmark; anna.flysjo@arlafoods.com

The dairy sector is responsible for 4% of global anthropogenic greenhouse gas (GHG) emissions (including meat by-products corresponding to 57% of the world's beef supply). When analysing how to mitigate climate change most efficiently, a system thinking is required. The difference in greenhouse gas emissions of raw milk between dairy farms has been found to be at least ±17% for relatively similar farm systems in Sweden, which indicates that there is scope for reducing GHG emissions by improving management practices. The present study gives an overview of different action to mitigate GHG emissions for milk production and addresses the complexity and some of the challenges related to it. For example, increased milk yield per cow is one strategy often presented to reduce GHG emissions for dairy. However, a higher milk yield per cow does not necessarily reduce the GHG emissions per kg milk, when also considering the alternative production of the beef by-product. Hence, a change in one part of the system can have an impact on other parts of the system, and it is thereby important to ensure a reduction in total net emissions. Other mitigation options discussed related to milk production are feed additives, biogas production, energy efficiency, renewable energy etc. To analyse different mitigation measures, models are required that account for the whole milk system, including inputs of e.g. feed, synthetic fertilisers, energy. Today the GHG emissions are calculated at many dairy farms to help identify 'hot spots' and how the specific farm can reduce emissions. When mitigating GHG emissions for milk at farm level there is no 'silver bullet'; instead many improvements which individually seem small can add up and together result in significant reductions. Mitigation strategies need to account for the individuality of farms, and the magnitude of reduction potential depends on the actual farm and its associated emissions.

Manure management and mitigation of greenhouse gases: opportunities and limitations
De Vries, J.W., Wageningen UR Livestock Research, Wageningen University and Research Centre, Environment, P.O. Box 135, 6700 AC Wageningen, Netherlands; jerke.devries@wur.nl

The emission of greenhouse gases (GHGs) from animal manure and its management (i.e., storage, processing and field application) in livestock and crop production systems remains a key factor in the life cycle impact of food production. Management of manure, including manure processing, contributes to impacts such as, climate change, acidification, eutrophication, and fossil fuel depletion. Manure processing technologies were developed with the potential to reduce the environmental impact, including greenhouse gas emissions and related costs, of manure management. Such technologies include: liquid and solid separation, filtration, drying and composting, and anaerobic digestion for bio-energy production. In this presentation the GHG mitigation potential of several processing technologies will be highlighted. Studies have shown that compared to conventional liquid manure management, greenhouse gases can be reduced substantially by implementing anaerobic digestion of manure (>100%) or by keeping urine and faeces separated in the housing system (>65%). However, other impacts, such as acidification and eutrophication, could increase as a result of processing. Future design of manure management systems, therefore, needs to consider available knowledge for creating optimal solutions. An outlook will be provided with a methodical approach for integrated design.

Greenhouse gas emissions from livestock food chains: a global assessment
Gerber, P.J.[1], Macleod, M.[1], Opio, C.[1], Vellinga, T.[2], Falcucci, A.[1], Weiler, V.[1], Tempio, G.[1], Gianni, G.[1] and Dietze, K.[1], [1]FAO, Animal Production and Health Division, Viale delle Terme di Caracalla, 00153 Rome, Italy, [2]Wageningen University, Livestock Research, P.O. Box 65, 8200 AB Lelystad, Netherlands; pierre.gerber@fao.org

Livestock food chains are a recognized source of greenhouse gas (GHG) emissions. A life cycle-based approach is used to generate emission estimates from production systems associated with main commodities, as a precondition to the initial identification and design of efficient mitigation measures. The model considers all significant direct and indirect GHG emissions (CH_4, N_2O and CO_2) across the food chain incorporating emissions from manufacture and use of production inputs such as fertilizer and feed; on-farm activities associated with animal management and feed production, manure storage and management; emissions related to land use change (soy-related deforestation and pasture expansion) and post-farm emissions related to the transport of animals from farm to processing units through to the distribution centre. The paper presents the methodological modelling framework developed for the quantification of greenhouse gas emissions and provide a comparative global and regional overview of the life cycle global warming potential disaggregated by species (chicken, pigs, cattle, small ruminants), commodity type (meat, eggs and milk) and farming systems. The assessment reveals the key factors contributing to the carbon footprint of production systems and related entry points for mitigation intervention along the livestock chains, in main geographical regions. In addition, it discusses the methodological innovations and challenges associated with the global modelling exercise (herd and feed baskets modelling on a global scale, computation in Geographical Information System – GIS – environment, data constraints and sensitivity analysis).

Determining the optimal use of by-products in animal production from an environmental perspective
Mollenhorst, H.[1,2], Van Zanten, H.H.E.[2], De Vries, J.W.[3], Van Middelaar, C.E.[2], Van Kernebeek, H.R.J.[2] and De Boer, I.J.M.[2], [1]Business Economics Group, Wageningen University, P.O. Box 8130, 6700 EW Wageningen, Netherlands, [2]Animal Production Systems Group, Wageningen University, P.O. Box 338, 6700 AH Wageningen, Netherlands, [3]Wageningen UR Livestock Research, Wageningen University and Research Centre, P.O. Box 135, 6700 AC Wageningen, Netherlands; erwin.mollenhorst@wur.nl

Using by-products in, e.g., animal feed may reduce the environmental impact. By-products are outputs from a multifunctional process of which the production volume is determined by another product, the main product. A theoretical framework to calculate the environmental load of the main product in a consequential life cycle assessment is described in literature. In this study we extended this framework to determine the optimal use of a by-product. It is illustrated with the example of manure management. In Europe, manure is a by-product, and is commonly used for crop fertilisation. An alternative use of manure is bio-energy production by means of anaerobic digestion. The environmental consequences of altered manure management were analysed with the theoretical framework ($D_1 - \Delta B_1 - I_1 + I_2 - D_{2(a+b)} + \Delta B_{2(a+b)}$). Manure used for bio-energy production cannot be used directly for fertilisation, requiring additional artificial fertiliser. This will increase environmental impact (D_1=52). Electricity and heat from bio-energy replace energy from fossil sources, which will reduce the environmental impact ($D2b$=38). Moreover, digestate is produced, that again replaces artificial fertiliser (D_{2a}=52). Finally, differences in emissions from the use of manure instead of artificial fertiliser (ΔB_1=41), digestate instead of artificial fertiliser (ΔB_{2a}=45) and emissions during storage of manure and the co-digestion process (I_1=39; I_2=42) need to be accounted for. In the end, using 1 ton of manure for bio-energy production instead of direct field application results in a decrease in greenhouse gas emissions of 31 kg CO_2-eq.

Reducing greenhouse gas emissions of pig production through feed production and diet composition

Meul, M.[1], Ginneberge, C.[1], Fremaut, D.[1], Van Middelaar, C.E.[2], De Boer, I.J.M.[2] and Haesaert, G.[1], [1]Faculty of Applied Bioscience Engineering, Valintin Vaerwyckweg 1, 9000 Ghent, Belgium, [2]Wageningen University, Animal Production Systems Group, P.O. Box 338, 6700 AH Wageningen, Netherlands; marijke.meul@hogent.be

We estimated greenhouse gas emission (GHGE) reduction potential of four diet scenarios for finishing pigs in Europe, directed at (1) improving crop production by increasing crop yields (CROP scenario) or decreasing inputs (e.g. fertilizer); (2) excluding soybean products and using European grown feed ingredients only (EU scenario); (3) maximizing use of by-products from the food and bio-energy industry; and (4) reducing crude protein content (N-LOW scenario). These scenarios were compared to a standard feed composition (STAND) based on a carbon foot print (CFP) assessment, using three different methodologies: (1) reference CFP excluding emissions from land use change (LUC); (2) CFP taking into account emissions from direct LUC; and (3) including emissions from both direct and indirect LUC. Our analysis showed that taking into account LUC has a major impact on the CFP associated with each scenario and results in a different estimation of the GHGE reduction potential and evaluation of the different scenarios. For example, reference CFP is lowest for CROP scenario with 6% GHGE reduction compared to STAND. When direct LUC is considered, EU is the best scenario with 18% reduction in GHGE compared to STAND because LUC emissions related to soybean products are avoided. However, when indirect LUC is considered N-LOW is the best scenario with 9% reduction in GHGE, because protein crops with high use of agricultural land are replaced by synthetic amino acids for which land use is neglected. Our analysis shows the urgent need for a concise methodology to estimate both direct and indirect land use change and their related GHG emissions. Until then, we suggest to consider land use as an important estimator of LUC risk in the CFP evaluation of agricultural products.

FeedPrint: insight in GHG emissions of the feed production and utilization chain

Vellinga, T.V.[1] and Blonk, H.[2], [1]Wageningen UR Livestock Research, P.O. Box 65, 8200 AB Lelystad, Netherlands, [2]Blonk Environmental Consultants, Gravin Beatrixstraat 34, 2805 PJ Gouda, Netherlands; theun.vellinga@wur.nl

The contribution of feed to the total emissions of livestock products is about 35 and 70% for ruminants and monogastrics, respectively. For this reason, the feed related industry wants to increase insight in the emissions and explore mitigation options. A database and calculation tool have been developed to cover all feed materials used in the Netherlands, sourced from different continents. To perform a cradle to farm gate LCA, this is combined with nutritional and farm models. The database is unique and contains new approaches in data collection and allocation methods. Transparency and wide acceptance are key words for industrial stakeholders, because the identification of mitigation options is considered a pre competitive issue. Due to a lack of good quality data, new methods have been developed to improve allocation procedures in industrial processing, to estimate the application of N from animal manure, to calculate the GHG emission of grass, to calculate GHG emissions related to land use change and to calculate uncertainty. A distribution type and range has been attributed to all data and emission factors. Monte Carlo simulation is used to gain insight in the overall uncertainty. A strong stakeholder involvement has been realized via working groups with industrial stakeholders and appeared to work in two directions. Researchers received feedback on data and methods, stakeholders became aware of the complexity of LCA and the related data requirements.

Environmental impact of milk production can be reduced using indicator traits and genomic selection
Hansen Axelsson, H.[1], Fikse, F.W.[1], Johansson, K.[2], Sørensen, M.C.[3], Sørensen, A.C.[3], Petersson, K.-J.[1] and Rydhmer, L.[1], [1]The Swedish University of Agricultural Sciences, Department of Animal Breeding and Genetics, Box 7023, 750 07 Uppsala, Sweden, [2]Swedish Dairy Association, Box 7023, 750 07 Uppsala, Sweden, [3]Aarhus University, Department of Genetics and Biotechnology, Box 50, 8830 Tjele, Denmark; Helen.Hansen@slu.se

We studied breeding strategies for Nordic Red Cattle aimed at reducing the environmental impact of milk production. The hypothesis was that specific indicator traits can be recorded in a few selected herds and this phenotype information can be used in breeding schemes with genomic selection. The stochastic simulation was performed. The breeding goal consisted of milk production (MY), functional traits (FT), and environmental impact (EI). The indicator traits for EI were categorized as large, medium or small depending on the scale how the traits were recorded. Two traits per category were selected to be analyzed as independent scenarios. These traits in the large category were stayability (STAY) and stature (STAT), in the medium category live weight (LW) and gases in the breath of the cow (BRH), and in the small category residual feed intake (RFI) and methane recorded in a gas chamber (METH). Additionally, one scenario without any records on indicator traits (No IT) and six scenarios each with an indicator trait were simulated. The traits in the large category could be recorded in the whole population while the traits in medium and small categories had a limited number of phenotype records. Despite this, the genetic gain was highest in scenarios BRH, RFI and METH resulting in 30-55% more improvement in EI compared to scenario No IT. The annual total genetic gain was €49 in scenario No IT and €54 in scenario METH. This proved that genetic improvement of mitigating CH_4 emissions in breeding schemes with genomic selection is possible even with a limited number of records on indicator traits. This improvement can be achieved without significant loss in milk production or in functionality.

Monitoring strategies to breed environment-friendly cows
De Haas, Y.[1], Crump, R.[1], Dijkstra, J.[2] and Ogink, N.[3], [1]Wageningen UR Livestock Research, Animal Breeding and Genomics Centre, P.O. Box 65, 8200 AB Lelystad, Netherlands, [2]Wageningen University, Animal Nutrition Group, P.O. Box 338, 6700 AH Wageningen, Netherlands, [3]Wageningen UR Livestock Research, Environment Dept., P.O. Box 135, 6700 AC Wageningen, Netherlands; Yvette.deHaas@wur.nl

In the Netherlands, agriculture produces 13% of the total greenhouse gas emissions, methane being one important contributor. At a constant feed intake and milk production level, there is variation among dairy cows for methane emission, opening up possibilities for breeding environment-friendly cows. This requires many individual cow records of real methane emission under practical conditions. However, these are currently unavailable, since most data originate from respiration chambers. We investigated practical measuring systems, calculating the accuracy of daily methane emission predicted in these systems compared with respiration chamber data. Data on 10 pairs of cows observed for 4 days in Wageningen respiration chambers were analysed and three scenarios were subsequently simulated: (1) measuring during milking (i.e., twice daily, for 15 minutes); (2) measuring in concentrate feeder (i.e., 5 times daily, 6 minutes); and (3) measuring in the cubicles (i.e., 4 hours continuously, 10 observations of 3 minutes per hour). For direct methane emission, calculated accuracies were 0.85, 0.89 and 0.96, respectively. However, under practical conditions full recollection of methane output of individual cows may be technically complicated and costly. Thus, we also investigated the accuracies that can be achieved by measuring the ratio between methane and carbon dioxide concentrations, which were 0.31, 0.33 and 0.39, respectively. Daily methane production can be predicted reasonably accurate by collecting samples of all cows during twice daily milking. This opens up the possibility of creating a large database of individual methane emission phenotypes for use in the identification of suitable indicator traits for the genetic merit of methane production.

Harmful gases concentration in broiler chicken halls during extremely low ambient temperature
Mihina, S.[1,2], Bodo, S.[1], Lendelova, J.[1], Galik, R.[1] and Broucek, J.[2], [1]Slovak University of Agriculture, Tr. A. Hlinku 2, 94976 Nitra, Slovakia, [2]Animal Production Research Centre Nitra, Hlohovecka 2, 951 41 Luzianky, Slovakia; stefan.mihina@uniag.sk

Concentrations and emissions of greenhouse gases (CH_4, N_2O, CO_2, and H_2S) and ammonia (NH_3) were monitored simultaneously in two broiler chicken halls during winter fattening period. In both halls, equipped with nipple drinker lines and tube-style pan feeder lines filled automatically, birds were kept on straw littered concrete floor. The houses were mechanically ventilated with regulated combined tunnel and cross two-sided ventilation. Fresh air inlets were placed on both side walls of the hall. Concentrations of gases were measured by the photoacoustic infrared detection method. Air samples for analysis of concentrations of gases were taken from air stream at ceiling fans, on animal level and from outdoor surroundings. Inside and outside air temperature, relative humidity of air and temperature of litter were measured permanently, too. Measurements were done during the winter season with extremely low ambient temperature. To ensure thermal comfort of chickens in the hall ventilators were running at very limited rate. It has influenced course of concentrations and emissions of gases as well as parameters of inside air and litter. Each hall has different feeding ratio, which influenced gases content in the halls, too.

The global research alliance livestock research group
Mottram, T.T., Defra, Food and Farming Group, Nobel House, 17 Smith Square, SW1P 3JR London, United Kingdom; toby.mottram@defra.gsi.gov.uk

The Global Research Alliance is a new intergovernmental organisation dedicated to coordinating and stimulating research into emissions of greenhouse gases from agriculture. The Livestock Research Group is looking at ways to improve the greenhouse gas intensity of livestock production systems and increase the quantity of carbon stored in those soils. Key emissions covered are methane from enteric fermentation and waste management, nitrous oxide from animal wastes and fertilisers, and soil carbon. The Group is co-chaired by New Zealand and the Netherlands. Immediate goals for the Livestock Group include: Collecting, collating and analyzing information on individual member efforts in livestock emissions research. Developing best practice guidance and standardized methodologies for measuring emissions from livestock production and making training and development opportunities available. Establishing networks and databases on key areas of activity, e.g. microbial genetics, manure management etc. Fostering research collaboration between member countries and with key partner organizations e.g. the CGIAR Climate Change and Food Security programme, the EU Joint Programming Initiative, the International Livestock Research Institute etc.

Quantification of dangerous gas emissions from pig fattening
Palkovičová, Z., Brouček, J. and Mihina, Š., Animal Production Research Centre, Hlohovecká 2, 951 41 Nitra, Slovakia; palkovicova@cvzv.sk

The aim of this study was to quantify emissions and emission factors of ammonia and greenhouse gases from pig fattening reared on the slatted floor during four fattening cycles. For this purpose the gas concentrations, airflow rate, housing and outdoor temperatures were monitored. In the experiment it was detected that both summer batches (batch 1 and 4) as well as autumn-winter batch (batch 2) and spring batch (batch 3) recorded a similar course in the amount of daily gas emissions. In summer batches the increase of daily gas emissions from the beginning to the end of fattening was mostly recorded. Conversely, in autumn-winter and spring batches the decrease of daily gas emissions from the beginning to the end of fattening was especially registered. Differences in daily emission rates of NH_3, CO_2, N_2O and CH_4 were significant between all batches except the batches: summer 1 and autumn-winter; summer 1 and summer 4; summer 1 and spring, summer 1 and summer 4; autumn-winter and spring, autumn-winter and summer 4, respectively. In the case of emitted airflow rate it was registered that differences were significant between all batches except the batches: summer 1 and spring, spring and summer 4. Temperature differences in animal area were also significant between all batches except the batches: summer 1 a summer 4. During all four batches there were produced 836 kg of NH_3, 266 138 kg of CO_2, 68.6 kg of N_2O, 4 176 kg of CH_4 and 431 428 kg of H_2O. The greatest emission rates of NH_3, CO_2 and N_2O were found in autumn-winter batch, CH_4 in summer batch 4, and H_2O in spring batch. The least emission rates of CO_2, CH_4 and H_2O were recorded in summer batch 1, and NH_3 a N_2O in summer batch 4. Calculated emission factors recorded the values 2.1 kg of NH_3, 654 kg of CO_2, 0.17 kg of N_2O, 10.3 kg of CH_4 a 1 073 kg of H_2O per animal and year. It was found that NH_3 emission factor calculated from the practical observations was lower by 0.79 kg per animal and year than assumed emission factor.

QTL detection for age at puberty and cyclicity resumption in a Holstein × Normande population
Lefebvre, R.[1], Larroque, H.[1], Barbey, S.[2], Gallard, Y.[2], Colleau, J.J.[1], Chantry-Darmon, C.[3], Laine, A.L.[4] and Boichard, D.[1], [1]INRA, Gabi, Domaine De Vilvert, 78350 Jouy en Josas, France, [2]INRA, UE326, Domaine Experimental Du Pin, 61310 Exmes, France, [3]Labogena, Domaine De Vilvert, 78350 Jouy en Josas, France, [4]INRA Laboratoire De Dosages Hormonaux, UMR PRC, Centre de Tours, 37380 Nouzilly, France; rachel.lefebvre@jouy.inra.fr

A QTL detection experiment was carried out in 'Le Pin' INRA experimental farm by crossing Holstein and Normande dairy cattle. The design included 1092 F2 as well as 561 'F3' (= F2 dams × F1 sires) females. Blood progesterone was assessed every 10 days until a positive assay, from 230 days old and 20 days post-partum, for 1096 heifers and 1039 primiparous cows respectively. Age at puberty and time between calving and cyclicity resumption were calculated with the date at the first positive assay. Mean age at puberty was 307 days (±51) and mean post-partum cyclicity resumption was 31 days (±12). Phenotypic correlation between both traits was very low (0.02). Heritability estimates were 0.22 for puberty and 0.15 for cyclicity resumption. All animals, including F1 parents and F0 grandparents, were genotyped with the Illumina 54k Beadchip. QTL detection was performed by linkage and linkage disequilibrium analysis (LDLA). For age at puberty, a total of 21 QTL were found on 12 chromosomes (1, 2, 3, 5, 8, 10, 11, 13, 14, 21, 27, 29), with the strongest ones on chromosomes 29 (26 and 44 Mb) and 13 (60 Mb). For time to cyclicity resumption, 14 QTL were found on 7 chromosomes (1, 5, 11, 21, 26, 27, 29), with the strongest ones on chromosomes 21 (26 Mb) and 26 (33 Mb). The latter region contains a gene encoding estradiol dehydrogenase involved in ovarian activity. Only one common region was detected for both traits (chromosome 21 around 26 Mb).

Comparative analysis of linkage disequilibrium in Fleckvieh and Brown Swiss cattle

Ertl, J., Edel, C., Neuner, S., Emmerling, R. and Götz, K.-U., Bavarian State Research Centre for Agriculture, Institute of Animal Breeding, Prof. Dürrwaechter-Platz 1, 85586 Poing-Grub, Germany; johann.ertl@lfl.bayern.de

The extent of linkage disequilibrium (LD) between genotyped single-nucleotide polymorphisms and unknown quantitative trait loci is assumed to be one factor influencing the accuracy of genomic breeding values. Additionally, the persistence of LD between breeds might influence the success of an across breed evaluation. For 54K genotypes (Illumina® 54K BeadChip) of 9,387 dual-purpose Fleckvieh bulls, mainly from Germany and Austria, and 4,068 Brown Swiss bulls, from Germany, Switzerland, Austria, Italy and USA, LD measures r and r^2 were calculated. In Fleckvieh, average r^2 is 0.32, 0.26, 0.17, 0.10, and 0.03 at genomic distances 10, 20, 50, 100, and 1000 kb, respectively. LD in Brown Swiss is higher than in Fleckvieh, especially at longer marker distances, with average r^2 of 0.32, 0.28, 0.20, 0.14, and 0.07. Analyzing German and Austrian subpopulations of Fleckvieh separately resulted in higher r^2 compared with values obtained from the complete population. The correlation of r (cor_r) was used as a measure of persistency of LD phase and calculated between Fleckvieh and Brown Swiss (born 1999-2005). For genomic distances 20, 50, 100, 200, 400, and 1000 kb, average cor_r is 0.86, 0.72, 0.54, 0.32, 0.15, and 0.05, respectively. Cor_r within breeds is much larger although results suggest that there is a certain genetic distance between subpopulations from different countries. Cor_r between German and Austrian Fleckvieh bulls is very close to 1 for small marker distances but decays to 0.96 and 0.92 at 200 and 1000 kb distance. Between German and US Brown Swiss, cor_r at these distances is 0.85 and 0.78, respectively. These genetic distances between subpopulations of Fleckvieh and Brown Swiss are also reflected by LD-based calculations of effective population size.

Genetic factors influencing coat color variation in two closely related Swiss cattle breeds

Flury, C.[1], Signer-Hasler, H.[1], Tetens, J.[2], Leeb, T.[3], Rieder, S.[4] and Drögemüller, C.[3], [1]Bern University of Applied Sciences, School of Agricultural, Forest and Food Sciences, Länggasse 85, 3052 Zollikofen, Switzerland, [2]Christian-Albrechts-University Kiel, Institute of Animal Breeding and Husbandry, Hermann-Rodewald-Str. 6, 24118 Kiel, Germany, [3]University of Bern, Vetsuisse Faculty, Institute of Genetics, Bremgartenstrasse 109, 3001 Bern, Switzerland, [4]Agroscope Research Station, Swiss National Stud Farm, Les Longs Prés, 1580 Avenches, Switzerland; christine.flury@bfh.ch

In the year 1885 the Swiss Evolener breed was separated from the Swiss Eringer breed due to the official order to breed single-colored Eringer animals. Since then the breeding standard is white spotted coat color for Evolener cattle, where such individuals were strictly excluded from the Eringer herdbook. Today, the Evolener herdbook contains 323 individuals in comparison to 7,367 in the Eringer herdbook. In total, 36 Evolener and 70 Eringer individuals were genotyped for the HD-Illumina-Beadchip. More than 627k SNPs remained after filtering. Based on weighted F_{ST}-values a strong selection signal was observed on BTA6 in the region of the KIT-gene. A genome wide association study revealed other genomic regions known to affect white spotting patterns in cattle. In both populations, different categories of basic coat colors are observed. It is known from other species that basic coat colors might interact with the expression of white markings. Respecting such epistasis in line with a more advanced correction for population stratification is expected to further enhance the results. The study allows for a better understanding of the genetic architecture of coat color variation in cattle, and will emphasize genetic differences between two closely related indigenous Swiss cattle breeds.

Signatures of selection in Holstein Friesian dairy cattle

Elferink, M.G.[1], Bovenhuis, H.[1], Veerkamp, R.F.[2], Coffey, M.P.[3], Wall, E.[3], Mc Parland, S.[4], Lunden, A.[5] and Bastiaansen, J.W.M.[1], [1]Wageningen University, Animal Breeding and Genomics Centre, P.O. Box 338, 6700 AH Wageningen, Netherlands, [2]Wageningen UR Livestock Research, Animal Breeding and Genomics Centre, Postbus 135, 6700 AC, Wageningen, Netherlands, [3]Scottish Agricultural College, Sustainable Livestock Systems Group, Easter Bush Campus, EH25 9RG Midlothian, United Kingdom, [4]Teagasc, Animal & Grassland Research and Innovation Centre, Moorepark, Cork, Ireland, [5]Swedish University of Agricultural Sciences, Department of Animal Breeding and Genetics, P.O. Box 7023, 75007 Uppsala, Sweden; Henk.Bovenhuis@wur.nl

The identification of genomic regions affected by past selection is of interest for understanding selection history and molecular pathways underlying breeding goal traits. The aim of the current study was to detect signatures of selection in the genome of Holstein-Friesian dairy cattle. Our data consisted of 2,029 cows from experimental herds in four European countries (Ireland, Netherlands, Scotland, and Sweden). These cows were genotyped with the Illumina Bovine SNP50 BeadChip as part of the RobustMilk project. To identify signatures of selection we applied the integrated haplotype score statistics averaged in 500 kb non-overlapping windows. We analyzed data from each country separately but also performed analyses based on a combined data set. In total 51 chromosomal regions were detected that showed evidence for selection in at least one of the four countries. Nine of these chromosomal regions showed evidence of selection in the combined Holstein-Friesian population. Of these regions, three – located on chromosome 10 (50-63 Mb), 13 (46-47.5 Mb) and 20 (24-38.5 Mb) – were independently detected in experimental herds from three countries. The large 14.5 Mb region on chromosome 20 includes the growth hormone receptor gene which has previously been associated with various production traits in cattle including milk yield.

G-by-E interactions at the SNP level in Holstein dairy traits recorded in China and UK

Pollott, G.E.[1], Wu, J.J.[2], Cooke, J.S.[1], Zhang, S.J.[2] and Wathes, D.C.[1], [1]RVC, VBS, London, NW1 0TU, United Kingdom, [2]HZAU, Animal Genetics, Wuhan, 430070, China; gpollott@rvc.ac.uk

Genotype-by-environment interactions (GEI) have been widely studied in livestock populations to improve the accuracy of genetic evaluations and to understand how genotypes are affected by differing environments. The use of SNP chips to genotype animals provides the opportunity to study GEI at the individual SNP genotype level. Two populations of Holstein dairy cows, one in SE England and the other in Hubei Province, China, were performance recorded and genotyped, using the Illumina SNP50 chip. The data from 482 cows, (221 from UK, 261 from China) and 34,952 informative SNPs on chromosomes 1 to 29 were analysed by Genomic GRAMMAR-GC in R, using genomic kinship coefficients, to obtain environmental residuals for each recorded animal. The environmental residuals were then analysed by fitting a regression coefficient across genotypes within each country for each SNP. A significant GEI for each SNP was investigated by testing the two regression coefficients using a Z score in PLINK. Three traits were analysed, heifer 305-d milk yield (MY305), age at first calving (AFC) and first calving interval (CI). Heritability estimates for the polygenic component of the model derived using the genomic kinship matrix were 0.27 ± 0.14 (MY305), 0.21 ± 0.15 (AFC) and 0.08 ± 0.07 (CI). For MY305, 84 SNPs indicated a GEI ($P<0.001$) with the region showing the greatest level of GEI around the DGAT1 gene on BTA14. For AFC, 42 SNPs indicated the presence of a GEI ($P<0.001$) with a prominent group of 5 SNPs on BTA26. The analyses of CI found 48 SNPs exhibiting a GEI ($P<0.001$) with groupings on chromosomes 8, 15 and 21. The use of a kinship matrix was particularly important in this study since it was not possible to link all the pedigrees back to common ancestors between the two countries. This initial study suggests that GEI can be detected at the SNP level and highlights a number of genomic regions that exhibit GEI for three key dairy traits in Holstein cows.

Fine mapping of QTLs of meat quality in three French beef breeds using the Bovine SNP50® chip
Allais, S.[1,2], Hocquette, J.F.[3], Levéziel, H.[4], Lepetit, J.[5], Rousset, S.[6], Denoyelle, C.[7], Bernard-Capel, C.[7], Rossignol, M.N.[8], Journaux, L.[1] and Renand, G.[2], [1]UNCEIA, 149, rue de Bercy, 75595 Paris, France, [2]INRA, UMR 1313, 78350 Jouy en Josas, France, [3]INRA, UMR 1213, 63122 Theix, France, [4]INRA/Univ. de Limoges, UMR 1061, 87060 Limoges, France, [5]INRA, UR 370, 63122 Theix, France, [6]INRA, UMR 1019, 63122 Theix, France, [7]Institut de l'Elevage, 149, rue de Bercy, 75595 Paris, France, [8]Labogena, Vilvert, 78350 Jouy en Josas, France; sophie.allais@unceia.fr

Without routine measurement of sensory meat quality traits, breeding companies wonder how to take advantage of the genetic variability of these traits at the molecular level. The project 'Qualvigène' was set in to find markers associated to beef quality therefore. A total of 1,059 Charolais, 1,219 Limousin and 947 Blond d'Aquitaine purebred young bulls from respectively 48, 36 and 30 sire families were genotyped with the Illumina Bovine SNP50® Beadchip. We measured ten traits related to quality of the longissimus thoracis muscle: shear force, tenderness, juiciness and flavour scores, intramuscular fat content, insoluble collagen content, collagen solubility, number and size of fibres, and lightness. QTLs were detected with a methodology combining linkage disequilibrium and linkage analysis on haplotypes. After quality control and with a minor allele frequency of 5%, about 37K SNP were used in each of the three breeds. If we consider the significant positions at the 1‰ threshold (without Bonferroni correction) and separated by a minimum distance of 4cM, we found about 170 QTLs in each of the three breeds, more precisely between 7 and 35 QTLs per trait. Compared to a previous QTL detection with microsatellites, we detected many more QTLs per trait. Some QTLs were common of two or three breeds as a QTL of shear force in the three breeds on the chromosome 3 between 16 and 21cM. Most of the other QTLs were breed-specific. The results were generally consistent with previous studies of candidate genes (Myostatin, Calpastatin and Calpain 1 genes).

Autozygosity by difference; a new method for locating autosomal recessive mutations
Pollott, G.E., RVC, VBS, Royal College Street, London NW1 0TU, United Kingdom; gpollott@rvc.ac.uk

This paper reports a novel method for analysing case/control SNP chip data from an autosomal recessive genetic disease to locate the region containing the causal mutation. The likely characteristics of such mutations are used as the basis for the method. One critical aspect of finding the causal mutation, not used in other methods, is to separate 'normal' background homozygosity from the autozygosity associated with the mutation. The proposed method is called 'autozygosity by difference' (ABD). The initial step involves identifying the most common homozygous genotype (CHG) in the cases at each SNP and giving each CHG a score of 1 if present or 0 for all other genotypes, for all SNPs and individuals. The total CHG per chromosome for each case and control is calculated. This dataset, comprising total CHG scores for each autosome for each individual, is then analysed by linear model using REML methodology to identify the chromosome which is significantly more homozygous in cases than in controls. The second stage of the analysis is to score each run of homozygosity (ROH) on that chromosome for each individual case and control animal. Each SNP in a ROH gets a score equal to the number of SNPs in the run. Each SNP on the chromosome is analysed by finding the mean score for cases and controls separately. The causal mutation is in the region with the highest score, calculated as the difference between SNP means for cases and controls. The probability of the SNP is calculated by permutation, randomising the genotypes within animals and reanalysing the data, ranking the results and estimating the probability of each result. The method was applied to data from a cattle genetic disease where segregation analysis of perinatal mortality indicated an autosomal recessive mode of inheritance. Using 13 cases and 42 controls, an area on chromosome 4 was located using ABD comprising 5 SNPs (P<0.001). The ABD method is viable for locating the potential region of the genome containing a causal mutation from a small number of cases and controls.

An imputation pipeline for cost effective genomic selection in commercial pig breeding

Cleveland, M.A.[1], Yu, N.[1], Foertter, F.[1], Deeb, N.[1] and Hickey, J.M.[2], [1]Genus plc, Hendersonville, TN, 37075, USA, [2]School of Environmental and Rural Science, University of New England, Armidale, Australia; matthew.cleveland@genusplc.com

The use of genomic selection in pigs can significantly increase genetic gain, but genotypes for young selection candidates are required for full potential. In pig breeding many candidates are routinely produced and the cost to generate the many genotypes desired for genomic evaluation is prohibitive. Imputation approaches have been developed to obtain high-density genotypes for animals that have genotypes for only few SNPs. We developed a very low-density marker panel to be used across traits and lines specifically to impute 60k genotypes. The panel SNPs were selected based on minor allele frequency in seven pig lines and for equal spacing across the genome. A pipeline was developed to perform imputation from this small panel using the software AlphaImpute. An automated process extracts raw genotypes from a custom database, prepares the input files for imputation, runs AlphaImpute on a high performance computing cluster, with jobs per chromosome distributed across multiple processors, and loads imputed genotypes back into the database. Imputed genotypes can then be rapidly extracted for use in genomic evaluation. More than 4,000 animals have been genotyped at 60k in each of multiple pig lines, including all sires and a large proportion of the dams. Selection candidates in each evaluated line are genotyped using the low-density panel. Tests using datasets with 2,500 to 3,500 high-density genotyped animals and 500 selection candidates resulted in imputation accuracies for candidates of 0.95-0.97 when both parents were genotyped. Total pipeline runtimes for these analyses ranged from 12 to 14 hours when using 20 processors. Continued optimization of the pipeline in the cluster is expected to result in stable or decreasing runtimes as data size grows. Imputation has now been implemented for several lines in routine evaluation to calculate genomic breeding values for young animals.

Structural equation models to study causal relationships between bovine milk fatty acids

Bouwman, A.C.[1], Valente, B.D.[2], Bovenhuis, H.[1] and Rosa, G.J.M.[2], [1]Wageningen University, Animal Breeding and Genomics Centre, P.O. Box 338, Wageningen, Netherlands, [2]University of Wisconsin, Department of Dairy Science, 1675 Observatory Drive, Madison, USA; aniek.bouwman@wur.nl

Bovine milk fat composition is determined by intrinsic biological pathways involving fatty acid synthesis and causal relationships among them. This study aimed to explore such causal networks using gas chromatography measurements of fatty acids on milk samples of 1,900 heifers. Data-driven methods were used to search for plausible causal structures, which were analyzed with mixed-effect structural equation models to estimate the magnitude of the causal relationships between fatty acids. The analysis comprised three steps: (1) residual covariance matrix was estimated with a Bayesian multi-trait animal model; (2) posterior residual covariance matrices obtained in step 1 were used as input for the inductive causation algorithm, which was applied to search for putative causal structures; (3) structural equation model was fitted conditionally on the causal structure obtained in step 2 to quantify causal relationships. Studying the residual distribution instead of the joint distribution of phenotypes allows for blocking the confounding resulting from correlated genetic effects. Residual correlations between traits should be present in order to find causal structures, however, fatty acids are genetically and residually highly correlated which complicates convergence of multi-trait models. Resulting causal structures resembled features of known pathways. Most fatty acids that are consecutive in the metabolic pathway (e.g. C4:0 and C6:0) were connected by an edge, and some unsaturated fatty acids were connected to their saturated equivalent (e.g. C18:0 and C18:1cis9). Not all detected connections could be directed from purely statistical information. Results in terms of selected structure and causal coefficients can contribute to improve our understanding of milk fatty acid synthesis, and to generate new hypotheses regarding the associations between them.

Molecular study for the sex identification in Japanese quails (*Coturnix Japonica*)
Vali, N. and Doosti, A., Islamic Azad University, Shahrekord Branch, Animal Sciences, Islamic Azad University, Shahrekord Branch, Post box, 166. 8813733395, Iran; nasrollah.vali@gmail.com

In many birds' species such as Japanese quail, sex determination in young and many adult birds is very difficult. Nowadays, sex identification of animals throughout their lives is possible by molecular genetics techniques such as polymerase chain reaction (PCR). The aim of this study was to determine the sex of Japanese quails (Coturnix japonica) by DNA analysis. Chromo helicase DNA (CHD) genes are preserved within avian Z and W sex chromosomes. The intron regions of the CHDW and CHDZ genes vary between male (ZZ) and female (ZW) individuals. The method used in this study was based on this difference. Genomic DNA was extracted from feathers instead of blood. The intron regions of CHDW and CHDZ genes were amplified by sex specific primers (Coja-F and Coja-R). PCR products were screened by agarose gel electrophoresis. These results show single (ZW) amplified fragments, about 320 bp for female, and no amplified PCR product for male (ZZ) Japanese quails. These results of this study show that CHD gene amplification is a convenient, inexpensive, safe, and simple technique for sex typing of Japanese quails and other avian species in the wild.

Nonlinear recursive models for weight traits in the Pirenaica Beef cattle breed
González-Rodríguez, A., Altarriba, J., Moreno, C. and Varona, L., Universidad de Zaragoza, Miguel Servet 177, 50013, Spain; lvarona@unizar.es

One of the main objectives of beef cattle selection is carcass weight. Live weights along different growth stages are frequently used as selection criteria under the hypothesis of a high and positive genetic correlation with weight at slaughter. However, the presence of compensatory growth may bias the prediction ability of early weights. Recursive models may represent an interesting alternative to understand the genetic and phenotypic relationship between weight traits along growth. We have performed three different analyses to study Weight at 120 days (W120) and at 210 days (W210), W120 and Carcass weight at slaughter (CW) and W210 and CW in the Pirenaica Beef Cattle Breed. The number of available data for each analysis was 8,592, 4,648 and 3,234, respectively. Further, we included a pedigree composed by 56,323 individuals. The statistical model includes sex, herd, year, season of birth and slaughterhouse for CW plus a recursive dependency between traits. This dependency was modeled as a polynomial up to the fourth degree and compared using a logCPO approach. Model comparison results suggest that the best models were the third degree polynomial for W120-W210 and W120-CW and the second degree polynomial for W210-CW, indicating a nonlinear relationship between traits. Moreover, the posterior mean estimates for heritabilities ranged between 0.29 to 0.44, in concordance with estimates of the literature. However, the posterior mean estimates of the genetic correlations between previous and subsequent weight traits were null or very low, showing that the relationship between traits is fully captured by the recursive dependency. These results imply that there is a low predictive ability of the performance of future growth by using only records of early weights and confirm the convenience of slaughterhouse records in beef cattle breeding evaluation.

Genetic variability of a South African ostrich breeding flock estimated from pedigree information

Fair, M.D.[1], Van Wyk, J.B.[1] and Cloete, S.W.P.[2,3], [1]University of the Free State, Department of Animal, Wildlife and Grassland Sciences, P.O. Box 339, 9301, Bloemfontein, South Africa, [2]University of Stellenbosch, Private Bag X1, 7602, Matieland, South Africa, [3]Institute for Animal Production, Private Bag X1, 7607, Elsenburg, South Africa; FairMD@ufs.ac.za

Pedigree records of 78,673 birds of a pair-breeding ostrich flock maintained from 1978 to 2005 at the Oudtshoorn Research Farm, South Africa were used to estimate the effective number of founders (f_e), the effective number of ancestors (f_a), the effective population size and the effective genome equivalents (f_g) under random mating, to assess the genetic variability present in the population. The average level of completeness of the pedigrees was high (99%) in the first generation and the average level of inbreeding (F) was 0.51%. The reference population was defined as the 78,347 birds hatched from 1990 to 2005. The estimated measures of variability were f_g=52.5, f_e=65 and f_a=64, with an f_e/f_a ratio of 1.02. The numbers of ancestors responsible for 100%, 50% and 20% of the genes in the reference population were 267, 22 and 6 respectively. The largest individual contribution to the population hatched from 1990 to 2005 was from a male, which was responsible for 4.85% of the genetic variability. The generation interval, calculated as the average age of parents when their offspring which were kept for reproduction were born was 7.70±0.14 years. The linear regressions of mean annual individual rate of inbreeding on year of birth for the two distinct periods, 1995-2000 and 2000-2005, were 0.10% and -0.03% per year respectively. The estimate of effective population size (N_e) computed via the increase in the individual rate of inbreeding was 103.7 animals. Estimates of N_e using the alternative methods of tracing the numbers of complete generations were 73.5, 181 and 95.1 for complete, maximum and equivalent complete generations. The results indicated that the population under study was at an acceptable level of genetic variability.

Time trends for performance traits of poultry included in the conservation programme in Poland

Calik, J., Krawczyk, J. and Szefer, M., National Research Institute of Animal Production, Department of Animal Genetic Resources Conservation, Krakowska 1, 32-083, Balice, Poland; mszefer@izoo.krakow.pl

Native breeds of poultry are closely associated with the agricultural landscape, tradition and culture of local communities. These populations are characterized by good health and resistance to adverse climatic conditions and uncontrolled conditions of backyard farming. They make efficient use of poor meadows and pastures and uncultivated land to produce flavoursome eggs and meat. The aim of the study was to evaluate the effectiveness of genetic resources conservation programmes using the example of poultry flocks maintained at the National Research Institute of Animal Production. Performance recording results from 2004-2008 were analysed in: 8 breeds/lines of laying hens (Z-11, Ż-33, G-99, H-22, R-11 S-66, K-22, A-33), 12 breeds of geese (Ga, Ka, Ki, LsD-01, Lu, Pd, Po, Ro, Ry, Sł, Su, Ku), 6 lines of ducks (K-2, KhO-1, LsA, P-8, P-9, P-33). In accordance with the genetic resources conservation programme, conservation flocks are not selected, which means that any changes in the values of the analysed traits are only due to genetic properties of a given population and environmental conditions. Analysis of the results revealed an upward tendency for the size of all populations evaluated. The production parameters and the indices of fertility (about 80-90%) and hatchability of chicks (about 70-80%) from five consecutive seasons were normal and did not differ significantly from the current breed patterns. In addition, the applied conservation programme ensures that the birds are maintained in full health. Work conducted as part of multiannual programme 08-1.31.9., financed by the Ministry of Agriculture and Rural Development.

Study on improvement of hair coat quality in scanblack mink

Wrzecionowska, M., Bielański, P. and Piórkowska, M., National Research Institute of Animal Production, Department of Animal Genetic Resources Conservation, Krakowska 1, 32-083, Balice, Poland; mmietlicka@izoo.krakow.pl

The most important traits of mink are animal size and hair coat quality traits. These include colour type, colour clarity, hair coat density and hair length. In mink, heritability of hair coat is high and ranges from 0.53 to 0.86. Hair length estimated for mink by Berg and Lohi (1991) depends on colour variety and ranges from 0.2 to 0.6. According to the same authors, the coefficient of heritability for animal size is in the 0.3-0.5 range. The aim of this study was to determine the effect of mating mink with different hair coat types and its influence on hair coat size and quality. Subjects were standard mink (scanblack, American) and their hybrids. The study used 20 female mink of each colour type and equal numbers of F1 and F2 progeny. The experiment was conducted in groups and divided into stages. Stage I: group I, standard American mink; group II, standard scanblack mink; group III, male American × female scanblack. Stage II: group I, male American mated to F1 American × scanblack; group II, male scanblack mated to F1 American × scanblack. Replacement crossing was carried out in the next stages. Parameters of reproductive and rearing performance were studied in young animals. Conformation was tested on live mink. Fur properties of hair coat and physical parameters of raw and dressed skins were evaluated in a laboratory. The evaluation involved 4 male skins and 4 female skins from each experimental group. The results were analysed statistically by two-factor analysis of variance and Duncan's D test, using Statgraphics Plus 4.0. In conclusion, the skins of mink derived from mating to American males were characterized by greater weight and length of raw and dressed skins. Skins obtained from animals mated to Scanblack males had greater area and lightness.

Conservation of the genetic resources of fur animals in Poland

Wrzecionowska, M., Bielański, P. and Kowalska, D., National Research Institute of Animal Production, Department of Animal Genetic Resources Conservation, Krakowska 1, 32-083, Balice, Poland; mmietlicka@izoo.krakow.pl

Acting in accordance with the principles of the Convention on Biological Diversity, in 1996 Poland started to implement the FAO's Global Plan of Action for Animal Genetic Resources. Since 2002, the National Research Institute of Animal Production has coordinated the conservation of the genetic resources of the native farm animal breeds in Poland. Some of the many species, breeds and varieties of fur animals were created or consolidated by Polish breeders and scientists. Special attention has been given to these animals because they are not always able to compete with high-producing breeds or varieties. The aim of the programme is to protect highly valuable species, breeds and varieties of fur animals that are threatened with extinction because of their small population. The programme covers the following fur-bearing carnivores and herbivores: white-necked fox, pastel fox, breeding polecat, Popielno White rabbit, Polish beige recessive chinchilla, and seven varieties of nutria (standard, black dominant, amber-golden, white non-albino, sable, pastel and pearl).One the goals of the conservation programme is to increase the population. The population should be increased to at least 700 foundation stock females in the case of nutria and to 200 foundation stock females in the case of the other species and breeds.To participate in the programme, animals have to be registered in the herd book of a given breed, as well as performance tested and estimated for breeding value by the National Animal Breeding Centre.The implementation of the above programme will result in the preservation of genetic biodiversity of the native breeds of fur animals created in Poland, which reflect sound breeding practices and have been well adapted to local environmental conditions and housing systems. Work conducted as part of multiannual programme 08-1.31.9., financed by the Ministry of Agriculture and Rural Development.

Genetic analysis of mortality during the strict lactation period in Gazella cuvieri
Cervantes, I.[1], Ibáñez, M.B.[2], Goyache, F.[3], Gutiérrez, J.P.[1] and Moreno, E.[2], [1]UCM, Puerta de Hierro s/n, 28035 Madrid, Spain, [2]CSIC, Ctra Sacramento, s/n, E-04120 Almería, Spain, [3]SERIDA, Cam Rioseco 1225, 33394 Gijón, Spain; gutgar@vet.ucm.es

Captive breeding of a threatened species is an important and, in some cases, successful tool for species conservation. In this study we used the information of the captive population of Gazella cuvieri located in 'La Hoya' Experimental Field Station (EEZA-CSIC) in Almería (Spain). This species, endemic to North Africa, is on the brink of extinction. The captive breeding programme began in 1975 under a European Endangered Species Programme. Population descends from 4 wild-born individuals (1 male and 3 females). Nowadays, the population size in Almería is 159 individuals. Genetic parameters of mortality during the strict lactation period from 673 calves belonging to 183 mothers were estimated defining the trait either as categorical or continuous. The total pedigree file contained 713 individuals. Each mortality record was assigned to the individual, to the mother, to the father or to both parents. Models included the mother age, mother age squared, year, mother experience and litter composition (interaction between the sex of the focal offspring and the sex of the littermate) as fixed effects. Maternal effect (random effect) was included when the mortality record was assigned to the individual. Putative differences in performance due to inbred mothers were ascertained including in the models fitted the following: coefficient of inbreeding (linear and quadratic) or individual increase in inbreeding of the calf. The inbreeding did not influence the trait. Heritability ranged between 0.03 and 0.27 and between 0.06 and 0.35, respectively, for the linear and categorical definition of the trait. We concluded that mortality to weaning is essentially a maternal-linear trait. The genetic trends showed that the dams with poorer genetic ability for mortality have been indirectly selected over the years. These results will be applied in the mating programme of this population.

Selective genotyping and logistic regressions to infer the genetic background of clinical mastitis
Bagheri, M.[1], Ashtiani, R.[2], Moradi-Shahrbabak, M.[2], Pimentel, E.C.G.[1] and König, S.[1], [1]University of Kassel, Animal Breeding, Nordbahnhofstr. 1a, 37213 Witzenhausen, Germany, [2]Tehran University, Animal Science, Daneshkade Avenue, Karaj, Iran; pimentel@uni-kassel.de

The objective of this study was to infer relationships between SNP-genotypes within candidate genes for clinical mastitis and EBVs for production traits in Holstein dairy cattle based on selective genotyping. The data set comprised in total 3,823 Holstein cows from two Holstein contract herds located in two regions in Iran. Data included EBVs for the production traits milk, fat, and protein yield, fat and protein percentage, and the phenotypic observation for the no. of cases of clinical mastitis per lactation. Data extraction for selective genotyping was based on extreme values for clinical mastitis residuals (CMR) from mixed model analyses. Two extreme groups of 135 cows were formed, and genotyped for two fragments of the genes TLR4 [= TLR4(1) and TLR4(2)] and CACNA2D1 [= CACNA2D1(1) and CACNA2D1(2)] using PCR-SSCP or PCR-RFLP. Associations between SNP-genotypes and traits of interest were carried out by applying logistic regression analyses, i.e. estimating the probability of the heterozygous genotype in dependency of the EBVs and of values for CMR. The heterozygous genotype was contrasted to both homozygous genotypes allowing in addition the estimation of effects for dominance. Allele G of TLR4(1) was associated with less cases for CM, and showed desired effects in production traits, e.g. higher milk yield and protein yield, and lower values for CMR. For CACNA2D1(2), the CM resistant group had substantially more G- than A- alleles, and accordingly, the value of genotype AG for CMR was 0.27 SD below the value of genotype AA. The favorable allele G for CMR was also associated with favorable EBVs in milk, fat and protein yield. Those SNPs are relevant for breeding objectives, because they simultaneously improve both antagonistic traits. Effects of dominance were of the same magnitude for the production trait (milk yield) and for the functional trait (CMR).

Analysis of CAPN2 and CAST genes expression in skeletal muscles in pig breeds raised in Poland
Bereta, A., Ropka-Molik, K., Różycki, M. and Tyra, M., National Research Institute of Animal Production, Department of Animal Genetics and Breeding, ul. Krakowska 1, 32-083 Balice, Poland; abereta@izoo.krakow.pl

Calpains are the cysteine endopeptidase which occur in all types of cells and their activity in a muscle cells is related with the cytoskeleton proteolysis. The activity of calpins and their inhibitor calpastatin is closely related with the changes in the muscle after slaughter – maturation of the meat post mortem and texture parameters. The aim of present study was to estimation of expression profile of calpain 2 (CAPN2) and calpastatin (CAST) genes in porcine skeletal muscles (semimembranosus and longissimus dorsi). The analysis was performed on four pig breed: Polish Landrace and Polish Large White (50 animals from each breed), Pietrain and Puławska (16 animals from each breed). Relative quantification of the transcript level of the genes studied was performed with 7500 Real-Time PCR System using TaqMan® MGB probes labeled with FAM, VIC or NED and with endogenous controls OAZ1 and RPL27. In longissimus dorsi muscle, the CAST mRNA abundance was the highest in Polish Landrace and Pietrain breeds and the lowest in Puławska pigs (P<0.05). The differences in CAST expression in ham muscle were similar, but not statistically significant. Interestingly, in both investigated muscles expression of CAPN2 gene was the lowest in Puławska breed (P<0.01) and at the similar level in the rest of breeds. The results obtained showed that expression of both genes is associated with pig breed, but it does not depend on a type of muscles – comparison between transcripts levels in muscles in each breed did not show significantly differences. Furthermore, the lowest CAPN2 and CAST expression in Puławska pigs may be related with the fact that this breed belongs to conservative breeds and is not undergo of selection for traits strongly associated with production economics, especially growth rate and percentage of meat in carcass. The study was supported by the Polish Ministry of Science and Higher Education (project no.NN311349139).

Effect of polymorphism in the leptin gene on reproduction performance in Pinzgau cattle
Moravčíková, N., Trakovická, A. and Kasarda, R., Slovak University of Agriculture in Nitra, Department of Animal Genetics and Breeding Biology, Tr. A. Hlinku 2, 949 76 Nitra, Slovakia; nina.moravcikova1@gmail.com

The aim of this study was to evaluate the relationship between the polymorphism in the leptin gene and reproduction performance in Pinzgau cattle. Leptin is a protein, which is synthesized by adipose tissue and which plays an important role in the regulation of feed intake, energy metabolism, growth and reproduction of cattle. A total of 85 Pinzgau cows were used to investigate how leptin gene polymorphisms affect reproduction traits. Evaluated reproductive parameters employed are indirect measurements of reproduction, also reflecting animal body conditions. The polymorphism of leptin gene in intron 2 on bovine chromosome 4 was genotyped by using the polymerase chain restriction fragment length polymorphism. A strategy employing PCR was used to amplify a 422 bp product from hair roots DNA samples. Digestion of PCR products with restriction enzyme Sau3AI revealed two alleles: allele A was 390, 32 fragments and allele B was 303, 88, 32. Three patterns were observed and frequencies were 0.447, 0.494 and 0.059 for AA, AB and BB, respectively. The statistical analysis shows significant effects on calving interval (P≤0.05) and on body weight at 210 days of age (P≤0.05) with B as a desirable allele. The allele A seems to increase calving interval and decrease body weight at 210 days of age. No significant associations were observed between the Sau3AI polymorphism and insemination interval and days open. The association of genetic markers with better reproductive performance is a very interesting finding and could be used in marker-assisted selection to improve reproduction in cattle.

Relationship between body condition score, locomotion and longevity in Polish Holstein-Friesian cows
Morek-Kopeć, M.[1] and Zarnecki, A.[2], [1]Agricultural University, al. Mickiewicza 24/28, 30-059 Krakow, Poland, [2]National Research Institute of Animal Production, ul. Krakowska 1, 32-083 Balice k. Krakowa, Poland; rzmorek@cyf-kr.edu.pl

The relations between longevity and the recently implemented body condition score (BCS) and locomotion (LOC) evaluations were examined using the Weibull proportional hazard model. Scores (1-9) for BCS of 146,441 cows and locomotion of 70,797 cows that calved for the first time between 2007 and 2011 were available. Mean true longevity (TL) was 1086 days from first calving to culling (uncensored records), and 952 days to the last available test day (censored records). Functional longevity (FL) was defined as TL adjusted for production. The statistical model included time-independent fixed effects of BCS or LOC scores, age at first calving, time-dependent fixed effects of year-season, parity-stage of lactation, annual change in herd size, fat yield and protein yield (in the FL model), and time-dependent random herd-year-season effect. Likelihood ratio tests showed highly significant effects of BCS and LOC on risk of culling. Both traits had a stronger effect on true than on functional survival; the impact of BCS was greater than that of locomotion. BCS displayed an intermediate optimum (score 5) for true and functional longevity. Cows with extremely low BCS were 3 times (for TL) and 1.6 times (for FL) more likely to be culled than cows with average score; similarly, cows with high scores were 1.7 (for TL) and 1.3 (for FL) times more likely to be culled. Low locomotion scores were associated with poor longevity. For TL the relative risk of culling (RRC) decreased more than twice with increasing locomotion score. With production included in the model, RRC was lowest for intermediate LOC score (6). Cows with this optimal score had a 40% better chance to survive than cows with the lowest locomotion score. RRC rose by 17% for the highest LOC scores as compared to the optimum.

Genetic parameter estimates of liver and kidney weights in Syrian hamsters selected for fertility
Satoh, M.[1] and Furukawa, T.[2], [1]National Institute of Livestock and Grassland Science, Ikenodai 2, Tsukuba, 3050901, Japan, [2]National Agricultural Research Center for Hokkaido Region, Hitsujigaoka 1, Sapporo, 0628555, Japan; hereford@affrc.go.jp

Genetic parameters were estimated for liver and kidney weights of 888 Syrian hamster males, sampled from four different strains in a selection experiment.. The experiment involved four lines. Traits were selected from combining litter size at birth (LS), litter weight at weaning (LW), and mature body weight (BW) from a base population (G0) through to generation 15 (G15) (line W). Selection traits were manipulated by combining LS and LW from G0 through G15 (line R), then a two-way-cross progeny between W and R at G15, and then maintained with relaxed selection through to G25 (line K). A randomly selected control line was established from G0 through to G25 (line C). Liver mass, kidney mass, and body mass data were randomly collected from 888 male hamsters from G8 through to G14 for lines, W, R, and C, and from G16 through to G18 for lines, K and C. Body weight was recorded at 8 weeks of age (BW8), and body weight, liver weight (LVW) and kidney weight (KDW) were recorded at 14 through to 21 weeks of age. BW8, LVW, KDW, and the relative weights of liver (LVW/BW) and kidneys (KDW/BW) were analyzed. Heritability estimates for LVW and KDW were moderate, but the estimate of KDW was higher than that of LVW. Heritability estimates for LVW/BW and KDW/BW were lower for both LVW and KDW, respectively. Estimated genetic correlation between LVW and KDW was moderate, and the genetic correlation between LVW/BW and KDW/BW was low. Estimated genetic correlations between LVW and LVW/BW and between KDW and KDW/BW were moderately high (0.61 and 0.40, respectively). High and positive genetic correlations for BW8 with LVW and KDW were estimated. Line W showed the heaviest BW8, followed by lines, K, R, C, in order of highest to lowest. The ranking of KDW for lines tended to be similar to that of BW8 and the differences between all lines were statistically significant.

Genome-wide association study for genetic heterogeneity for milk yield and somatic cell score
Fikse, W.F.[1], Rönnegård, L.[1,2], Mulder, H.A.[3] and Strandberg, E.[1], [1]Swedish University of Agricultural Sciences, Department of Animal Breeding and Genetics, P.O. Box 7023, 750 07 Uppsala, Sweden, [2]Dalarna University, Statistics Unit, School of Business and Technology, Rödavägen 3, 781 70 Borlänge, Sweden, [3]Wageningen University, Animal Breeding and Genomics Centre, P.O. Box 338, 6700 AH Wageningen, Netherlands; Erling.Strandberg@hgen.slu.se

Recently, genetic variation in residual variance was reported for both Swedish Holstein and Swedish Red. The aim of this study was to perform a genome-wide association study for this genetic heterogeneity. Breeding values for residual variance for milk yield and somatic cell score were available for 842 Swedish Red bulls. In addition, marker data were available for 701 bulls using the Illumina Bovine SNP50 BeadChip, which includes 54,001 single nucleotide polymorphisms (SNP) markers. After edits on minor allele frequency, call rates and GenCall scores more than 48,000 markers were available to be included in the analyses. A multi-locus Bayesian stochastic search variable selection model was used for the analysis. Here, allele effects follow a priori a mixture distribution, where a small fraction of the markers (prior probability of 5%) has a large effect and the remaining markers have virtually no effect. For milk yield in the Swedish Red breed, two regions with a Bayes factor larger than 10 were identified and a further five regions had a Bayes factor larger than three. For somatic cell score, the corresponding numbers were one and four, respectively. In conclusion, we found a few SNPs associated with residual variance of milk yield and somatic cell score in the Swedish Red breed.

PyPedal, an open source software package for pedigree analysis
Cole, J.B., Agricultural Research Service, USDA, Animal Improvement Programs Laboratory, 10300 Baltimore Avenue, Beltsville, MD 20705-2350, USA; john.cole@ars.usda.gov

The open source software package PyPedal (http://pypedal.sourceforge.net/) was first released in 2002, and provided users with a set of simple tools for manipulating pedigrees. Its flexibility has been demonstrated by its used in a number of settings for large and small populations. After substantial revisions and feature additions, an alpha version of PyPedal 2 became available in 2006. The production version of PyPedal 2.0.0 was released in 2010, and featured a completely rewritten object model and many tools for pedigree manipulation and analysis. Many measures of genetic variation can be calculated from pedigrees, including effective population sizes; effective founder and ancestor numbers; and coefficients of regular, ancestral, and partial inbreeding. The software has been used on pedigrees of up to 600,000 animals in several species, including dogs, dairy cattle, and beef cattle. Data can be loaded from, and saved to, plain-text files, GEDCOM 1.5 and GENES 1.20 binary files; and MySQL, Postgres, SQLite and databases. Version 2.1.0, which is currently undergoing beta testing, adds several new tools for manipulating pedigrees by treating them as sets. Pedigrees can be merged by taking the union of two or more pedigrees, for example. The intersection of a two or more pedigrees is the set of animals common to all of them, and union and intersection operations can be used together to perform operations analogous to subtraction. Fast algorithms for computing coefficients of inbreeding also have been added. Unknown parent groups can be constructed based on a description provided by the user. Missing values are now handled more consistently, and missing values within individual records can be accommodated. Support for a data frame class provides a much more efficient framework for calculating summary statistics, such as demographic measurements. PyPedal is provided under an open source license, and may be freely modified, distributed, and used.

Genetic parameters of conformation traits of young Polish Holstein-Friesian bulls

Otwinowska-Mindur, A.[1], Ptak, E.[1] and Zarnecki, A.[2], [1]Agricultural University, al. Mickiewicza 24/28, 30-059 Krakow, Poland, [2]National Research Institute of Animal Production, ul. Krakowska 1, 32-083 Balice k. Krakowa, Poland; rzmindur@cyf-kr.edu.pl

The objective of this study was to estimate the genetic parameters of conformation traits of Polish Holstein-Friesian bulls, evaluated as required for registration in the herd book and for entering progeny testing. Data were 8 linearly scored (1-9 scale) and 6 descriptive (scored from 50 to 100) conformation traits of 2762 young bulls born between 2000 and 2011. A multi-trait animal model was used to estimate genetic parameters. (Co)variance components for 14 traits were estimated by a Bayesian method via Gibbs sampling. The linear model included fixed linear regressions on age at evaluation (from 10 to 23 months), fixed effects of year of birth, fixed effects of herd-classifier, and random animal effect. Estimates of heritabilities for all analyzed traits were within the range of 0.14-0.37. Among the 6 descriptive type traits the heritabilities were greatest for size (0.37) and for overall conformation score (0.31). The lowest heritability (0.16) was for feet and legs. The range of estimated heritabilities was similar for linearly scored traits. Heritabilities were lowest for rear legs side view (0.14) and chest width (0.18), of moderate magnitude for body depth, rump width, foot angle and rear legs rear view (0.20-0.26), and highest for rump angle (0.36) and muscularity of front end (0.33). The heritabilities of the same traits assumed for routine progeny testing are in most cases similar to those obtained in this study.

Fine mapping of production and meat quality QTL in Large White pigs using the PorcineSNP60 Beadchip

Sanchez, M.P.[1], Tribout, T.[1], Iannuccelli, N.[2], Bouffaud, M.[3], Servin, B.[2], Dehais, P.[2], Muller, N.[3], Mercat, M.J.[4], Estelle, J.[1], Bidanel, J.P.[1], Rogel-Gaillard, C.[1], Milan, D.[2] and Gilbert, H.[2], [1]INRA, UMR1313 GABI, 78350 Jouy-en-Josas, France, [2]INRA, UMR444 LGC, 31326 Castanet-Tolosan, France, [3]INRA, UE450 TP, 36651 Le Rheu, France, [4]IFIP, Genetics, 35651 Le Rheu, France; marie-pierre.sanchez@jouy.inra.fr

A total of 495 Large White pigs distributed in 106 sire families were genotyped using the PorcineSNP60 Beadchip and tested for feed intake, growth, carcass composition and meat quality in a French control test station between 2007 and 2009. Of the 64,432 SNP of the chip, 44,412 were used for genome-wide association analyses (GWAS) with an animal mixed model including a regression coefficient for the tested SNP (FAmily-based Score Test for Association option, GenABEL R package). A genomic kinship matrix was used in this analysis. Haplotypic effects were tested using a mixed model with multiple regression on the haplotypes. A total of 45 regions with significant effects ($P<10-4$) have been identified for SNP distributed on all chromosomes (SSC) except on SSC5 and SSC12. Among these regions, 18 with a size ranging from 7 to 1,251kb presented from 2 to 6 SNP with significant associations on traits related to growth and feed intake (3 regions on SSC3, 6 and 10), carcass composition (10 regions on SSC1, 3, 7, 10, 14, 16, 17 and 18) and meat quality (5 regions on SSC1, 4, 9). Depending on the trait and on the region, effects varied from 0.3 to 0.6 phenotypic standard deviation. Eight of these regions have never been described before. The chromosomal region which had the greatest number of significant effects (a region of 183kb on SSC1 for meat quality) was submitted to haplotype analyses. Three haplotypes of 6 SNP were found in the studied Large White population. They presented significant effects on all meat quality traits included in the study. These results show that GWAS analyses with the PorcineSNP60 Beadchip allow QTL to be mapped with a good accuracy in commercial pig populations.

Utilization of whole blood cells culture for chromosome preparation in the sterlet

Parkanyi, V., Masar, J., Ondruska, L. and Rafay, J., Animal Production Research Centre Nitra, Institute of Small Farm Animals, Hlohovecka 2, 951 41 Luzianky, Slovakia; ondruska@cvzv.sk

The aim of the present study was the use of whole blood of the sterlet (Acipenser ruthenus) for cytogenetics analysis. Whole blood cells, from vena caudalis, were incubated at 20 °C for 120 hours. A suitable and repeatable method is proposed to gain dividing lymphocytes in defined culture conditions without the need of killing the fish and with a good yield of metaphases. The study of chromosomes is important from a molecular and a selection standpoint. In the lymphocytes of peripheral blood diploid number of chromosomes, 2n=118, was found. Different chromosome numbers were recorded ranging from 116 to 120. The cells with 116-117 chromosomes are in the hypodiploid and the cells with 119-120 chromosomes are in the hyperdiploid category. The variability of chromosomes in c- metaphases is caused by microchromosomes mainly.

A polymerase chain reaction technique for fetal sex determination using ovine amniotic fluid

Jaayid, T.A.[1], Hasan, E.F.[1] and Al-Allak, Z.S.[2], [1]College of agriculture, Animal production, Basrah university, Basrah, Iraq, [2]College of Education, Biology, Misan university, Misan, Iraq; taleb1968@yahoo.com

Prenatal sexing was undertaken using 10 samples of ovine amniotic fluid obtained through amniocentesis from pregnant uteruses of animals killed in the slaughterhouse. The aim of the study was to evaluate the applicability and efficiency of the polymerase chain reaction technique (PCR). For these purposes multiplex PCR were carried out with male-specific sequence (SRY) gene on the Y chromosome to give a 116 bp product and X amelogenin (Aml-X) gene on the X chromosome to give a 300 bp product. The amplified samples were plotted in 2% electrophoresis agarose gel and viewed using an ultraviolet gel documentation. Six samples (52.63%) were identified as male(116 bp and 300 bp) and four samples (47.37%) as female (300 bp); this information was confirmed by a morphological analysis of all fetuses. The results allow to conclude that PCR, when using SRY and Aml-X markers on ovine amniotic cells, is efficient and reliable to fetal sex identification. This article describes the applicability of PCR primers for the detection of fetal sex. For the first time completed research such specifications in Iraq using molecular biology in sex determination, for the first time used gene concept in the selection and this is one of the best success in the selection.

Genetic analysis of calf survival in Dutch Holstein calves

Van Pelt, M.L. and De Jong, G., CRV, AEU, P.O. Box 454, 6800 AL Wageningen, Netherlands; mathijs.van.pelt@crv4all.com

A durable and animal welfare-friendly dairy industry wants vital calves and heifers in the rearing period. Loss of calves is not only costly for the farmer from an economic point of view, but it also causes a negative image of the sector and decreased work satisfaction of the farmer. Little is known about calf survival between the period just after birth and when a heifer starts lactating. In this study, the genetics of calf survival between day 3 and 365 was examined. The data set for the parameter estimation comprised 522,335 heifer calves for replacement. Calves were at least 75% Holstein Friesian and born between 2002 and 2009. Early death in the first month of life has different causes than death during the rearing period. Therefore, the first year of rearing was divided in four periods: day 3 to 14, day 15 to 30, day 31 to 90, and day 91 to 365. The complete period day 3 to 365 was analysed as well. Genetic parameters of binary traits were estimated with a linear sire model. Fixed effects included in the model were herd, year × month of birth and parity number of the dam. In the first year of rearing 6.45% of the calves died. The heritability was lowest for day 15 to 30 with 0.001 and highest for the complete period day 3 to 365 with 0.011 (0.040 on the underlying scale). The genetic standard deviation was 2.5% for the complete period. It is possible to distinguish bulls with good and bad calf survival based on their progeny groups. Genetic correlations between subsequent periods were high (between 0.66 and 0.90) and very high with the complete period (between 0.85 and 1.00). Despite the low heritabilities, reliabilities are at an acceptable level, due to large progeny groups. The trait day 3 to 14 has a high correlation with day 3 to 365 and is a good predictor for this trait. Bull calves could be included in trait day 3 to 14 as well.

Genotype by environment interactions when environments are defined from Herd-Test-Day profiles

Huquet, B.[1,2], Leclerc, H.[2] and Ducrocq, V.[1], [1]INRA, UMR 1313 GABI, Domaine de Vilvert, 78352 Jouy-en-Josas, France, [2]Institut de l'Elevage, 149 rue de Bercy, 75012 Paris, France; berenice.huquet@jouy.inra.fr

The objective of this study was to assess genotype by environment interactions on dairy cattle in France using an innovative definition of environments based on Herd Test Day (HTD) profiles. HTD profiles estimated using a Test-Day Model are thought to reflect herd management: they represent the part of the daily production (milk yield, fat and protein contents) only due to month-to-month variations in environmental conditions, mainly related to feeding and climatic conditions. Herd clustering of 15,576 French herds of the three dairy breeds (Normande, Montbéliarde, Holstein) based on their HTD profiles led to the definition of three contrasted environments. Herd cluster 1 was characterized by high milk yield HTD profiles (+2 kg/day above the mean), herd cluster 2 by high protein and fat contents HTD profiles (+1.25 and +0.69 g/kg/day) and herd cluster 3 by low HTD profiles for milk yield and protein contents (-2.7kg/day and -0.59 g/kg/day). The (co)variance components of 305d first lactations were estimated with an animal model for each combination of breed and trait (milk yield, fat and protein yields and contents). The datasets included 43,611 lactations from 309 herds, 125,465 lactations from 796 herds and 246,740 lactations from 1690 herds for Normande, Montbéliarde and Holstein, respectively. Heterogeneous variances according to the environment were taken into account for genetic and residual random effects. Genetic correlations between environments were extremely high (close to 1) whatever the breed and trait, showing no reranking of animals between environments. However, a clear gradient of heritabilities was found (e.g., from 0.30 to 0.40 for milk yield in the Normande breed) showing a scale effect: animals with good performances in herds characterized by less productive management were penalized. A reaction norm model based on the first principal component(s) of the HTD profiles confirmed these results.

Is modernisation beneficial for the welfare of dairy cattle?

Tuyttens, F.[1,2], Ampe, B.[1] and Sonck, B.[2], [1]Ghent University, Salisburylaan 133, 9820 Merelbeke, Belgium, [2]ILVO, Scheldeweg 68, 9090 Melle, Belgium; frank.tuyttens@ilvo.vlaanderen.be

Modernisation has clearly improved the efficiency of milk production, but the consequences for animal welfare are poorly documented. We, therefore, compared dairy cattle welfare housed in modern versus traditional farms. This experiment took place on 19 modern and 20 traditional dairy farms shortly before the 2011 outdoor period. Traditional farms were defined as tying stalls in which no structural changes had taken place during the last 20 years; modern farms were defined as cubicle houses of maximum 20 years old. Cattle welfare was quantified using the Welfare Quality® protocol. This protocol allows the overall welfare status to be categorized based on 32 welfare indicators grouped into 11 welfare criteria, which in turn are grouped into 4 welfare principles. The scores for the welfare criteria and principles were compared using the two sample T-test. Not a single farm received either the lowest ('not classified') or highest ('excellent') welfare category. All farms were categorized as 'acceptable' or 'enhanced'. The overall welfare category hardly differed between modern (53% enhanced) and traditional (50% enhanced) farms. Detailed inspection, however, did reveal some differences. Modern farms scored better for the welfare principle 'good housing' because cows have greater freedom of movement. On the other hand, traditional farms scored better for the welfare principle 'appropriate behaviour' because cows spend more time on pasture, perform fewer undesirable social behaviours (chasing, head butting) and are less fearful of humans. Assuming that our sample of farms were indeed representative of traditional and modern dairy farming and assuming that the Welfare Quality® method quantifies cattle welfare correctly, our results indicate that the effect of modernisation on cattle welfare is limited and not unequivocal. Despite the improved ease of movement for cows housed in modern farms, scores for other aspects of welfare were comparable or even worse.

Subsequent effects of the artificial pastures on winter performance of beef cattle

Bozkurt, Y.[1], Turk, M.[2] and Albayrak, S.[2], [1]Suleyman Demirel University, Faculty of Agriculture, Department of Animal Science, Isparta, 32260, Turkey, [2]Suleyman Demirel University, Faculty of Agriculture, Department of Field Crop Science, Isparta, 32260, Turkey; yalcinbozkurt@sdu.edu.tr

The aim of this study was examine the subsequent effects of pasture types with different botanical composition on the winter performance of beef cattle grown under Mediterranean climatic conditions. The experiment was conducted at university farm in Isparta province located in the west Mediterranean region of Turkey. For this purpose, 20 Holstein beef cattle with an average age of 9 months old were assigned to two outdoor paddocks evenly (10 animals per paddock) and fed similar ration for 7 months in late 2011. The animals were previously grazed separately on a 3 ha artificial pastures divided into two pasture lands (AP1 and AP2) and characterised by two different botanical compositions: AP1, mixtures of *Medicago sativa* L. (20%) + *Bromus inermis* L. (40%) + *Agropyron cristatum* L. (30%) + *Poterium sanguisorba* (10%), and AP2, mixtures of *M. sativa* L. (15%) + *Onobrychis sativa* Lam. (15%) + *A. cristatum* L. (35%) + *B. inermis* L. (35%), respectively. The average initial weights of AP1 and AP2 group animals were 284 and 281 kg respectively. At the end of the experiment, AP1 and AP2 animals reached mean final weights of 480 and 499 kg, respectively. However, there were no significant (P>0.05) differences in total gain (196 vs. 216 kg) and daily liveweight gains (0.934 vs. 1.025 kg). The results indicated that although AP2 group animals showed better performances there were no subsequent effects of the artificially established pasture types on winter performance of beef cattle under the Mediterranean conditions.

Comparison of Holstein bulls and crossbreed Holstein-Belgian Blue: 1. the growing period
Robaye, V., Knapp, E., Istasse, L., Hornick, J.L. and Dufrasne, I., Liège University, Nutrition Unit, Veterinary Medecine Faculty, Boulevard de Colonster 20, 4000 Liege, Belgium; listasse@ulg.ac.be

Meat from male dairy cattle is usually considered as a 'by product' of the dairy industry. In Belgium, most of the dairy males are used by multinational companies to produce veal calves in integrated systems. As opposed to other countries, there is thus nearly no meat production from steers or bulls originating from the dairy cattle owing to rather poor performances in terms of killing out or carcass composition. Crossbreeding between dairy cows and a Belgian Blue bull was suggested as an alternative system to produce beef meat. Ten Holstein and 10 Holstein × Belgian Blue calves were compared during the growing period from birth to the end of month 13. The animals were housed in group in a free stanchion barn with straw as bedding. Live weight was recorded at regular intervals in order to adapt food intakes to maintain daily live weight gain between 0.7-0.9 kg/d. There were 3 phases during the growing period. During phase 1, which lasted for 78.5 days, all calves were given daily 6 L milk and they consumed an average of 0.71 kg first age compound feedstuff and 1.5 kg hay. During the 223.5 days phase 2, another compound feedstuff with sugar beet pulp was offered along with hay. The respective intakes were 2.1 and 3.1 kg/d. A mixed diet made of grass silage (0.40), maize silage (0.40), barley (0.05) dehydrated lucerne (0.08), sugar beet pulp (0.035) and bran (0.035) was given during the 2 last months at a rate of 14 kg fresh food or 6.52 kg DM/d. No significant differences were observed between the 2 types of calves on the animal performances which were presented as averages for the Holstein, the Holstein × Belgian Blue and SEM. Birth weight was 45 and 48 (0.73 kg). The average daily gain was 0.86 and 0.80 (0.04 kg/d), 0.78 and 0.82 (0.02 kg/d) and 0.84 and 0.84 (0.06 kg/d) during phases 1, 2 and 3 respectively. In such conditions, the final live weight was 363 and 368 kg at an average age of 392 days.

Comparison of Holstein bulls and crossbreed Holstein-Belgian Blue: 2. the fattening period
Robaye, V., Knapp, E., Istasse, L., Hornick, J.L. and Dufrasne, I., Liège University, Nutrition Unit, Veterinary Medecine Faculty, Boulevard de Colonster 20, 4000 Liege, Belgium; listasse@ulg.ac.be

Beef meat is usually obtained in Belgium by the fattening of bulls from the beef herd and culled cows from both the dairy and beef herds, Belgian Blue being the main beef breed used in Belgium. The bulls are offered diets high in energy at a rather young age so that the system is called 'growing fattening'. The 10 Holstein and 10 Holstein × Belgian Blue young bulls described in the companion paper were offered *ad libitum* a fattening diet made of sugar beet pulp (0.44), rolled barley (0.10), rolled spelt (0.09), crushed maize (0.08), bran (0.08), soya bean meal (0.085), linseed meal (0.085), molasses (0.03) and mineral mixture (0.01). Straw was available in a rack. The initial live weights were 363 and 368 kg (SEM 10.7, NS) and the final live weights 600 and 617 kg (SEM 14.48, NS) so that the average daily gains were 1.71 and 1.79 kg/d (SEM 0.05, NS), values to be considered as extremely high. The average food intake was 12.8 kg with the Holstein and 12.5 kg with the Holstein × Belgian Blue. The corresponding feed conversion ratio was 7.5 and 7.0 kg/kg. The slaughter weights were 587 and 606 kg (SEM 14.0, NS). There were significant differences between the 2 groups in terms of warm carcass weights (309 and 353 kg, SEM 8.52, $P<0.05$) and of killing out percentage (52.5 and 58.3%, SEM 0.38, $P<0.001$). The composition of the carcass was also significantly different, the crossbreed bulls carcasses being characterized by a large proportion of muscle (60.9 and 68.2%, SEM 0.42, $P<0.001$), a lower proportion of fat and connective tissues (19.3 and 16.0%, SEM 0.47, $P<0.01$) and a lower proportion of bones (19.9 and 15.8%, SEM 0.32, $P<0.001$). When the comparison was made in terms of muscle mass, there was a difference of 52 kg corresponding to an improvement of 29%. It can therefore be concluded that cross breeding with a Belgian Blue bull largely improved carcass production and composition.

Comparison of Holstein bulls and crossbreed Holstein-Belgian Blue: 3. meat quality and composition
Robaye, V., Knapp, E., Istasse, L., Hornick, J.L. and Dufrasne, I., Liège University, Nutrition Unit, Veterinary Medecine Faculty, Boulevard de Colonster 20, 4000 Liege, Belgium; eknapp@ulg.ac.be

Over the last years, growing attention has been paid to the dietetic aspect of bovine meat. The fat content and the fatty acids composition of meat are indeed of interest for the consumers who attach also great importance to sensory traits such as tenderness, flavour or colour. Two days after slaughter, meat samples were taken from the carcasses of the 10 Holstein and 10 Holstein × Belgian Blue young bulls described in the two companion papers for meat quality and chemical composition analysis. Meat quality was assessed on a sample of the Longissimum thoracis (LT) muscle. There were no differences in meat colour parameters both on d2 and 7 after slaughter except that meat of the Holstein bulls was less red on d7 (a* value of 17.8 and 19.3, SEM 0.3% P<0.01). The drip loss was lower with the Holstein (2.4 and 3.5, SEM 0.15% P<0.01). The opposite was found for the cooking loss at d7 (34.2 and 29.9, SEM 0.22% P<0.001). Tenderness was not different. The chemical composition was assessed on samples of the LT and Rectus abdominis (RA) muscles. In the LT, the dry matter content was higher in the Holstein (25.0 and 24.2, SEM 0.13% P<0.01) and a similar effect on ether extract was observed (9.0 and 4.7, SEM 0.03%/DM, P<0.05). There was a tendency for Se content to be higher in the Holstein (587 and 533, SEM 13.5 µg/kg/DM, NS). The animal type did not affect the Zn content in meat (average of 260 mg/kg/DM). The effects were similar for the RA muscle. One has to note that the Se content was lower in the RA than in the LT (560 and 493 µg/kg/DM). The fatty acids content was measured in both muscles. On the whole, the individual fatty acids content was lower in the Holstein × Belgian Blue bulls than in the Holstein, the difference being significant only in the LT. It can be concluded that crossbreeding in Holstein cows with a Belgian Blue double muscled bull improved meat quality and meat composition for the consumer.

Nutritional value and technological suitability of milk from cows of 3 Polish local breeds
Litwinczuk, Z.[1], Barlowska, J.[2], Chabuz, W.[1] and Brodziak, A.[1], [1]University of Life Sciences in Lublin, Department of Breeding and Protection of Genetic Resources of Cattle, Akademicka 13, 20-950 Lublin, Poland, [2]University of Life Sciences in Lublin, Department of Commodity Science and Processing of Raw Animal Materials, Akademicka 13, 20-950 Lublin, Poland; zygmunt.litwinczuk@up.lublin.pl

The aim of the present study was to evaluate the nutritional value and technological suitability of milk from cows of 3 Polish breeds included in the programme of genetic resources conservation. The study included 976 samples of milk obtained from cows of 3 native cattle breeds, i.e. White-Backed (191), Polish Red (168) and Polish Black and White (69), maintained in the traditional husbandry system. The reference group consisted of milk from cows of Polish Holstein-Friesian breed – PHF (219) maintained in the intensive system and milk from cows of Simmental breed (329) kept under traditional husbandry system. Following parameters were determined: content of fat, protein, lactose and dry matter and casein; acidity; heat stability; rennet coagulation time and content of selected whey proteins. Selected samples were investigated for fatty acid profile as well as content of macro- and microelements. The obtained results were analyzed statistically using StatSoft Inc. software STATISTICA ver. 6. Cows of native breeds produced milk of a higher nutritional value (higher content of whey protein and polyunsaturated fatty acids, including CLA) and more suitable for processing as compared to the PHF cows. Milk from cows of Polish Red breed was the most valuable in relation to aforementioned parameters, which can be associated with a distinctive phylogenetic origin of this breed. In relation to the obtained results, milk from Polish Red cows seems to be remarkably valuable, as the high content of fat (4.35%) was concomitant with especially important high protein content (3.61%), including casein (2.68%). The highest proportion of CLA was observed also in milk from Polish Red cows (2.24%).

Pedigree analysis of Slovak Holstein cattle

Pavlík, I., Hazuchová, E., Kadlečík, O. and Kasarda, R., Slovak University of Agriculture in Nitra, Department of Animal Genetics and Breeding Biology, Tr. A. Hlinku 2, 949 76 Nitra, Slovakia; ivan.pavliktn@gmail.com

The aim of the study was to evaluate the genetic diversity in Slovak Holstein cattle population by the methods of pedigree analysis. The analyzed pedigree population consisted of 275,058 individuals of that 267,390 dams and 7,668 sires. The 94,528 individuals (94,359 dams and 169 sires) born from 1997 to 2009 and registered in SHA set up the analysed reference population (RP). They were living cows and sires in 2011 as well as those sires with frozen semen doses in the AI centres. For calculation of genetic diversity parameters the program Endog v.4.8 (Gutiérrez and Goyache, 2005) was used. An average maximal number of generations traced was 9.44±1.43, 2.96±0.89 complete generations and equivalent number of generations traced was 5.59±0.96. An average individual inbreeding intensity was 1.92% (2.48% in purebred cows and bulls), individual increase in inbreeding was 0.40% (bulls 0.53% and purebred cows 0.50%). The average relatedness was 2.38% (bulls 2.72% and purebred cows 2.69%). The 83.0% individuals from the RP were inbred, the ratio of inbred animals was higher in sire population (98.82%) than in dams. The effective population size computed via individual increase in inbreeding was 125.26 and 48.35 via regression on equivalent generations. The effective number of founders was 129, effective number of ancestors 48 and only 22 ancestors described 50% of diversity. From these results, we can conclude that the genetic diversity of Holstein population in Slovakia is reduced by more factors (inbreeding, bottlenecks). Further reducing of diversity could lead to degradation of production and the other important traits. Obtained results point out the need to improve mating strategy as well as include maintaining of genetic diversity into breeding goals of breed development.

Genetic diversity in the Slovak Spotted breed using pedigree information

Hazuchová, E., Kadlečík, O., Pavlík, I. and Kasarda, R., Slovak University of Agriculture in Nitra, Department of Genetics and Breeding Biology, Tr. A. Hlinku 2, 94976 Nitra, Slovakia; hazuchova.eva@gmail.com

The aim of this study was to evaluate genetic variability based on pedigree information in Slovak Spotted breed. The pedigree information was available from The Breeding Services of the Slovak Republic, s.e. The analyzed pedigree population consisted of 109,534 individuals (105,124 dams and 4,410 sires). The 36,896 individuals (36,834 dams and 62 sires) born from 1990 to 2009 and registered in herdbook set up the analysed reference (RP) populations. The 41.71% of animals in the RP were inbred. For monitoring of genetic variability in populations was used ENDOG v4.8 software (Gutiérrez and Goyache, 2005). The maximum number of generations traced was 8.06±1.44, number of fully traced generations 2.107±0.64 and equivalent complete generations 4.45±0.78. The average inbreeding coefficient for reference population was Fi=0.355% the average individual increase of inbreeding was ∆Fi=0.093%, effective population size computed via individual increase in inbreeding Ne=528.33±104.9. The average individual relatedness coefficient was AR=0.8%. Total number of founders was 42,398, effective number of founders was 346, the effective number of ancestors was 87 and 44 individuals explained 50% of genetic diversity. The average relatedness coefficient in the RP was higher than inbreeding coefficient. Therefore we can assume increasing inbreeding in the next generation. Results will be used in genetic management of breeding work in Slovak Spotted and monitoring of parameters characterizing genetic diversity and its development, as well.

Competitiveness of different cattle breeds in milk production

Szabó, F., Buzás, G., Konrád, S., Kovácsné Gaál, K., Pongrácz, L. and Gulyás, L., University of West Hungary, Department of Animal Science, Vár 2., 9200 Mosonmagyaróvár, Hungary; szf@mtk.nyme.hu

During the study income and cost of milk production and milk processing for different breeds such as Holstein, Simmental and Jersey producing milk with different fat and protein content was evaluated based on the data collected in various farms. The competitiveness of the studied breeds was compared based on the difference between the income and feeding costs. The income per cow of Holstein, Simmental, and Jersey was 100%, 53% and 89%, respectively. Feeding cost per kg milk was 100%, 127% and 112%, per butterfat kg 100%, 120% and 78% while that of per kg of milk protein kg 100%, 115% and 94%, respectively. The results show that milk production usually gives deficit in the case of present milk price; however, there is a small profit in some farms due to the subsidization. The Simmental breed is the last in competition behind the Holstein and Jersey by profitability of milk production. No meaningful difference can be found between the latter two breeds, despite Jersey needs less but more expensive feedstuffs than Holstein. During the processing more butter and cheese can be produced from a given amount of Jersey's higher fat and protein content milk than of Holstein milk, however no difference in the per cow productivity.

Importance of polymorphism in the promoter region of GH gene for beef production and quality

Sugita, H.[1], Ardiyanti, A.[1], Hirayama, T.[2], Suzuki, K.[1], Shoji, N.[1], Yonekura, S.[3], Roh, S.-G.[1] and Katoh, K.[1], [1]GSAS, Tohoku University, Animal Science, Aoba-ku, Sendai, 981-8555, Japan, [2]University of Ryukus, Animal Science, Nishihara, Okinawa, 903-0213, Japan, [3]Shinshu University, Animal Science, MInami-minowa-mura, Nagano, 399-4598, Japan; kato@bios.tohoku.ac.jp

Single nucleotide polymorphism (SNP) in the genes for growth hormone (GH), fatty acid synthase (FASN) and stearoyl-coA desaturase (SCD) are massively involved in performance of meet quality of Japanese Black cattle. For GH SNP, cattle with the allele C showed lower carcass weight (CW) but richer intramuscular oleic acid % (C18:1) than those with other alleles (A and B). The SNPs preferable for high-fat meet production have also been reported for FASN and SCD genes. The aims of the present study were 1) to investigate SNPs at the promoter region of GH gene, and 2) to analyze the interaction between GH promoter SNPs and meet production and quality, in Japanese Back cattle. We found three SNPs at the promoter region: -396 (T/C), -346 (T/C) and -147 (T/C). Finally we classified 4 haplotypes according to the promoter SNPs: P ~ S. The allele frequency of the promoter haplotypes was 0.14, 0.13, 0.05 and 0.68 for P ~ S, respectively, in Miyagi prefecture. The cattle with allele P showed a larger CW and lower C18:1 %, while the cattle with allele Q showed smaller CW, BMS No and C18:1 %. In addition, mRNA expression of SCD and ELOVL genes in the muscle for P type was smaller, but that for S type was greater, than that for the each other types. These results indicate that SNPs at the promoter region of GH gene significantly influences CW and meet fatty acid compositions, partly due to the alteration in related gene expression.

Comparison between determined and calculated energy balance for dairy cows in early lactation

Van Knegsel, A.T.M.[1], Heetkamp, M.J.W.[1], Van Den Brand, H.[1], Kemp, B.[1] and Dijkstra, J.[2], [1]Wageningen University, Adaptation Physiology Group, De Elst 1, 6708 AW Wageningen, Netherlands, [2]Wageningen University, Animal Nutrition Group, De Elst 1, 6708 AW Wageningen, Netherlands; ariette.vanknegsel@wur.nl

The Dutch net energy system for dairy cows (VEM system; Van Es, 1975) was developed and evaluated for dairy cows mainly in a positive energy balance. The objective of this study is to evaluate the correspondence between energy balance determined in climate-respiration chambers (EB) and energy balance calculated according to Dutch net energy system (EBc) for dairy cows in a negative energy balance (NEB). Twelve Holstein-Friesian dairy cows were housed in climate-respiration chambers within a few days after parturition till week 9 postpartum (pp). The EB and EBc were calculated weekly for in total 6 chambers (2 cows each): 4 chambers from week 2 to 9 pp, and 2 chambers from week 1 to 9 pp. Cows were fed a total mixed ration with grassilage, cornsilage, wheat straw and concentrate (27:27:6:40 on DM basis). Data are presented as means (±SD). Cows produced 39.3 (±3.7) kg of milk per day, with 4.62±0.62% fat and 3.15±0.31% protein. Dry matter intake was 20.4 (±2.6) kg/d. The mean EB was -156.2±238.0 $kJ/kg^{0.75}$ per day. The mean EBc was -252.2±139.6 $kJ/kg^{0.75}$ per day. In addition, EBc overestimated the EB when cows where in severe NEB (<-360 $kJ/kg^{0.75}$ per day), whilst EBc underestimated the EB when cows experienced a mild NEB (>-360 $kJ/kg^{0.75}$ per day). In conclusion, these data indicate that the VEM system underestimates the energy balance of cows in a negative energy balance, except for cows in a severe NEB. In addition, it can be discussed that besides the origin of the VEM system, also low activity profile of cows in the chambers, difference between efficiencies to produce milk out of feed vs. body reserves and limited animal numbers contribute to the observed difference between EB and EBc in this study.

Modelling the impact of heat stress on female fertility in three production systems

Brügemann, K.[1], Gernand, E.[2] and König, S.[1], [1]University of Kassel, Animal Breeding, Nordbahnhofstr. 1a, 37213 Witzenhausen, Germany, [2]Thuringian State Institute of Agriculture, August-Bebel-Straße 2, 36433 Bad Salzungen, Germany; sven.koenig@uni-kassel.de

The aim of this study was to assess the impact of heat stress in Holstein dairy cows on the female fertility trait 'conception rate (CR)' in three different production systems in the state of Lower Saxony, Germany. Production systems were defined as follows: A production system characterized by intensive crop production (= indoor housing), a pasture based production system, and a maritime region at the coast comprising 11,660, 7,173, and 13,061 insemination records, respectively. Heat stress was assessed by daily temperature-humidity indices (THI) modelled as random regression coefficients using Legendre polynomials of order 3 in an analysis of variance. Further fixed effects were herd-year-season of calving, the level of test-day milk yield, and the interaction of parity × DIM. The additive animal genetic effect, the effect of the permanent environment, and the effect of the service sire were modelled as random. Genetic analyses were performed using the program THRGIBBS1F90 that allows the application of random regression methodology for binary data. In all production systems, THI=60 was identified as a general threshold denoting a substantial decline in CR. Best results for CR were observed for THI<45; especially for high yielding cows which responded sensible to increasing THI. Same trends were observed on the phenotypic scale when comparing different production systems. Additive genetic variances and heritabilities for CR were relatively low and constant in the course of THI, and ranging between h^2=0.01 and h^2=0.05 on the underlying liability scale. An increase of additive genetic variances for CR was only observed for THI>70 indicating that genetic differentiation in female fertility is more accurate in 'challenging' environments.

Pre-weaning and fattening performances of five genotypes of cattle breeds in Morocco

El Fadili, M.[1], El Ouardi, L.[1] and Francois, D.[2], [1]Institut National de la Recherche Agronomique, Animal Production, BP.415, Avenue de la Victoire, Rabat, 10060, Morocco, [2]INRA, Genetique Animale, SAGA, 31326 Castanet Tolosan, France; Dominique.Francois@toulouse.inra.fr

The study was carried out to evaluate pre-weaning and fattening performances under the same management conditions of five genotypes sired by the local Oulmes and European improved breeds. A total of 107 veal progeny of Oulmes (OxO=39), Tarentais (TxO=25), Pie-Noire (PNxO=21), Simmental (SxO=7) and Limousin (Lx(TxO, PNxO, SxO)=15) breeds were weighted from birth to weaning at 9 month. The veal average fattening starting age and body weights were 248 day and 113 kg, respectively for O purebred and 243 day and 156 kg for the two and tree way crosses veal. The feed intake was controlled individually and the conversion index was computed. The results showed that purebred O realized significantly lower birth (20.34 kg) and weaning (128.13 kg) weights, while those sired from improved breeds T, S, PN and L showed superior weights at birth and weaning: 26.12 and 162.82, 30.00 and 191.44, 28.56 and 185.49, and 27.99 and 187.18 kg, respectively. During the first three months, veal sired by PN and S bulls realized superior daily gains, 639 and 646 g/d respectively compared to those sired by T (550), L (516) and local O (447 g/d). However, from 6 to 9 month veal sired by L bulls realized similar to superior daily gain compared to veal sired by S and PN. During fattening, veal sired by L bulls realized the highest daily gains (982 g/d), dry matter intake (6.01 kg) and the best conversion index (6.12), while veal sired by PN (877), S (896) and T (866 g/d) realized intermediate average daily gain. Their conversion index and dry matter intake were intermediate as well. However, O veal realized the lowest average daily gain (734 g/d) and the highest conversion index (8.59), while the progeny of PN and T sires showed comparable performances. These results indicate that the utilization of the L as terminal sire breed could be considered in crossbreeding to improve growth and bovine meat production in Morocco.

Effect of calf fatness on further grazing and feedlot performance

Simeone, A.[1], Beretta, V.[1], Elizalde, J.[2] and Caorsi, J.[1], [1]Facultad de Agronomía-Universidad de la República, Animal Science, Ruta 3 km 363, 60000, Uruguay, [2]Consultant, Rosario, 2000, Argentina; beretta@fagro.edu.uy

An experiment was conducted to evaluate the effect of calf fatness after winter feedlot and type of pasture on further grazing and finishing feedlot performance. Forty-eight Hereford calves (238.8±33.5 kg; 11 months old) were divided in 2 groups according to subcutaneous backfat depth (SBFi<3mm or >3mm) and randomly allocated to a native pasture (NP) or an improved pasture (IP, *Festuca arundiancea, Lotus corniculatus, Trifolium repens*) in a factorial arrangement 2×2. Cattle grazed from spring to fall, followed by a winter finishing feedlot period (91 days). Liveweight (LW) was recorded every 28 days and LW gain (LWG) estimated by regression. *Longisimus dorsi* area (LDA) and SBF were measured at the end of grazing season, feed:gain ratio (F:G) was estimated for the feedlot phase, and carcass traits were recorded at slaughter. Statistical model included main effects, the two-way interaction and initial LW as covariate. Mean pasture DM biomass was 1,089±192 and 2,786±329 kg/ha for NP and IP, respectively. The effect of SBFi on grazing performance was independent of pasture type (P>0.05). Calves with SBFi>3mm showed lower spring-LWG (0.78 vs. 0.88 kg/d, P<0.05), but by the end of autumn they were heavier (404 vs. 373 kg, P<0.01), had more SBF (5.7 vs. 4.2 mm, P<0.01) and larger LDA (51 vs. 46 cm², P<0.01) compared to SBFi<3 mm. Grazing IP increased spring-LWG and final grazing LW (P<0.05), but had no effect on SBF or LDA at same date (P>0.05). Treatments did not affect feedlot LWG or F:G (P>0.05). At slaughter, animals with SBFi<3mm had less SBF (P<0.01) and carcasses of animals from this group that grazed NP were the lightest (P<0.01) compared to other treatments which did not differ (P>0.05). Results suggest that differences in animal fat retention during early stages of growth may still be observed at slaughter, but the effect on carcass weight could be modulated by grazing management prior to finishing feedlot.

Veal composition of Holstein calves fed by different ration
*Vavrišínová, K., Bučko, O., Margetín, M., Haščík, P., Juhás, P. and Szencziová, I., Slovak University of
Agriculture in Nitra, Tr. A. Hlinku 2, 94976 Nitra, Slovakia; klara.vavrisinova@uniag.sk*

The veal composition of two groups of Holstein breed calves was observed. The animals of 1st group were
fed with feed mixture and hay after weaning (5 head). The animals of 2nd group were fed with hay, feed
mixture and maize silage (5 head) after weaning. From birth to weaning all animals were fed with the dairy
feed mixture and starter feed mixture. The total length of fattening (from birth to slaughter) was 150 days.
The experiment was carried out on ordinary dairy farm under operating conditions. Evaluation shows that
the calves of the 2nd group achieved lower average daily gains and significantly lower body weight at the age
of 150 days (P <0.05). We have found a lower weight of the main parts of the meat in 2nd group (P <0.05)
in carcass value indicators too. Animals in 2nd group had a higher proportion of kidney fat compared with
1st group, but the difference was not significant. From the analysis of meat samples (MLT) in calves of 2nd
group we have found: higher water content (75.25 g/100g compared to 74.77 in the 1st group), higher content
of intramuscular fat (1.35 g/100g compared to 1.25 in 1st group) and lower total protein content (22.40
g/100g compared to 22.99 in 1st group). Differences in chemical composition of meat were not statistically
significant. Significant differences between groups in values of pH1 and pH24 were not found. The observed
traits did not showed any qualitative variations of meat. We have not found any differences between groups
in meat colour (L* value: 1st group 43.03; 2nd group 42.30) too. On the contrary statistically significantly
lower value of percentage drip loss was found in animals of group 1 (P<0.05). This work was supported by
projects VEGA 1/0493/12, VEGA 1/2717/12 and ECOVA Plus – ITMS: 26220120032.

The relationship between US and CT fat thickness of rib joints measured by Angus crossbred animals
*Somogyi, T.[1], Holló, I.[1], Kovács, A.Z.[1], Anton, I.[2] and Holló, G.[1], [1]Kaposvár University, Faculty of Animal
Science, 40. Guba Sándor str., 7400, Kaposvár, Hungary, [2]Research Institute for Animal Breeding and
Nutrition, 1. Gesztenyés str., 2053, Herceghalom, Hungary; somogyitms@gmail.com*

The aim of this study was to determine the relationship in fat thickness measured with ultrasound (US)
and computerised tomography (CT) among 12th and 13th rib joint in Angus crossbred heifers and bulls.
Altogether 16 heifers and 10 bulls were kept in two groups under the same condition. The diet of animals
was based on maize silage and high level of concentrate. The average slaughter age was the same, 437.3±21
d for heifers and 451±48 d for bulls, whereas slaughter weight differed, 486.9±20.6 kg for heifers and
603.1±26.3 kg for bulls (P<0.001). The in vivo US fat thickness (USFT) was measured with ANISCAN
100 device before the transport to the slaughterhouse, between the 12th and 13th rib on the right side. For
CT-analysis, rib joint were taken between the 11-13th rib, from the right half carcass, after 24 hours chilling.
Each right half carcass was dissected into lean, fat and bone. The following data were evaluated: USFT, CT
fat thickness at spine (CTS), at centre (CTC), at rib (CTR) and CT fat % in rib joint (CTFAT). Statistical
analysis was used by SAS 9.1 package program. The carcasses of heifers had significant (P<0.001) more
fat (20.5±2.0% vs. 13.2±1.5%) and less lean (57.4±2.5% vs. 64.6±1.1%) than bulls. The average USFT of
bulls and heifers were 12.21. The values of US measurements did not differ between genders. The USFT
in heifers moderately correlated to CTS (r=0.63, P<0.05), CTC (r=0.62, P<0.05), and CTFAT (r=0.60,
P<0.05). The USFT in bulls correlated only to CTS (r=0.67, P<0.09). According to our results, ultrasound
and CT are useful tools for the estimation of fat content of carcass, however further experiments are needed
to their combined use.

Prediction of beef carcass meat and fat content with the joint usage of US and CT methods
Somogyi, T., Kovács, A.Z., Holló, I. and Holló, G., Kaposvár University, Faculty of Animal Science, 40. Guba Sándor str., 7400, Kaposvár, Hungary; somogyitms@gmail.com

The main purpose of this study was to estimate the lean and fat content of beef carcasses with the combined use of ultrasound technique (US) and computed tomography (CT). Altogether, 50 animals, purebred Holstein and Hungarian Simmental bulls, and crossbred Charolais × Angus bulls and heifers were used to the evaluation, slaughtered in average 567.48±57.24 kg live weight and in 498.36±50.6 d final age. *In vivo* US fat thickness measurements (USFT) were carried out with ANISCAN 100 portable US device, on the right side, between the 12th and 13th rib. From kidney fat, fat samples were collected for CT fat density (CTFD) evaluation. After 24 hours chilling, rib samples were taken between the 11-13th rib, from the right half carcass, Thereafter, each right half carcass was dissected into tissue components. On cross sectional CT scans the following data were measured and evaluated: area (CTA), width (CTW) and density (CTD) of longissimus muscle. The SAS 9.1 version, corr and stepwise procedure were used for statistical analysis. With carcass muscle properties USFT correlated negatively ($P<0.01$, $r=-0.47 - -0.68$), while with CT measurements positively ($P<0.01$, $r=0.56-0.83$). As expected, carcass fat properties correlated in opposite signs with USFT ($P<0.01$, $r=0.82$), and CT measurements ($P<0.01$, $r=-0.49 - -0.88$), and surprisingly negatively with CTFD ($P<0.01$, $r=-0.88$). Using US and CT data, stepwise linear regression was applying to determine lean weight in carcass (CTW+CTD+CTA, $R^2=0.78$), lean % (CTD+USFT+CTA, $R^2=0.80$), fat weight in carcass (CTFD+CTD+USFT, $R^2=0.92$) and fat % (CTFD+CTD+CTW+USFT, $R^2=0.93$). The exclusion of USFT from regression equations resulted lower regression coefficients in the case of CFW ($R^2=0.87$) and CF% ($R^2=0.92$). These results indicate that the co-use of US and CT imaging methods can improve the predictability of beef carcass tissue composition, thereby to make more accurate the slaughter value estimation.

Effect of supplementation with linseed and CLA on adipose tissue cellularity of Holstein young bulls
Gómez, I.[1], Mendizabal, J.A.[1], Beriain, M.J.[1], Albertí, P.[2], Sarries, M.V.[1], Arana, A.[1], Insausti, K.[1], Soret, B.[1] and Purroy, A.[1], [1]ETSIA Universidad Pública de Navarra, Campus de Arrosadía, 31006 Pamplona, Spain, [2]CITA de Aragón, Avda. Montañana, 930, 50059 Zaragoza, Spain; inma.gomez@unavarra.es

The aim of this work was to study the effect of whole linseed and conjugated linoleic acid (CLA) diet supplementation on adipocyte cellularity in bulls. Forty eight Holstein bulls were distributed into four feeding groups: Control (C, n=12), fed on corn, barley and soybean meal concentrate; Linseed (L, n=12), receiving the same concentrate but including 10% of whole linseed; Conjugated linoleic acid (CLA, n=12), including 2% of synthetic CLA; Linseed plus conjugated linoleic acid (L+CLA, n=12), including 10% of whole linseed and 2% of synthetic CLA. The four concentrates were isoenergetic (3.34 McalEM/kg) and isoproteic (16.9% CP). Animals were fattened from 239.8±6.61 to 458.6±9.79 kg body weight (322±6.0 d old at slaughter). Bulls of the four groups had similar growth, carcass, and fattening parameters. Subcutaneous adipocyte diameters were similar in the four groups of young bulls (160.0, 160.0, 159.7, 169.4 µm for C, L, CLA and L+CLA, respectively; $P>0.05$). In the same way, intramuscular adipocyte diameters were similar in the four groups (45.6, 48.1, 43.6, 48.8 µm for C, L, CLA and L+CLA, respectively; $P>0.05$). The frequency distribution of subcutaneous adipocyte diameters showed a normal distribution in the four groups of animals. However, the frequency distribution of intramuscular adipocyte diameters showed a not normal distribution (skewness coefficients: 0.8, 1.2, 0.9, 0.8 for C, L, CLA and L+CLA, respectively; $P<0.05$), indicating a possible adipocyte proliferation in this adipose tissue. In conclusion, the supplementation with linseed and CLA had not effect on the cellularity of subcutaneous and intramuscular adipose tissues, which showed a different pattern of development.

The effect of the F94L substitution in myostatin on the live weight of cattle raised on the pasture

Tomka, J., Polák, P., Krupa, E., Vašíček, D., Bauer, M. and Vašíčková, K., Animal Production Research Centre Nitra, Hlohovecká 2, 95145 Lužianky, Slovakia; tomka@cvzv.sk

Myostatin is well known for its mutations and consequent uncontrolled development of muscles. Most of studies on the effect of the myostatin mutation were performed using commercial herds where animals are raised in the intensive feeding regimes. This research was aimed on identification and evaluation of presence of myostatin mutation F94L in the purebred and crossbred limousine herds in Slovakia. These herds are kept in the extensive conditions (pasture) in different parts of Slovakia. In our research, 79 randomly chosen animals from 3 herds were used. Two copies of mutation were found in two animals and one copy of mutation was found in 8 animals. The effect of genotype on the live weight was assessed using linear models. The coefficient of determination of model with only genotype included was 0.19. After including the sex of animal into the model the coefficient of determination raised to 0.53. After including the age of animal into the model the coefficient of determination raised to 0.59. The age and sex alone in the model explained 6% and 16% respectively.

Effect of pH on fermentation and microbiota in a dual-flow in-vitro simulation of ruminal digestion

Pinloche, E.[1,2], Auclair, E.[1] and Newbold, C.J.[2], [1]Lesaffre Feed Additives, 137 rue Gabriel Peri, 59700 Marcq-en-Baroeul, France, [2]Institute of Biological, Environmental and Rural Sciences, Penglais Campus, SY23 3DA Aberystwyth, United Kingdom; ea@lesaffre.fr

The aim of this study was to evaluate the impact of a drop in pH (simulate the onset of acidosis) on fermentation (VFAs) and the associated microbiota. Dual-flow bioreactors (1,600 ml) were inoculated with rumen fluid sampled from 3 canulated dry cows. The diet consisted of barley (45%), hay (40%), soya (10%) and molasses (5%) and was added at a rate of 4 g/hr. During the adaptation period (7d), the pH was maintained at 7 and then was dropped by 1 unit every 4 days and VFAs and bacterial population (T-RFLP and 454 pyrosequencing) analysed. The change from pH7 to pH6, did not alter the acetate concentration but decreased both propionate (-13%) and N-butyrate (-25%). At pH5 changes in all VFAs were observed and at pH 4 VFAs production dramatically decreased. Fingerprint analysis (T-RFLP) showed that the inoculated bacterial population grouped with the bioreactors population when they were maintained at pH7 and pH6 but that there was a clear shift in the population for bioreactors maintained at pH5 or pH4. The bacterial diversity remained stable at pH7 (Shannon index of 3.13) and pH6 (3.30) but drastically decreased at pH5 (1.54) and 4 (1.77). Out of the 60 genera detected in the bioreactors 13 were highly correlated with pH (linear or quadratic). Some cellulolytic bacteria were not detected in the bioreactors (*Fibrobacter, Ruminococcus*) after the adaptation period and were replaced by others (*Butyrivibrio*). *Streptococcus* was found at high levels at pH7 and 6 but disappeared at pH5. *Bacillus* and *Lactobacillus* represented more than 95% of the bacterial population at pH4 and 5 against 5% of the total population at pH6 and 7. This study provided information on the impact of a drop in pH on the rumen microbiota and demonstrates that *in vitro* continuous culture may be a useful tool to study and manipulate the rumen ecosystem.

Effect of including greenhouse gas emission costs into economic values of traits for beef cattle
Åby, B.A.[1], Aass, L.[1], Sehested, E.[2] and Vangen, O.[1], [1]Norwegian University of Life Sciences, Department of Animal and Aquacultural Sciences, Box 5003, 1432 Ås, Norway, [2]GENO Breeding and AI Association, Box 5003, 1432 Ås, Norway; bente.aby@umb.no

Genetic improvements of livestock are permanent and cumulative and thus have large potentials to reduce emissions. Indirect selection through correlated traits is currently the best option. In breeding goals, traits are usually weighed by their respective economic value (EV), calculated from e.g bio-economic models. Green house gas (GHG) emission costs may be estimated using a shadow price (politically negotiated value for CO_2), and included in the calculation of EV. In this study, EV were calculated for two breed groups (intensive and extensive breeds) under two production conditions: (1) semi-intensive using considerable amounts of concentrates; and (2) completely roughage based. Other production conditions were also investigated, but not presented here. EV were estimated for seven functional traits: herd life of cow (HL), age at first calving (AFC), calving interval, stillbirth (S), twinning rate (T), calving difficulty, limb and claw disorders, and for seven production traits: birth weight, carcass weight, carcass conformation, carcass fatness, growth rate from birth to 200 days (weaning), growth rate from 200 to 365 days and growth rate from 365 days to slaughter. Inclusion of GHG emission costs decreased the relative economic importance of the functional traits HL, AFC, S and T while increasing the importance of the production traits. However, the overall effect of including GHG emission was small and similar for both of the included production conditions. In addition, little reranking between the traits was observed. A sensitivity analysis for increased shadow price also showed small effects. The results suggest that the economic values are robust towards the inclusion of GHG emission costs into the estimation of EV and increased shadow price. Thus, broad breeding goals for beef cattle do not need to be changed considerably to include the emission of GHG.

Effect of ACTH injection on acute phase and immune response in heifers
Sgorlon, S., Colitti, M., Gaspardo, B. and Stefanon, B., Dipartimento di Scienze Agrarie e Ambientali, Università degli Studi di Udine, via delle Scienze, 208, 33100 Udine, Italy; sandy.sgorlon@uniud.it

The effect of ACTH challenge (AC) on acute phase and immune response has been evaluated in dairy heifers. Fifteen dairy Friesian heifers, homogeneous for age and body conditions, with no clinical signs of disease were used. During the trial, animals were fed twice a day a basal diet (concentrate and forage). After 21 days (T21), all heifers were given 0.5 mg of ACTH (Synachten, Novartis, Varese) every 12 hours for 5 consecutive days. Blood was collected from jugular vein in the morning before (days 19 and 22) and during (days 24 and 26) AC, and analysed for cortisol, employing an immunoenzymatic kit, and for glucose, ceruloplasmin (Cp), haptoglobin (Hp), Zn and paraoxonase (PON) by an automatic analyser (ILab 600, Instrumentation Laboratory) according to Bertoni et al.. Real Time PCR (Sybr® Green chemistry) was performed on total RNA isolated from whole blood to analyse the transcriptional pattern of TNF-a, IFN-g, IL-2 and IL-6. Biochemical data were analysed with ANOVA model with fixed effects for time of sampling (SPSS, 1997). Biomolecular data were expressed as relative expression (n-fold), in comparison to the average values measured at days 19 and 22. The n-fold variations before and during AC were analysed with T-test. In all the groups after AC, Hp was unaffected, cortisol, glucose and Cp increased (P=0.000), whereas Zn (P=0.000) and PON (P<0.01) decreased as well as the proinflammatory cytokines expression (P<0.05). These results indicate that ACTH-induced cortisol release has strong effects on acute phase response and on inflammatory response, with a down regulation of the main proinflammatory cytokines. Thus, ACTH challenge can be efficiently used as experimental model to mimic stressful conditions in cattle. The research was supported by SAFEWASTES, EU Project number 513949, Bruno Stefanon.

The management of buffalo breeding in Hungary

Barna, B. and Holló, G., Kaposvár University, Faculty of Animal Science, Guba S. street 40, 7400 Kaposvár, Hungary; brighitte@vipmail.hu

The management of buffalo breeding and the main features of keeping and feeding conditions of Hungarian buffalo populations are presented. The buffalo population contains about 700-800 buffalo cows in Hungary. Especially the National Parks breed them, but there are some private milking and meat stocks. Observations were made on three different breeding farms: a National Park (Kápolnapuszta), a dairy farm (Virágoskút, with 72 heads), and a farm for meat purpose (Elekmajor, with about 70 heads). On the farms, the animals are kept on the pasture throughout the most part of the year and most of them are kept outdoors without building in the winter, too. The buffalo is well adapted to extreme weather conditions (Virágoskút-dry, hot summer), and especially suitable for use in wet marshy areas (Kápolnapuszta). In winter supplementary feed (alfalfa hay, grass hay, oat hulls) is required for the animals. The milking stocks were fed with grass-hay, alfalfa-hay and concentrate before milking. The length of productive life of buffalo is 10-15 years; the main culling causes are reproductive and locomotion problems. The behaviour and temperament of buffalos affects the habitat and keeping conditions, too. The wallowing is a typical behaviour form of buffalo in the wet marshy areas. The temperament of dairy herd seems to be calmer than that of non-dairy animals. No significant physical behaviour response of buffalo to machine milking was detected, but the presence of calves affects the milk let down. The milk has an organic qualification. The price of buffalo milk is twice as much as the price of cow milk. This suggest that the buffalo breeding has some advantages in Hungary.

The composition of buffalo milk using an ultrasonic milk analyser

Barna, B. and Holló, G., Kaposvár University, Faculty of Animal Science, Guba S. street 40, 7400 Kaposvár, Hungary; brighitte@vipmail.hu

The aim of this study to determinate the main composition of buffalo milk using an ultrasonic milk analyser. For this purpose, 101 milk samples from 18 dairy buffalo were collected during a 3-day consecutive period. In Hungary, there is only one buffalo farm for dairy purposes. The animals were fed with grass-hay, alfalfa-hay and concentrate (in winter) and were kept loose housing system. Milking was practiced at 5 am and 5 pm twice a day, milk samples were taken in both cases. Milk composition was recorded by using an ultrasonic milk analyser (Ecomilk ultra pro, Bulteh 2000), the following data were recorded: fatness, solids non fat, milk-density, protein, conductivity, freezing point. The statistical analysis was carried out using SPSS 11.5 program, the significant differences were calculated by ANOVA using Tukey test. Buffalos were divided into three classes according to their age (>10 yr, 6-10 yr, <6 yr) and three lactation stage classes (<90 d, 91-180 d, >181 d) were used in this study. Milk composition of milk samples from morning and evening milking did not differ significantly. The milking time was 8.91 minutes on average. 1.67 litres of milk was led down. The mean density of milk samples and percentages of fat, solids non fat, and protein in the milk of buffaloes were found to be 30.16 ± 2.33 g/cm^3, $10.86\pm1.77\%$, $11.07\pm0.87\%$ and $5.16\pm0.75\%$, respectively. The average freezing point of the milk samples were 572.4 milicelsius, while the conductivity 3.27 mS/cm. The effect of age was significant on the fat of milk. The fattiest milk was produced by the 6-10 yr group. The stage of lactation was significant on all milk composition parameter except for conductivity. The highest level of fat, solids non fat, and protein were found in the third lactation group. These data indicated that the effect of lactation stage is higher on milk composition of buffalo than age.

In vivo performance of Italian Heavy Draught Horse foals fed two diets with different protein levels
Mantovani, R.[1], Guzzo, N.[1], Sartori, C.[1] and Bailoni, L.[2], [1]Dept. of Agronomy Food Natural Resources Animals and Environment, Viale dell'Universita', 16, 35020 Legnaro (PD), Italy, [2]Dept. of Comparative Biomedicine and Food Science, Viale dell'Universita', 16, 35020 Legnaro (PD), Italy; roberto.mantovani@unipd.it

This study aimed at evaluating growth rate and feed intake in 47 young foals of Italian Heavy Draught Horse breed (IHDH). Foals (274 ± 35 d of age and 339 ± 52 kg of BW) were allocated in 8 pens in relation to sex, age and BW. Pens (2 of females, F; and 6 of males, M) were then randomly assigned to 2 diets (low protein, LP; and high protein, HP) given in 2 subsequent periods of about 4 mo. each (10.5 and 12.3% CP on DM for LP; and 12 and 15% CP on DM for HP in 1st and 2nd period, resp.). Diets were given ad lib. and formulated to be isoenergetic. About half of animals (n=25) were slaughtered at 13 mo. of age (1st period); the remaining (n=22) at 18 mo. (2nd period). Animals were weighted at 3 weeks interval. Feed intake in each pen was weekly measured. Growth patterns were estimated within period for diet, sex and diet×sex by modeling BW as a function of age using a 4th degree Legendre polynomial. ADG and feed intake were analyzed with the same previous effects adding the pen (diet×sex) or the week plus diet×week effects, resp. During the 1st period, different growth patterns (i.e., significant Legendre coefficients) were observed comparing females fed with HP or LP diet, or males in the two different diets. However, no significant differences were detected on ADG (1.02, 0.93, 1.07 and 1.15 kg/d for LP-M, LP-F, HP-M, and HP-F, resp.). In the 2nd period the remaining foals had equal growth patterns and again no differences on ADG were found (1.12, 1.37, 1.24, and 0.99 ordered as above). The DM intake was influenced by diets in both periods. In conclusion, a small dietary protein restriction (i.e. on average 2% of DM) did not change the ADG in IHDH foals both up to 13 or to 18 months of age with possible positive effects on costs of diet and on N pollution.

Insulin, metabolites and arachidonic acid in horses offered linseed and sunflower oils
Patoux, S., Robaye, V., Dotreppe, O., Istasse, L. and Hornick, J.L., Liège University, Nutrition Unit, Veterinary Medecine Faculty, Boulevard de Colonster 20, 4000 Liege, Belgium; listasse@ulg.ac.be

During the training and race seasons, sport horses are offered diets high in concentrate. Fat has been suggested as an alternative compounds to cereals to provide energy and to decrease the level of starch, and associated disturbances. Oil increases the energy but also the essential fatty acids supplies. In this study, eight 6 months exercised adult horses, were used in 3 3×3 latin squares, one being incomplete. The diet was made of 50% grass hay and 50% compound feedstuff with 47.5% whole spelt, 47.5% rolled barley, 3% molasses and 2% mineral mixture in the control diet. In the linseed oil diet, first pressure oil was incorporated at a rate of 8% and substituted to a similar proportion of rolled barley. In the sunflower oil diet, the substitution was similar. The fat content was low in the control concentrate at 15.4 g/kg DM. This rose to 94.1 g/kg DM when the oils were included. The horses remained healthy over the 6 months of the experiment. There were no effects on plasma glucose and on beta hydroxybutyrate concentrations, while plasma cholesterol increased with the oil supplementations. Plasma insulin concentration rose after the meal to reach a plateau within 2 hours. The extends of the increase and of the plateau were reduced when both oils were added. The arachidonic acid (C20:4n-6) content was reduced in the triglycerides fraction (P<0.001) both with linseed oil (1.06 mg/100 ml) and with sunflower oil (0.91 mg/100 ml) as compared with the control diet (1.15 mg/100 ml). It can be concluded that the reduced responses in insulin associated to the supplementation with linseed oil or sunflower oil suggest that these oils may be fed to reduce the negative effects of a high concentrate diet in the developmental diseases of the young horse intensively trained. Similarly, the reduction in the plasma C20:4n-6 concentration can induce less proinflammatory reactions.

Duration and frequency of suckling by foals during the first 3 months of age

Kaić, A.[1], Boštic, T.[2] and Potočnik, K.[2], [1]University of Zagreb, Faculty of Agriculture, Department of Animal Science, Svetošimunska 25, 10000 Zagreb, Croatia, [2]University of Ljubljana, Biotechnical Faculty, Department of Animal Science, Groblje 3, 1230 Domžale, Slovenia; klemen.potocnik@bf.uni-lj.si

The objective of this research was to find out duration and frequency of suckling by foals. In the experiment we observed two foals with their mares. The animals were set up in a stable, in individual boxes. For observation we used two cameras attached in the middle, between two boxes. Observation lasted from birth to ninetieth day of age of foals. Duration and frequency of suckling were based on observation of recorded material. Individual measurements were defined as sucking only if the foal had muzzle at udder 5 seconds or more. All together 1969 individual suckling's were observed. All data were processed by analysis of variance using SAS software package. Regression coefficients for the stage of lactation and influence of mares on duration and frequency of suckling were estimated. First day after birth an average duration of suckling was 87.4 seconds. In second and third month individual duration of suckling varied between 70 and 90 seconds and generally decreased. Total daily time of suckling was maximal first day after birth with an average of 258 minutes. This time was shortened with the progress of age and after second month of life was less than 60 minutes per day. In the first day after birth frequency of suckling was 150 times per day and, compared to the other observed lactation days, was the highest one. From second to fifteenth day frequency of suckling rapidly decreased from 125 to 66 times per day. After that, frequency of suckling in the second month decreased slowly from 60 times per day and at the end of the third month was stabilized at 40 times per day. The analysis of variance revealed that stage of lactation significantly affected duration and frequency of suckling. Influence of mares on duration and frequency of suckling in this research was not statistically significant.

Factors influencing stallion reproductive success in Swedish warmblood riding horses and trotters

Eriksson, S., Johansson, K. and Jönsson, L., Swedish University of Agricultural Science, Department of Animal Breeding and Genetics, P.O. Box 7023, 750 07 Uppsala, Sweden; Susanne.Eriksson@slu.se

Stallion fertility is important for the economy of horse breeders. The reproductive success is strongly influenced by several environmental factors in addition to physiological and genetic factors. The aim of this study was to analyse per season data on mating and foaling outcomes in Swedish Warmblood (SWB) and Swedish Standardbred Trotters (ST). Data were recorded in years 2000-2006 and included 537 SWB stallions and 313 ST stallions, mated with 10-1,656 mares each during that time period. In total, the data comprised 72,468 mare-season combinations. Testicle size was available for 142 ST stallions. Stallion age at mating varied between 3-30 years. Information on conceptions confirmed by veterinarians was not available, but number of conceptions was estimated based on number of reported abortions, foals born (adjusted for twins), and mares that died during pregnancy. The average conception rate was 71.4% for SWB and 74.7% for ST. Average foaling rate was 63.5% for SWB and 63.3% for ST. Four reproductive techniques were used: natural mating or artificial insemination (AI) using fresh, chilled or frozen semen. Logistic regression was used to analyse number of successes (conceptions or foalings) per trials (mated mares) per season for stallions. The fixed effects of year, reproductive technique and stallion age were significant. AI with fresh semen gave the highest probability of conception and foaling, and frozen semen the lowest. For ST-stallions, regression on inbreeding coefficient was also included in the model, and higher inbreeding coefficients lowered the reproductive success. In a subset of data including stallions with testicle measurements, the fixed effect of testicle size was not significant. Estimated repeatabilities for reproductive success of stallions were generally low: 0.04-0.07 for SWB and 0.05-0.08 for ST. The project will continue with further studies including genetic analysis on more recent data and per cycle data.

Impact of different health disorders on length of reproductive life in warmblood stallions

Bernhard, V. and König V. Borstel, U., University of Göttingen, Animal Science, Albrecht-Thaer-Weg 3, 37075, Germany; koenigvb@gwdg.de

Breeding records and health data of 455 warmblood stallions serving between 1975 and 2010 at Marbach state stud, Germany, were collected and submitted to survival analysis. Mean length of service increased over the decades from 6.7±0.5 years (1970s) to 9.9±0.7 in the 1980s. Estimates suggest a further increase for the 90ties and 00s, but owing to a high proportion of censoring (64 and 97% of the records, respectively) results are not yet reliable for the last two decades. Of all health disorders, hazards were highest for fertility problems and other issues leading to castration (hazard ratio (HR)=4.27), followed by lameness (HR=2.19), respiratory diseases (HR=1.72), accidents (HR=1.63), disorders of the digestive system (HR=0.99), and disorders of the cardiovascular system (HR=0.63; all compared to hazards for stallions to leave the stud for non-breeding purposes such as use in sport). However, with all health disorders resulting in the ending of the breeding career, the proportion of affected horses declined over the decades. The most marked decline could be seen in issues leading to castration (affecting 40/19/4/1% of the stallions) and deaths due to respiratory diseases (affecting 11/11/0/0% of the stallions in the 70ties, 80ties, 90ties and 00s, respectively). In addition to decade and culling reason (both P<0.0001), coat colour tended to influence hazards (P=0.056) with liver chestnuts having a lower risk of culling than chestnuts (HR=0.39) and bay horses a higher risk than liver chestnuts (HR=2.99) or grey (HR=2.37) horses. Furthermore, the stud fee influenced survival such that the mortality rate declined with rising fees (HR=0.98; P<0.0001). Overall, there is a clear, positive trend in time for the length of reproductive life, probably based on improved management conditions. However, factors such as coat colour and stud fee, which are linked to demand for individual stallions also influence length of reproductive life.

Biotechnologies of reproduction in the horse: what has changed?

Reis, A.P.[1,2], Palmer, E.[2] and Nakhla, M.[1], [1]AgroParisTech, 5, rue Claude Bernard, 75005 Paris, France, [2]Cryozootech SA, 16, rue André Thome, 78120 Sonchamp, France; alline@cryozootech.com

Development of biotechnologies of reproduction was initially encouraged to promote rapid population's growth and genetic diffusion. This study was proposed to compare the model of development, purposes and target markets of different biotechnologies in the horse industry (artificial insemination (AI), embryo transfer (ET), OPU/ICSI and cloning) in order to understand the evolution of the innovative process. We applied the methodology of study of multiple cases. We built a chronology of events from a review of literature and identified the model of development, methods of transfer to the field and target markets; finally we compared data collected. Biotechnologies progressed from simple techniques, involving small teams and high efficiency (i.e. AI) to complex systems, involving multiple actors and high degree of uncertainty (i.e. cloning). The market of each technology is different and it evolves from a wide market (i.e. AI) to limited, extremely demanding markets: ICSI and cloning. The purpose of use changed from fast spreading desired genes (AI) to individual solutions for subfertility or infertility (ICSI; cloning). The model of development and transfer to the field is similar with minor variations for AI and ET: large scientific exchanges and long periods of knowledge gaining before commercialization (>30 years for AI and 20 years for ET). This interval is reduced in recent technologies (6-10 years), resulting in commercialization before technological maturity and reduced exchange of information. Knowledge is now greatly kept inside institutions as a factor of leadership and a marketable product in an environment highly competitive to obtain private funds. In conclusion, biotechnologies and related innovative processes were influenced by the model of research funding, expectations of users and evolution of horse status. Recent technologies are related to the clinical aspect of fertility and not only genetic selection.

Horse clones registration: historic, scientific and rationale basis
Palmer, E. and Reis, A.P., Cryozootech, 16, rue Andre Thome, 78120 Sonchamp, France;
ericpalmer@crytozootech.com

The first cloned horse was born in 2003, the first registration of a clone in a studbook was in 2005, the first approved cloned stallion in 2007, the first foal by a cloned stallion was born and registered as 'known origin' in 2008 and in a full studbook in 2010. At least 18 clones are registered in EU studbooks. No cloned horse has participated to a competition. The scientific background should be considered before deciding rules on horse clones. The clone has a full copy of the nuclear DNA of the model, but the mitochondrial DNA of the recipient oocyte and possible epigenetic differences due to environment or imperfect reprogramming. The phenotype of the clone can differ, because of mitochondria, epigenetics and environment. As a stallion the clone is equal to the donor, as he does not transmit mitochondria and his genome is reprogrammed at meiosis and fertilization. In the female, the mitochondria are transmitted, but their importance is still unknown. Analyses can confirm the similar nuclear DNA and the different mitochondrial DNA, if the donor is known. If no supposed donor is available, no biological parameter can detect a clone. AI and ET took longer from success to application in breeding. They were also discussed and were accepted after successive steps: (1) rejection; (2) acceptation with restrictions such as number of offspring, quality of the animal and mention of 'AI' or 'ET' on their passport; (3) full acceptation. Some breeds such as thoroughbred are still at step 1. Clones will probably follow the same pathway but should keep the mention of clone, at it has a genetic signification. For rules, a distinction exists between law, which applies to everyone and can be enforced, and rules within studbooks or competitions organizations. Some laws can be used against cloning, such as human food safety or others, but some can be used for cloning such as trade freedom. Rules must be applicable in practice: if no test can prove that a horse is a clone, a ban is impossible. At last, one must question the utility of making specific rules.

Horse welfare and behavior in therapeutic riding
Miraglia, N.[1], Cerino, S.[2], Pierni, E.[3] and Gagliardi, D.[4], [1]University of Molise, Animal, Vegetables
and Environmental Sciences, Via De Sanctis, 86100 Campobasso, Italy, [2]F.I.S.E., Therapeutic Riding,
Viale Tiziano 74, 00197 Rome, Italy, [3]Professional Veterinary, Via Ferrari 12, 86100 Campobasso, Italy,
[4]Professional Veterinary, Via Ferrari 12, 86100 Campobasso, Italy; s.cerino@alice.it

In every equestrian activities the horse welfare is, of course, paramount. A particular interest have the horse welfare and behavior in the field of therapeutic riding.In the present paper the authors would like to point out how important they are in the above mentioned case, where the horses are mounted by disabled people with a general different approach to the riding techiniques and to the horse itself. To understand the horse behavior support the cognitive and meta-cognitive approach useful to disabled people to structure the empathic relationship with the animal, which is the basis for a good therapeutic work. At the same time the horse welfare results to be one of the crucial knots we cannot ignore in Therapeutic Riding. To train horses to be able to work with disabled people, it is essential to check periodically their attitudes and conducts. The authors propose the employ of a detailed card to measure the horse linear behavior, when they work in therapeutic riding sessions. In the card they propose the evaluation of the horse linear behavior (both at rest and work) by a score among 1 and 5, togheter with a clinical examination of the animal. The results of the card submission during a therapeutic riding project with schizophrenic patients will be discussed here.

Residual feed intake divergent selection in sheep

Francois, D.[1], Bouvier, F.[2], Ricard, F.[1], Bourdillon, Y.[2], Weisbecker, J.L.[2] and Marcon, D.[2], [1]INRA, UR 631 SAGA, BP 52627, 31326 Castanet Tolosan, France, [2]INRA, UE 332 Domaine de Bourges, la Sapinière, 18390 Osmoy, France; Dominique.Francois@toulouse.inra.fr

Meat sheep selection programmes concern meat traits such as growth and body composition. One way to improve lamb production may be to increase the efficiency of feed transformation by lambs. Measurement of feed intake in sheep has been performed with automatic feeders on young rams as they are submitted to individual test for selection on liveweight (LW), growth and body composition. The test is following a national scheme process and lasts 8 weeks from about 100 to 156 days of age. LW and ultrasound scan (US) are recorded at start (LW only), mid (both LW and US) and end (both Lw and US) of the test. Feeders provided detailed data for every meal of the ram (identity, intake, duration) allowing computation of daily feed intake (DFI). Since DFI is strongly correlated with LW, growth and body composition; linear regression of DFI on these traits was performed and residual feed intake (RFI) was expressed as the residue of the regression. A demonstrating one-generation divergent selection was carried on. Sires have been selected among a trial of 151 Romane (INRA-401) rams. The 10 more favorable (FA) with the most negative residual feed intake rams and 10 less favorable (LF) with the most positive residual feed intake ones were selected. 7 of each group were mated to Romane ewes and procreated the next generation. FA sires had a RFI mean of -211 g/day, as LF sires had + 204 g/day. The selection differential was 415 g (=3.7 residual standard deviation). Among the offspring, 160 males entered the individual test protocol, 149 achieved it, 82 from FA sires and 67 from LF sires. They were measured at the same age following the same protocol as the sires. On average the RFI was -21 g/day for FA offspring as + 26 g/day for LF offspring, differential was 47 g/day. Realized heritability was 0.23. Correlated responses on DFI, LW, growth and body composition will be presented.

Effect of mixing grass silage with concentrate on feed intake in ewes and live weight gain in lambs

Helander, C.[1], Nadeau, E.[1], Nørgaard, P.[2] and Arnesson, A.[1], [1]Swedish University of Agricultural Sciences, Department of Animal Environment and Health, P.O. Box 234, 532 23 Skara, Sweden, [2]University of Copenhagen, Faculty of Life Sciences, Department of Basic Animal and Veterinary Sciences, Grønnegårdsvej 3, 1870 Fredriksberg C, Denmark; Carl.Helander@slu.se

The aim was to study the effects of chopping grass silage and of mixing grass silage with concentrate on intake, body weight (BW) and body condition score (BCS) in pregnant and lactating ewes and on LWG of lambs until weaning. Two similar experiments (Exp. 1 and Exp. 2) were carried out over two consecutive years. Each year, 21 ewes were allocated equally to three treatments; unchopped silage *ad libitum* and 0.8 kg concentrate daily, fed separately (US), chopped silage *ad libitum* and 0.8 kg concentrate daily, fed separately (CS) and chopped silage and concentrate in a mixed ration *ad libitum* (CM) with the same forage to concentrate ratio as in CS. The silages in Exp. 1 and Exp. 2 contained 10.9 and 11.4 MJ metabolisable energy /kg DM and 580 and 483 g NDF /kg DM, respectively. The ewes increased their silage DM intakes from late pregnancy to lactation by 47 and 64%, in Exp. 1 and Exp. 2, respectively, when averaged over treatments (P<0.001). In Exp. 2, the daily DM intake of lactating ewes was, on average, 13% higher for CM than for US and CS, which did not differ (4.75 vs. 4.21% of BW, P<0.05). Neither BW nor BCS of ewes were affected by chopping silage or by mixing silage and concentrate. The ewe BW was 97.8 and 97.3 kg in pregnancy and 87.2 and 91.7 kg in lactation in Exp. 1 and Exp. 2, respectively, when averaged over treatments (P<0,001). The ewe BCS was 3.1 and 3.4 kg in pregnancy and 2.7 and 3.0 kg in lactation in Exp. 1 and Exp. 2, respectively, when averaged over treatments (P<0.001). In Exp. 2, CM increased daily LWG from birth to weaning by, on average, 16% compared to US and CS, respectively (454 vs. 392 g, P<0.01). Mixing silage and concentrate increased DM intake in lactating ewes and LWG of lambs from birth to weaning.

A novel approach to the prediction of carcasses composition of lambs
Cadavez, V.A.P.[1] and Henningsen, A.[2], [1]Mountain Research Center (CIMO), ESA – Instituto Politécnico de Bragança, Animal Science, Campus de Santa Apolónia, Apartado 1172, 5301-855 Bragança, Portugal, [2]University of Copenhagen, Institute of Food and Resource Economics, Rolighedsvej 25, 1958 Frederiksberg C, Denmark; vcadavez@ipb.pt

The objective of this study was to apply the simultaneous equations model technique to predict lamb carcasses composition. Forty male lambs, 22 of Churro Galego Bragançano Portuguese local breed and 18 of Suffolk breed were used. Lambs were slaughtered and carcasses were weighed approximately 30 min after slaughter in order to obtain hot carcass weight (HCW). After cooling at 4 °C for 24 h, the subcutaneous fat thickness measurement (C12) was taken between the 12th and 13th ribs. The left side of all carcasses was dissected into muscle, subcutaneous fat, inter-muscular fat, bone, and remainder (major blood vessels, ligaments, tendons, and thick connective tissue sheets associated with muscles). The carcasses tissues proportions were calculated, and a simultaneous equations model was fitted. The model fitting quality was evaluated by the coefficient of determination and by the root mean square error. This study shows that simultaneous equations models can improve the accuracy of the predictions of lamb carcasses tissues proportions.

Investigation on specific combining abilities of Merinos and several sire sheep breeds for lamb meat
Henseler, S., Preuss, S. and Bennewitz, J., Institute of Animal Husbandry and Animal Breeding, University of Hohenheim, Garbenstraße 17, 70599 Stuttgart, Germany; stefanie.henseler@uni-hohenheim.de

Lamb meat production in the south-west of Germany is often practised by performing a one way cross. Merino breed is used as a fertile mother line and is mated with a sire sheep breed that is selected for meat and growth traits. In this study the special combining ability of five sire breeds with the Merinos is investigated. The sire breeds were Texel, Suffolk, Ile de France, Schwarzkopf, and Charolais. Additionally, Merinos purebreds were considered. Around 250 lambs were generated for each cross (in total 1,600 lambs). Paternity control was done by genotyping 768 SNPs. They were housed and fattened on one central farm. Several growth, carcass and meat quality traits were recorded at slaughter. These are weight at slaughter, kidney fat, carcass length, pH1, pH24, conductivity, drip loss, meat colour, meat loss after cooking, and usual carcass classification parameter. Additionally, meat quality was investigated by a sensoric test. The statistical analysis was done in order to find out which breed is most suited as a sire breed for the production of high quality lamb meat. First results showed that the large size of the experiment resulted in a high experimental power to detect breed differences. The final results will be available during 2012.

Genetic improvement in the dairy goat sector in Germany

Hamann, H.[1], Herold, P.[1], Wenzler, J.G.[2], Mendel, C.[3], Krogmeier, D.[3] and Götz, K.U.[3], [1]State Office for Geo-Information and Rural Development Baden-Württemberg, Stuttgarter Straße 161, 70806 Kornwestheim, Germany, [2]Goat Breeders Association Baden-Württemberg, Heinrich-Baumann-Straße 1-3, 70190 Stuttgart, Germany, [3]Bavarian State Research Center for Agriculture, Institute for Animal Breeding, Prof.-Dürrwaechter-Platz 1, 85586 Poing, Germany; Henning.Hamann@lgl.bwl.de

The goat breeding structure in Germany is decentralised with no breeding value estimation, and neither planned breeding nor artificial insemination is available. Aim of the present study is to evaluate a breeding program incorporating two regional goat breeding organisations. Objectives of the study are the investigation of alternative breeding plans based on buck circles or nucleus breeding. Furthermore, genetic connectedness of the two dairy goat populations is evaluated and genetic parameters are determined. Finally, a breeding value estimation scheme is developed. In the beginning the traits milk yield, fat and protein yield, fat- and protein% and somatic cell score (LSCS) are analyzed. Data sets contain 19,385 lactations and 7,075 goats in Baden-Württemberg and 8,603 lactations and 2,729 goats in Bavaria. Average milk yield is 2.7±0.8 kg milk/day and 2.4±0.8 kg milk/day, respectively. Heritability was estimated to be 0.30, 0.35, 0.35, 0.45, 0.50 and 0.20 for milk yield, fat yield, protein yield, fat-%, protein-% and LSCS, respectively. A first attempt in modelling breeding plans showed that in terms of genetic gain a breeding plan with controlled buck circles is superior to the actual breeding program or a breeding plan with AI if the population size is small (2,000 or 7,500 goats). With growing population size a breeding plan with AI achieves highest genetic gain (12,000 or 50,000 goats). Generally, it can be shown that every change of the actual breeding program improves the situation. Therefore, starting out with a regional breeding program seems promising for improving the sustainability and efficiency of goat breeding in Germany.

Estimates of genetic parameters for growth traits in Arabi sheep

Shakrollahi, B.[1], Baneh, H.[2] and Karimi, K.[3], [1]Sanandaj Branch, Islamic Azad University, Department of Animal Science, 6617983851, Iran, [2]Young Researchers Club, Sanandaj Branch, Islamic Azad University, 6617983851, Iran, [3]Varamin-Pishva Branch, Islamic Azad University, Department of Animal Science, 12 Varamin, Iran; Borhansh@iausdj.ac.ir

Arabi sheep is highly valued for meat production and kept in southwestern Iran, nearby regions of Iraq and Saudi Arabia. The aim of our study was to estimate genetic parameters for birth weight (BW), weaning weight (WW) and weight at six months of age (6 MW), which are essential to design a selection program for this sheep. Data collected over a period of 10 years (1999-2009). Number of correct records for the traits consisted of 277 for BW, 2,002 for WW and 1,885 for 6MW. Analyses were performed by restricted maximum likelihood procedures (REM L), by fitting six animal models with various combinations of direct and maternal effects. All three weight traits were significantly influenced by birth year, sex and birth type. Age of dam only significantly affected BW. Log–likelihood ratio tests were conducted to determine the most suitable model for each growth trait in univariate analyses. Direct and total heritability estimates for BW, WW and 6MW based on the best model were 0.42 and 0.16, 0.38 and 0.13 and 0.14 and 0.14, respectively. Estimation of maternal heritability for BW and WW were 0.22 and 0.18, respectively. Genetic and phenotypic correlations among these traits were positive. Phenotypic correlations among traits were low to moderate Genetic correlations among traits were positive and higher than corresponding phenotypic correlations. WW had a strong and significant correlation with 6MW (0.99). The results suggested that selection based on WW could be more effective than the other traits on improvement of growth traits in Arabi sheep.

Fatty acids profile of intramuscular and extramuscular fats in lambs raised under grazing system
Margetín, M.[1], Apolen, D.[1], Polák, P.[1], Zaujec, K.[1], Debrecéni, O.[2], Vavrišinová, K.[2], Kubinec, R.[3], Blaško, J.[3] and Soják, L.[3], [1]Animal Production Research Centre Nitra, Hlohovecká 2, 951 41 Lužianky, Slovakia, [2]Slovak University of Agriculture Nitra, Faculty of Agrobiology and Food Resources, Tr. Andreja Hlinku 2, 949 76 Nitra, Slovakia, [3]Comenius university, Institute of Chemistry, Faculty of Natural Sciences, Mlynská dolina, 842 15 Bratislava, Slovakia; margetin@cvzv.sk

We evaluated carcass data and took meat samples (Musculus longissimus lumborum et thoracis – MLLT) from 20 lambs of Ile de France breed with average empty live weight 27.3 kg and age 107 days, raised under mothers on pasture, to assess carcass data and to determine physical-chemical meat properties. We took parallel meat samples from MLLT (intramuscular fat – IMF) and fat samples from root of tail (extramuscular fat – EMF) to determine profile of fatty acids (totally 69 FAs) by gas chromatography. By means of ANOVA we detected significance of differences among individual FAs or FAs groups in dependence on the fat type (IMF or EMF) and sex. Dressing percentage of lambs was on average 48.9%. Average amount of total water in MLLT sample was 1.26; of total proteins 21.7 and total fat 1.93 g /100 g. We found significant differences ($P<0.001$) in amount of SFA (48.7 or 55.9 g/100 g FAME), PUFA (16.9 or 9.0), essential FAs (8.8 or 4.2), omega-6 FAs (8.5 or 3.2) and omega-3 FAs (4.6 or 2.0) when we compared FAs in IMF and EMF. Ratio of n-6/n-3 was 1.85 in IMF and 1.61 in EMF ($P<0.001$). When we compared IMF and EMF fatty acids we found significant differences ($P<0.001$) with linoleic acid (C18:2 n-6; 6.41:2.84 g 100 g-1 FAME), GLA (C18:3 n-6; 0.05:0.02); α-linolenic (C18:3 n-3; 2.38:1.38), arachidonic acid (1.83:0.28), EPA (0.82:0.08), DPA (0.91:0.36), DHA (0.29:0.09). Differences between trans-vaccenic acid (11t-C18:1; 4.05:4.04) and rumenic acid (9c11t-18:2 CLA; 1.82:1.96) were not significant between IMF and EMF. Factor sex had significant influence on amount of α-linolenic and rumenic acids ($P<0.05$).

Fattening performance and carcass traits of local a Romanian breed crossed with meat rams
Ghita, E., Lazar, C., Pelmus, R., Dragomir, C., Mihalcea, T., Ropota, M. and Gras, M., National Research Development Institute for Animal Biology and Nutrition, Animal Biology, Calea Bucuresti nr.1, 077015 Balotesti, Ilfov, Romania; elena.ghita@ibna.ro

Previous studies have shown that although Teleorman Black Head (TBH) have better aptitudes for meat production than other Romanian sheep breeds (Tsurcana, Tsigai), it is still inferior to the specialised meat breeds. Therefore, the opportunity of crossing TBH with Suffolk rams as a solution to increase the competitiveness of TBH fat lamb carcasses was investigated. Thus, the fattening performance and carcass traits of the hybrid lambs (TBH × Suffolk) were compared with those of TBH, in a fattening trial, for 97 days, on 20 hybrids and on 20 TBH lambs (similar conditions of maintenance and feeding). At the beginning of the trial the lambs (83 days of age) had an average weight of 18.57 and 16.98 kg, respectively ($P>0.05$) and displayed a previous average daily weight gain of 0.251 kg (hybrid lambs) and 0.237 kg (TBH lambs) ($P>0.05$). The final weight, at the age of 180 days was 42.95 kg (hybrid lambs) and 39.99 kg (TBH lambs) ($P<0.05$). At the end of the experiment 8 lambs from each group were slaughtered in order to determine the slaughter and commercial outputs, the proportion of the different carcass parts, the proportion of butcher parts, the meat to bone ratio (for parts/for entire carcass), specific measurements of the carcass, the chemical composition of the meat including the fatty acids and cholesterol level. No significant difference between hybrid and local lambs were found for the slaughter output or for the commercial output. On the other hand, width and depth dimensions were higher or tend to be higher for the hybrid lambs carcasses ($P<0.001$). TBH × Suffolk hybrid lambs tend to have better dressing with muscles, particularly at the leg, loin and rack. In conclusion, crossing TBH local breed with Suffolk rams allowed higher rate of growth of the hybrid lambs and improvement of some parameters describing carcass quality.

Dietary NSC levels and its effect on rumen development of Döhne Merino lambs
Le Roux, M.[1,2], Van De Vyver, W.F.J.[2] and Hoffman, L.C.[2], [1]Department of Agriculture, land reform and rural development, Animal production, Vaalharts research station, Private bag X9, Jan Kempdorp 8550, South Africa, [2]Stellenbosch University, Department of Animal Sciences, Private bag X1, Matieland, Stellenbosch 7602, South Africa; wvdv@sun.ac.za

The following study investigated the effect of the non structural carbohydrate (NSC) fraction in creep feeds on the development of the rumen of lambs. Fermentation of NSC in the rumen gives rise to higher proportions of buytric- and propionic acids which in turn stimulates rumen development. The current study aimed to determine the extent of papillae development in the rumen of suckling lambs due to different NSC fractions in creep feeds. Three creep feed rations were formulated at NSC levels of 455, 477 and 508 g/kg to represent the treatments CFC, CF1 and CF2 respectively. Another treatment (control; CON) was included to represent lambs with no access to creep feed but acces to grazing. Twin lambs averaging a birth weigth of 4.42 kg entered the trial at an average age of seven days and were slaughtered at the average age of 69 days at a live weight of 23.6 kg. The rumen development characteristics which include papillae length, width, mucosa thickness and rumen wall thickness were measured from microscopic slides in micrometers. Rumen wall samples from lambs on creep feed treatments had longer papillae than samples from CON lambs. Papillae width, in turn, decreased as the NSC level increased. It was concluded that papillae length responds to increasing levels of NSC and that the longer a papilla grows the thinner it becomes. Absorptive area for nutrient uptake is therfore increased as the NSC fraction of the feed increases.

Chemical composition of meat from lambs fed 'in nature' or hydrolyzed sugar cane
Endo, V., Da Silva Sobrinho, A.G., Zeola, N.M.B.L., De Almeida, F.A., Lima, N.L.L. and Manzi, G.M., Faculdade de Ciências Agrárias e Veterinárias, FCAV, Unesp, Zootecnia, Via de Acesso Prof. Paulo Donato Castellane, s/n, 14884-900 Jaboticabal, SP, Brazil; nivea.brancacci@ig.com.br

The objective was to evaluate chemical composition of meat from lambs fed 'in nature' or hydrolyzed sugar cane. Ile de France lambs (n=24, from 15 to 32 kg of body weight) were subjected to one of three treatments: 'in nature' sugar cane + concentrate; hydrolyzed sugar cane with 0.6% calcium oxide (CaO) in aerobic environment + concentrate; or hydrolyzed sugar cane with 0.6% CaO + concentrate in an anaerobic environment, in a completely randomized design. The values of chemical composition of meat were not affected (P>0.05) by treatments, averaging for moisture 75.96%, CP 21.26% and ash 1.21%. For the ether extract (EE), it was observed that lambs fed hydrolyzed sugar cane in anaerobic environment had a higher value (2.77%) than lambs fed hydrolyzed sugar cane in aerobic environment (2.07%). These results were not expected, because the diets showed little variation in composition, ranging from 1.92% to 2.14% EE, and did not show differences in intake and digestibility of EE. The slaughter age of lambs is another factor that could influence the chemical composition, since intramuscular fat is the last to be deposited in the carcass, but the ages were similar, averaging 153 days. The results showed that hydrolysis of sugar cane with alkaline agents was not of any value added.

Use of ultrasound to predict carcass composition and evaluation of carcass and meat traits in lambs

Esquivelzeta, C., Casellas, J., Fina, M. and Piedrafita, J., Universitat Autònoma de Barcelona, Facultat de Veterinària, Campus Universitat Autònoma s/n, 08029 Cerdanyola del Vallès, Spain; ceciliaer@gmail.com

Real time ultrasound techniques are useful tools for predicting body composition in livestock, having the advantage of assess carcass traits in live animals. The aim of this study was to evaluate ultrasound measurements to predict carcass composition in 124 Pascual-type lambs (13-16 kg), and compare carcass and meat traits between breeds (Ripollesa, R; Lacaune, L and its crossbreed; RL). Images were taken between 12th and 13th ribs. Skin thickness (ST), subcutaneous back-fat thickness (BFT) and depth (DLM), width (WLM) and area of longissimus dorsi (ALM) were obtained with the ImageJ 1.42q software. After slaughter BFT (0.2 ± 0.01 cm), DLM (2.5 ± 0.03 cm), WLM (4.5 ± 0.04 cm) and ALM (10 ± 0.12 cm^2) were measured on lamb carcasses, and right leg (1.4 ± 0.01 kg), rack (3 ± 0.03 kg), shoulder (1.1 ± 0.01 kg) and neck (0.8 ± 0.01 kg) were weighted. Correlation coefficients between ultrasound and direct measurements were moderate for BFT (0.6) and high for DLM (0.9), WLM (0.7) and ALM (0.7). Carcass pieces were also correlated (0.2-0.4) with ultrasonic measurements. The meat fatty acid (FA) analysis showed that the FA mostly represented in muscular tissue were: oleic (C18:1n-7, 38.8%) being greater for L, palmitic (C16:0, 21.7%) without differences between breeds, and stearic (C18:0, 14.3%) lower percentage for L. Differences between breeds for saturated (SFA; 39.0-41.2%), monounsaturated (MUFA; 48.1-49.6%) and polyunsaturated FA (PUFA; 6.0-7.5%) proportions were found. Differences for nutritional ratios were only found for PUFA/SFA ratio, being greater for L (0.19), followed by R (0.16) and RL (0.15) breeds. Nutritional ratios for n-6/n-3 varied from 6.27 to 6.96 and from 7.85 to 8.57 for C18:2n-6/C18:3n-3 ratio. In conclusion, ultrasound measurements are useful for predicting accurately DLM, WLM, ALM and BFT. All these information can be included in sheep genetic improvement programs as a new selection procedure to improve meat traits of sheep breeds.

Possibility of using ultrasound measurements for *in vivo* assessment of heavy lambs

Polák, P., Tomka, J., Krupová, Z., Oravcová, M., Margetín, M., Apolen, D. and Zaujec, K., Animal Production Research Centre Nitra, Department for Animal Breeding and Product Quality, Hlohovecká 2, 95115 Lužianky, Slovakia; polak@cvzv.sk

Fattening, carcass quality characteristics and ultrasound measurements of 40 heavy lambs of Tsigai and Ile de France breeds were analysed in order to judge possibilities of ultrasound measurements for *in vivo* assessment of meat production traits. Sonograms of transversal cut of musculus longissimus thoracis et lumborum (MLTL) were obtained one day before slaughter on last thoracic vertebra by echocamera Aloka PS 2 and ultrasound probe UST 5820 - 5. Muscle width, thickness and area and fat thickness were measured on digitalised sonograms in special software for video image analysis NIS – Elements. Tsigai lambs had significantly higher slaughter age, weight and muscle width. However Ile de France lambs had significantly higher muscle thickness, weight of meat in shoulder or proportion of round in half carcass. It means that here is significant difference between carcass composition between Tsigai and Ile de France. Lambs of specialised meat breed Ile de France in lower age and slaughter weight showed better meat production in comparison to Tsigai – dual purpose one. When we used ultrasound measurements with combination of daily gain or live weight before slaughter, coefficients of determination of the models were on level 0.09-0.24. Muscle width and daily gain had significant coefficient of correlation with carcass weight but the highest coefficient of determination for carcass weight was found in combination of live weight and subcutaneous fat layer (0.24). Despite of limited number of animals and quite large difference between carcass composition of analysed breeds in this preliminary study, the findings indicate that ultrasound can be successfully utilized for *in vivo* assessment of carcass weight.

Determination of the most appropriate growth curve model and growth characteristics in Norduz Kids
Özel, D. and Aygün, T., Yüzüncü Yıl University, Department of Animal Science, Faculty of Agriculture, 65080 Van, Turkey; taygun@yyu.edu.tr

The aim of this study was to examine the effects of some major environmental factors such as gender, birth type and dam age on the body weight and some body measurements at various periods of 14 Norduz kids, and to determine the most suitable growth model. For this aim, the growth models of Monomolecular, Gompertz, 3 parameters Logistic, and Richards were used. Birth weight, live weight, and body measurements such as withers height, body length, chest circumference, leg circumference, chest depth, and width of chest behind shoulders were fortnightly recorded from birth to 180 days of age. Corresponding the averages at 180 days of age were estimated as 2.94±0.14 kg, 20.30±0.74 kg, 56.41±0.54, 58.60±0.80, 76.09±1.82, 59.49±1.09, 24.44±0.59, and 13.62±0.32 cm, respectively. Only birth type on body weight, body length, chest circumference, and chest depth at weaning age had significant effect (P<0.05). Significant effect of gender on leg circumference at weaning time was found (P<0.05). The effect of birth type on withers height, chest depth, and leg circumference at 180 days of age was statistically significant (P<0.05).It was concluded that the most suitable growth model describing the body weight and the body measurements of Norduz kids was Logistic model with 3 parameters.

Evaluation of reproductive parameters in sheep females inseminated with Lacaune ram semen
Zeola, N.M.B.L., Da Silva Sobrinho, A.G., Sena, L.S., Columbeli, A.C., Da Silva, L.M., Silva, F.U., Santana, V.T., Endo, V. and Lima, N.L.L., Faculdade de Ciências Agrárias e Veterinárias, FCAV, Unesp, Zootecnia, Via de Acesso Prof. Paulo Donato Castellane, s/n, 14884-900 Jaboticabal, SP, Brazil; nivea.brancacci@ig.com.br

It was used 50 Ile de France females, being 25 ewes with age between 4 and 5 years old and 25 hoggets with age between 1 and 2 years old, submitted to the estrus synchronization, using a hormonal protocol, being later inseminated with fresh semen of Lacaune ram (Fixed Time Artificial Insemination – FTAI). During the gestation the sheep females stayed in tifton-85 pasture and received a diet with 14% of crude protein, containing 40% of corn silage and 60% of concentrate.The ½ Lacaune ½ Ile de France lambs were submitted to a suckling period of 42 days and in the creep feeding from the 10 days of age, received concentrated with 20.0% of crude protein. The body weight of sheep females before of the lamb's birth (73.08 and 71.94 kg) and after (63.67 and 61.34 kg), the weight losses after the lamb's birth (9.42 and 10.60 kg; 13.43 and 14.77%), the gestation period (146.5 and 147 days), the body weight at 28 days of age (8.80 and 12.10 kg), the body weight at 42 days of age (12.13 and 15.60 kg) and the weight gain from birth to 42 days of age (0.22 and 0.26 kg) were not influenced (P>0.05) for the sheep females age, however the lamb's body weight at the birth and at 14 days of age was influenced (P<0.05), ewes with age between 4 and 5 years old propitiated lamb's weight of 3.02 and 7.15 kg, respectively, while hoggets with age between 1 and 2 years old, propitiated higher weights of 4.50 and 9.79 kg, respectively. The Ile de France sheep females presented zootechnical indexes, related to the reproductive parameters, appropriate regardless of their age.

Ojinegra sheep local breed: identifying factors of variability of performance

Ripoll-Bosch, R.[1], Villalba, D.[2], Blasco, I.[1], Congost, S.[3], Falo, F.[4], Revilla, R.[3] and Joy, M.[1], [1]CITA, Av. Montañana 930, 50059 Zaragoza, Spain, [2]UdL-ETSEA, Av. Rovira Roure 191, 25198 Lleida, Spain, [3]CTA, Apdo. 617, 50080 Zaragoza, Spain, [4]AGROJI, Plaza del Ayuntamiento 1, 44556 Molinos, Spain; rripoll@aragon.es

In local and endangered livestock breeds it is necessary to study their productive performance in order to be able to promote their utilization and conservation. The lack of reliable data may lead to erroneous replacement criteria and on-farm decisions. This is specially the case of the Ojinegra sheep breed from Teruel (Spain). The aim of this study was to characterize the productive parameters of this breed and to determine the variability in results. Live weight (LW) and body condition scores (BCS) at lambing and weaning of ewes and live weight at born (LWb) and average daily gain (ADG) of lambs, from lambing to weaning, were controlled. Data was recorded in 8 conventional farms along 1 year. Data was analyzed with the MIXED procedure of SAS, with the farm being the random effect. Ewe's LW and BCS at lambing (43.3±7.4 kg and 2.55±0.45) and weaning (42.3±7.3 kg and 2.73±0.45) were influenced by the age of the ewe, the season of the year where lambing took place and the farm (P≤0.05). BCS at lambing was also influenced by the number of lambs born (P<0.05). LWb and ADG (3.5±0.67 kg and 168±51 g/day) was influenced by lamb's gender, lambs reared per ewe, age of the ewe, BCS of ewe at lambing and the farm (P≤0.05). ADG was also affected by the season of lambing and LWb by the interaction 'season of lambing*number of lambs reared' (P≤0.05). The farm was decisive in results, representing up to 48% of variability in some parameters studied. Variability between farms could be explained either by on-farm management or by genetic heterogeneity of the breed. The farm should be strongly considered when characterizing productive parameters of a breed.

Genetic parameters for milk traits using fixed regression models for Istrian sheep in Croatia

Špehar, M.[1], Barać, Z.[1], Kastelic, M.[2], Mulc, D.[1], Jurković, D.[1] and Mioč, B.[3], [1]Croatian Agricultural Agency, Ilica 101, 10000 zagreb, Croatia, [2]University of Ljubljana, Groblje 3, 1230 Domžale, Slovenia, [3]University of Zagreb, Svetošimunska 25, 10000 Zagreb, Croatia; mspehar@hpa.hr

Istrian sheep is a Croatian autochthonous breed originated from Istrian peninsula and neighbouring Karst plateau in Slovenia and Italy. The objective of this study was to estimate genetic parameters for daily milk, fat and protein yield and fat and protein content using test-day records of the Istrian sheep in Croatia. Data consisted of 13.101 test-day records for 2.320 ewes. Production data was recorded using the AT4 method and taken from the central database of the Croatian Agricultural Agency. The number of animals in pedigree was 3.588. A single-trait repeatability fixed regression test-day model was used to estimate genetic parameters. Fixed class effects in the model were: parity, litter size, lambing season, and flock. Days in milk and age at lambing were treated as covariates. For yield traits, the effect of days in milk was fitted using the Ali-Schaeffer lactation curve and nested within parity and litter size. Age at lambing was modelled as linear regression nested within parity. This effect was not included in the model for fat and protein content, while the effect of days in milk was nested within parity. Direct additive genetic effect, flock-test-day, and permanent environment effect over lactations were included in the model as random effects. Variance components were estimated using Residual Maximum Likelihood as implemented in the VCE-6 program. The estimated ratios for daily milk, fat and protein yields (kg), and fat and protein contents were: 0.15, 0.07, 0.013, 0.07, and 0.15 for additive genetic, 0.29, 0.31, 0.28, 0.34, and 0.18 for flock-test-day, and 0.21, 0.20, 0.21, and 0.05, and 0.07 for permanent environment effect over lactations. Results provide genetic parameters for the application of genetic evaluation for milk traits evaluation of the Istrian sheep in Croatia.

Analysis of milk production traits in dairy goats

Wensch-Dorendorf, M., Schafberg, R. and Swalve, H.H., Martin-Luther-University Halle-Wittenberg, Institute of Agricultural and Nutritional Sciences, Theodor-Lieser-Str. 11, 06120 Halle, Germany; hermann.swalve@landw.uni-halle.de

Data from one of the largest dairy goat farms in Germany collected within a period of 9 years were available for investigations. About 71,000 test day records from about 4,000 goats were included in the analysis of 4 lactations of milk (kg), fat percent, protein percent, somatic cell count (SCC, log10 transformed) and fat-protein-ratio. Raw averages were 3.48 kg for milk, 4.09 for fat percent, 3.47 for protein percent, 2.92 for SCC and 1.18 for fat-protein-ratio. Days in milk (DIM, limited to 600 days) were modeled as classes of a main fixed effect since the Ali-Schaeffer regression function, widely used and proven in analysis of dairy cows, showed no better fit. In the first 100 days in milk, a dim-class was defined for every ten days in lactation. For the following days in milk, a dim class was defined for every 50 days. The results show that when comparing parity numbers, in the first 100 days mostly only minor differences for the LSmeans of the traits exist, with SCC being an exception. Goats in their first lactation show lowest somatic cell counts. The further the stage of lactation progresses, the more differences between parity numbers for the LSmeans of milk(kg) and protein percent were observed, with an advantage for the first lactation goats. LSmeans of fat percent and SCC increase with increasing parity number. Heritabilities as well as genetic correlations between the lactations, estimated using a multiple-trait test-day model, were moderate to high.

Improvement of productive and reproductive traits in Cyprus Chios sheep and Damascus goats

Hadjipavlou, G.[1], Papachristoforou, C.[2], Mavrogenis, A.P.[1] and Koumas, A.[1], [1]Agricultural Research Institute, P.O. Box 22016, 1516 Lefkosia, Cyprus, [2]Cyprus University of Technology, P.O. Box 50329, 3603 Lemesos, Cyprus; georgia.hadjipavlou@arinet.ari.gov.cy

In Cyprus, the breeding structure of sheep and goat populations is based on a closed nucleus flock system with three tiers: the nucleus flocks, the multipliers and the commercial producers. Since the 1980s, most efforts have been concentrated on the genetic improvement of pure-bred Chios sheep and Damascus goats. For over thirty years, the Agricultural Research Institute (ARI) and the Department of Agriculture have maintained nucleus flocks of about 600 Chios sheep and 500 Damascus goats. Males are mated as lambs or kids, with individual hand-matings, following detection of animals in heat by vasectomized males. Evaluation procedures are based on selection indices for milk and growth rate, utilizing information on both pedigree and individual performance. Selection intensity is 5-10% for males and female replacement rate is 25-30% (average productive life of 3.5 parities). The main breeding goals are to improve milk production and quality and to increase meat output. The chosen goal traits are based on genetic parameter estimates and correspond to fat-corrected part-lactation (60-day) milk production, 60- and 90-day, post-weaning growth rate for lambs and kids, respectively, and twinning rate. Adjustments are made for season, dam parity and type of birth. In the past thirty years, along with advancements in flock management and disease prevention, genetic evaluation procedures at ARI resulted in significant improvement of productive and reproductive characters for both species. Milk yield per animal per year increased by 70 kg for Chios ewes and by 170 kg for Damascus goats. Litter size at birth increased for mature females (from 1.8 to 2.1 lambs and from 1.9 to 2.2 kids). Weaning weight per litter increased by 1.3kg for lambs and by 3.5kg for kids. Additionally, a higher respective post-weaning growth rate of 100 and 80g per day was observed for lambs and kids.

Genome wide SNP -data for the assessment of genetic diversity in Swiss sheep breeds
Burren, A.[1], Signer-Hasler, H.[1], Tetens, J.[2], Kijas, J.[3], Flury, C.[1] and Drögemüller, C.[4], [1]Berne University of Applied Sciences, School of Agricultural, Forest and Food Science, Länggasse 85, 3052 Zollikofen, Switzerland, [2]Institute of Animal Breeding and Husbandry, Christian-Albrechts-University Kiel, Hermann-Rodewald-Str. 6, 24118 Kiel, Germany, [3]CSIRO Livestock Industries, 306 Carmody Road, St Lucia QLD 4067, Australia, [4]University of Berne, Institute of Genetics, Bremgartenstrasse 109, 3001 Bern, Switzerland; alexander.burren@bfh.ch

In the context of the global sheep hapmap project 24 individuals from each of 7 indigenous Swiss sheep breeds were genotyped using Illumina's OvineSNP50 BeadChip. These 168 animals were subjected to a detailed analysis of genetic diversity and selection signatures. After filtering, a total 45,508 informative SNPs (93%) remained for final analysis. The results for different k from cluster analysis are presented. These results support the already known proximity between populations such as Valais Blacknose and Valais Red or Swiss Mirror and Swiss White Alpine, respectively. For the derivation of possible selection signatures the genotypes were grouped according to obvious phenotypic differences (i.e. base color, specific color patterns of the head, and horned vs. polled). F_{ST}-values were derived. Beside others, peaks of the weighted F_{ST}-values were observed on chromosome 6 (specific color patterns of the head) and on chromosome 10 (horned vs. polled). The results allow a better understanding of the genetic diversity within and between Swiss sheep breeds and indicate or confirm associated genomic candidate regions for qualitative phenotypic differences.

The use of the reverse real-time PCR for the FecB gene detection in Merinolandschaf sheep
Czerneková, V., Milerski, M. and Kott, T., Institute of Animal Science, Přátelství 815, 104 00 Prague 10 – Uhříněves, Czech Republic; czernekova.vladimira@vuzv.cz

A highly prolific line consisting of the carriers of a major gene FecB (Booroola) was created within the Merinolandschaf breed of sheep. For a detection of the FecB gene carriers, a new method of genotypization based on reverse real-time polymerase chain-reaction (PCR) using new primers and Tag-Man probes was developed. The proposed method enables fast, accurate and specific detection of genetic variants of the FecB gene. The PCR is prepared in an environment of genomic DNA, Taq polymerase, primers and probes, the buffer and magnesium chloride. The amplification is carried out in the device for the real time PCR, when periods of denaturation (95 °C), and anelation with elongation (60 °C), are changing while measuring the fluorescence signal of marked probes from which genotypes are then identified. High impacts of prolific allele on litter size (+0.86 lamb in litter) and number of lambs weaned (+0.55 lamb) were confirmed. In 43% of cases the FecB carriers had triplets or more numerous litters. The effects of the prolific allele on lamb mortality and lamb liveweight at the 100 days of their age were not statistically significant. FecB carriers showed slightly lower longissimus muscle depth and backfat thickness measured by ultrasound. By the use of computer simulations, the economic effectiveness of prolific line was compared with the rest of Merinolandschaf population.

Analysis of melatonin receptor 1A gene polymorphisms in the population of dairy sheep in Slovakia
Vasickova, K., Margetin, M., Bauer, M., Apolen, D. and Vasicek, D., Animal Production Research Centre Nitra, Hlohovecka 2, 951 41 Luzianky – Nitra, Slovakia; apolen@cvzv.sk

Melatonin receptor 1A (MNTR1A) plays a key role in the control of the photoperiod-induced seasonality, which is mediated by the circadian levels of melatonin. The objective of this study was to investigate MNTR1A gene polymorphisms in synthetic population of dairy sheep (Improved Valachian × Lacaune crossbreds). One hundred and two blood samples were collected and genomic DNA was isolated using Maxwell 16. A part of exon 2 of the MNTR1A was amplified and PCR product was digested with Mnl I (SNP612) and Rsa I (SNP606) restriction endonuclease. Restriction digestion allowed the determination of three genotypes in each SNP with frequencies MM 0.73, Mm 0.25, mm 0.02 and CC 0.35, CT 0.51, TT 0.14 respectively. Possible association between polymorphism and reproductive cyclicity is under investigation.

Myofibrillar activity from two different muscles of Gentile di Puglia lambs
Nicastro, F.[1], De Marzo, D.[1], Facciolongo, A.M.[2], Nicastro, A.[3] and Toteda, F.[1], [1]Dipartimento di Produzione Animale, Università degli Studi di Bari, Via Amendola 165/A, 70126 Bari, Italy, [2]C.N.R., Istituto di Genetica Vegetale, Via Amendola 165/A, 70126 Bari, Italy, [3]Dipartimento di Scienze Mediche di Base, Università degli Studi di Bari, Piazza Giulio Cesare 1, 70100 Bari, Italy; nicastro@agr.uniba.it

Different studies suggest that the disparity in functional properties among various muscles or fiber types can be attributed mainly to myofibrillar proteins. The myosin is different in istochemical, biochemical, enzymatic properties and degradation on postmortem storage. Proteolytic enzyme activity participates in myofibrillar breakdown and continues to be active in muscle post-mortem influencing tenderization of meat. Age and type of muscle are two factors which are involved in the proteinases activity. In this study activity of cathepsin D and Warner- Bratzler Shear (W.B.S.), were measured in the longissimus lumborum (LL) and in the semimembranosus (SM)muscles of 24 lambs slaughtered in two different ages (70 and 110 days). Samples from each muscle were prepared in the lysing buffer (sodium citrate buffer, EDTA and Triton) and pH adjusted to 5. Haemoglobin was used as substrate at the 0.6 and 2.0 final concentrations (in %, w/v). The extraction of cathepsin D was also determined with Lowry method. Cathepsin D activity increased with substrate concentrations, reaching the maximum value in the LL muscle with 2% of haemoglobin (P<0.01) in sample of animals 70 days old. Values of the specific activity (37.12 vs 23.15 mg of tyrosine/g protein/h at 45 °C) and total activity (mg of tyrosine/g muscle/h at 45 °C) of cathepsin D showed the same trend observed above. These data are confirmed by the rheological (W.B.S.) values of the muscle LL compared to SM (2.21 vs 3.25 kg/cmsq; P<0.01) that correlate the amount of the cathepsin D with tenderization of the meat. Between the two ages the younger lambs showed better, but not significant, results for amount of cathepsin D and for shear values.

Monitoring of bluetongue disease on the sheep farm

Lacková, Z., Bíreš, J., Kočišová, A., Smitka, P. and Smaržik, M., University of Veterinary Medicine and Pharmacy, Clinic of Ruminants, Komenského 73, 041 81 Košice, Slovakia; lackova_z@azet.sk

The aim of this study during years from 2008 to 2011 was to gain information about the incidence and epidemiological situation of an economically important orbivirus transmissive disease of bluetongue in the sheep farm. Monitoring of BT including entomological survey, clinically-biochemical examinations and serological diagnostics was realized. Entomological survey consisted of capturing midges and their classification based on the characteristical marks. Clinical examination of animals was carried out monthly on the basis of general inspection of animals (clinical examination of the herd and infected animals). Biochemical examination of blood consisted of hematological profile (Hb, RBC, Htk, Le, differential blood picture), enzymatic activity (ALP, AST, GGT, CPK), concentration of total bilirubin, total protein, albumin, creatinin, total immunoglobulins, urea, beta-hydroxybutyrate and minerals (Ca, P, Fe, Cu, Zn). Serological examination for the detection of anti-BTV antibodies (vp 7 protein) took place in the State Veterinary Institute in Zvolen monthly using ELISA method (ID VET Kit for the detection of anti-VP7 antibodies by competitive ELISA). During years from 2008 to 2011 in the sheep farm totally 42,884 midges were captured. From this number complex Culicoides Obsoletus was 63.43% (27,202), complex C. Pulicaris was 5.15% (2,206) and complex C. Nubeculosus were 0.12% (52). From other Cullicoides spp. were 31.3% (13,424). During monitoring period we did not detect the presence of Culicoides imicola regarded as the main vector of bluetongue disease. Each year we noticed three peaks of midges occurence that were associated with climatic conditions suitable for life and development of the vector. Findings of clinical and biochemical examinations that were detected are not typically present during bluetongue disease and they pointed to other diseases. During serological diagnostics no anti-BTV antibodies were detected.

Relationship between adult feed intake and litter weight in mice selected for litter size

Rauw, W.M., INIA, Departamento de Mejora Genética Animal, Crta. de la Coruña, km. 7,5, 28040 Madrid, Spain; rauw.wendy@inia.es

Animal breeders have taken advantage of extensive variation in resource allocation systems when selecting for high production combined with a low (efficient) feed intake. Whereas most production traits are selected upwards (more meat, more milk, more eggs, more wool), traits that are related to the animals' energy balance, such as feed intake and body fatness, are selected downwards. The first because of the high costs associated with feed intake and the second because of the consumers' desire for edible lean with a minimum of visible fat. Whereas no apparent selection limit exists for traits selected upwards, the selection limit for traits selected downwards is apparent, namely at a value of zero. Several undesirable correlated effects of selection for high production efficiency have become apparent at the current production levels, many related to metabolic imbalance. Selection for increased leanness and improved food efficiency in pigs has resulted in a decreased voluntary food intake, which is particularly challenging for sows in lactation. The phenotypic correlation of growth intake and mature intake with maximum lactation food intake was investigated in a mouse model in 179 mice of a selection experiment for litter size at birth. Feed intake was measured from 3 weeks of age until 21 days in lactation. Females with a higher growth intake also had a higher mature intake. Females selected for high litter size with a higher growth and mature feed intake supported larger litter weights, but no significant relationship existed in the non-selected control line. It is suggested that lactating control line females eat to support a given litter size, while females selected for high litter size support the maximum litter weight that is allowed for by their intake capacity. Since lactating sows mobilize body reserves, the relationship of growth intake with lactation intake may be reflected in the relationship between growth intake and body condition.

Genetic parameters of reproduction traits in hungarian pig populations
Nagyné Kiszlinger, H.[1], Farkas, J.[1], Kövér, G.Y.[1], Malovrh, S.[2], Czakó, B.[1] and Nagy, I.[1], [1]Kaposvár University, Kaposvár Guba S. u. 40., 7400 Kaposvár, Hungary, [2]University of Ljubljana, Groblje 3., 1230 Domzale, Slovenia; nagy.istvan@ke.hu

Genetic parameters were estimated for the pregnancy length and the number of piglets born alive for the Hungarian Large White (HLW), Hungarian Landrace (HL) and their reciproc cross. The analysis was based on the national database from field tests (2001-2010) using bivariate animal models. Gestation length and number of piglets born alive records of the purebred and crossbred pigs were considered as separate traits. Thus HLW-HLW x HL and HL – HLW x HL were separately analyzed. Altogether four runs were performed. The data consisted of 167865, 56743 and 163980 records for HWL, HL and the cross, respectively and the total number of animals in the pedigree was 126340. Estimated heritabilities for the gestation length were 0.30 ± 0.02, 0.22 ± 0.04 and 0.25 ± 0.03, 0.25 ± 0.02 in HWL, HL and their cross, respectively. Estimates for permanent environment effect were 0.06 ± 0.01, 0.04 ± 0.02 and 0.07 ± 0.02, 0.07 ± 0.02, in HWL, HL and their cross, respectively. Estimated heritabilities for number of piglets born alive were 0.09 ± 0.03, 0.06 ± 0.04 and 0.07 ± 0.02, 0.07 ± 0.02 in HWL, HL and their cross, respectively. Estimates for permanent environmental effect were 0.06 ± 0.02, 0.06 ± 0.03 and 0.06 ± 0.02, 0.06 ± 0.02 in HWL, HL and the cross, respectively. The genetic correlations between purebreds and crossbreds were high, for pregnancy length they were 0.96 ± 0.02 and 0.81 ± 0.06 and for number of piglets born alive they were 0.82 ± 0.03 and 0.93 ± 0.04. Based on the results the use of the purebred breeding values are satisfactory for both traits.

Heritability of sow longevity and life time production in Finnish Large White and Landrace pigs
Sevón-Aimonen, M.-L.[1], Haltia, S.[2] and Uimari, P.[1], [1]MTT Agrifood Research Finland, Biotechnology and Food Research, Genetics Research, Myllytie 1, 31600 Jokioinen, Finland, [2]Figen Ltd., P.O. Box 40, 01301 Vantaa, Finland; marja-liisa.sevon-aimonen@mtt.fi

Replacement costs in piglet production can be decreased by increasing life time production of sows. Aim of this study was to estimate heritabilities of traits related to longevity and life time production. Studied traits were age at culling (AC), total number of parities (TNP), life timeproduction measured as total number of piglets born (LTNB), total number of piglets born alive (LTBA), total number of still born (LTSB), total number of weaned (LTW), total number of piglets died before weaning (LTPM), and proportion of non-productive days to total herd days from first mating to culling (HTNP). Data were obtained from Figen Ltd. containing records of 27,103 Landrace (LR) and 23,078 Large White (LW) sows born between years 2000 and 2006. (Co)variances were estimated using an animal model REML and DMU program package. The statistical model contained herd and birth month as fixed effects and dam, additive animal and error as random effects. Heritabilities for AC,TNP,LTNB,LTBA,LTSB, LTW, LTPM,HTNP in LR (LW) were 0.08, 0.08, 0.09, 0.09, 0.07, 0.08, 0.08 and 0.06 (0.11, 0.11, 0.12, 0.12, 0.12, 0.10, 0.16 and 0.06), respectively. Estimated heritabilities were higher in LW than in LR but were at a reasonable level in both breeds to be used in selection program. Results will be further used in developing genomic evaluation for longevity and life time production traits.

Development of new selection strategies to decrease piglet mortality
Brandt, H.[1] and Henne, H.[2], [1]Department of Animal Breeding and Genetics, Ludwigstr. 21 B, 35390 Giessen, Germany, [2]BHZP GmbH, An der Wassermühle 8, 21368 Dahlenburg-Ellringen, Germany; horst.r.brandt@agrar.uni-giessen.de

With the increased litter size in pigs up to 14 piglets total born per litter and more an increased piglet mortality is observed which could be explained by a possible negative genetic correlation between litter size and piglet mortality. On two large data sets with individual piglet birth weights (n=103,010 and n=15,407 piglets) and a third data set with detailed information about piglet losses (n=11,779 piglets) genetic parameters for litter quality traits and their correlation to litter size were estimated using animals models including maternal and litter effects. For piglet losses heritabilities between 8 and 15% were estimated while the average birth weight in kg or as subjective score from 1 to 4 show heritabilities of 25 and 20%, respectively. All litter quality traits show strong antagonistic genetic correlations to litter size at birth ranging from 0.25 to 0.32. New selection criteria on litter quality without measuring individual piglet birth weights are discussed which should be included in the breeding goal for reproduction traits in pigs.

Genetic of residual feed intake in growing pig: relationships with nitrogen and phosphorus excretion
Saintilan, R.[1], Mérour, I.[2], Brossard, L.[3], Tribout, T.[1], Dourmad, J.Y.[3], Sellier, P.[1], Bidanel, J.[2], Van Milgen, J.[3] and Gilbert, H.[1,4], [1]INRA, UMR 1313 GABI, 78350 Jouy-en-Josas, France, [2]IFIP, Institut du porc, 35650 Le Rheu, France, [3]INRA, UMR 1348 PEGASE, 35590 Saint Gilles, France, [4]INRA, UMR 444 LGC, 31320 Castanet-Tolosan, France; romain.saintilan@jouy.inra.fr

The aim of this study was to evaluate residual feed intake (RFI), which is the difference between observed daily feed intake (DFI) and predicted feed intake from production and maintenance requirements, as a selection criterion to improve feed efficiency in commercial pig breeds, and its potential to reduce the proportions of nitrogen (Nratio) and phosphorus (Pratio) excreted relatively to intake. Data were collected in French central test stations between 2000 and 2009 for four pig breeds with different selection objectives: Large White (LWD, n=10,694) and French Landrace (LR, n=6,470) dam breeds and Large White (LWS, n=2,342) and Piétrain (PP, n=2,807) sire breeds. The RFI equations were computed separately for each breed. Nratio and Pratio during the test period (35-110 kg) were estimated from feed intake, body composition and body weight gain. Genetic parameters were then computed. The RFI accounts for about 29% of the phenotypic variability of DFI for LWD, LWS and LR and 20% for PP. Heritability estimates for RFI range from 0.22±0.03 (LWD, LWS, LR) to 0.33±0.05 (PP). The RFI shows favorable genetic correlations with feed conversion ratio (FCR), Nratio and Pratio. Predicted responses to a 5 years selection program based on current selection indexes are an increase of RFI in dam breeds (+0.24 to +0.35 SD) and a moderate reduction in sire breeds (-0.06 to -0.11 SD). Moreover, Nratio and Pratio are expected to decrease in all breeds to varying extent according to the breed (-0.38 to -1.16 SD for Nratio; -0.38 to -0.98 SD for Pratio). New selection indexes using RFI rather than FCR as a selection criterion can improve feed efficiency, growth and carcass traits while reducing nitrogen and phosphorus excretion.

Stability of genetic parameters for average daily gain and lean meat % in Hungarian Landrace pigs
Nagy, I., Kövér, G.Y., Nagyné Kiszlinger, H., Czakó, B. and Farkas, J., Kaposvár University, 40 Guba S., 7400 Kaposvár, Hungary; nagy.istvan@ke.hu

Stability of genetic parameters and breeding values for average daily gain (ADG) and lean meat percentage (LMP) was analyzed for 13,2004 and 68,062 Hungarian Landarce pigs born between 1994 and 2004. The dataset was divided into 7 successive 5 year long periods (1: 1994-1998, 2: 1995-1999, 3: 1996-2000, 4: 1997-2001, 5: 1998-2002, 1999-2003 and 2000-2004) then, after selecting the appropriate part of the pedigree for these sub-datasets, genetic parameters and breeding values were estimated for ADG and LMP using REML and BLUP methods using bivariate animal models. In the applied models sex, year-month, herd, animal and random litter effects were considered. Estimated ADG heritabilities for periods 1 to 7 ranged between 0.16±0.01 and 0.19±0.01. Magnitudes of LMP heritabilities for the same periods showed higher variability (11±0.02 and 0.31±0.01). After breeding value estimation the adjacent datasets were merged pair wise (1-2, 2-3, 3-4, 4-5, 5-6, 6-7) using inner join. Thus in the merged datasets only the common records of the datasets representing the adjacent 5 year long periods were included. In these merged datasets each pig had two breeding values for ADG and LMP based on two different periods. Linear regression coefficients (and R) were calculated between the breeding values to characterize their stability. The estimated regression coefficients among the breeding values predicted for different year groups ranged between 0.73-0.86 and 0.70-1.24 for ADG and LMP, respectively. The obtained R values for ADG and LMP ranged between 0.83-0.90 and 0.75-0.91, respectively. It can be concluded that the breeding values showed moderately high stability for both traits.

Effect of MC4R genotype on growth, fat deposition, appetite and feed efficiency in Canadian pigs
Jafarikia, M.[1], Fortin, F.[2], Maignel, L.[1], Wyss, S.[1] and Sullivan, B.[1], [1]Canadian Centre for Swine Improvement, Ottawa, ON, Canada, [2]Centre de développement du porc du Québec, Québec City, QC, Canada; mohsen@ccsi.ca

The melanocortin-4 receptor (MC4R) gene plays a vital role in energy homeostasis. A missense G to A mutation in amino acid Asp298Asn of the MC4R gene has been shown in many studies to affect growth rate and fat deposition. The objectives of this study were to estimate the frequency of this mutation in major Canadian pig breeds and to investigate the effect on average daily gain (ADG), backfat thickness (BFT), average daily feed intake (ADFI) and feed conversion ratio (FCR). A total of 722 barrows, 162 females and 138 males evaluated at the Deschambault test station in Quebec over four trials were genotyped for MC4R. The numbers of pigs by breed (and G allele frequency) were 277 (0.33), 304 (0.77) and 441 (0.48) for Duroc, Landrace and Yorkshire, respectively. Animals entered the nursery at the test station at an average age of 17.9±4.5 days and average weight of 5.3±1.4 kg. Pigs were transferred to finishing pens at an average age of 69.2±7.0 days and average on-test weight of 29.4±7.1 kg. On average, animals spent 92.5±15.0 days on test and had an average off-test weight of 118.0±8.7 kg. The statistical model included fixed effects of MC4R genotype and the interaction of sex, breed and trial, as well as on-test age and weight as covariates. Pen within trial, farm of origin within trial and litter were included as random effects. MC4R genotype had a significant effect on ADG, BFT and ADFI (P<0.01) but not on FCR. The least square means by genotype showed that AA pigs had 30±1 g/day greater average daily gain, 0.6±0.2 mm thicker backfat and consumed 90±2 g/day more feed than their GG counterparts. Including MC4R genotype in BLUP evaluation would increase EBV accuracy and genetic progress in traits influenced by MC4R genotype. Direct use of MC4R genotypes may also have application to breed for increased appetite or higher fat without adversely affecting feed efficiency.

Global gene expression analysis of liver for androstenone and skatole production in the young boars

Neuhoff, C.[1], Pröll, M.[1], Große-Brinkhaus, C.[1], Frieden, L.[1], Becker, A.[2], Zimmer, A.[2], Tholen, E.[1], Looft, C.[1], Schellander, K.[1] and Cinar, M.U.[1], [1]University of Bonn, Institute of Animal Science, Animal Breeding and Husbandry group, Endenicher Allee 15, 53115 Bonn, Germany, [2]University of Bonn, Institute of Molecular Psychiatry, Sigmund Freud Str. 25, 53127 Bonn, Germany; christiane.neuhoff@itw.uni-bonn.de

Boar taint is the nasty odour or taste that can be evident during the cooking or eating of pig or pig products derived from non-castrated male pigs. Although boar taint has genetic basis, genetic background of boar taint formation is not fully understood. Therefore, the objective of this study was to identify differentially expressed genes and genetic pathways associated with divergent levels of androstenone and skatole in porcine liver. For this purpose, total RNA of liver tissue from 20 Pietrain × F2 cross boars with divergent androstenone and skatole level were isolated. Liver gene expressions pattern were produced using 20 Affymetrix GeneChip Porcine Array. The analysis of the microarrays showed that 107 genes differentially expressed in the comparison of high and low skatole, of which 49 were up and 58 down regulated in high level of skatole compared to the low level skatole. Two genes were identified in the high vs. low level androstenone comparison with FDR$\geq$0.8. Moreover, statistical analysis revealed that 197 genes differentially expressed in the androstenone and skatole interaction group in which 94 genes were down regulated and 103 genes were up regulated. 84 genes were found common for high vs. low skatole and for the interaction group. Two genes were found in common for high vs. low androstenone and the interaction group. The most relevant pathways for skatole production and for the interaction of androstenone and skatole were retinol-, fatty acid- and steroid metabolism pathways. Our study identified candidate genes which are potentially influence on pathways and in connection with the metabolism of androstenone and skatole production.

Slaughter traits of hybrid pigs as influenced by terminal sire line and gender

Lukic, B., Djurkin, I., Kusec, G., Radisic, Z. and Maltar, Z., Faculty of Agriculture, Department of Special Zootechnics, Kralja Petra Svacica 1d, 31 000 Osijek, Croatia; idurkin@pfos.hr

The present study was conducted on 90 pig carcasses (44 gilts and 46 barrows) originating from two PIC terminal sire lines in order to analyze carcass and meat quality traits and evaluate their suitability for further processing into dry-cured products. At the slaughter line and in laboratory following carcass and meat quality traits were determined: carcass weight, fat and muscle thickness according to 'TP' method, leanness, carcass lengths 'a' and 'b', ham length and its circumference, pH_{45} and pH_{24} in ham and LD muscle, drip loss, CIE-Lab, cooking loss and WBSF. The data was analyzed using analyses of variance (ANOVA) to test influence of terminal sire line and gender, as well as their interaction on the carcass and meat quality traits examined. Results show that the carcass traits as backfat thickness, muscle thickness, lean meat percentage and carcass length –'a' were affected by gender with significant differences (P<0.01), while the carcass length – 'b', ham length and ham circumference were affected both by terminal sire line and gender (P<0.01). Meat quality was influenced by sire line only in pH_{24} measured in LD muscle and CIE-a (P<0.01) while the WBSF was influenced by gender (P<0.05).

Analysis of relationships between fattening and slaughter performance of pigs and the level of IMF
Tyra, M. and Żak, G., National Research Institute of Animal Production, Department of Animal Genetics and Breeding, ul. Krakowska 1, 32-083 Balice, Poland; mtyra@izoo.krakow.pl

The aim of the study was to determine the level of basic fattening and slaughter traits (growth rate, level of meatiness and fatness, age at slaughter) on account of different levels of intramuscular fat that determine different sensory perceptions of consumers. Subjects were 4,430 gilts from pedigree farms, which were tested in performance stations. The breed composition of the animals was as follows (head): Polish Large White – 1,240, Polish Landrace – 2,083, Puławska – 104, Hampshire – 35, Duroc – 152, Pietrain – 208, line 990 – 608. Animals were kept in individual pens and fed standard diets. Intramuscular fat (IMF) content of the longissimus dorsi muscle was determined by Soxhlet using the SOXTHERM SOX 406 system (Gerhardt). The level of IMF served as a basis for dividing the test animals into three groups: below 2% (group I), between 2% and 3% (group II) and above 3% (group III). Animal breed had the highest and highly significant effect on the level of all traits analysed. As regards age at slaughter and carcass meat %, an interaction was found between animal breed and the group factor determined based on IMF level ($P \leq 0.001$). The factor expressed as IMF group had no effect on the level of analysed traits ($P > 0.05$). Therefore, the results of this analysis concerning the parameters obtained from live evaluation do not permit these data to be used in selection for improved IMF levels. The high rate of lean deposition in the modern breeds prevented genetic differences in the level of IMF to fully manifest themselves at a slaughter weight of about 100 kg. This unfavourable information leads one to look for other factors that determine variation of this trait.

Effect of genotype on n-3 and n-6 fatty acid content in pig meat and subcutaneous fat
Ribikauskiene, D.[1,2], Ribikauskas, V.[2], Stuoge, I.[2] and Macijauskiene, V.[3], [1]Lithuanian Endangered Farm Animal Breeders Association, R. Zebenkos 12, Baisogala, Radvilliskis distr., 82317, Lithuania, [2]Institute of Animal Science of Lithuanian University of Health Sciences, R. Zebenkos 12, Baisogala, Radvilliskis distr., 82317, Lithuania, [3]Siauliai University, Vilniaus 88, Siauliai, 76285, Lithuania; daiva@lgi.lt

The amounts of n-3 (C18:3, C22:6) and n-6 (C18:2, C20:3, C20:4, C22:4) fatty acids in the meat and subcutaneous fat of various pig breeds were defined. Three groups of different pig breeds were formed: group 1 comprised of Lithuanian White (commercial type), group 2 was made of Yorkshire and group 3 Landrace pigs. The content of fatty acids was analysed at the Analytical Laboratory of the Institute of Animal Science of LUHS. The subcutaneous fat of the Yorkshire breed was biologically more valuable. It contained, respectively, 0.6 and 4.3% ($P < 0.001$) more indispensable polyunsaturated fatty acids (PUFA) compared with the subcutaneous fat of Landrace and Lithuanian White (commercial type) pigs. However, the intramuscular fat in M. Longissimus dorsi of the Yorkshire breed had the lowest (12.3%) PUFA amount. The amount of PUFA in the lipids of Landrace pig meat was 3.6 and 3.9% ($P < 0,025$) higher than that in Lithuanian White (commercial type) and Yorkshire intramuscular fat. The Landrace breed showed a more favourable n-6/n-3 ratio than the Yorkshire and Lithuanian White, and it exceeded current nutritional recommendations. According to the total amount of saturated and polyunsaturated fatty acids in intramuscular fat of M. Longissimus dorsi and subcutaneous fat, Lithuanian White had more saturated ($P < 0.05 – P < 0.025$) and less polyunsaturated ($P < 0.025 – P < 0.005$) fatty acids than Yorkshire and Landrace breeds. The generalisation of the data indicated, that the meat of Landrace pigs is more valuable if compared with other breeds and the subcutaneous fat is more valuable than that of the Yorkshire pig breed.

Responses to divergent selection for residual feed intake in growing pigs, consequences on pork

Gilbert, H.[1,2], Bidanel, J.P.[1], Billon, Y.[3], Meteau, K.[4], Guillouet, P.[5], Noblet, J.[6], Sellier, P.[1], Gatellier, P.[7], Sayd, T.[7], Faure, J.[6] and Lebret, B.[6], [1]INRA, UMR1313, 78320 Jouy-en-Josas, France, [2]INRA, UMR444, 31320 Castanet-Tolosan, France, [3]INRA, UE367, 17700 Surgères, France, [4]INRA, UE1206, 17700 Surgères, France, [5]INRA, UE88, 86480 Rouillé, France, [6]INRA, UMR1348, 35590 Saint-Gilles, France, [7]INRA, UR370, 63122 Saint-Genès-Champanelle, France; helene.gilbert@toulouse.inra.fr

Data were collected during eight generations of divergent selection on residual feed intake (RFI) during growth in a Large White population. The selection criterion was measured on males fed *ad libitum* from 35 to 95 kg body weight (BW) (1,118 pigs). Growth, feed intake, carcass composition and meat quality were recorded on sibs slaughtered at 110±7 kg BW (1895 pigs). After eight generations, line difference for RFI was 2.0 genetic standard deviation (σ_g), i.e. 97 g/d. Correlated responses for daily feed intake and feed conversion ratio were 1.0 and 1.8 σ_g, respectively, and not significant for growth rate. Low-RFI pigs had higher lean meat content (1.5 σ_g), killing out percentage (1.0 σ_g) and loin weight (1.7 σ_g). Early generations of selection induced correlated responses on meat quality traits (ultimate pH and colour), from 1.2 to 2.1 σ_g. These differences did not increase in the last 3 generations. In the seventh generation of selection, right loins of 117 pigs from both lines were sampled. Low-RFI pigs had higher muscle glycolytic potential (+11.9 μmol/g, P=0.005) and lower intramuscular fat content (-0.22 g/100 g muscle, P=0.006) but lipid (TBARS) and protein (carbonyl group) oxidation levels of raw and cooked meat aged 4 days were not different between lines. Visual assessment of raw pork showed greater homogeneity and intensity of red colour and marbling in the high-RFI line (P<0.01). However, among sensory traits recorded on cooked meat, only flour sensation after mastication (P<0.05) and juiciness (P<0.10) were significantly different between lines (favourable in the high-RFI line).

SNP's in different genes involved in fat storage are not associated with boar taint in Belgian pigs

Schroyen, M.[1], Janssens, S.[1], Stinckens, A.[1], Brebels, M.[1], Bertolini, F.[1], Lamberigts, C.[1], Millet, S.[2], Aluwé, M.[2], De Brabander, D.[2] and Buys, N.[1], [1]KU Leuven, Biosystems, Livestock Genetics, Kasteelpark Arenberg 30, 3001 Heverlee, Belgium, [2]Institute for Agricultural and Fisheries Research (ILVO), Animal Sciences Unit, Scheldeweg 68, 9090 Melle, Belgium; martine.schroyen@biw.kuleuven.be

Boar taint represents an unpleasant odour and flavour in the cooked meat of some entire male pigs (5 to 8%). This offensive phenotype is mainly caused by two constituents, skatole and androstenone. Because levels of both compounds are elevated when sexual maturity is reached, castration of pigs eliminates boar taint, but castration without anaesthesia is no longer accepted because of animal welfare issues. A sustainable alternative is raising entire male pigs and genetically selecting against high levels of skatole and androstenone. Only, selection against high levels of androstenone in particular is challenging since a high genetic correlation between this boar taint component and reproduction-related hormones is observed. However, both skatole and androstenone are metabolised in the liver and stored in adipose tissue. Hence, to circumvent the co-selection of undesired reproduction characteristics, one can choose to focus genetic selection on the storage of both components in fat tissue. Therefore in our study, genes that are associated with fat percentage in the pigs' carcass, apolipoprotein M (APOM), leptin (LEP) and lipin-1 (LPIN1), were examined in relation to boar taint in pig breeds mostly used in crossbred in Belgian pig farms, such as Landrace, Large White and Piétrain pigs. The single nucleotide polymorphisms (SNP's) APOM-intron2-g.2289G>C, LEP-exon3-g.3469C>T and LPIN1-exon2-c.93C>T were genotyped using the Eco130I, HinfI and HincII PCR-RFLP method respectively. However, no associations between these SNP's and the presence of boar taint were observed, indicating that these SNP's cannot be of use in selection against boar taint in typical Belgian pig breeds.

LEP and LEPR gene polymorphisms in the populations of polish maternal breeds

Szyndler-Nędza, M., Tyra, M. and Piórkowska, K., National Research Institute of Animal Production, Department of Animal Genetics and Breeding, ul. Krakowska 1, 32-083 Balice, Poland; mszyndle@izoo.krakow.pl

Leptin (LEP) and its receptor (LEPR) play a key role in the regulation of energy metabolism in mammals. Numerous studies with pigs have shown significant differences in lean meat content and fatness of animals resulting from the polymorphism of the above genes. Through their effects on metabolism, appetite and fatness level of animals, these genes may indirectly regulate reproductive processes. The aim of the study was to analyse LEP and LEPR gene polymorphisms in the Polish population of Polish Large White (PLW) and Polish Landrace (PL) pigs. Subjects were 121 PL sows and 109 PLW sows. The maternal breeds differed in the frequency of LEP gene polymorphisms. In the PL breed, there were almost exclusively animals of LEP^{TT} genotype (frequency of >98%). Three polymorphic forms were found in the PLW breed, with most animals having LEP^{TT} genotype (>77%), followed by LEP^{CT} heterozygotes (>20%) and only one animal of LEP^{CC} genotype (1.83%). In the case of leptin receptor (LEPR) gene, a greater diversity of polymorphic forms was observed in both maternal breeds. Among PL pigs, most animals were LEPR^{BB} homozygotes (49.6%), followed by LEPR^{AB} heterozygotes (43.8%) and animals of LEPR^{AA} genotype (6.6%). Among PLW pigs, animals of LEPR^{AB} genotype formed the largest group (52.3%). Like in the PL breed, LEPR^{AA} homozygotes were by far the least frequent (5.5%).

Effect of EGF, AREG and LIF gene polymorphism on litter size in pigs

Mucha, A., Ropka-Molik, K., Piórkowska, K., Tyra, M. and Oczkowicz, M., National Research Institute of Animal Production, Department of Animal Genetics and Breeding, ul. Krakowska 1, 32-083 Balice, Poland; amucha@izoo.krakow.pl

The aim of the experiment was to use the DNA mutations in the EGF, AREG, LIF1 and LIF3 genes to determine associations between the genotype and litter size in Polish Large White and Landrace sows. Reproductive traits investigated were: number of piglets born alive (NBA) and number in 21 day (N21). The polymorphisms in EFG gene were genotyped by PCR methods. The polymorphism in AREG gene were genotyped by PCR-RFLP method, with specific primers and the restriction enzymes StyI. The relationships between the EGF, AREG, LIF1 and LIF3 genotypes and NBA and N21 were analysed. The polymorphisms in LIF1 and LIF3 genes were genotyped by PCR-SSCP and PCR-RFLP methods, the restriction enzyme used in this analyze was DraIII. Two different alleles of EGF, AREG, LIF1 and LIF3 gene were identified: alleles A (0.81) and B (0.19) of the EGF gene, alleles A1 (0.83) and A2 (0.17) of the AREG gene, alleles A (0.65) and B (0.35) of the LIF1 gene and alleles A (0.35) and B (0.65) of the LIF3 gene. In the first parity, statistically significant differences were found at the EGF gene locus in favour of sows of BB genotype compared to the other sows for the number of pigs born alive (NBA) and litter size on day 21 (N21) ($P<0.01$ and $P<0.05$); at the LIF1 gene locus for sows of AA compared to BB genotype for NBA ($P<0.01$); and at the LIF3 gene locus for sows of BB compared to AA and AB genotype for NBA ($P<0.01$ and $P<0.05$). In parities two to four, statistically significant differences were observed between sows of A1A1 and A1A2 genotype of the AREG gene for NBA ($P<0.05$) and between AA and BB genotypes of the LIF1 gene for NBA and N21 ($P<0.01$ and $P<0.05$). In parities five to eight, there were no statistically significant differences in NBA and N21 between sows of different genotypes of the analysed genes. This study was supported by the Polish Ministry of Science and Higher Education (Project no. NN311220938).

Changes in peripheral blood leukocytes in swine selected line for resistance to MPS

Shimazu, T.[1], Liu, S.Q.[1], Katayama, Y.[1], Li, M.H.[1], Sato, T.[1], Kitazawa, H.[1], Aso, H.[1], Katoh, K.[1], Suda, Y.[2], Sakuma, A.[3], Nakajo, M.[3] and Suzuki, K.[1], [1]Tohoku University, Sendai, 981-8555, Japan, [2]Miyagi University, Sendai, 982-0215, Japan, [3]Miyagi Prefecture Animal Industry Experiment Station, Iwadeyama-cho, Miyagi, 989-6445, Japan; tomoyuki-shimazu@m.tohoku.ac.jp

Mycoplasma pneumonia of swine (MPS), a highly prevalent chronic respiratory disease, is known to cause significant economic losses in the swine industry. One approach to reduce the risk of MPS is genetic improvement. In this study, we characterized the immunophenotype of a novel Landrace line genetically selected to reduce the incidence of pulmonary lesions caused by MPS. Twelve castrated males of the selected line and 12 control Landrace pigs were used. The mean body weight and age of the pigs at the start of the experiment were 65 kg and 16 weeks, respectively. Blood samples were collected by jugular venipuncture on days -14, -7, 0, 2, 7, and 14. MPS vaccine was injected twice intramuscularly on days -7 and 0 of blood collection. The samples were subjected to white blood cell (WBC) count analysis, phagocytosis assay, and cell population analysis (B, T, and myeloid cells). The mitogenic activity of the blood cells was also analyzed before and after the first sensitization. The statistical analyses were performed using the SAS MIXED procedure. We found significant changes in WBC count between the 2 lines. The MPS-resistant line showed reduction in B cell count and increase in the myeloid cell ratio after the MPS vaccination. In contrast, the control line showed opposite results. Furthermore, the percentage of CD4+ T cells significantly increased in the control line even though the total T cell ratio remained unchanged. Moreover, lymphocyte proliferation was significantly lower in the MPS-resistant line. These results suggest that the MPS-resistant phenotype is associated with differences in immunological profile and that the magnitude of MPS is certainly influenced by the host immunophenotype.

Correlated response of cytokines with selection for reduced MPS pulmonary lesions in Landrace pigs

Sato, T., Tohoku University, Graduate School of Agricultural Science, Amamiya 1-1, Sendai, 981-8555, Japan; b1am1116@s.tohoku.ac.jp

Mycoplasma pneumonia of swine (MPS) is responsible for great financial losses in the swine industry. Currently, antibiotic feed additives and vaccines are used as preventive measures. However, they are rendered ineffective because of food safety issues and emergence of drug-resistant bacteria. We used a novel swine line selected for reduced MPS pulmonary lesions through 5 generations. Several immune traits of the line have already been measured. In this study, we measured cytokine concentrations (IL10/IL13/ IL17, TNFa, and IFNg) and estimated their correlation with MPS lesions. Landrace pigs (approximately 600) from each generation were used. MPS vaccine was injected twice intramuscularly at 70 and 95 kg of body weight. Blood serum samples were collected at 105 kg body weight. The samples were analyzed for cytokine concentration by ELISA. Statistical analyses were performed using the VCE and PEST programs to estimate breeding values and the SAS GLM procedure for multiple comparison tests. The heritability of IL10, IL13, IL17, TNFa, and IFNg was 0.20±0.06, 0.12±0.06, 0.27±0.07, 0.20±0.10, and 0.05±0.03, respectively. Genetic correlations of IL17 and TNFa with pulmonary lesions in MPS were -0.86±0.13 and 0.69±0.29, respectively. Medium-to-high correlations indicated that the cytokines were related to MPS lesions. Through selection, the breeding values of IL17 and IFNg increased substantially and those of TNFa decreased. These results suggest that natural and cellular immunity are more important for the suppression of pulmonary lesions in MPS than humoral immunity, such as antibody response. These results and conclusions support those of previous studies on the MPS-selected line.

Selection of reduced MPS pulmonary lesions influences the production of soluble factors in blood

Katayama, Y.[1], Li, M.H.[1], Sato, T.[1], Liu, S.Q.[1], Shimazu, T.[1], Kitazawa, H.[1], Aso, H.[1], Katoh, K.[1], Suda, Y.[2], Sakuma, A.[3], Nakajo, M.[3] and Suzuki, K.[1], [1]Tohoku University, Graduate School of Agricultural Science, Sendai, 981-8555, Japan, [2]Miyagi University, Department of Food, Agriculture and Environment, Sendai, 982-0215, Japan, [3]Miyagi Prefecture Animal Industry Experiment Station, Iwadeyama-cho, Miyagi Prefecture, 989-6445, Japan; cronohousebox@yahoo.co.jp

Mycoplasma hyopneumoniae (Mhp), the primary pathogen of mycoplasma pneumonia of swine (MPS), is a major cause of economic losses in the swine industry. Breeding for disease resistance is beneficial for preventing the disease. Previously, we established a novel swine line with reduced MPS pulmonary lesions having a different immunophenotype from that of the control group. In this study, we further characterized the selected line by focusing on blood-soluble factors after sensitization with commercial MPS vaccine. The Landrace line, selected for resistance to MPS, and a control line (comprising 12 castrated males each) were used. The mean body weight and age of the pigs at the start of the experiment were 65 kg and 16 weeks, respectively. Blood samples were collected on days -14, -7, 0, 2, 7, and 14. The MPS vaccine was injected twice intramuscularly on days -7 and 0 of blood collection. The blood samples were analyzed for endocrine hormones (cortisol, insulin, leptin, and IGF-1), antigen-specific IgG, and cytokines (IL-10, 13, 17, IFNg, and TNFa) by ELISA or RIA. Cytokine mRNA expressions were analyzed by real-time PCR. Statistical analyses were performed using the SAS MIXED procedure. We found no differences in the endocrine hormone and antigen-specific IgG levels between the 2 lines after sensitization. However, IL-10 and IFNg concentrations increased after MPS vaccine inoculation and were significantly higher in the MPS-selected line, which was reflected in the mRNA levels. At the moment, we are conducting dry chemical analysis. These results will provide insights on the underlying mechanisms of MPS resistance and accelerate the establishment of a new Mhp infection-resistant pig line.

Greenhouse gases emissions and energy consumption in French sheep for meat farms (1987-2010)

Benoit, M.B.[1] and Dakpo, H.D.[2], [1]INRA, SAE2, Theix, 63122 F-St Genes-Champanelle, France, Metropolitan, [2]CERDI, Bd François Miterrand, 63000 Clermont-Ferrand, France, Metropolitan; marc.benoit@clermont.inra.fr

Livestock production is seen as one of the major contributor to GHG emissions. Focusing on French sheep for meat breeding systems, this study sheds light on the main factors that can influence greenhouse gases emissions (GHG: CO_2, CH_4, N_2O) and non-renewable energy consumption. These two indicators (GHG expressed in equivalent CO_2 and energy in Mega Joules) are studied through the life cycle assessment method, taking into account a large number of inputs such as feeding, fertilizers, fuel, equipment, etc. They are analyzed for a sample of 1180 observations (about 49 farms per year) over the period 1987-2010. A focus is made to identify specificities of farming systems located in plain or mountain areas, managed on conventional or organic way. Preliminary results show average GHG gross emissions of 32 kg CO_2 eq for 1 kg of carcass(CW) and average net emissions (after carbon sequestration in soils) of 28 kg CO_2 eq per CW. CH_4 represents 61% of the total emissions, CO_2 21% and N_2O 18%. In average, for each gas the main emission factor was respectively enteric fermentation (77%) for CH_4, feed (34%) for CO_2 and manure emissions on pasture (61%) for N_2O. For energy consumption, with an average of 80 MJ per CW the main factors are feeding (26% of the total), fuel (25%) and fertilizers (17%). There is a high variability between farms, for GHG emissions or energy consumption. For GHG emissions, the mean in the first quartile group is 24 kg CO_2 eq per CW and for the forth group it is about 42. For energy, they are respectively 56 and 108 MJ per CW. The main variable that explains the observed variability in the GHG emissions appears to be ewe productivity (number of lambs weaned per ewe per year). Variability of energy consumption is linked (inversely) to forage self-sufficiency. The trends observed over the 24 years will also be discussed in relation with farm functioning and technic performances.

Environmental impact of the pork supply chain depending on farm performances

Reckmann, K. and Krieter, J., Christian-Albrechts-University Kiel, Institute of Animal Breeding and Husbandry, Olshausenstraße 40, 24098 Kiel, Germany; kreckmann@tierzucht.uni-kiel.de

The aim of the study is to figure the effect of changes in farm performances on the environmental impact of pork production. The assessment was performed using data representing typical Northern German pork production in 2010. Further, data of pig farms with 25% highest and lowest efficiency in terms of net profit (economic success) was used. Data for the farms was gathered from an extension service. The database of the feed and slaughtering stage was composed of own data collection and of literature. The system boundaries of the Life Cycle Assessment cover feed production, pig housing and slaughtering. Infrastructure, packaging, retail and consumption were excluded. Results were expressed per '1 kg pork'. Three impact categories were considered: Global warming potential (GWP), Eutrophication potential (EP) and Acidification potential (AP), expressed in equivalents (eq). The average pork production results in a GWP of 3.62 kg CO_2-eq, an EP of 42 g PO_4-eq and an AP of 89 g SO_2-eq per kg pork. Feed production is the main contributor to GWP (80%), followed by pig housing (15%) and slaughtering. The largest part of Eutrophication is caused by pig housing (51%) and feed production (48%). In case of the AP, pig housing plays a key role with an amount of 72%, feed production is responsible for 27% of the AP. Over all, slaughtering has only a marginal share (1-5%) to the environmental impacts. A higher efficiency on farm level reduces the GWP (4%) as well as the EP (1%) and AP (5%). A lower efficiency results in a 1% increased GWP and EP, but in a 0.5% decreased AP. Different scenarios with varying parameters on farm level (e.g. litter size, daily weight gain, FCR) will be used for further calculations. Results will be compared to identify performance parameters with a high effect on the overall impacts. In order to illustrate the variation of these impacts, Monte Carlo methods will be applied additionally.

Biotic components of sustainability: assessing ecosystem services in livestock farming systems

Marie, M.[1,2] and Merchier, M.[1], [1]INRA, SAD-ASTER, 662 avenue Louis Buffet, 88500 Mirecourt, France, [2]Université de Lorraine, ENSAIA, 2 avenue de la Forêt de Haye, 54505 Vandoeuvre lès Nancy, France; michel.marie@mirecourt.inra.fr

In order to assess ecosystems services, a set of indicators have been designed and implemented in a dairy mixed farming system (MFS) and a grassland-based farming system (GFS). The support function is represented by an ecologic regulation area indicator and earthworm abundance, the regulation function by ground beetles abundance, bird community specialization index, and pollinator abundance, while the supply function is represented by the wild vegetal diversity. The percentage of the surface occupied by hedges, bushes, isolated trees, brooks or ponds was higher in GFS than in MFS (20.1 vs 9.6%), as well as the % of centers of plots closer than 50 m from an ecological regulation area (53% vs 26%), giving an indicator value of respectively 8.0 and 4.8. Earthworm abundance ($133/m^2$ in MFS and $205/m^2$ in GFS) was higher in permanent than in temporary meadows, and variable for different cultures. The indicator for carabid beetles is higher for MFS (6.2) than for GFS (1.8), the number of beetles trapped in one week being 3.7 for permanent meadows, and 61.5 for cultures. The total number of birds identified in 20 observation periods was 393 and 396 and the community specialization index 0.67 and 0.71 respectively for GFS and MFS, giving close scores for the 2 systems. The values are also close for the two systems for the floristic diversity of grassland (6.8 and 7.1 for MFS and GFS), but with lower scores for alfalfa-Orchardgrass than for temporary or permanent meadows. The diversity of the habitats of a farming system is a positive factor for the biodiversity, as evidenced by the link between cultures and ground beetles, or by the ecological regulation areas development. Propositions to improve the biotic resources of the surveyed farming systems are done on the basis of these observations.

Identification of on-farm innovations in sheep and beef systems

Ingrand, S.[1], Pailleux, J.Y.[1] and Devun, J.[2], [1]INRA, SAD, UMR1273 Metafort, 63122 Saint Genes Champanelle, France, [2]Institut de l'Elevage, UMT PASF, 9, Allée Pierre de Fermat, 63170 Aubière, France; stephane.ingrand@clermont.inra.fr

Innovation in agriculture should be seen differently by researchers and farmers. For a researcher, innovation is often seen as something new dealing with technology, genetics, facilities....as for a farmer, it is something dealing with the integration in the farming system of something new according to the system, and not obviously a novelty considering the scientific community. The present paper concerns on-farm innovation in beef cattle and meat sheep systems. 226 questionnaires were sent to experts of livestock farming systems and we obtained 50 answers. The 7 questions in the questionnaire were: (1) which are the domains concerned by changes/innovation in your region? (2) could you describe one example? (3) could you describe one practice dramatically different from the 'average' practices in your region? (4) could you describe some practices which do not fit with your recommendations? (5) what are the reasons in your opinion? (6) what are interesting unusual practices, from your point of view? (7) what about innovations in the future? The results are compared to the results of surveys carried out in private farms (24 meat sheep farms and 25 beef cattle farms). The main result is that very few 'innovative changes' observed in farms can be considered as novelties, but those changes are very numerous: 64 in sheep farms and 86 in beef farms, concerning herd, buildings, equipments, crops, energy, workforce. Sheep and beef cattle farmers are rather different according to what they consider as useful in innovations: easier work (respectively 67% and 32% of beef and sheep farmers), increase income (63% and 12%), improvement of work organisation (33% and 8%), improvement of social aspects (17% only for beef farmers).

A comparison of two milk production systems from an economic point of view

Kunz, P.L.[1], Frey, H.J.[2] and Hofstetter, P.[3], [1]Bern University of Applied Sciences, School of Agricultural, Forest and Food Sciences, Laenggasse 85, 3052 Zollikofen, Switzerland, [2]Vocational Education and Training Centre for Nature and Nutrition, Sennweidstrasse, 6276 Hohenrain, Switzerland, [3]Vocational Education and Training Centre for Nature and Nutrition, Chlosterbuel 28, 6170 Schuepfheim, Switzerland; peterkunz@bfh.ch

The aim of this study was to compare indoor feeding (IF) and pasture-based feeding (PF) systems in dairy farming. From 2008 to 2010, two separate herds were kept under the same conditions and within an equal agricultural area: 15.8±0.4 ha for the IF herd and 15.7±0.7 ha for the PF herd at the same location in Hohenrain, Switzerland. The IF herd consisted of 24 Swiss Dairy cows. These cows were kept in a free-stall barn and fed a mixed ration composed of corn silage, grass silage, and 1,094±150 kg of concentrates per cow and year. The PF herd consisted of 28 Swiss Dairy cows. This herd received during wintertime hay and 285±26 kg of concentrates per cow and year. From March to November, the cows were on pasture. Body weight of both herds was measured every four weeks, milk yield and composition 22 times per year. For statistical analysis the Equal-Variance t-Test (normal distribution) or the Wilcoxon Rank-Sum test were used. Energy corrected milk yield was considerably higher for the IF cows: 9,607±2,304 kg, compared to the PF cows: 5,681±1,233) kg per lactation (P<0.01). The IF herd produced 202'300 kg energy corrected milk (ECM) per year, whereas the PF herd produced 161,900 kg ECM per year. Total costs were SFr 34,759 higher for the IF herd as compared to the PF herd. This was due to higher expenditures for conservation of feeds, concentrates, veterinary services, seeds, fertiliser, wages and machines. Working time was 285 hours higher for the IF operation. This resulted in an hourly wage of SFr. 7.90 for the IF farm and SFR. 13.20 for the PF farm. Within the conditions given in Switzerland and their related costs, it seems to be economically more interesting to reduce costs of production than to increase milk yield of cows.

Economic and grazing resources analysis of extensive livestock farming systems (dehesas) in Spain
Gaspar, P., Escribano, M., Mesías, F.J., Pulido, A.F. and Escribano, A.J., Universidad de Extremadura, Escuela de Ingenierías Agrarias, Ctra. Cáceres s/n, 06071 Badajoz, Spain; pgaspar@unex.es

The main commercial value of dehesa systems today is extensive livestock farming. The present work analyzes the economics of livestock farming over two years in a sample of dehesas in Extremadura. Dehesa is an agroforestry system (a combination of grazing, woodland and cropping lands) whose main production consists of extensive livestock farming. Principal Components and Cluster analyses were used to establish a dehesa farm typology, and a study was made of the technical and economic characteristics linking livestock farming with the use of the territory's grazing resources. Four groups of dehesa farms were distinguished by the statistical analyses applied to 52 technical and economic indicators. Overall livestock production was found to obtain 71.18% of its energy needs from the environment. By groups, the grazing resources covered 76.04% of the needs of livestock in Type 1, 74.07% in Type 2, 60.01% in Type 3, and 59.18% in the mixed production farms with Iberian pig. By species, cattle obtained 82.09% of their energy needs from grazing, small ruminants 76.87%, sows 48.70%, and acorn-fed pigs 79.67%. Finally, an economic evaluation was made of the feed requirements of these groups that were covered by grazing or by supplements. The importance of this work lies in the analysis of the use of grazing resources and the income generated by these. This procedure improves the understanding about the technical and economic performance of extensive livestock systems, since usage and market value of natural resources is not often taken into consideration.

Farmed fish homeostasis and neurohormonal status under integrated production conditions: a review
Papoutsoglou, S.E., Agricultural University of Athens, Animal Science and Aquaculture, Iera Odos 75 Votanikos Athens, 11855, Greece; sof@aua.gr

Farmed Fish Homeostasis and Neurohormonal Status under Integrated Production Conditions: A review Sofronios E. Papoutsoglou Abstract Several factors characterizing our presentday world are conducing to the increasing application of integrated methods/techniques within the now well-established field of fish farming production, this also involving the sustainable utilization of water reserves in a world of dwindling water resources. These factors are, in particular, the ever mounting global demands for farmed fish production, the destructive human impact on the aquatic environment and various global climate changes, According to well-documented data that continue to accumulate, the most critical consideration during the rearing period of farmed fish is scrupulous maintenance of their homeostasis and, particularly, of their neurohormonal status. This comprises the key biological issue which, via optimal management, may ensure achievement of the highest levels of fish farming goals. The aim of the present review is to emphasize the importance of attaining insight into specific essential factors. One of these concerns the interactions between the homeostatic mechanisms (e.g. neurohormonal functions-interactions/responses to external or internal or of both origin stimuli) of farmed fish. The other relates to the status of their rearing environment as regards the managing of the biological, chemical, physical and hydrological parameters during the application of different production systems under integrated conditions. It is thus to be concluded that the achievement of optimum results is dependent upon the degree to which the application of fish production systems is carried out within the framework of a concept of integration, the most indispensable component of this framework being maintenance of the highest possible level of homeostasis in farmed fish.

Development of integrated aquaculture in zone Z1, Zadar County, Croatia

Župan, I.[1], Peharda, M.[2], Bavčević, L.[3], Šarić, T.[1] and Kanski, D.[1], [1]University of Zadar, Department of Ecology, Agronomy and Aquaculture, M. Pavlinovica, 23000 Zadar, Croatia, [2] Institute of Oceanography and Fisheries, Š.I.Meštrovića, 21000 Split, Croatia, [3]Agricultural Chamber, Zadar, 23000 Zadar, Croatia; tosaric@unizd.hr

Integrated aquaculture (IMTA) is recognized worldwide as a solution for mitigation of environmental impact and economical and social sustainability. Several studies in the Mediterranean suggested a great potential for development of sustainable aquaculture under IMTA principles. Almost 50% of Croatian marine fish farm production is situated in Zadar County which makes it the main aquaculture area in the Croatia. In planning of further development of this county, integrated coastal management principles were used and zones (Z1 – Z4) for aquaculture were proclaimed. Zone Z1 was selected as area where aquaculture is a top priority and where integrated aquaculture combining fish and bivalves is planned. Study conducted in Z1 zone demonstrated that mussel (Mytilus galloprovincialis Lamarck, 1819) has the same production cycle as in traditional areas such as Mali Ston Bay, coupled with adequate meat quality. Further more, preliminary data show that cca 5kg of mussel spat per 1 meter of submerged rope can be collected in vicinity of the fish farm installations. Increased condition indices on locations close to fish farms in comparison to natural populations were noticed also for Noah's ark (Arca noae Linnaeus, 1758), while preliminary research shows great potential for production of two valuable bivalves – oyster *Ostrea edulis* Linnaeus, 1758 and pectinid *Mimachlamys varia* (Linnaeus, 1758). All producers in the area have shown interest for diversification of production trough shellfish farming. Two EU projects (IPA, FP7) are under evaluation for examing all aspects of the potential for IMTA in Z1 zone. These projects will evaluate production capacity and health standards for bivalves. The final goal is to classify zone Z1 as a shellfish production area according to EU health standards.

Effect of ruminal cobalt-EDTA infusion on milk fat composition and mammary lipogenic gene expression

Vilkki, J.[1], Viitala, S.[1], Leskinen, H.[1], Taponen, J.[2], Bernard, L.[3] and Shingfield, K.J.[1],[1]MTT Agrifood Research Finland, Jokioinen, 31600, Finland, [2]University of Helsinki, Helsinki, 00014, Finland, [3]INRA, Saint Genès-Champanelle, 63122, France; johanna.vilkki@mtt.fi

Intravenous or ruminal administration of cobalt as Co-EDTA or Co-acetate is known to alter bovine milk fat composition but the underlying mechanism is not known. Four Finnish Ayrshire cows in mid-lactation fitted with rumen cannulae were used in a 4×4 Latin square to examine the effects of ruminal infusions of incremental amounts of Co-EDTA on milk fat composition and mammary lipogenic gene expression. Each experimental period comprised a 18 d continuous infusion of Co-EDTA in the rumen supplying 0, 1.5, 3.0 or 4.5 g Co/d followed by a 10 d washout. Mammary tissue samples for gene expression analysis were obtained on d 16 of each period. Administration of Co-EDTA had no effect ($P>0.05$) on dry matter intake, milk production or milk fat secretion, but induced dose-dependent changes in milk fatty acid composition. Changes in milk fat composition to Co-EDTA were evident within 3 d of infusion and characterized by linear or quadratic decreases ($P<0.05$) in the concentration of fatty acids containing a cis-9 double bond, an increase in 4:0 concentration and linear decreases in milk 8:0, 10:0, 12:0 and 14:0 content. However, linear or quadratic decreases ($P<0.05$) in milk fat cis-9 14:1/14:0, cis-9 16:1/16:0, cis-9 18:1/18:0 and cis-9, trans-11 18:2/trans-11 18:1 concentration ratios to Co-EDTA were not accompanied by alterations in mammary stearoyl-CoA desaturase (SCD1) mRNA abundance. In contrast, changes were detected in the expression of several genes involved in de novo fatty acid synthesis and triglyceride synthesis. In conclusion, changes in milk fat composition to ruminal infusions of Co-EDTA do not arise from down-regulation of the SCD1 gene in the mammary gland, but rather from other mechanisms leading to the inhibition of enzyme function and correlated changes in the regulation of the lipogenic gene networks.

Follistatin, muscle development, puberty and fertility in Merino ewe lambs

Rosales, C.[1,2,3], Ferguson, M.[1,3,4], Macleay, C.[1,3], Briegel, J.[1,3], Hedger, M.[5], Martin, G.[2] and Thompson, A.[1,3,4], [1]The department of agriculture and food of Western Australia, Sheep Industry, 3 Baron-Hay Court, Kensington, 6151 WA, Australia, [2]The University of Western Australia, Institute of Agriculture, 35 Stirling Highway, Crawley, 6009 WA, Australia, [3]CRC for Sheep Industry Innovation and the University of New England, Sheep Industry, Armidale, 2351 NSW, Australia, [4]Murdoch University, School of Veterinary and Biomedical Sciences, 90 South Street, Murdoch, 6150 WA, Australia, [5]Monash University, Monash Institute of Medical Research, 27-31 Wright Street, Clayton, 3168 Vic, Australia; nieto_cesar@hotmail.com

It has been thought that females must accumulate a certain mass of body fat before they can go through puberty and maintain fertility but this theory is being challenged because muscle development also seems to be linked to fecundity. Body fat and reproduction are linked physiologically by leptin, but a link among increased muscling and reproduction has not been explored. One possible link is follistatin (FS), a key player in the regulation of both muscle development and follicle-stimulating hormone secretion. We assessed the relationships between breeding values for muscling, the onset of puberty and fertility, and circulating FS concentrations in Merino ewe lambs (n=136). To detect onset of puberty, testosterone-treated wethers were run with the lambs from 6 to 8 months of age and then replaced by entire rams. Blood FS concentrations determined by RIA decreased as puberty approached and increased as conception approached. The proportion of ewe lambs that attained puberty decreased as FS values increased (P<0.05), but FS concentration had no effect on fertility. We conclude that FS secretion is related to the onset of puberty but not fertility of Merino ewe lambs. The data presented do not imply a cause-effect relationship; however, further research is necessary to clarify the effect of circulating FS on reproductive traits in Merino ewe lambs.

Maternal undernutrition during midpregnancy affects muscle cellular characteristics of lambs

Kuran, M.[1], Sen, U.[1], Sirin, E.[2] and Ulutas, Z.[2], [1]Ondokuz Mayis University, Animal Science, Kurupelit, 55139, Samsun, Turkey, [2]GOP University, Animal Science, Taslicifilik, 60250 Tokat, Turkey; mkuran@omu.edu.tr

It has been reported that maternal nutrition during mid-pregnancy affects birth weight and muscular development of lambs. The aim of this study was to investigate the effect of maternal undernutrition during mid-pregnancy on muscle cellular characteristics in lambs. Mature Karayaka ewes were fed between day 30 and day 80 of pregnancy as follows; 100% of daily requirements (control group, C; n=9) or 50% of daily requirements (under-fed, UF; n=15) or 175% of daily requirements (over-fed, OF; n=7). Following lambing, lambs were weaned at day 90 of age and fattened until slaughter at 150 days. Samples from the longissimus dorsi (LD) and semitendinosus (ST) muscles were excised from the right side of the carcasses and frozen in liquid nitrogen immediately after slaughter. Total protein content in muscles samples was determined on the basis of total nitrogen content. Myofibrillar and sarcoplasmic protein contents in muscles samples were determined by biochemical analyses. The amount of connective tissue protein was calculated by subtracting the amounts of myofibrillar and sarcoplasmic proteins from the amount of total protein. Total cellular DNA content in muscles samples was determined by spectrophotometry. Daily weight gain during fattening period of lambs in UF group were lower than C and OF groups (P<0.05). Lambs in OF group had a lower content of total protein and connective tissue protein than those in C and UF groups (P<0.05) in LD and ST muscles. Protein to DNA ratio in the samples from lambs in OF group were lower than C and UF groups (P<0.05). There were no significant differences between groups in terms of myofibrillar and sarcoplasmic proteins in both muscles. These results show that maternal undernutrition during mid-pregnancy results in altered muscle cellular characteristics in lambs. The study was supported by TUBITAK (TBAG-U/148).

Effects of mannan oligosaccharide on blood biochemical parameters of broiler chickens

Karkoodi, K.[1], Mahmoodi, Z.[1] and Solati, A.A.[2], [1]Department of Animal Science, Saveh Branch, Islamic Azad University, Saveh, Iran, [2]Department of Veterinary Science, Saveh Branch, Islamic Azad University, Saveh, Iran; karkoodi@yahoo.com

Effects of mannan oligosaccharide (MOS) supplementation in the diet of broiler chickens on blood biochemical parameters were examined. A number of eighty Ross-308 day-old male broiler chicks were randomly allocated to two treatment diets (four replicates with ten sub-replicates). Broilers were fed corn-soybean meal based diets without MOS (control group) and control diet plus 2 g/kg MOS (MOS group). At the end of the experiment (Day 42) blood samples were collected from three chickens of each replicate. Results of the present study indicated that the diet containing MOS significantly decreased Alanine aminotransferase activity, Glutathione peroxidase, Low Density Lipoprotein concentration, and leukocytes numbers ($P<0.05$). Supplementation of MOS in the diet significantly increased Alkaline Phosphatase, Aspartate aminotransferase, Albumin, and Hemoglobin concentration ($P<0.05$). No significant difference in Superoxide dismutase, total protein, triglycerides andHigh Density Lipoprotein concentrations, erythrocyte, monocyte, lymphocyte, neutrophil, eosinophil numbers, and haematocrit counts were noticed between the MOS treatment and control.

Monoclonal antibodies as a tool for evaluation of the protein changes during bull sperm capacitation

Cupperová, P., Michalková, K., Simon, M., Horovská, Ľ. and Antalíková, J., Institute of Animal Biochemistry and Genetics, Slovak Academy of Sciences, Immunogenetics, Moyzesova 61, 90028, Ivanka pri Dunaji, Slovakia; petra.cupperova@savba.sk

Capacitation is an important physiological pre-requisite before the sperm cell can acrosome react and fertilize an oocyte. This process involves an increase in membrane fluidity, cholesterol efflux, ion fluxes, reorganization of membrane proteins and increased tyrosine phosphorylation of proteins. In order to understand the molecular basis underlying capacitation, it is very important to characterize proteins and phosphoproteins involved in signal transduction pathways. In our study a set of 34 anti-sperm monoclonal antibodies were used to detect changes of bovine sperm surface proteins reaction patterns after capacitation induced by TL Sperm cell capacitation medium (Minitube). Monoclonal antibodies (mAbs) were produced by hybridoma cell lines obtained after intrasplenic immunization of BALB/c mice with intact bull sperm. The changes in the reaction patterns were evaluated by indirect immunofluorescence, PAGE-SDS and two-dimensional gel electrophoresis followed by western blot analysis with anti-sperm mAbs and anti-phosphotyrosine α-PY antibody. In the indirect immunofluorescence a two-fold increase of sperm reactivity with seven mAbs was detected after capacitation, while another three mAbs detected a decreased reactivity (approximately twice). Western blot analysis showed changes in molecular weight of proteins detected by five mAbs. Two-dimensional PAGE and western blot analysis detected by four mAbs showed a shift of the reactive region to less acidic region after capacitation. Further analysis of detected changes in protein composition after capacitation can contribute to a better understanding of the basic biochemical processes of fertilization. This work was supported by grants VEGA 2/0006/12 and APVV-0137-10.

The effect of maternal undernutrition during mid-gestation on some placental characteristics in ewes
Sen, U.[1], Sirin, E.[2] and Kuran, M.[1], [1]Ondokuz Mayis University, Agriculture Faculty, Department of Animal Science, 55139, Samsun, Turkey, [2]Gaziosmanpasa University, Agriculture Faculty, Department of Animal Science, 60250, Tokat, Turkey; ugur.sen@omu.edu.tr

There is evidence that maternal undernutrition at specific points of gestation can affect the development of the ovine placenta and the fetus. The aim of this study was to determine the effects of maternal undernutrition during mid-gestation on some placental characteristics in ewes. Estrus of 3 to 5 years old Karayaka breed ewes was synchronized and mating was monitored to determine the day 0 of gestation. The ewes had similar body weights (47.8±0.7 kg) and eye muscle values (loin thickness; 20.9±1.0 mm and fat thickness; 4.7±0.5 mm) at mating. The ewes were allocated to two treatment groups at day 30 of gestation; under-fed (UF; n=12) and well-fed (WF; n=13) groups. The ewes in UF group were fed with a diet to provide 50% of their daily requirement from day 30 to day 80 of gestation and 100% of their daily requirement during the rest of the gestation period. The ewes in WF group were fed at least 100% of their daily requirement throughout gestation. Results showed that the singleton bearing ewes in UF group had lower (P<0.05) placental weight (354.1 vs. 378.3 g), average cotyledon weight (1.50 vs. 1.82 g) and lamb birth weight (3.8 vs. 4.2 kg) compared to the singleton bearing ewes in WF group. Additionally, the pattern of weight gain/loss was significantly different (P<0.05) between the two groups. Ewes in UF group lost body weight progressively from day 30 of gestation until day 80. The results of the present study show that maternal undernutrition during mid-gestation may cause an insufficient placental development and hence alter fetal development resulting in a reduced birth weight from singleton pregnancies. This study was financed by TUBITAK (TBAG-U/148).

Tissue accumulation and urinary excretion of chromium in lambs supplemented with chromium picolinate
Dallago, B.S.L.[1], Lima, B.A.F.[1], Mustafa, V.[1], Mcmanus, C.[2], Paim, T.P.[3], Campeche, A.[3], Gomes, E.[3] and Louvandini, H.[3], [1]Universidade de Brasília-UnB, Faculdade de Agronomia e Medicina Veterinária, Campus Universitário Darcy Ribeiro – ICC Sul, Brasília/DF, 71910-900, Brazil, [2]Universidade Federal do Rio Grande do Sul, Departamento de Zootecnia, Av. Bento Gonçalves 7712, Porto Alegre/RS, 91540-000, Brazil, [3]Universidade de São Paulo, Centro de Energia Nuclear na Agricultura-CENA, Av. Centenário, n° 303, Caixa Postal 96, Piracicaba/SP, Brazil., 13400-970, Brazil; dallago@unb.br

Chromium concentrations in liver, kidney, spleen, heart, lymph node, skeletal muscle, bone, testis and urine were measured to trace the biodistribution and bioaccumulation of Cr after oral supplementation with CrPic. Twenty-four Santa Inês lambs were treated with four different concentrations of CrPic: placebo, 0.250, 0.375 and 0.500 mg CrPic/animal/day for 84 days. The basal diet consisted of Panicum maximum cv Massai hay and concentrate. Cr concentrations were measured by ICP-MS using 52Cr as collected mass. There was a positive linear relationship between dose administered and the accumulation of mineral in the heart (P=0.0162), lung (P=0.0049) and testis (P=0.04). Urinary excretion of chromium occurred in a time and dose-dependent manner, so the longer or more dietary Cr provided, the greater excretion of the mineral. Thus, there is a risk of bioaccumulation and biomagnification due to Cr offered in the CrPic form.

Transcription of IL-6 and IFN-γ in chicken lymphocytes stimulated with synbiotics *in vitro*

Slawinska, A.[1], Siwek, M.[1], Brzezinska, J.[1], Zylinska, J.[2], Bluyssen, H.[3], Bardowski, J.[2] and Bednarczyk, M.[1], [1]University of Technology and Life Sciences, Mazowiecka 28, 84-085 Bydgoszcz, Poland, [2]Institute of Biochemistry and Biophysics, Pawinskiego 5a, 02-106 Warszawa, Poland, [3]University of Adam Mickiewicz, Umultowska 89, 61-614 Poznan, Poland; siwek@utp.edu.pl

Synbiotic is a substance obtained by combining prebiotic (e.g. raffinose family oligosaccharides RFO) and probiotic (e.g. LAB – lactic acid bacteria). Synbiotic supplements in chicken diet stimulate gut microflora and immune system through gut-associated lymphoid tissue. The aim of the presented study was to assess the reaction of the chicken lymphocytes to *in vitro* stimulation with synbiotics on the molecular level. Chicken lymphocytes were isolated from spleens. In the next step lymphocytes were separated by density gradient centrifugation. Lymphocyte culturing was performed in 24-well plates by applying 1 ml of cell suspension containing 10^7 viable cells and 0.5 ml of a substance that stimulated an immune response. Lymphocytes were stimulated for 3, 6, 15 and 18 hrs. Five experimental groups were defined for *in vitro* culture: (1) Synbiotic 1 (RFO + Lactococcus lactic subsp. lactis); (2) Synbiotic 2 (RFO + Lactococcus lactic subsp. cremoris); (3) Synbiotic 3 (commercial synbiotic Duolac); (4) prebiotic (RFO); and (5) cell culture medium (negative control). Impact of the pre-/ and synbiotics on the chicken lymphocytes was assessed by qRT-PCR. Two genes coding chicken cytokines were used as target genes: IL-6 (gen ID 395337) and IFN-γ (gen ID 396054). Expression level was defined based on relative quantification of the target gene compared to the reference gene (ACTB, gen ID 396526). Preliminary results point towards Synbiotic 1 and Synbiotic 2 as the best stimulators of the immune system genes during *in vitro* culturing of chicken lymphocytes. This research project was supported by grant No. N N311 623938 from the National Science Centre, Poland.

The role of *Escherichia coli* in the pathogenesis of coliform mastitis in sows

Kemper, N.[1], Gerjets, I.[2], Looft, H.[3] and Traulsen, I.[2], [1]Martin-Luther-University, Institute of Agricultural and Nutritional Science, Animal Hygiene, Theodor-Lieser-Str.11, 06120 Halle, Germany, [2]Christian-Albrechts-University, Institute of Animal Breeding and Husbandry, Olshausenstr. 40, 24098 Kiel, Germany, [3]PIC Germany GmbH, Ratsteich 31, 24837 Schleswig, Germany; nicole.kemper@landw.uni-halle.de

Coliform mastitis (CM) as one main symptom of puerperal disorders in sows subsumed under the term Postpartum Dysgalactia Syndrome (PDS) affects both the sow and the piglets seriously. Even though it is a multifactorial disease, the causative agents are bacteria, and *Escherichia coli* has often been isolated from diseased animals. However, in previous studies, the isolated strains have not been further investigated for their virulence-associated genes. Bacteriological analysis of *E. coli* was performed from milk samples of five farms. 1,271 *E. coli* isolates from milk samples of 979 healthy sows, and 1,132 isolates from 1,026 diseased sows were identified. These isolates were further investigated with multiplex PCR for 27 virulence genes. SAS and R were used for statistical analysis and to generate heat maps to illustrate possible correlations. *E. coli* was found in 70.6% of the milk samples of CM-affected and in 77.9% of the milk samples of non-infected sows. 1,132 *E. coli* isolates from CM-positive samples and 1,271 isolates from CM-negative samples were further examined. Both in isolates from healthy and diseased animals, the median number of virulence genes was two. Four virulence genes (hra, chuA, iroN, kpsMTII) occurred significantly more frequently in isolates of diseased animals. However, no specific virulence gene profiles for isolates from either diseased or healthy sows were detected. The association between farm and the occurrence of virulence genes was significant, indicating farm specific *E. coli* isolates. Futhermore, seasonal effects were detected. In conclusion, these results support the theory that any given *E. coli* strain can cause CM in sows, if further adverse factors are present.

Chewing behaviour of pregnant and lactating ewes fed long or chopped grass silage

Helander, C.[1], Nadeau, E.[1] and Nørgaard, P.[2], [1]Swedish University of Agricultural Sciences, Department of Animal Environment and Health, P.O. Box 234, 532 23 Skara, Sweden, [2]University of Copenhagen, Faculty of Life Sciences, Department of Basic Animal and Veterinary Sciences, Grønnegårdsvej 3, 1870 Fredriksberg C, Denmark; Carl.Helander@slu.se

The objective was to study the effects of silage particle length and feeding strategy on feed intake and chewing behaviour in pregnant and lactating ewes. In each of two experiments (Exp. 1 and Exp. 2), seven ewes were assigned to one of three dietary treatments: (1) unchopped grass silage *ad libitum* and 0.8 kg concentrate daily, fed separately (US); (2) chopped silage *ad libitum* and 0.8 kg concentrate daily, fed separately (CS); or (3) chopped silage mixed with concentrate *ad libitum* (CM). The silages averaged 10.9 and 11.4 MJ metabolisable energy kg^{-1} DM and 580 and 483 g NDF /kg DM in Exp. 1 and Exp. 2, respectively. The intake was higher in lactation than in pregnancy for all treatments in both experiments, but the difference was greater in Exp. 2 (P<0.001). In Exp. 1, the daily silage DM intake in late pregnancy was 2.0, 1.8 and 1.8 kg and in lactation 2.5, 2.4 and 2.7 kg for US, CS and CM, respectively. In Exp. 2, the daily silage DM intake in late pregnancy was 1.8, 1.9 and 1.9 kg and in lactation 3.4, 3.3 and 3.4 kg for US, CS and CM, respectively. Thus, intake was not affected by silage particle length or feeding strategy. Chopping silage increased effective rumination time kg^{-1} DM intake by 31 and 59% in Exp. 1 (156 vs. 119 min, P<0.05) and Exp. 2 (154 vs. 97 min, P<0.01), respectively. Effective eating time (min/day) was 32% shorter in CS and CM compared with US during pregnancy in Exp. 1 (264 vs. 349 min, P<0.01). In Exp. 2, the effective eating time (min/day) was 33% shorter in CS and CM compared with US (262 vs. 349 min, P<0.05). Conclusively, effective rumination time in relation to feed ingested was increased due to chopping of grass silage in both experiments, whereas effective eating time per day was decreased by chopping silage.

Behavioural differences in pigs raised in outdoor and indoor systems

Juska, R., Juskiene, V. and Leikus, R., Lithuanian University of Health Sciences, Institute of Animal Science, R. Zebenkos 12, 82317, Lithuania; violeta@lgi.lt

The growth performance and behaviour of intensively raised pigs were compared under outdoor and indoor rearing systems. Three groups of Norwegian Landrace and Norwegian Landrace × Pietrain crossbred pig, analogous by parentage, age, weight, condition score and gender were used in the study. Each group was formed of 14 animals. The pigs were housed in either indoor pens (ID) or outdoor enclosures (OD-1 and OD-2). The pigs in all groups were given compound feed *ad libitum*. The diet of OD-2 group was additionally supplemented with red clover grass. The study indicated that the pigs in OD-1 and OD-2 consumed daily 5.6 and 5.2% less compound feed respectively and compound feed consumption per kg gain was 4.5 and 11.5%, respectively, lower in comparison with the ID group. Although compound feed consumptions of OD-1 and OD-2 were almost same, the consumption per kg gain was 7.3% lower in pigs additionaly supplemented with red clover grass (OD-2). Nevertheless, average daily gain of three groups did not differ significantly (P>0.05). The behavioural observations indicated that the pigs in both OD-1 and OD- 2 groups spent more time for rooting 100.9 min (P=0.014) and 60.9 min (P=0.017) respectively. In general, the pigs in both outdoor groups were more active than indoor pigs (278.0% (P=0.005) and 219.2% (P=0.010), respectively). The OD-2 pigs additionally supplemented with red clover spent less 40,2% (P=0.052) time for environmental exploration than those fed only compound feed (OD-1). There were no significant differences between the groups regarding aggressiveness, nutrition and other behavioural elements, however, pigs of both OD-1 and OD-2 groups spent somewhat less time for eating and drinking in comparison with ID group (P>0.050). It is concluded that outdoor raising possitively affected pigs by increasing their activity, their motivation to explore environment. However, no data was found confirming lower aggressiveness of outdoor raised pigs.

Network analysis: interruption of the chain of infection by removal of the most central premises

Büttner, K., Krieter, J. and Traulsen, I., Institute of Animal Breeding of Husbandry, Christian-Albrechts-University, Olshausenstr. 40, 24098 Kiel, Germany; kbuettner@tierzucht.uni-kiel.de

The structure of trade networks determines the spread of infectious diseases. With detailed knowledge about this structure, disease spread can be predicted and possibly prevented. The aim of this study is to analyse the change in the network structure by a successive removal of the most central premises in the network of pig production. From 2006 to 2009 contact data from a producer community in Northern Germany were analysed. The data contain information on 4,635 animal movements between 483 premises. For a static network analysis repeated trade connections between two premises were aggregated to a single one (926 trade contacts). If two premises are connected by at least one direct or indirect connection they are part of the same component. In terms of disease spread it is preferred to have many but small components instead of one giant component (largest component of the network). To realize a reduction of the giant component size of more than 95% all premises with more than 3 outgoing contacts (out-degree) have to be removed (9.7%). The number of components increases from 1 to 380 with a median number of nodes per component of 1. If we look at the parameter degree (in- and outgoing contacts) all premises with a degree higher than 6 (11.4%) have to be removed to get the same reduction. The number of components increases to 368 with a median number of nodes per component of 1. The removal of the premises with the most ingoing contacts (in-degree) is not a suitable parameter for the fast reduction, because 95.7% of the premises have to be removed to achieve the 95%-reduction as with out-degree and degree. To interrupt the chain of infection the strategic removal following the ranking of out-degree or degree, e.g. using vaccination, is an appropriate measure because only about 10% of the premises have to be removed to get a reduction of the giant component of more than 95%, so a further disease spread can be prevented.

Transfer of trace elements from feed to pig tissues: management of feed and food limits

Royer, E.[1] and Minvielle, B.[2], [1]Ifip-institut du porc, 34 bd de la Gare, 31500 Toulouse, France, [2]Ifip-institut du porc, La Motte au Vicomte, 35651 Le Rheu, France; eric.royer@ifip.asso.fr

Cadmium (Cd) and lead (Pb) are environmental contaminants resulting from human activity. Presence of these heavy metals in the animal production chain may result from contamination of agricultural soils or pollution of minerals added to feed. Most notifications to the European Union (RASFF) rapid alert system concerning high levels of heavy metals are related to mineral raw materials. A pig study was undertaken to evaluate the effects of long-term dietary exposure to cadmium or lead at levels slightly below regulatory limits for feed on tissue concentrations of these elements. Cadmium and lead could not be detected (<1 or <5 µg/kg) in any muscle samples. Metal concentrations in the liver were increased by exposure, but all samples conformed to the regulatory limit (500 µg Cd or Pb per kg on a wet weight basis). In the kidney, Pb concentration was also below the limit (500 µg/kg ww) but Cd concentration exceeded the maximum value for human consumption (1000 µg/kg ww). Effects on the tissue content were equivalent irrespective of whether the source was minerals or crop material. We conclude that compliance with feed regulatory levels does not mean that cadmium concentration in kidneys will be below regulatory limits in pigs exposed during the whole fattening period. This may partly explain why surveys in Member States have shown that offal occasionally exceeds limits for human consumption. The preventive roles of feeding practices and awareness of the feed industry must be emphasised. The relationship between the regulatory limits for trace elements in feed and those in animal products is discussed.

Effect of housing system during pregnancy on lameness and claw health of sows in farrowing crates

Calderón Díaz, J.A.[1,2], O'doherty, J.[1] and Boyle, L.[2], [1]University College Dublin, Animal Science Dept., Belfield, Dublin 4, Ireland, [2]Teagasc Moorepark Research Centre, Pig Production Dept., Fermoy, Co. Cork, Ireland; julia.calderon-diaz@teagasc.ie

The aims of the study were to evaluate the effect of gestation housing system on lameness in sows and claw lesion (CL) scores and to study the association between CL and lameness. 42 sows housed in stalls (S) and 43 loose (L) housed sows fed by an ESF during pregnancy were transferred to farrowing crates on day 110 of pregnancy (d-5) and kept on 2 floors: tri-bar steel (n=25 L; n=23 S) or plastic coated iron (n=18 L; n=19 S). Lameness was scored (0 to 5) on d-5. While sows were lying in the farrowing crate on d-5 and prior to weaning (d28) the hind claws were inspected for CL: heel overgrowth/erosion (HOE); heel sole crack (HSC); white line (WL) damage; wall cracks (WC) and dew claw injuries (DCI). CL were scored using a modified version of the FeetFirst™ scale. PROC MIXED of SAS was used to analyze the relationship among CL and housing system, the model included housing, period, floor and their interactions. Lameness was analyzed using PROC LOGISTIC. The model included housing and HOE, HSC, WL, WC and DCI as covariates. All sows had at least one CL. S sows had higher scores for WL (3.99 S vs. 3.15 L±0.22; P<0.01) and DCI (5.59 S vs. 4.55 L±0.22; P<0.01) while L sows had higher HOE scores (8.31 L vs. 7.36 S±0.25; P<0.01). HOE, WL and DCI scores increased (P<0.01), whereas HSC scores decreased (P<0.035) between d-5 and d28. There was no interaction between floor in the farrowing crate and gestation housing system (P>0.05). Floor had an effect on HSC (P<0.001) and DCI (P<0.05). L sows had a lower odds of being scored 0 or 1, and higher odds of being scored 2 or ≥3 for locomotory ability (P<0.05). There was no significant relationship between lameness and any of the CL (P>0.05). Although S sows had higher scores for CL potentially associated with pain, L sows were more likely to be lame on entry to the farrowing crate. Claw health deteriorated in the farrowing crates.

Animal welfare in animal oriented Dutch middle level vocational education

Van Der Waal, M.E.[1] and Hopster, H.[2], [1]Wageningen University, Education and Competence Studies Group, Hollandseweg 1, 6706 KN, Netherlands, [2]Van Hall Larenstein University for Applied Sciences, Animal Management, Animal Welfare Group, Agora 1, 8934 CJ, Leeuwarden, Netherlands; waalspaans@hetnet.nl

As animal welfare is of increasing relevance in society, policy and animal practice, it is important to know how animal welfare issues are dealt with in animal oriented vocational education. In the Netherlands, however, no such information was available. Therefore, 20 schools across the country with programmes in animal care, horse husbandry, veterinary nursing, and livestock management were selected for a study of animal welfare education. 105 teachers and 376 pupils filled in the surveys, 26 teachers, 85 pupils and 21 team leaders participated in an interview. The questions in surveys and interviews involved 6 topics: 1. general view on animal welfare at school, 2. existing culture, attitude and priority regarding animal welfare in the curriculum and classes 3. animal welfare in qualification dossiers and examination 5. competences associated with animal welfare 6. need for support. Results indicate that in general animal welfare has nowhere lead to school wide views on the importance of animal welfare in the curriculum and how it should preferably be taught. Teachers mostly implicitly combine animal welfare topics with established courses as in animal housing, animal health, animal nutrition and animal breeding. Teachers and pupils within the educational programme of livestock management displayed a relatively more anthropocentric attitude, where in the other programmes attitudes tended to be more zoöcentric. Although animal welfare is 'mentioned' frequently in qualification dossiers and examination standards, only a few schools developed explicit courses in animal welfare. The research has led to recommendations for a more clearly defined, structured and explicit treatment of animal welfare issues in the different educational programmes, internships, teacher training courses and care for animals used as 'educational material' at schools.

Resistance of six commercial laying hen strains to an Ascaridia galli infection

Kaufmann, F., Das, G., Moors, E. and Gauly, M., University of Göttingen, Animal Science, Albrecht-Thaer-Weg 3, 37075, Göttingen, Germany; gdas@gwdg.de

Six genotypes of commonly used commercial laying hens, namely Lohmann Brown (LB), Lohmann Silver (LSi), Lohmann LSL classic (LSL), Lohmann Tradition (LT), Tetra SL (TETRA) and ISA Brown (ISA), were compared for their ability to resist an experimental Ascaridia galli infection. Laying performance, feed intake, change in the integument and faecal egg counts were determined during the experiment. The hens were infected at the beginning of laying period and slaughtered 105 d after infection i.e., at an age of 35 weeks, to determine their worm counts. No large differences were observed among the genotypes for the performance parameters. However, significant differences in average worm counts of the genotypes were quantified (P=0.008). LSL hens had the highest (25.8) and LT hens had the lowest (12.9) worms per hen. Although worm burden of LSL hens did not differ than those of TETRA and ISA (P>0.05), they had higher worm burdens than LSi, LT and LB hens (P<0.05). ISA hens also had higher worm burdens when compared with LT and LB hens (P<0.05). LSL and ISA hens had higher number of larva than LSi, TETRA, LT and LB hens (P<0.05). Number of female worms, length of the worm and worm sex ratio did not differ significantly between the genotypes (P>0.05). The results suggest that there is a considerable variation in the responses of most commonly used chicken genotypes to the nematode infection. Although no large differences were observed in the performance of the genotypes, LT and LB hens were more resistant than LSL and ISA hens when exposed to an experimental A. galli infection. It is concluded that genetic differences in the responses of the breeds to the nematode infection may contribute to the efforts for selecting more suitable breeds for alternative floor production systems, where hens face poor bio-security.

Relational database system for pig research farms

Karsten, S.[1], Stamer, E.[2] and Krieter, J.[1], [1]Institute of Animal Breeding and Husbandry, Christian-Albrechts-University, Olshausenstr. 40, 24098 Kiel, Germany, [2]TiDa Tier und Daten GmbH, Bosseer Str. 4c, 24259 Brux, Germany; jkrieter@tierzucht.uni-kiel.de

Quantity of collected data on farms increases due to higher stock numbers, use of sensor techniques and legal requirements. Beyond that, on research farms even more traits and parameters are recorded. However, herd management software on the farm is rarely linked with sensor techniques and cannot process additional traits, for example observation of fundament. Thus the recorded information cannot be used in an optimal way for monitoring of the animals as well as of the process technology. For integration of the data a central relational database system has been developed. Records of parentage, fertility and health are transferred automatically from the herd management program to the database. The same applies for automatically recorded data like feed intake or body weight. Sporadically recorded data like body condition scores and observations of fundament can be entered on-line via an electronic form. Consideration of house climate in stables as well as feeding or housing system is possible as well. The flexible design offers options for an extension of the system for future traits. All records are checked for plausibility and consistency during the import process. In case of errors a protocol is sent to the responsible person who can react promptly. An on-line user interface enables realtime monitoring of animal and herd performance as well as of the process technologies. Every animal (even piglets) is stored individually, which enables tracking its path of life from birth via rearing, fattening up to slaughtering. At the moment the central database system comprises the data of two sow holding research farms in Germany.

The role of culture media on infectivity of *Capillaria obsignata* eggs

Tiersch, K.M.[1], Das, G.[1], V. Samson-Himmelstjerna, G.[2] and Gauly, M.[1], [1]University of Göttingen, Animal Science, Albrecht-Thaer-Weg 3, 37075, Göttingen, Germany, [2]Free University Berlin, Institute of Parasitology and Tropical Veterinary Medicine, Königsweg 67, 14163 Berlin, Germany; gdas@gwdg.de

This study investigated whether infectivity of *Capillaria obsignata* eggs depends on media culture used for embryonation of the eggs. Intact female worms were kept in one of following four media: 0.5% formalin, 2% formalin, 0.1% potassium dichromate and 0.1N sulfuric acid. Embryonation rates of the eggs were quantified either daily in intact females for 16 days, or weekly in disrupted females. To test infectivity of the eggs embryonated in different culture media, an experimental infection of chickens with a single dose of 250 eggs/bird was performed. Vast majority of the eggs (>82%) in the first two-thirds of uteri were able to complete embryonation, irrespective of the culture media used for incubation. However, only 32.6% of total eggs could be harvested after disruption of the intact females. Embryonation rates of the eggs from disrupted worms were different among four culture media, with 0.1N sulfuric acid resulting in highest embryonation rate (44.2%). All the experimentally infected birds harbored mature worms, with varying establishment rates of the eggs depending on the culture media (P<0.001). Incubation of the eggs in potassium dichromate 0.1% resulted in lower establishment rate of the eggs (10.2%) when compared with formalin (70.5% and 47.9% for concentrations at 0.5% and 2%, respectively) or with 0.1N sulphuric acid (57.5%). It can be concluded that most of the eggs in first two-thirds of uteri in the intact females have the potential to complete embryonation without being influenced by the incubation media. However, disruption of the intact females results in lower number of harvestable embryonated eggs with a considerable variation due to culture media used. With the exception of 0.1% potassium dichromate, any of the three media, particularly 0.1 N sulfuric acid, can be suggested for embryonation of *C. obsignata* eggs.

Combined effects of humic acids and probiotic on health and performance of Japanese quail

Hanusová, E.[1], Pospišilová, D.[2], Hanus, A.[1], Árvayová, M.[2] and Peškovičová, D.[1], [1]Animal Production Research Centre Nitra, Hlohovecká 2, 951 41 Lužianky, Slovakia, [2]Vetservis,s.r.o., Kalvária 3, 949 01 Nitra, Slovakia; peskovic@cvzv.sk

Combined effects of the humic acids (Humac Natur) and probiotic (Propoul) on performance and health in Japanese quail, as a model animal for gallinaceous poultry, were investigated. A total 111 birds were fed either a standard control (CON) diet (ME=11.7 kJ/kg, CP≥200 g/kg) or an experimental diet, which was made of control plus humic acids (3 g Humac/kg feed). The diets were fed *ad libitum* for 44 weeks. Birds fed experimental diet additionally received a *Lactobacillus fermentum* based probiotic (Propoul) for one week at monthly intervals (0.06 g/bird). Feeding started at day 1 with both additions. The birds were weighed at ages of 0, 28 and 42 days. Microbial compositions of the intestinal contents were examined for the presence of clostridia and lactobacillus spp. at 33 (H), 40 (H+P) and 42 (H) weeks. Cumulative mortality at day 10 was lower in birds fed diet supplemented with humic acids and probiotic (3.3%) than the birds fed the standard diet (10.9%). Although birds receiving humic acids and probiotic were heavier (P=0.011) than control fed birds on day 28 (101.6±1.86 and 94.5±2.04 g, respectively), the difference between groups was not significant on day 42 (P=0.19). Effects of humic acids + probiotic on the reproductive characters of the birds were more pronounced. Birds given both additions laid heavier (P=0.04) eggs and showed a higher (P=0.03) laying performance (165 and 151 eggs/201 days) when compared with the birds on the standard diet. The results of microbiological composition of the intestinal contents confirmed the absence from clostridia in the birds given Humac+Propoul. Lactobacillus content raised after propoul addition. It is concluded that humic acids and probiotics, incorporated into a standard diet, reduce mortality at the early life, improve body weight development to some extend, and favour laying performance of Japanese quails.

Evaluation of indirect methods to estimate body composition of goats

Silva, T.S.[1], Busato, K.C.[1], Chizzotti, M.L.[1,2], Rodrigues, R.T.S.[1], Chizzotti, F.H.M.[3], Yamamoto, S.M.[1] and Queiroz, M.A.A.[1], [1]Univasf, Colegiado de Zootecnia, Centro de Ciências Agrárias, 56304-917 Petrolina PE, Brazil; [2]UFLA, Depto de Zootecnia, cx. postal 3037, 37200-000, Brazil; [3]UFV, Depto de Zootecnia, 36571-000 Viçosa MG, Brazil; mariochizzotti@dzo.ufla.br

Aiming to evaluate two methods for obtaining body composition of the Brazilian indigenous breeds Canindé and Moxotó and F1 Boer crossbreds with non descript goats, a comparative slaughter trial was conducted with 60 goats (20 of each breed group), averaging 15 kg of initial body weight (BW). The baseline group consisted of five animals of each breed group. The remaining goats were allocated on three treatments: *ad libitum* intake and fed 50 or 75% of the *ad libitum* intake, and were fed 90 d. Diet consisted of 40% of elephant grass and 60% of concentrate. Evaluations of ribeye area (REA) were performed on day 90 using ultrasound. The REA was also determined in the carcass between the 12th and 13th ribs (REAmea) using a caliper for further evaluation and correlation with REA obtained by ultrasound (REAus). After the experimental period all animals were slaughtered. The cleaned gastrointestinal tracts, organs, carcasses, heads, hides, tails, feet, blood, and tissues were weighed to measure empty BW (EBW). These parts were frozen, ground, homogenized and subsampled for chemical analyses. The left side of the neck of each animal was dissected into muscle, bone and fat and subjected to chemical analysis. Measurements of REAus were highly correlated and thus equations for determinations of EBW, meat (kg), bone (kg), % meat, % bone, crude protein, ether extract, ash, water were proposed. To predict body composition through analysis of the chemical composition of the neck, equations were developed for EBW, Water (kg), CP (kg), EE (kg) and ash (kg). Among the indirect methods tested, analysis by ultrasound showed better results. This associated with its lower costs turns it into a reliable tool in predicting body composition. Funded by CNPq, FACEPE and FAPEMIG.

Influence of exposure time to low cadmium contamination in feed on retention in pig liver and kidney

Royer, E.[1] and Lebas, N.[2], [1]Ifip-institut du porc, 34 bd de la Gare, 31500 Toulouse, France; [2]Ifip-institut du porc, Les Cabrières, 12200 Villefranche-de-Rouergue, France; eric.royer@ifip.asso.fr

Kidney and liver are the critical animal organs affected by dietary exposure to cadmium (Cd). A total of 36 female pigs (LWxLD × LWxPiétrain) were used to evaluate tissue accumulation following low dietary contamination during the weaning and growing period or the whole fattening period. After weaning, pigs were blocked (mean initial weight 9.2 kg) and housed in plastic and stainless steel pens until slaughtering. Following a standard phase 1 diet, three groups (n=12) received *ad libitum*: (1) non contaminated phase 2, growing and finishing diets (control); (2) contaminated feeds for phase 2 and growing periods before being returned to control finishing feed; (3) contaminated diets for all periods. In groups 2 and 3, Cd was added as $Cd(NO_3)_2$ or by introducing contaminated wheat and sunflower meal in feeds (final dietary concentrations 0.30 to 0.54 µg Cd/kg). The average total Cd intakes of groups 1, 2 and 3 were 14±7, 58±12 and 102±26 µg per pig, respectively. Cd levels were significantly higher in liver and kidney of exposed pigs. Cd concentrations of groups 1, 2 and 3 were 66±47, 245±56, 359±70 µg per kg in liver and 182±86, 896±375 and 1,119±376 µg/kg in kidney, respectively. For all muscle samples, concentrations were below the detection limits (1 to 5 µg/kg). Transfer from dietary cadmium was calculated for kidney and liver and relationships are compared with equations from literature.

Nutritional strategies to reduce enteric methane emissions in ruminants

Newbold, C.J., Aberystwyth University, Institute of Biological, Environmental and Rural Sciences, Penglais Campus, Aberystwyth, Ceredigion, SY23 3DA, United Kingdom; cjn@aber.ac.uk

Methane produced during anaerobic fermentation in the rumen represents an energy loss to the host animal as well as contributing to emissions of greenhouse gases into the environment. The microbial ecology of the rumen ecosystem is complex and the ability of this system to efficiently convert complex carbohydrates to fermentable sugars is in part due to the effective disposal of hydrogen through reduction of CO_2 to methane by methanogenic archaea. A wide range of approaches aimed at decreasing methane emission from the rumen based on direct inhibition of methanogens (vaccination, biocontrol using bacteriophage or bacteriocins, chemical or plant based inhibitors etc), redirection of hydrogen towards alternative sinks (organic acids, nitrate, promotion of acetogenic populations, high sugar grasses and other novel plant material etc) and decreasing overall hydrogen production (ionophores, plant extracts, lipids, defaunation etc) in the rumen have been discussed. However the ecology of the rumen microbial system is such that it frequently reverts back to initial levels of methane production though a variety of adaptive mechanisms. Effective strategies for decreasing methane emissions from ruminants will require the further characterisation of hydrogen generation and utilisation in the rumen and a greater understanding of the factors influencing rumen microbial ecology. It is likely that a multi-factorial approach to methane mitigation involving inhibition of methanogens, provision of alternative electron acceptors and development of low methane-emission diets may be required to bring about a meaningful reduction in methane emissions from ruminants.

Greenhouse gas emissions from feed production and enteric fermentation of rations for dairy cows

Mogensen, L.[1], Kristensen, T.[1], Nguyen, T.L.T.[1], Knudsen, M.T.[1], Brask, M.[2], Hellwing, A.L.F.[2], Lund, P.[2] and Weisbjerg, M.R.[2], [1]Aarhus University, Dept. of Agroecology, Blichers Allé 20, 8830 Tjele, Denmark, [2]Aarhus University, Dept. of Animal Science, Blichers Allé 20, 8830 Tjele, Denmark; Lisbeth.Mogensen@agrsci.dk

Farmers can reduce carbon footprint (CF) of milk by choice of feed. However, a low CF from feed production might be offset by a high enteric methane emission or conversely. CF of different feedstuffs were calculated taking into account the contribution from growing, processing, transport, land use (LU) and land use change (LUC). Subsequently, CF from each feedstuff was added to the CF from methane from enteric fermentation when the feed was combined to typical rations. Roughages have relative high dry matter yields per ha, and only little contribution from transport. CF of roughage is therefore lower than CF of concentrated feeds, especially for maize silage. However, when taking LU into account, CF of maize silage becomes higher than that of grass silage due to carbon sequestration in grass lands. Rapeseed cake and soybean meal are both co-products from oilseed production. The CF contribution from growing and processing rapeseed cake and soybean meal are quite similar, but due to transport of soybean meal from mainly South America, CF of soybean meal is nearly double that of domestically produced rapeseed cake, and much higher if also contribution from LUC is added. Replacing soybean meal with rapeseed cake reduced CF from feed production (by 39%) and CF from enteric emission (by 4%). A ration based on grass clover compared to maize silage increased CF from enteric emission (by 14%), but reduced CF from feed production if LU was taken into (by 36%). A ration with a roughage:concentrate ratio of 80:20 vs. 50:50 caused 24% higher enteric emission. However, depended on the type of the concentrated feed this could be counterbalanced by a lower GHG from feed production.

Role of the nature of forages on methane emission in cattle

Doreau, M.[1], Nguyen, T.T.H.[1,2], Van Der Werf, H.M.G.[2] and Martin, C.[1], [1]INRA / VetAgro Sup, UMR 1213 Herbivores, St-Genès Champanelle, 63122, France, [2]INRA / Agrocampus Ouest, UMR1069 Soil Agro and hydroSystem, Rennes, 35000, France; michel.doreau@clermont.inra.fr

We compared the effect of the nature of forages in diets on enteric methane (CH_4) emission determined by the SF6 method in three experiments in which animals were fed according to their requirements. In Exp. 1, dairy cows were fed 4 diets with 50% hay or 60% maize silage, with increasing amounts of linseeds in concentrate. Exp. 2 was carried out in dairy cows fed diets with 45% grass silage or 45% maize silage and concentrates differing in protein source, soybean meal or dehydrated lucerne. Exp. 3 in young bulls compared diets with 50% hay or 65% maize silage and concentrates made of maize grain and soybean meal. Diets based on maize silage emitted less CH_4 than diets based on hay in exp. 1 (by 11-20%), and less than diets based on grass silage in exp. 2 (by 9-11%), when expressed per kg of dry matter intake (DMI). In exp. 3, maize silage diet emitted 12% more CH_4 per kg DMI than hay diet. The decrease in CH_4 per kg milk when diets were based on maize silage compared to hay or grass silage was 7-25% for exp. 1, and 12-13% for exp. 2. In exp. 3, CH_4 per kg liveweight gain (LWG) did not vary with the nature of forages. These results are consistent with variation in volatile fatty acid pattern in the rumen: an increase in (acetate + butyrate)/propionate ratio is positively related to CH_4 emission. The impact of diets on climate change expressed in CO_2 equivalents per kg of LWG was evaluated by life cycle assessment in exp. 3. It was lower by 8% for maize silage diet than for hay diet, partially due to the higher proportion of concentrate and the lower LWG in hay diet. This difference is compensated for by accounting carbon sequestration by permanent grasslands for hay diet. In conclusion, a trend to a lower CH_4 emission was observed with maize silage diets compared to hay or grass silage diets, but the impact of the nature of forages on climate change requires a global evaluation of greenhouse gases balance.

Assessment of archaeol as a molecular proxy for methane production in cattle

Mccartney, C.A.[1], Bull, I.D.[2], Yan, T.[3] and Dewhurst, R.J.[1], [1]Teagasc, Grange, Dunsany, Co. Meath, Ireland, [2]University of Bristol, School of Chemistry, Cantock's Close, Bristol BS8 1TS, United Kingdom, [3]Agri-Food and Biosciences Institute, Large Park, Hillsborough BT26 6DR, United Kingdom; christine.mccartney@teagasc.ie

Archaeol was previously identified as a potential faecal marker for methanogenesis. This study compared faecal archaeol concentration and methane production for 16 lactating heifers offered a standard diet (30/70 grass silage/concentrates on a dry matter (DM) basis) to assess potential to use the marker to study between-animal variation. Total collections of faeces and methane measurements, using calorimetric chambers, were made at 71 (SD=7.67) and 120 (SD=8.56) days in milk for two groups of 8 animals. Archaeol was analysed according to McCartney *et al.* DM intake was higher for the mid-lactation group (17.3 vs. 14.9 kg/day; SED=0.84; P=0.011). The average coefficient of variation for triplicate analysis of archaeol was 13.59% (SD=3.17). Mean archaeol concentration in faeces (A) was 24.95 mg/kg DM (SD=7.06), and mean methane production (M) was 20.87 g/kg DM intake (SD=3.23), with M = 13.8 + 0.28A (R^2=0.34; P=0.011). Further analysis showed an effect of stage of lactation on this relationship (P=0.03). The low concentration of archaeol in faeces corresponds with the values obtained by McCartney *et al.* for cattle offered high-concentrate diets. Whilst there was a positive relationship, some unaccounted variation remains, which could relate to differences in the activity levels of methanogens and the selective retention of archaeol in the rumen. The effect of lactation period may reflect differences in the rate of passage of archaeol from the rumen and be related to the differences in feed intake.

Trade-offs between methane emission reduction and nitrogen losses

Bannink, A.[1], Ellis, J.L.[1], Sebek, L.B.J.[1], France, J.[2] and Dijkstra, J.[1], [1]Wageningen UR, Wageningen UR Livestock Research, P.O. Box 65, 8200AB Lelystad, Netherlands, [2]University of Guelph, Canada, Centre for Nutrition Modelling, 50 Stone Road East, Building #70, ON N1G 2W1, Guelph, Canada; andre.bannink@wur.nl

Differences in feed intake, feed nutritive value, feed N content and production level primarily determine the variation in average level of enteric methane produced and N excreted per unit of product or feed. However, within a certain production condition significant variation remains due to nutritional factors affecting methanogenesis and N excretion. Understanding of underlying mechanisms is required to benefit from this variation. Rumen degradation of feed OM is the main determinant of rate of fermentation and methanogenesis. The level of N intake and production determine N excretion and associated N losses. Dietary lipids (resistant to fermentation; no N), a higher resistance to rumen fermentation of starch and protein (less fermentation), a change in dietary carbohydrate content and composition and in protein content (changed methanogenesis and N intake), and changed intraruminal fermentation conditions may all affect rumen fermentation rate and methanogenesis, production and N excretion. Some well-established trade-offs need to be taken into account when evaluating the effects of nutrition on methane emission and N excretion. All factors mentioned are involved with optimization of feeding strategies (e.g. fertilization and harvest management, forage production, grazing management, feeding of concentrates and by-products) and have to be considered when aiming to mitigate methane and N excretion on farm. The same trade-offs need to be considered when evaluating the effectiveness of methane reducing compounds added to feed.

Effect of supplementation of whole crushed rapeseed on methane emission from heifers

Hellwing, A.L.F., Sørensen, M.T., Weisbjerg, M.R., Vestergaard, M. and Alstrup, L., Aarhus University, Animal Science, Post box 50, 8830 Tjele, Denmark; Lene.Alstrup@agrsci.dk

It is important to minimize the negative impact from cattle production on environment and climate. The aim of this investigation was to evaluate the effect of supplementing the diet with fat on methane emission from heifers. Sixteen growing heifers were assigned to two different diets fed *ad libitum* from four month of age. The control (CONT) and fat supplemented (FAT) diet compositions in % of dry matter (DM) were: grass-clover silage (37 and 29), maize silage (38 and 40), barley straw (20 and 25), rapeseed meal (3 and 0) and crushed rapeseed (0 and 5), resulting in 487 and 492 g NDF, 30 and 47 g crude fat and 130 and 108 g crude protein per kg DM, respectively. The methane emission was measured by means of open-circuit indirect calorimetry for four days. Body weight of heifers was similar for FAT (291 ± 20 (SEM) kg) and CONT (286 ± 10 kg) at the time of measurements. The daily production of CH_4 was 172 l for FAT which was lower (P=0.04) than the 200 l for CONT. However, the daily intake of DM was 5.5 kg for FAT compared to 6.3 kg for CONT (P=0.08). Thus per kg DM intake there was no difference in methane emission (P=0.34). Nevertheless loss of gross energy as methane was 6.5% for the FAT diet which tended (P=0.08) to be lower than the CONT diet (6.9%). This difference shows that there may be a potential for reducing methane emission from heifers by supplementing fat to the diet as we expect that the higher fat concentration in the FAT diet is the main driver for the reduced methane emission. We conclude that supplementing fat to heifer diets may reduce methane emission, however further studies are warranted to confirm this.

Milk production and carbon footprint in two samples of Italian dairy cattle and buffalo farm
Carè, S.[1], Terzano, M.G.[2] and Pirlo, G.[1], [1]CRA-FLC, via Porcellasco 7, 26100 Cremona, Italy, [2]CRA-PCM, Via Salaria 31, 00016 Monterotondo, Italy; sara.care@entecra.it

Milk production systems are a source of greenhouse gas (GHG) emissions. While there is a good knowledge about carbon footprint (CF) of cattle milk, there is hardly any information about CF associated to the production of buffalo milk. The CF of one kilo of buffalo milk was estimated in four farms in the 'Mozzarella di bufala campana-DOP' production area (Caserta, Italy). The estimated values were compared with those obtained from a sample of nine intensive dairy cattle farms in northern Italy (Lombardia). Direct GHG emissions were CH_4 from enteric fermentation and decomposition of organic matter in manure; N_2O from denitrification and nitrification of organic N of manure and urine and of N of chemical fertilizers; CO_2 emitted with the fuel combustion. Indirect GHG emissions were those associated to the production of purchased feeds, animals, fuels, fertilizers, seeds and other off-farm items. Direct CH_4 and N_2O emissions were estimated according to ISPRA (2010); the other emissions were estimated through an extensive bibliography review and by considering specific Italian condition. Linear regression was used to determine a relation between milk production and CF for buffalo and cattle milk. Buffalo and dairy cattle herds averaged 244±146 and 157±48 mature buffaloes and cows respectively; average milk productions were 2,560±740 kg/buffalo/year and 9,160±1,030 kg/cow/year. Estimated CF were 4.08±1.16 (kg CO_2eq) and 1.16±0.20 (1 kg FPCM) in buffaloes and cows respectively. Regression coefficient (b1) between CF and FPCM production were -0.65 (P<0.1) for buffalo, and -0.15 (P<0.05) for cows. The lower regression coefficient for dairy cows can be explained by the higher productivity in comparison to buffaloes.

Methane generating potential of *Lotus uliginosus* var. Maku harvested in three consecutive dates
Marichal, M.D.E.J.[1], Crespi, R.[1], Arias, G.[1], Furtado, S.[1], Guerra, M.H.[1] and Piaggio, L.[2], [1]Facultad de Agronomía – Universidad de la República, Producción Animal y Pasturas, Garzón 780, 12900 Montevideo, Uruguay, [2]Secretariado Uruguayo de la Lana, Baltasar Brum 3764, 11700 Montevideo, Uruguay; mariadejesus.marichal@gmail.com

Methane generating potential of *Lotus uliginosus* var.Maku harvested in three consecutive dates was investigated. Three replicate plots were seeded in August at the Experimental Research Center of the Secretariado Uruguayo de la Lana (S 33°52', W 55°34'). When plants were 20±5cm, forage was harvested in October(LM1), November(LM2) and December(LM3). Pastures were analyzed for NDF, ADF and H2SO4 soluble lignin (Lig$_{sa}$). An *in vitro* gas production procedure was followed, rumen contents of two fistulated whethers fed alfalfa hay were collected two hours after morning feed. Three batches of 33 of bottles (3 bottles by experimental forage, 3 with alfalfa as standard, and 3 blanks) were incubated for 24 h. Cumulated gas was collected at 8 hs and from 8 to 24 hs in separated bottles and methane was measured by gas chromatography. Fiber fractions and methane production at 8, 8 to 24 and 24 h were analyzed in a complete randomized design (PROC GLM, SAS;Tukey test). LM2 presented smaller (P<0.05) NDF, ADF and Ligsa (369, 282 and 114 g/kg DM, respectively), than LM1 (478, 363 and 201 g NDF, ADF and Lig$_{sa}$ /kg DM, respectively); similar (P>0.10) and intermediate values were observed in LM3 (419, 324 and 139 g NDF, ADF and Ligsa /kg DM, respectively). Methane in gas accumulated in 24h was greater (P=0.027) in LM2 than in LM3, presenting LM1 intermediate and similar (P>0.20) values (19, 16 and 18 mg CH_4 /g OM for LM2, LM3 and LM1, respectively). Methane measured in gas accumulated up to 8 h was greater (P<0.02) in LM1 and LM2 than in LM3 (8, 7 and 5 mg CH_4 /g OM for LM1, LM2 and LM3, respectively). No differences (P=0.26) in CH_4 production was registered in gas collected from 8 to 24 h of incubation (11 mg CH_4 /gOM). Results suggest that pastures regrowth may vary in methanogenic potential.

Session 41

Theatre 1

New perspectives and risks of precision livestock farming systems

Faverdin, P., INRA, Agrocampus Ouest, UMR PEGASE, 35000 Rennes, France; philippe.faverdin@rennes.inra.fr

The farmer knows his animals through the information system that he develops. If observation was the major route of acquisition of information, development of new technologies has grown rapidly in recent years and provides access to an increasing number of parameters on animals and their environment. Biological parameters, morphological and behavioral of animals and of their products can be better known and with a higher frequency, thus offering new perspectives in animal tracking. Beyond the technical performance associated to these new technologies that open up promising challenges for high-throughput phenotyping research in livestock animals, the question of their use in farming systems is a complex challenge with different expectation. This information should indeed find a place in the global information system and decision processes of the farmer, otherwise overloading him by information flow too cumbersome. The expected benefits concern increasing labor productivity, but also improved performance or better efficiency of production systems by better and earlier diagnosis and decision-making assistance. However, farmers have to adapt their organization based on these new technologies, often not compatible with each other, and requiring their own monitoring and maintenance. Moreover, it is possible to show that the benefit of the same additional information varies with the livestock farming system. The best systems will probably use a limited number of sensors, if possible simple, with complementary views of animals (behavior, biology, morphology), but providing multiple outputs emerging from dynamic and simultaneous processing of these data within a large herd information system. The process of data management will have to take into account the process of decision to effectively support the farmer in the livestock management. A lot of progress in the structure and treatment of data within the information system and in the integration of the results in management is still required to really capitalize on these new technologies.

Session 41

Theatre 2

Innovation in animal feeding, a key driver in the concept of sustainable precision livestock farming

Den Hartog, L.[1] and Sijtsma, R.[2], [1]Nutreco R&D and Quality Affairs, P.O. Box 220, 5830 AE Boxmeer, Netherlands, [2]Wageningen University, Animal Nutrition Group, P.O. Box 338, 6700 AH Wageningen, Netherlands; Leo.den.Hartog@nutreco.com

Sustainable precision livestock farming is a way forward in a world in which animal production will increasingly be affected by external factors. These include surging demands for animal products and struggling supplies of feed raw materials. Simultaneously, there is growing concern about food and the impact of production methods on animal welfare and environment. Throughout, producers are facing reduced profit margins. Sustainable precision livestock farming integrates the technological approach of precision livestock farming with the social and ecological aspects related with consumer and societal demands. Optimization of productivity and efficiency within the constraints of the concept are important objectives, as well as maximization of the profit for all stakeholders. Animal feed plays a crucial role as it is usually the largest cost factor in animal production and at a critical junction in the chain. Several indicators demonstrate that further optimization of animal feeding is potentially still possible. The genetic potential is only partially utilized, the utilization of most nutrients appears to be low and there is a huge variation in performance among farms and, within farms, among animals. New science and technologies seem to offer many opportunities for innovation in animal feeding. Future innovation will be driven by (gen)omics, microsystem- and nanotechnology and information and communication technology. These mainstream technologies are the foundation of many application technologies of relevance for animal feeding. However, technology access and acceptance by consumers and society need to be managed in a proper way. In conclusion, sustainable precision livestock farming may become a 'license to produce' and will maximize profits. Animal feeding is essential in this concept and innovation in animal feeding is a key driver for realization and optimization.

Precision livestock farming: review and case studies
Halachmi, I., Agricultural Research Organization (ARO), Institute of Agricultural Engineering, The Volcani Center, P.O. Box 6, 50250 Bet Dagan, Israel; Halachmi@volcani.agri.gov.il

The dairy industry reached a stage that Precision livestock farming (PLF) technology can raise dairy management to a new level of significant added-value to the farm profitability and animal wellbeing. In order to materialize this process, companies and research institutions combine existing sensors, models and management programs. To apply PLF we need to define key indicators and corresponding golden standards. A key indicator is continuously, fully automated, monitored by a sensor (such as image, sound, movement) directly at animal level. This review paper will introduce five successful case studies. Then, zoom-in to the development of the latest two sensors and their associated business model that are now being under development in the ARO Israel in cooperation with commercial companies. (Case study 1) Automatic early detection of calving diseases; (case study 2), automatic cow lameness detection, (case study 3) approaching calving sensors; (case study 4) applications of on-line real-time milk analyzer with robotic milking; (case study 5) automatic body condition squaring. This review paper will include both (1) companies' new developments, and a large EU-FP7 industry-academia project on the development of electronic sensors for animal welfare. The description of each case study will specify: the objectives, the technology and experimental methods, statistical analyses during the sensor validation phases, and then synthesis of the results.

ATOL: an ontology for livestock
Meunier-Salaün, M.C.[1], Bugeon, J.[2], Dameron, O.[3], Fatet, A.[4], Hue, I.[5], Hurtaud, C.[1], Nedellec, C.[6], Reichstadt, M.[7], Vernet, J.[7], Reecy, J.[8], Park, C.[8] and Le Bail, P.Y.[2], [1]INRA-AgroCampus Ouest, UMR 1348 Pegase, 35590 Rennes, France, [2]INRA, UR 1037 LPGP, 35000 Rennes, France, [3]INSERM-Université de Rennes I, U936, 35000 Rennes, France, [4]INRA, UMR 85 PRC, 37380 Nouzilly, France, [5]INRA, UMR 1198 BDR, 78352 Jouy-en-Josas, France, [6]INRA, UM 1077 MIG, 78352 Jouy-en-Josas, France, [7]INRA, UMR 1213 UMRH, 63122 Saint-Genès Champanelle, France, [8]Iowa State University, Ames, Iowa, USA; Marie-Christine.Salaun@rennes.inra.fr

The development of ontology in biology is necessary to better organize and exploit the knowledge on the individual characteristics like his phenotype (result of the expression and regulation of the genome). The computerized data processing for the characterization of each phenotype highlights the need of standardized and explicit language, defining without ambiguity the traits used and referred by various users: geneticists, physiologists, biochemists, producers, archivists. This can be reached trough the structured collection of terms and concepts within an ontology built to meet the user's needs: ATOL for 'Animal Trait Ontology for Livestock', has been specifically dedicated to the animals breeding (fish, poultry, mammals). However, this generic ontology of phenotypic traits provides characters whose aggregation can explain the components of the quantity and quality of cattle productions: meat, milk, effective nutritional and fertility, as well as those taken into account for the assessment of animal welfare. ATOL should help to annotate phenotypic databases in order to establish more precise genotype/phenotype relationships while considering environmental effect on phenotype using EOL (Environment Ontology for Livestock). Moreover ATOL might contribute to a systemic, predictive approach of animal performance and using the semantics tools an access to more relevant and precise information in document content.

A mobile automatic milking system used both indoors and on pasture: data from pasture
Dufrasne, I., Robaye, V., Knapp, E., Istasse, L. and Hornick, J.L., Liège University, Nutrition Unit, Veterinary Medecine Faculty, Boulevard de Colonster, 20, 4000 Liege, Belgium; eknapp@ulg.ac.be

In most farms equipped with an automatic milking system (AMS), the cows remain indoors without any grazing opportunity. The benefits from grazing are lost. In order to carry on grazing, a mobile AMS, built at the University of Liege, was made of 2 trailers. The robot and its equipment were located in the first trailer while the milk tank and the refrigerating system were in the second. A herd of 48 Holstein cows was thus milked with an AMS located on a permanent pasture. The cows grazed on a rotational system with 11 paddocks. The areas were comprised between 1.33 and 4 ha. The distance between paddocks and the AMS varied from 100 up to 425 m. The AMS was located in a paddock of 1.33 ha. From 12/05 until 12/07 the cows were fetched either twice or once a day but they could also reach the robot freely. The sward height was measured at the entry and exit of each paddock. The data about daily milk yield and milking number were analysed according to a GLM procedure including the effect of animal, days in paddock, distance between AMS and paddock, rotation cycle number and complementation. The cows produced daily 24.3 kg when fetched twice and only 20.7 kg when fetched once (P<0.001). The corresponding number of milkings per day was 2.3 and 1.8 (P<0.001). The milking number was reduced when grass height decreased (P<0.001; r^2=0.53). The models explained 76 and 28% of the milk yield and milking number variations respectively. Amongst the parameters studied, the animal effect explained 77% of the milk yield and 53% of the milking number variations, respectively (P<0.001). The distance explained a weak but significant part of the variation in milk yield and milking number (2.3% and 3.8% ; P<0.001). There were no clear relationships between milk yield or milking number and distance. It is concluded that milking with an AMS located in pasture away from the farm is possible and therefore of interest for dairy organic farms.

The potential for direct measurement of rumen pH in commercial dairy farming
Mottram, T.T. and Nimmo, S.B., eCow Ltd, Innovation Centre, University of Exeter, Rennes Drive, EX4 4RN, Exeter, United Kingdom; toby@ecow.co.uk

The detection and correction of acidosis is an important element in managing high producing dairy cows. The most direct way of detecting the pH status of the rumen is to use a bolus that measures rumen pH and telemeters the data to a base station. The first reports of rumen pH monitoring by wireless have been associated with cattle in research situations. This paper presents new data about the relationship between measurements made in different locations in the rumen. It also discusses methods of extending the life of rumen boluses and extracting them by endoscope to develop techniques to reduce the cost of use of a bolus to make it suitable for commercial farmers to monitor acidosis on a routine basis. New definitions of Sub-Acute Rumen Acidosis will be proposed and data presented on the combination of pH, redox and temperature. The variability of pH value due to location within the ventral sac indicates that the reticulum is the most reliable location for rumen moinitoring, data will be presented to support this. The eCow bolus has a specific gravity greater than 1.5 to retain itself in the reticulum and to ensure that the sensor is in a downward orientation to extend its life.

Comparison of 3 methods of multivariate analysis to study economic efficiency in dairy cattle farms

Atzori, A.S.[1], Steri, R.[1], Tedeschi, L.O.[2] and Cannas, A.[1], [1]Dipartimento di Agraria, Sezione di Scienze Zootecniche, Università di Sassari, via De Nicola, 9, 07100, Sassari, Sardinia, Italy, [2]Department of Animal Science, TAMU, Kleberg Center, College Station,TX, USA; asatzori@uniss.it

The objective of this work was to compare the ability of three different methods of multivariate analysis to explain the variability of the farm income over feed cost (IOFC), used as profitability indicator. A stochastic approach was used to study the relationships between routinely available farm information and farm profitability. Two databases were used: a development database (dDB), based on 18 technical and economical variables of 135 dairy cow farms, and a simulated database (sDB), with 5,000 dairy farms, developed by applying a Monte Carlo technique on the dDB data distribution and relationships. The IOFC was calculated for both the dDB and the sDB as the difference between milk, calves, and culling revenues minus the estimated herd feeding costs. Three multivariate methods were used to reduce the number of variables and to identify data clusters that could be translated into technical recommendations. The sDB was firstly analyzed with: (1) Principal Component (PC) Analysis; and (2) Factor (F) Analysis, after excluding for both the IOFC data; selected PC and extracted F were then regressed on the observed IOFC to quantify their contribution to the profitability; then, 3) a Canonic Discriminant (CD) analysis was performed on sDB after dividing the farms in 3 classes of IOFC level. The first 4 selected PC (associated with nutritional efficiency, milk quality and payment, poor management, and reproductive efficiency) explained 74.5% of the IOFC variability; the 4 extracted F (related with nutritional efficiency, management, milk fat and protein content), explained 68.5% of IOFC variability; the first CD was able to identify the most important variables related to IOFC (herd genetic level, milk price, and management related variables) with the highest explained IOFC variability (92.7%) among the three methods of multivariate analysis.

Systems for determining carcass lean meat yield% in beef and lamb

Gardner, G., Anderson, F., Williams, A., Ball, A.J., Hancock, B. and Pethick, D.W., Murdoch University, School of Veterinary and Biomedical Science, Murdoch, WA 6150, Australia; G.Gardner@murdoch.edu.au

Carcass lean meat yield percentage (LMY%) is a key profit driver for beef and lamb processors. Currently LMY% is poorly assessed in Australia, as processors rely on carcass weight and a single point measure of fatness to estimate LMY%. This approach has a limited capacity to predict carcass LMY%, and almost no capacity to describe variation in the distribution of lean between regions of the carcass. In response to this the Australian lamb industry has been assessing and developing a number of different high-throughput technologies to predict LMY% within abattoirs. This activity is being driven at two levels, the first targeting simpler and less expensive devices that will deliver single-site measurements of tissue depth. These devices predict LMY% with less precision (R^2 range from 0.2-0.4), and at the whole carcass level only. They include mechanical tissue depth probes, ultrasound, and boning room vision systems that can determine tissue depth at the GR site (11 cm from mid-line over the 12th rib), or muscle and fat depth at the C-site (5 cm from the mid-line over the 12th rib). Secondly, we are assessing more expensive whole carcass systems that will enable more accurate determination of LMY% (R^2 range from 0.4-0.7) as well as determining lean distribution between different regions of the carcass. These include carcass vision systems, dual energy x-ray absorptiometry, and computer aided tomography scanning (CTscan). In all cases these LMY% prediction devices are trained upon a central 'gold-standard' dataset generated using CTscan of carcasses scanned in 3 sections (fore, saddle and hind). This central CTscan dataset has the advantage of generating consistent and repeatable data, not subject to human bias. Thus processors can select an LMY% prediction technology that best optimises their trade-off between cost, speed, and precision. This model is now being adapted to the beef industry, with obvious constraints being the size and expense of working with beef carcasses.

Modelling for the prediction of beef sensory quality
Chriki, S.[1], Legrand, I.[2], Journaux, L.[3], Picard, B.[1], Pethick, D.[4], Polkinghorne, R.[5] and Hocquette, J.F.[1], [1]INRA, UMR1213, 63122 Theix, France, [2]Idele, Qualité des viandes, Limoges, France, [3]UNCEIA, 149 avenue de Bercy, Paris, France, [4]Murdoch University, Veterinary School, Perth, Australia, [5]Merringanee, Murrurundi, NSW 2338, Australia; jfhocquette@clermont.inra.fr

Beef is characterized by a high and uncontrolled variability of its palatability, which contributes to consumer dissatisfaction. In Europe, there is no reliable tool to predict beef quality in order to deliver consistent quality beef to consumer. Faced to the number of factors influencing beef quality, partners of the ProSafeBeef programme have brought together all the data they have collected over 20 years. The resulting data warehouse contains available data of animal growth, carcass composition, muscle tissue characteristics and beef quality. This database was useful to analyze relationships between intramuscular fat level and flavour, and also to determine which are the most important muscle characteristics related to a high tenderness. In addition, Australia has developed the Meat Standards Australia (MSA) grading scheme to predict beef quality by untrained consumers for each individual 'muscle × cooking method' combination using various information on the corresponding animals and meats. The results of our work indicate that it would be possible to manage a grading system in Europe similar to the MSA system. Finally, tenderness, juiciness, flavor, overall liking and the final MSA scores were positively correlated with intramuscular fat level ($r>0.49$, $P<0.01$) and dry matter ($r>0.29$, $P<0.01$). In addition, the 'soluble collagen content/total collagen content' ratio was positively correlated to tenderness score ($r>0.23$, $P<0.05$) for all samples. In conclusion, this is the first large study in Europe which related biochemical parameters through different muscle types to quality scores determined by a large number of untrained consumers. This study confirmed the importance of intramuscular fat level for beef quality and of collagen solubility for tenderness across muscles.

Growth breeding value redistributes weight to the saddle region of lamb carcasses
Gardner, G.E., Anderson, F., Williams, A., Kelman, K.R., Pannier, L. and Pethick, D.W., Murdoch University, School of Veterinary and Biomedical Science, Murdoch, 6150, WA, Australia; F.Anderson@murdoch.edu.au

Increased growth rate and carcass lean meat yield % are key profit drivers for the lamb industry, however redistributing lean tissue to more highly priced parts of the carcase will also increase its value. Faster growing lambs are known to be leaner and less mature at slaughter. Therefore we hypothesised that selection for growth using the Australian Sheep Breeding Value (ASBV) for greater post weaning weight (PWWT) would increase whole carcase lean weight, when animals are compared at the same carcass weight. Lamb carcases (n=1,218) from the Sheep CRC Information Nucleus were scanned in sections (fore, saddle, and hind) using Computed Tomography (CT) to determine fat, lean and bone weights. Data was analysed using the log-linearised allometric equation $\log y = \log a + b.\log x$. Fixed effects were site-year, sex, sire type, birth-type rear-type and kill group within site-year, with random terms sire and dam by year. For the same carcass weight, PWWT caused no composition differences, except in female lambs which had 3.3% more carcase lean ($P<0.01$) across the 25 unit PWWT range. Alternatively for the same fat, lean or bone weight, these tissues were all proportionately heavier in the saddle region of the high PWWT lambs by 3%, 7%, and 16% across the PWWT range. Aligning with our hypothesis, PWWT was associated with increased total carcase lean, although only in females. Unexpectedly, PWWT caused a redistribution of carcass tissues to the saddle region, particularly for bone and lean, implying an altered conformation in these high growth lambs. Conflicting with the premise of our hypothesis, these effects appear to be independent of maturity as there was no whole body increase in bone weight. Furthermore loin muscle myoglobin concentration in the high PWWT lambs was increased (by 0.03±0.018 mg/g tissue) rather than decreased as would be expected in a less mature animal. In conclusion, PWWT redistributes carcase weight to the saddle region of lambs.

Mineral contents in herbage over the grazing season

Schlegel, P. and Bracher, A., Agroscope, Tioleyre 4, 1725 Posieux, Switzerland;
patrick.schlegel@alp.admin.ch

The objective of this survey was to verify if mineral contents in herbage was influenced by time over the grazing period in intensively managed pastures with rotation paddocks. Seventy two grass samples were collected on the days before dairy cows entered the paddock on two Agroscope research farms (Posieux, 650 m and Sorens 860 m a. s. l.) between 2008 and 2010. Date, growth stage and botanical composition were determined and samples were chemically analyzed for dry matter (DM) and mineral (Ca, P, Mg, K, Na, S, Co, Cu, Fe, Mn, Zn, Se) contents. Data were analyzed using GLM procedures including month (April to October), growth stage and of botanical composition as fixed effects. Year and location (research farm) were used as covariables. Contents of P (4.7 ± 0.5 g/kg DM), K (34.6 ± 4.0 g/kg DM), Na (0.22 ± 0.09 g/kg DM), Cl (8.2 ± 2.9 g/kg DM), S (2.5 ± 0.5 g/kg DM), Cu (9.9 ± 1.6 mg/kg DM), Mn (78.7 ± 42.8 mg/kg DM), Zn (32.7 ± 5.1 mg/kg DM) and Se (<0.025 mg/kg DM) were not influenced ($P>0.05$) by date. Contents of Ca and Mg were lower ($P<0.001$) during April, Mai and October, and contents of Co and Fe were higher ($P<0.001$) during September and October than during the other months. Pastures with lower proportions of grasses presented higher contents of P ($P<0.001$), K ($P<0.05$) and Cu ($P<0.05$) and higher contents of Ca and Mg ($P<0.001$) than pastures with high grass proportions. Growth stage did not influence ($P>0.05$) mineral contents. The present data indicate that (1) mineral contents of forage from intensively managed pastures were variable; (2) part of this variation was explained by the botanical composition (P, K, Cu) and by the month (Ca, Mg, Co, Fe).

Basis for a tool for transporters to self-control the welfare of animals during transport

Mounaix, B.[1], De Boyer Des Roches, A.[2], Mirabito, L.[1] and David, V.[1], [1]Institut de l'Elevage, Service Bien-être Santé Taaçabilité Hygiène, 149 rue de Bercy, 75595 Paris Cedex 12, France, [2]VetAgroSup, 1 Avenue Bourgelat, 69280 Marcy L'Etoile, France; a.de-boyer@vetagro-sup.fr

To the request of professionals, a study was conducted to select and to test several animal-based welfare indicators which could be used by transporters to self-control the quality of transport. Welfare indicators were first reviewed from recent international scientific literature. Then animal-based indicators were selected upon scientific relevancy and were tested by experienced observers in various transport contexts. All criteria and animal-based indicators were tested for reproducibility (Kappa index) within 3 trained and experienced observers. Two prototype tools have been proposed to transporters: a handbook to describe each criteria and its measurement, and reminders of current European regulation on transport and good practices recommendations; 3 grids to observe and to qualify animal welfare, loading and unloading areas and handling during transport; different grids were proposed for adult cattle, calves and sheep. These prototypes are now tested for feasibility by several voluntary transporters, including drivers. They are a basis for an operating tool for professionals to self-control the quality of their work regarding animal welfare.

Evaluation of cow comfort index and stall usage index in different cooling systems for dairy cows
Lendelová, J.[1], Botto, Ľ.[2], Pogran, Š.[1] and Reichstädterová, T.[1], [1]Slovak University of Agriculture in Nitra, Tr. A. Hlinku 2, 949 76, Slovakia, [2]Animal Production Research Centre Nitra, Hlohovecká 2, 951 41 Lužianky, Slovakia; JanaLendelova@gmail.com

The aim of this study was to investigate the changes in cow comfort index (CCI) and stall usage index (SUI) in groups with different cooling systems in summer period. In the first group, was used sprinkling system for animal cooling (S) and in the second group was disposable sprinkling system and diagonally ordered vents (SV). There were evaluated two identical groups of Holstein dairy cows with the same number of free-stalls (2x43) with separated manure solids bedding. Behaviour of animals was noticed by 10 minute intervals during 24 hour cycle. The data were analyzed using a General Linear Model ANOVA by the statistical package STATISTIX 9.0. Significance of differences was determined by Tukey HSD test (at $\alpha \leq 0.05$). The CCI was higher in group SV with sprinklers and vents compared with that in group S with sprinklers but without vents (82.66% compared to 76.51%) in daily period one hour before evening milking. CCI was more than 5% higher in both groups at night period – computed one hour before morning milking and cow comfort index was at night higher in group SV than in group S again (87.82% compared to 82.66%). It was found by monitoring of whole night period (from end of evening milking to start of morning milking) that free-stalls were most used at last 5 hours before morning milking. Animals from group S without vents, which were less cooled, were significantly more lying down in alleys (2.52 h/d/cow vs. 0.56 h/d/cow, $P<0.001$). Total time spent by lying in stalls and in alley represented in group SV 11.31 /d/cow, and in group S 10.22 /d/cow. There was found positive effect of animal enhanced cooling using sprinkling system with increased air movement by vents. It reflected in significant prolongation of whole lying time and shortening of time, when animals were standing.

Genome-wide association mapping using single-step GBLUP
Misztal, I.[1], Wang, H.[1], Aguilar, I.[2], Legarra, A.[3] and Muir, B.[4], [1]University of Georgia, 425 River Rd, Athens 30602, USA, [2]INIA, Las Brujas, 90200 Canelones, Uruguay, [3]INRA, BP 52627, 32326 Castanet-Tolosan, France, [4]Purdue University, Animal Science, W Lafayette 47907, USA; ignacy@uga.edu

The purpose of this study was to extend single-step GBLUP (ssGBLUP) to genome wide association analysis (GWAS). ssGBLUP is a procedure that calculates GEBVs based on combined pedigree, genomic and phenotypic information. In this study, GEBVs were converted to marker (SNP) effects. Unequal variances for markers were incorporated by deriving weights from SNP solutions and including the calculated weights into a new genomic relationship matrix. Improvements on the SNP weights were obtained iteratively either by recomputing the SNP effects only or also by recomputing the GEBV. Efficiency of the method was examined using simulations for 10 replications with 15,800 subjects across 6 generations, of which 1,500 were genotyped with 3,000 SNP markers evenly distributed on 2 chromosomes. Heritability was assumed 0.5 all due to 30 QTL effects. Comparisons included accuracy of breeding values and cluster of SNP effects of ssGBLUP and BayesB with several options for each procedure. For genomic evaluation, an accuracy of prediction of 0.89 (0.01) was obtained by ssGBLUP after only one iteration, which was slightly higher than BayesB of 0.88 (0.02), but required only a small fraction of time. Power and precision for GWAS applications was evaluated by correlation between true QTL effects and the sum of m adjacent SNP effects, where m varied from 1 to 40. The highest correlations were achieved with m=8 and were 0.82 (0.02) for ssGBLUP, and 0.83 (0.07) for BayesB with m=16 according to marker density and extent of linkage disequilibrium in simulated population. Computing time for ssGBLUP took about 2 min while BayesB took about 5 hrs. ssGBLUP with marker weights is fast, accurate and easy to implement for GWAS applications. In particular, ssGBLUP is applicable to GWAS with complex models including multitrait, maternal and random regression.

Estimation of dominance effects in paternally genotyped populations

Boysen, T.J., Heuer, C., Tetens, J. and Thaller, G., Institute of Animal Breeding and Husbandry, Christian-Albrechts-University, Olshausenstrasse 40, 24098 Kiel, Germany; tboysen@tierzucht.uni-kiel.de

To apply genomic selection in cattle breeding, large cohorts of bulls have been genotyped at a high number of single nucleotide polymorphisms (SNP). Aggregated phenotype information for estimating genomic breeding values is mainly based on performance of female offspring while marker information is predominantly available for males. To estimate non-additive effects directly, however, both genotypes and phenotypes of animals must be available. In addition, a huge number of observations is required to achieve reliable estimates. Here we propose an approach to overcome these restrictions by deriving genotype probabilities for cows given the genotypes of sires, sires of the cows and the respective allele frequencies in the population. Thereof coefficients for additive and dominance effects can be calculated and used in regression models for genome wide association analyses. Properties such as power of method and bias of estimates for a broad range of parameters mode of inheritance, effect of size, allele frequencies, number of animals, and, linkage disequilibrium between SNP and QTL were investigated using extensive simulation studies. Results indicate that number of cows in the range between 150,000 and 1,000,000 are needed to achieve a power of 90% for realistic QTL-scenarios. Preliminary results from analyses based on 2,400 bulls genotyped for 54k SNP and 374,000 daughters showed unrealistic high power values indicating that further research is essential to account for possible stratifications and to develop appropriate sampling strategies.

Consequences of genomic imprinting in livestock genetic evaluation

Varona, L., Casellas, J., Moreno, C. and Altarriba, J., Universidad de Zaragoza, Miguel Servet 177, 50013, Spain; lvarona@unizar.es

Genetic evaluation by mixed models becomes the basis for its genetic improvement in livestock. The most widespread methodology accounts for direct polygenic additive genetic effects plus some systematic effects and a residual source of variation. However, the genetic determinism of many phenotypic traits must be also linked with maternal and paternal genetic effects, such as paternal and maternal genomic imprinting. Genetic evaluation models must capture all potential sources of genetic variability. A complete model must include direct additive genetic effects and parental, maybe imprinted, genetic effects. However, when partial imprinting is present, it originates covariance between direct and parental genetics effects that cannot be inferred from the phenotypic and pedigree data. Given that the complete model cannot be properly addressed, the aim of this study was to analyse the consequences of ignoring some of the genetic effects (i.e., direct, maternal and paternal genetic effects) in terms of variance component estimation and breeding value prediction by using beef cattle data sets and simulated populations under several scenarios. Simulated and real data were analysed using a Bayesian approach with several models that include: (1) direct additive effect; (2) direct and paternal effects; (3) direct and maternal effects; (4) paternal and maternal effects; (5) direct, paternal and maternal effects with null correlations between them. The conclusion of this study indicates that paternal or maternal effects may play a relevant role in some relevant traits in livestock production. Further, its absence in the genetic evaluation models may lead to biased parameter estimation and to erroneous ranking of the candidates of selection, with important consequences in the phenotypic performance of future generations. In particular, the cause of negative correlations between direct and maternal effects may be caused by the presence of not considered paternal effects.

Effect of linkage disequilibrium, haplotypes and family relations on accuracy of genomic prediction
Wientjes, Y.C.J.[1,2], Veerkamp, R.F.[2] and Calus, M.P.L.[2], [1]Wageningen University, Animal Breeding and Genomics Centre, P.O. Box 338, 6700 AH Wageningen, Netherlands, [2]Wageningen UR Livestock Research, Animal Breeding and Genomics Centre, P.O. Box 65, 8200 AB Lelystad, Netherlands; yvonne.wientjes@wur.nl

Our objective was to investigate the effect of linkage disequilibrium (LD), haplotypes and family relationships on the accuracy of direct genomic values. A formula based on selection index theory was used to predict accuracies, using genomic relationships and information of a reference population (RF) of 529 genotyped cows. Four groups of selection candidates were simulated using increasing amounts of information: (1) allele frequency of RF (FREQ); (2) allele frequencies and LD pattern of RF (LD); (3) randomly drawing haploid chromosomes from RF (HAP); (4) animals from RF, thereby being the only group with real family relationships to RF (FAM). Accuracy of FAM was predicted using the remaining 528 animals as RF. At a heritability of 0.6, accuracies were on average 0.093 ± 0.003 (FREQ), 0.168 ± 0.006 (LD), 0.355 ± 0.015 (HAP) and 0.577 ± 0.064 (FAM). At a heritability of 0.1, relative differences between accuracies across groups were similar. FREQ used the same assumptions as the deterministic formula to predict accuracies of Daetwyler *et al.*, namely; no LD between loci, no relationships with RF and all loci have an effect. As a result, accuracies of this group were equal to predictions with the Daetwyler-formula. Variance of the accuracy of FAM was much higher compared to the other scenarios, due to much higher variances in relationships with animals in RF. Accuracies of FAM were on average more than 50% higher than the accuracy of HAP. It is concluded that level of relationship with RF has a much higher effect on the accuracy of direct genomic values compared to linkage disequilibrium per se. Furthermore, increasing length of haplotypes shared with animals in RF improves prediction accuracy, especially when multiple haplotypes are shared with one or more reference animals due to family relationships.

A software pipeline for animal genetic evaluation
Truong, C.V.C.[1], Krostitz, S.[2], Fischer, R.[2], Mueller, U.[2] and Groeneveld, E.[1], [1]Institute of Farm Animal Genetics, Mariensee, FLI, Höltystr 10, 31535 Neustadt, Germany, [2]Saxon State Office for Environment, Agriculture and Geology, Köllitsch, LfULG, Am Park 3, 04886 Köllitsch, Germany; cong.chi@fli.bund.de

Genetic evaluation of animals is essential in any modern breeding program. This procedure usually involves many tools to deal with different aspects. Most of existing tools are independent programs and disconnected from others in terms of usage. Linking them into a complete process is challenging since each requires its own configuration in a specific format. Moreover, for large datasets the repeated manual parameterization is inefficient and error-prone. To address this issue, we have developed 'GEPipe', a software pipeline for genetic evaluation. Our objective is to automate the input parameterization and provide the integration of all involved components. GEPipe is a web-based application easily accessible from a web browser in the network. It is based on the principle 'less input more output' requiring from the user only a minimum set of data to maximize the output. Thus, with one single upload of PEST data and configuration files, six sub-systems (AGen, AfterBLUP, EvolveBLUP, PopRep, OptBS and ZwISSS) are automatically executed to generate different reports. As a result, GEPipe produces a complete documentation on the process of genetic evaluation: computing aggregate genotypes (AGen), analyzing and estimating BLUP (AfterBLUP), monitoring BLUPs over time (EvolveBLUP), creating population reports (PopRep), modeling and optimizing the structure of breeding programs (OptBS) and publishing breeding values (ZwISSS). This process helps breeders or breeding organizations with the quality management of their breeding programs. GEPipe and its sub-systems are freely released under the GPL license. The project is supported by funds of the Federal Ministry of Food, Agriculture and Consumer Protection (BMELV) based on a decision of the Parliament of the Federal Republic of Germany via the Federal Office for Agriculture and Food (BLE) under the innovation support programme.

Sparse structures for mixed model equations for different animal models

Masuda, Y., Obihiro University of Agriculture and Veterinary Medicine, Nishi 2-9, Inada, 0808555, Obihiro, Japan; masuday@obihiro.ac.jp

Most of computing time in residual maximum likelihood (REML) is spent on the sparse factorization and sparse inversion for a coefficient matrix of mixed model equations (MME). When the sparse patterns are properly used, computing time would be greatly saved. The objective of this study was to suggest an optimal design for sparse solver to utilize sparse characteristics of MME. The eight systems of equations were formed by applying different animal models, including single-trait, multiple-trait and random regression model, to actual datasets from Holsteins in Japan. The orders of coefficient matrices were 100 thousands or more. A supernode was defined as a maximal block of contiguous columns of a Cholesky factor whose diagonal block was full lower triangular, and whose off-block-diagonal column sparsity structures were identical. To quantify the properties for a factor, following statistics were calculated: the most frequent number of columns within supernodes (ModeNC) the number of columns within the largest dense lower triangular (NT) and NT to the order of MME (NT%). Two of ordering algorithms, approximate minimum degree (AMD) and multilevel nested dissection (METIS), were compared. ModeNC was almost same to the order of the (co)variance matrix of additive genetic effects. For all MME, more than half of the supernodes contained few columns. NT (NT%) ranged from 3,311 to 29,173 (from 2.5% to 3.6%) for AMD and from 1,590 to 19,440 (from 1.5% to 2.4%) for METIS. The dense triangular matrix was reduced by METIS, but the size remained relatively larger. The factorization and inversion of the dense matrix would be bottlenecks. Sparse package requires a technique for efficient storage of the large dense matrix and optimized subroutines for dense operations.

Imputation from lower density marker panels to BovineHD in a multi-breed dataset

Schrooten, C.[1], Van Binsbergen, R.[2,3], Beatson, P.[2] and Bovenhuis, H.[3], [1]CRV BV, P.O. Box 454, 6800 AL Arnhem, Netherlands, [2]CRV AmBreed, NZ P.O. Box 176, Hamilton 3240, New Zealand, [3]Wageningen University, Animal Breeding and Genomics Centre, P.O. Box 338, 6700 AH Wageningen, Netherlands; chris.schrooten@crv4all.com

Starting in 2007, genomic selection based on 50k genotypes has been applied in breeding programs of CRV Ambreed for Friesians, Jerseys, and crossbreds. Reliability of genomic breeding values could increase if reference populations of these breeds could be combined into one multi-breed genomic evaluation. One of the prerequisites is that high density genotypes are available. This can be accomplished by genotyping animals with high impact on the population with a high density chip, and deriving high density genotypes for the remaining animals genotyped at lower density. The objective of this study was to investigate the error rate for imputation from lower density marker panels to BovineHD (777k) in a multi-breed dataset. BovineHD genotypes were obtained for 465 Friesians, 227 Jerseys and 57 crossbreds. Imputation using Beagle version 3.3 was studied for 5 alternatives, each with animals from a different combination of breeds serving as reference set for imputation. In all alternatives, the same set of 25 Friesians, 25 Jerseys, and 10 crossbreds was considered as validation animals. For these animals, only the genotypes for the relevant lower density chip SNP were retained, and genotypes for the remaining SNP were imputed. Each alternative was replicated 4 times, with different validation and reference animals per replicate. Imputation errors were computed as the number of differences between imputed alleles and observed alleles, divided by the number of compared alleles. First results for the alternative with all breeds in the reference set showed average imputation errors of 0.89%, 0.68%, and 0.44% for crossbreds, Friesians, and Jerseys, respectively, when imputing from 50k to HD. It was concluded that this error rate was low enough to apply this strategy to obtain HD genotypes for all animals genotyped at lower density.

From the 50K chip to the HD: imputation efficiency in 9 French dairy cattle breeds

Hoze, C.[1,2], Fouilloux, M.N.[3], Venot, E.[2], Guillaume, F.[3], Dassonneville, R.[2,3], Journaux, L.[1], Boichard, D.[2], Phocas, F.[2], Fritz, S.[1], Ducrocq, V.[2] and Croiseau, P.[2], [1]Union Nationale des Coopératives agricoles d'Elevage et d'Insémination Animale, 149 rue de Bercy, 75012 Paris, France, [2]INRA, UMR 1313 GABI, Domaine de Vilvert, 78350 Jouy-en-Josas, France, [3]Institut de l'Elevage, 149 rue de Bercy, 75012 Paris, France; chris.hoze@jouy.inra.fr

Since 2008, genomic selection based on the Bovine SNP50 BeadChip® (50K) is implemented in France for the three main dairy breeds. The major challenge for the other breeds is to build large reference populations. New perspectives are offered with the BovineHD BeadChip® (HD). Indeed the density of this chip (1 Single Nucleotide Polymorphism per 4 kilobases) should be high enough to detect conserved linkage disequilibrium across breeds and therefore combine reference population with an appropriate methodology. With this aim, 2102 bulls from 9 breeds (52 to 551 per breed) were genotyped with the HD chip in the ANR-APISGENE funded GEMBAL project. 549 Holstein bulls genotyped in EUROGENOMICS consortium were added to this population leading to a total of 2,651 bulls genotyped on the HD chip. This sample will be used to impute the 50K-genotypes from the national databases. Imputation error rates were computed in each breed with a validation set of 20% of the animals, selected either by age (youngest cohort) or by randomly choosing complete families. Markers were masked for validation population in order to mimic 50K genotypes. Imputation was done using the Beagle 3.3.0 software. Mean allele imputation error rates ranged from 0.34% to 5.1% depending on the breed. Results were better for large breeds than for regional French breeds due to differences in effective population size and, above all, on number of animals HD-genotyped. Accuracies were lower in Brown Swiss and Simmental because foreign ancestors were not genotyped, they could be improved through genotypes exchanges. Multi-breed imputation may also be a way to improve efficiency of imputation for related breeds.

Significance and bias: choosing the best model for variance composition

Ilska, J.J.[1], Kranis, A.[2], Burt, D.[1] and Woolliams, J.A.[1], [1]The Roslin Institute, Easter Bush, EH25 9RG, United Kingdom, [2]Aviagen Ltd., Newbridge, EH28 8SZ, United Kingdom; joanna.ilska@roslin.ed.ac.uk

BLUP methodology has accelerated the dissemination of genetic progress, enabling the expansion of poultry production. The progress achieved relies on the accuracy of breeding values (BVs), which in turn depends on the validity of models used to estimate variance components. Aside from significance testing, the fit of models can be assessed by bias and prediction error variance (PEV) of EBVs. Analyses involved estimation of variance components of hen housed production (HHP), body weight (BWT) and egg weight (EWT) and identification of the best model using log-likelihood ratio test. The population consisted of 1.3M chicks over 24 generations. For HHP, the only significant random term was the direct genetic effect (σ^2_A) of the chick (M1). A typical model used in commercial setting (M2) includes σ^2_A and permanent environment ($\sigma^2_{p.e}$) effects. For BWT, the best model included σ^2_A, maternal genetic effects and $\sigma^2_{p.e}$ (M3), for EWT it also included covariance between σ^2_A, and maternal genetic effects (M4). Prediction errors through bias from the models were estimated by regressing phenotypes on BVs predicted from ancestral information only. Slopes for BWT and EWT were significantly different from 1. There was no statistical difference between slopes for HHP due to large SE, caused by small numbers (n=1,800) and low h^2. The slopes for BWT predictions from the best model M3 (β=0.91, SE 0.02) and M2 (β=0.94, SE 0.02) were not statistically different (n=70,614). For M1 the slope was 0.77 (SE 0.02). The slopes for EWT were similar for the best model M4 (β=0.9, SE 0.09) and M2 (β=0.91, SE 0.08). Slopes from models M1 and M3 were estimated at 0.89 and 0.88 (SE 0.08) respectively. Differences in PEV were small, with lowest values found for M3 in BWT, M2 in HHP and M1 in EWT. Although best models for BWT and EWT were not statistically different from M2, they hold the potential benefit of estimating maternal BVs and thus expanding breeding objectives.

Estimation of genetic parameters for longitudinal measurements of feed intake in Piétrain sire lines

Dufrasne, M.[1,2], Jaspart, V.[3], Wavreille, J.[4] and Gengler, N.[2], [1]FRIA, 5 rue d'Egmont, 1000 Brussels, Belgium, [2]University of Liège – Gembloux Agro-Bio Tech, 2 Passage des Déportés, 5030 Gembloux, Belgium, [3]Walloon Pig Breeding Association, 4 rue des Champs-Elysées, 5590 Ciney, Belgium, [4]Walloon Agricultural Research Center, 9 rue de Liroux, 5030 Gembloux, Belgium; marie.dufrasne@ulg.ac.be

The aim of this study was to estimate the genetic parameters for longitudinal measurements of feed intake (FI) in a crossbred population of pigs to develop a genetic evaluation model for the estimation of breeding values for FI of Piétrain boars. Data were collected on crossbred pigs in test station in the context of the genetic evaluation system of Piétrain boars in the Walloon Region of Belgium. Trait analyzed was daily FI (DFI). Because there were no facilities to record individual DFI in the Walloon test station, individual DFI were assumed to be the total pen FI divided by the number of pigs per pen. The edited dataset consisted of 3,902 measurements of DFI recorded on 1,975 crossbred pigs from 75 purebred Piétrain sires and 150 Landrace dams from the hyperprolific Landrace K+ line. A random regression animal model with fixed effects of sex and pen, and random effects of additive genetic, permanent environment and residual was developed in this study. Random additive genetic and permanent environment effects were modeled with linear splines with knots located at 75, 100, 175 and 210 d. The mean DFI was 1.979 kg/d with a SD of 0.479 kg/d. Estimated heritability for DFI increased with age from 0.02 at 75 d to 0.30 at 210 d. Estimated genetic correlation between age decreased when age interval increased. These preliminary results are consistent with literature. However, additional research are ongoing to test alternative random regression models that should be better than using splines for longitudinal performance of DFI. Furthermore, genetic relationships between DFI and other production traits, like growth and carcass traits, must be analyzed.

Marker assisted selection for milk production traits in Churra dairy sheep

Sanchez, J.P.[1], Garcia-Gamez, E.[2], Gutierrez-Gil, B.[2] and Arranz, J.J.[2], [1]IRTA, Avd Rovira Roure 191, 25198, Spain, [2]Universidad de Leon, Campus de Vegazana, s/n, 24071, Spain; juanpablo.sanchez@irta.cat

The aim of this work was to assess the predictive ability of a marker assisted selection (MAS) procedure for daily milk yield (MY), fat yield (FY), fat percentage (FP), protein yield (PY), protein percentage (PP) and somatic cell score (SCS) in Churra sheep. The data set comprised 1681 yield deviations (YDs) for these traits. Animals were genotyped with the Illumina Ovine SNP50Chip. After quality control 43779 SNPs were retained for reconstructing chromosome-wise haplotypes (DualPHASE). The complete data set was divided in two: 90% of the records were used as training set (TS) and the remaining 10% constituted the validation set (VS). An LDLA genome scan based on 6-SNPs haplotypes was carried out on the TS and the significant associations ($P\leq0.05$) were retained. The VS was used for correlating their actual YDs to their predictions, calculated from the predicted breeding values and haplotype effects obtained from a model applied to the TS. This procedure was repeated 10 times, in every repetition a different tenth of the animals became the VS and average results across repetitions were estimated. For PP the average of the correlations (SD) between real and predicted YDs rose from 0.48 (0.02), when no haplotypes information was considered, to 0.52 (0.004) when this information was included. For PY the correlation increased from 0.21 (0.03) to 0.22 (0.07). For the remaining traits the averages from the pedigree index were higher than those from the MAS model. These results might be explained by the low accuracy when declaring the SNPs to be associated to the traits. In most of the cases not clearly associated loci were detected; while for PP, systematically, a QTL explaining around 10% of the genetic variance was mapped on OAR3. We speculate that the inaccuracy in QTL detection is mainly due to not properly considering the covariance structure between YDs. Further research on this regard and testing alternative methods still needed.

Combining approaches for the analysis of the genetic structure of Avileña-Negra Ibérica cattle breed

Martín-Collado, D.[1], Toro, M.A.[2], Abraham, K.J.[3], Carabaño, M.J.[1], Rodriguez-Ramilo, S.T.[4] and Díaz, C.[1], [1]INIA, Mejora Genética Animal, Ctra. Coruña km. 7,5, 28040, Madrid, Spain, [2]E.T.S.I. Agrónomos, UPM, Producción Animal, Ciudad Universitaría, 28040, Madrid, Spain, [3]Faculdade de Medicina de Ribeirão Preto, Universidade de São Paulo, Programa de PósGraduação em Genética, Ribeirão Preto SP, São Paulo, Brazil, [4]Facultad de Biología, Universidad de Vigo, Bioquímica, Genética e Inmunología, Campus Lagoas Marcosende, 36310, Vigo, Spain; martin.daniel@inia.es

The inference of the genetic structure of domestic animal populations has important implications in the design of breeding and conservation programs. Several tools have been developed to determine it. In general terms, they are grouped in distance-based and the model-based methods. Both approaches have proved to be useful tools to infer the structure of simulated populations and real populations with relatively simple gene flow. But problems arise when populations show more complex relationships like is usually the case of domestic animal populations. The Spanish local cattle breed Avileña-Negra Ibérica is an example of this complexity. We used three different tools to analyze its structure. On the one hand we used the widely known model-based method Structure software. On the other hand two distance-based methods were applied. One maximize the Nei's minimum genetic distance among subpopulations while the other, developed in this study, uses of the molecular coancestry matrix to determine the most genetically isolated individuals and built on them the subpopulations. 13.415 animals from 60 herds were genotyped for 17 microsatellites. We compared and combined the results of the three approaches to infer the genetic structure of the breed. We show that although initially results of the different methods get different results, when combining all of them the hidden genetic structure breed shows up. Finally, implications of the genetic structure of the breed on its genetic management are discussed.

Developments in dairy worldwide, from a dairy farmers perspective

Beldman, A.C.G.[1], Daatselaar, C.H.G.[1] and Prins, B.[2], [1]LEI- Wageningen UR, P.O. Box 65, 8200 AB Lelystad, Netherlands, [2]President Global Dairy Farmers, Jan Altinkstraat 18, 9791 DM Ten Boer, Netherlands; alfons.beldman@wur.nl

Dairy farmers in EU have become more exposed to the world dairy market because of a trend towards globalisaton. A result of this exposure is the increasing volatility of milk prices since 2008 compared to a period with relatively stable prices before. To anticipate, it is becoming more important to know what is happening elsewhere in dairy husbandry in the world. Global Dairy Farmers (GDF) is an international platform of dairy farmers and companies with the goal to exchange knowledge and strategies with members from all over the world. Based on information from this network, some major developments in dairy in general will be pictured, like the development of the market and in dairy production structures worldwide. From a farmers perspective it is interesting to see what these developments mean for their own situation. The GDF network describes the predominant farming system for each region, the most important developments in these regions (market, economy, society) and how these developments are incorporated into the farmers strategy for that region. This picture will be presented for USA (large scale, purchased feed), South America (Large scale producing for local market), New Zealand\Australia (Pasture based; utilizing international market), and China (from dairy village to large scale) in comparison with the European situation. Also some lessons derived from these developments will be translated to the dynamics of the dairy sector in Europe.

The study of the functioning of the meat market for EU consumers
Gbur, P. and Theelen, M., European Commission, DG Health and Consumers, rue Froissart 101, 1040 Brussels, Belgium; margareta.theelen@ec.europa.eu

The study of the functioning of the meat market for EU consumers covers both market conditions and consumer decision making within the market for fresh meat (beef, chicken, pork, veal, turkey and lamb) and meat products. The results show that 43% of EU households eat meat or meat products 4 times per week or more. One third of consumers would like to buy meat less often, half of them for health reasons. Consumer awareness of specific types of meat, such as organic or origin certified, does not translate into purchases. For example, 49% of consumers know organic meat but only 16% say they buy it. Shares of consumers who buy meat with nutrition claims or environmental certificates are even smaller: 15% and 5% respectively. However, 41% of EU consumers would like to buy more often organic meat, 31% meat with nutrition claims and 29% meat with environmental certificates. What prevents them from doing so is mainly a high price, but also a limited choice of such products or insufficient information they have. Indeed, a mystery shopping revealed that organic meat was in general over 60% more expensive than the regular one, and meat with environmental or animal welfare certificates was hardly or not available in a number of countries. Consumers look at labels when buying meat but they do not always understand the information. For example, 53% thought that it was not safe to use a meat product after a best before date and 36% indicated that it would lose some of its quality but still could be consumed. In comparison to EU15, consumers in countries that joined EU recently are less aware of different meat types (which are also less often available in Eastern Europe), use less information sources and information aspects when buying meat, are less satisfied with safety or hygienic conditions in the market. In contrast, they are more satisfied with availability of meat in general.

Dairy chain relation development in Slovakia
Stefanikova, M., Slovak Association of Dairy Farmers, Vystavna 4, P.O. Box 1, 94910 Nitra, Slovakia; stefanikova@agrokomplex.sk

This paper presents the dairy chain development in Slovakia from the point of view of all stakeholders. The dairy farming sector is presented in terms of development of dairy farmer numbers, milk prices, qualitative parameters (fat and protein content), including comparisons on the European Union level. Economic efficiency of milk production is presented based on the latest figures on milk production costs and revenues by the European Dairy Farmers model. Dairy Chain Relation Development in Slovakia is explained based on consumer price construction of 1,5% UHT milk. Generally said, consumer price consists of raw material – milk price (paid by dairies to dairy farmers), dairy processing margin, retail margin and VAT. Currently, from the consumer price of 1,5% UHT milk, – the dairy farmer receives a capital consumption allowance (cca) of 44%, dairy gets cca 15%, retail gets cca 24% and VAT gets cca of 17% (VAT tax is 20%). The paper explains in detail all stakeholders price developments from 2008 to 2012. The last part of paper is dedicated to milk and milk products consumption including promotion programmes for milk consumption as one of the possible ways to improve the situation in the dairy chain.

Optimisation of economics in dairy cattle and sheep farms in Slovakia
Krupová, Z.[1], Huba, J.[1], Michaličková, M.[1], Krupa, E.[2] and Peškovičová, D.[1], [1]Animal Production Research Centre Nitra, Hlohovecká 2, 951 41 Lužianky, Slovakia, [2]Institute of Animal Science, P.O. Box 1, 104 01 Prague, Czech Republic; krupova@cvzv.sk

The objective of this study was to evaluate efficiency of dairy cattle and sheep farms in Slovakia and to propose optimal parameters important for economic sustainability of the farms. The main production and economic characteristics of dairy cattle (15 farms) and dairy sheep (9 farms) for the year 2010 were analyzed separately for intensive and extensive ones. In cattle as well in sheep the most important items were costs for feeds (39 and 23%), other direct costs (26 and 41%) and labour costs (8 and 16%). Depreciations represent also higher costs proportion (19 and 16%), however, they are nonfinancial i.e. implicit costs item. The total annual costs without depreciation were of 1,823 and 1,491 € per cow and 264 and 153 € per ewe in the intensive and extensive farms, respectively. The total annual revenues per cow were based on milk yield (6,268 and 4,741 kg), milk marketability (94 and 91%) and the average milk price (0.299 and 0.300 €/kg) for studied type of farms in 2010. Appropriate values important for revenues in dairy sheep farms were milk yield (220 and 67 kg), price of milk (0.963 and 0.827 €/kg), price of cheese (3.80 €/kg in extensive farms only), sold lambs (1.47 and 1.12 lamb) and price of lambs (11 and 22 €/lamb). Total profit (-62 and -198 € per cow and -36 and -35 € per ewe) calculated for intensive and extensive farms considered revenues from milk (1,761 € and 1,293 €) in cattle and from lambs (16 € and 23 €), milk and cheese (212 and 95 €) in sheep. Based on prognoses of costs development in dairy farms and prices of the cattle and sheep commodities in 2012 it is recommended to achieve milk yield 8,323 kg and 5,451 kg per cow and year in the intensive and extensive farms. Corresponding values for dairy sheep farms are 250 and 140 kg of milk and 1.47 and 1.21 lambs put on the market per ewe and year.

Grazing livestock in Baltic countries and development paths of dairy farmers in Lithuania
Stalgiene, A.[1], Preisegolaviciute-Mozuraitiene, D.[2] and Jankauskas, I.[3], [1]Lithuanian Institute of Agrarian Economics, V. Kudirkos st. 18-2, Vilnius, 03105, Lithuania, [2]Lithuanian Cattle Breeders Association, Kalvarijos g. 128, Kaunas, 46403, Lithuania, [3]Lithuanian Agricultural Advisory Service, Stoties g. 5, Akademija, 58343, Lithuania; aldona@laei.lt

In Lithuania number (nr) of cattle decreased in period 2003-2010 by 17%, dairy cows 22%, goats 62%. In this period 108,000 farms stopped keeping dairy cows (-56%). In Estonia nr of cattle decreased by 12%, dairy cows 20% and goats 17%. Nr of dairy farmers decreased by 72%. In Latvia situation was different: nr of cattle increased by 4%, while nr of dairy cows decreased by 8% and goats by 21%. In all three Baltic countries nr of sheep doubled in this period. Despite a decrease in dairy cows, production level in all countries increased more than 25%, explained by structural changes in management. In frame of 'a Leonardo da Vinci in combination with Wageningen UR Eastern Europe entrepreneurship study', future strategies and attitudes of dairy farmers were examined. As tool a questionnaire was filled in by 300 farmers, which was divided over the various size groups of farms to represent the whole farming community. 71% of farms in the sample are characterized by 'a specialized family dairy farm', 8% by 'a specialised large scale dairy farm with personnel' and, 14% by 'a diversified farm in agricultural activities'. On average more than 75% of income comes from dairy. Focus in development is on 'Expanding of dairy production' (50% of farmers), '(Further) specialization in dairy farming' (18%), 'Diversify into other agricultural activities than dairy' (10%), and 'Wait & see' (11%). Major farming goals are 'Earn enough money to support my family' (74%), 'Maximize profit' (69%), 'Keep costs as low as possible' (62%). Some changes in farm management and structure suggested in next 5 years are increase in milk level (54% of farms), invest in modern technology (44%) and start with organic farming (32%), the latter being a remarkable high percentage of the farms.

Livestock production in West Balkan countries and development paths of dairy farmers in Slovenia
Klopcic, M.[1], De Lauwere, C.[2] and Kuipers, A.[3], [1]Biotechnical Faculty, Dept. of Animal Science, Groblje 3, 1230 Domžale, Slovenia, [2]LEI Wageningen UR, P.O. Box 35, 6700 AA Wageningen, Netherlands, [3]Expertise Centre for Farm Management and Knowledge Transfer, P.O. Box 35, 6700 AA Wageningen, Netherlands; Marija.Klopcic@bf.uni-lj.si

Livestock structure in West Balkan countries will be presented – number and size of farms, cooperative and processing structures, and level of self-sufficiency and consumption of animal products. This provides a view in the past.To gain insight in future development paths from farmers' view point, a case study will be presented about Slovenia. In frame of the 'Leonardo da Vinci in combination with Wageningen UR Eastern Europe entrepreneurship study', strategies and attitudes of dairy farmers are examined. As tool a questionnaire with 50 questions was used filled in by 365 dairy farmers, divided over the various size groups of farms to represent the whole farming community. 66% of farms in the sample are characterized by a specialized dairy farm, 24% by a diversified farm in agricultural activities and 10% in non-agricultural activities. On average 72% of income comes from dairy. Focus in development is on expansion and specialisation in dairy production (61% of farms), diversify into other agr. activities than dairy (10%) and in non-agr. activities (5%), and cooperate/chain integration (9%). Major farming goals are keep costs as low as possible, maximize profit, earn enough money to support family, breed sustainable cows, improve animal welfare, produce best quality/ safe product. Minor attention will be towards providing employment to others. Possible changes in farm management suggested in next 5 years are buy/rent additional land (79% of farms), improve feeding (45%), keep costs low (45%), improve sustainable traits of cows (42%), increase number of cows (40%), invest in larger barn (34%), and increase milk yield/cow (34%). Highest ambition can be found in group of younger and higher educated farmers. These results are compared to previous studies questioning dairy, suckler cow and Cika (rare breed) farmers.

Dairy sector developments and farm strategies in Poland
Malak-Rawlikowska, A.[1] and Żekało, M.[2], [1]Warsaw University of Life Sciences – SGGW, Faculty of Economic Sciences, ul. Nowoursynowska 166, 02-787 Warsaw, Poland, [2]Institute of Agricultural and Food Economics, Farm Accountancy Department, ul. Świętokrzyska 20, 00-002 Warszawa, Poland; Marcin. Zekalo@ierigz.waw.pl

In recent decade dairy sector in Poland has been a stage for thorough and dynamic changes. This paper presents general trends in the Polish dairy production sector in recent decade and describes farmers' attitudes towards the future in a context of strategy formulation. During period 2004-2010 number of dairy farms decreased by 54.7%. At the same time the average milk production per farm doubled. Since 2000 the annual milk production in Poland has remained at the level of 11.5-12.0 million tons. Milk yield/cow has been steadily increasing by almost 30% in 2000-2010. This trend was accompanied by a decline in dairy cow number by 18% and an increase in milk deliveries to dairy processing by 33%. In order to gain more insight in farmer's attitudes towards the future, a questionnaire about farm strategic management was realised within the Leonardo da Vinci in combination with Wageningen UR Eastern Europe entrepreneurship project. The questionnaire was in 2011 directed to 300 dairy farmers in Mazovian region in Poland, structured according to farm size represented in the region. Most of farmers declared to focus their development strategy on further specialisation (87%) and expanding dairy production (81%). Significant was also focus on vertical (39%) and horizontal (48%) integration. However, about 59% of farmers were uncertain about the future milk price and only 24% expected it to rise. For about 82% of farmers 'profit maximisation' was an important or very important goal. Highly ranked were also 'to run the farm efficiently' (79%) and 'to breed cows with high milk production' (78%). Almost 73% appreciated to be independent. 54% of farmers increased milk production volume after accession and about half is expecting to continue this trend in near future. Interesting is that almost 60% of farmers declared to focus on keeping costs as low as possible in the future.

Strategies, innovation and entrepreneurship of dairy farmers in Eastern Europe
De Lauwere, C.[1], Beldman, A.[1], Lakner, D.[1] and Kuipers, A.[2], [1]LEI Wageningen UR, P.O. Box 35, 6700 AA Wageningen, Netherlands, [2]Expertise centre for farm management and knowledge transfer, P.O. Box 35, 6700 AA Wageningen, Netherlands; carolien.delauwere@wur.nl

Farmers in Eastern Europe face challenges since the fall of the communist regime in the 1990s and the accession to the EU in 2004 and 2007. It is assumed that improved capacities of farmers on strategic management and entrepreneurship enable them to better anticipate towards the continuous changes and keep their farms viable. Therefore a study was started in 2011 which aims (1) to give insight into the strategies of dairy farmers in Eastern Europe and its influencing factors, and (2) to give insight into the effects of a facilitated interactive learning methodology on the capacities of farmers with regard to strategic management and entrepreneurship. A survey amongst about 300 dairy farmers in Poland, Lithuania and Slovenia was carried out in 2011 to get insight into the strategies of farmers and its influencing factors. Most farmers interviewed characterized their farms as a specialized dairy farm and strived for expansion and specialization. They qualified the farming goals maximize profit, earn enough money to support the family, run the farm efficiently, improvement of animal welfare and breed sustainable dairy cows (regarding longevity and fertility) as (very) important. Perceived opportunities were EU subsidies and accession to the EU and perceived threats were future reduction of direct payments, future milk quota abolition and new EU agricultural policy. The farmers interviewed perceived the possibilities to perform their preferred strategies and their knowledge to do so as neutral or a little positive. They were quite positive about their entrepreneurial competences and abilities for strategic reflection. They appeared to be reasonably positive about their future. Differences in outcome between the countries will be discussed. The results of the survey serve as basic measurement for trainings on strategic management, innovation and entrepreneurship which will be carried out in the spring of 2012.

Production cost, an indicator for the profitability of beef cattle operations
Sarzeaud, P., French Livestock Institute, Monvoisin, BP 85225, 35652 Le Rheu cedex, France; patrick.sarzeaud@idele.fr

The recent evolution of the french beef farms marked by the continuation of the extension of the structures and the increase of the labor productivity. In the exploitations of the French beef farms network coordinated by the Livestock Institute, the beef production per unit of labor, approximately increased 3% per year over the period 2000-2009. But in the same time the economic efficiency, evaluated by the profit margin on global farm returns dropped down by 12 points since 2005. In order to help farmers in improving this efficiency and their competitiveness, the French livestock Institute proposed a new method of calculation of production costs. Available for beef and dairy farms, this new method is in correspondence with those used at the international level (agri benchmark). This study is based on work concerning economic efficiency in the 570 French beef farms, which were monitored in 2009. Results 2009 were treated and developed in order to describe the costs and their variability. Statistic approach aimed to describe relationship between costs and income. This observation illustrates a great variability in the farms costs. The averages by type of beef enterprise vary from 1 to 2 for comparable situations: between 208 € for 100 kg live weight for fatteners to 451 €/100 kg for breeders. It's mainly due to the productiveness of those enterprises, their involvement in inputs and equipment consumption and their efficiency in using factors such as labor, land and capital. Actually, productiveness, costs and returns interfere together to contribute to the profitability of those beef enterprises. In conclusion, it seems that the implementation of the calculation of the production costs in beef farms provides new references in order to judge the economic farm efficiency. However, others factors interfere in the profitability such as productiveness and efficiency of inputs. As far as enlargement of farms, this will probably carry on in the future (2015-2020 projections), therefore efforts should focus on economic efficiency.

The reindeer herders are interested in selective breeding

Muuttoranta, K. and Mäki-Tanila, A., MTT Agrifood Research Finland, Biotechnology and Food Research, Myllytie 1, 31600 Jokioinen, Finland; kirsi.muuttoranta@mtt.fi

Reindeer is a semi-domesticated animal. However, reindeer herders value selection very highly in improving the production. In the vast Finnish reindeer herding area, selection may be strong, as there is quota system due to the limited pasture resources. The herders make selection decisions independently, thus the breeding aims vary. We wanted to study how consistent the selection decisions are across the reindeer herding units (cooperatives), comprising the herding area. The managers of cooperatives were interviewed on their selection decisions. For analysing the regional differences, the reindeer herding area was divided to six subregions. The analysis relied on R software, using ANOVA. Official statistics on the cooperatives were also used in studying the differences. The latitude largely determines the area for pasturing and the number of animals (herd size) in the cooperative. The selection of calves in the autumn gatherings is based on their phenotype and also on the information on their dams when available. Despite the differences among the cooperatives, the breeding aims are similar throughout the herding area. The most important traits, such as calf size, muscularity, health and dam performance are highly valued in all the cooperatives. According to the cooperative managers, hair and antlers are good indicators of health and vigour. The proportion of selected calves varies from 20 to 40% of all calves across the six subregions (with more variation within them, i.e. from 4 to 75%). The selection intensity is correlated with predation pressure, animal density and herd size. The herders are also aiming at an optimum age distribution to maximise the reproduction rate. The benefits from the basic elements of selection, including pedigree information and accurate weight recording, interest the reindeer herders, while they are difficult to be accomplished.

Survey of genetic selection on pasture-based dairy farms in the USA and Romania

Schutz, M.M.[1], Maciuc, V.[2], Gay, K.D.[1] and Nennich, T.D.[1], [1]Purdue University, 125 S. Russell St., West Lafayette, IN 47907, USA, [2]University of Agricultural Sciences and Veterinary Medicine, Aleea Mihail Sadoveanu nr. 3, Iasi 700490, Romania; mschutz@purdue.edu

Graziers from US and Romania (RO) were surveyed to determine management and genetic selection practices. The overall aim was to inform development of genetic selection indexes. Respondents included 80 farmers in 23 US states and 23 farmers in Romania. Producers were questioned about their herd's grazing history and milk and component production. They were also questioned about breeding practices such as use of seasonal grazing and breeds utilized. Producers were asked to rank traits by their expected genetic importance. Traits were ranked from -5 to +5 with negative being selection against and positive for a trait. Production was 21.3±5.1 kg (US) and 17.9±13.5 kg (RO) of milk per cow per day, 4.0±0.4% (US) and 3.97±0.3% (RO) milk fat, and 3.3±0.3% (US) and 3.5±0.3% (RO) milk protein. Also, 47.5% (US) and 52.2% (RO) of producers participated in seasonal calving, defined as 75% of cattle calving in any 3 mo window. Further, 72.5% (US) and 82.7% (RO) utilized crossbreeding to the extent that at least 10% of the herd was crossbred. Holstein, Jersey, Ayrshire, and Milking Shorthorn were most common breeds used in US, while Simmental, Brown Swiss, and Holstein were most common in RO herds surveyed. The average ranks of traits were (US; RO): productive life (3.83; 4.14), udder composite (3.56; 4.08), somatic cell score (-3.18; -2.90), feet and legs (3.16; 3.27), fat percentage (3.06; 3.48), calving ability (2.97; 3.86), daughter pregnancy rate (2.95; 4.11), protein percentage (2.84; 3.53), fat yield (2.69; 3.84), body size (-2.66; 3.00), protein yield (2.51; 3.34), and milk yield (2.23; 4.24). US graziers place more emphasis on traits relating to longevity and fertility and less on production traits than the most widely used US selection indexes. Graziers in RO emphasized milk yield more and productive life and somatic cell counts less than US graziers.

Cost efficiency of dairy sheep organic systems in Spain applying a translog cost function
Angón, E.[1], Toro-Mújica, P.[1], Perea, J.[1], García, A.[1], Acero, R.[1] and De Pablos, C.[2], [1]University of Córdoba, Animal Production, Campus de Rabanales, 14071, Córdoba, Spain, [2]University Rey Juan Carlos, Economía de la Empresa, Paseo de los Artilleros s/n, 28032 Madrid., Spain; z82anpee@uco.es

The aim of this study was to evaluate the cost efficiency of the dairy sheep organic systems in Castilla La Mancha. The cost function model was elaborated by applying a translog equation and four variables were selected to determine the model. The obtained function allowed evaluating the systems globally and according to typology. In addition, some variables explaining farm's profitability were identified. The average efficiency was of 62% and a 26% of the sample farms showed high level of efficiency. The familiar farms presenting a commercial profile, with medium-sized herds managed in family labor semi-intensive systems, reached the highest levels of efficiency. In contrast, the semi-intensive commercial systems, reported poorer results than the before mentioned. This system was caracterized with the following inputs: keeping sheep housed most of the year, high percentage of foreign labor, high levels of investment and poor utilization of supplementary feeding.

Development of costs efficiency in dairy cattle in Slovakia
Michaličková, M., Krupová, Z., Huba, J. and Polák, P., Animal Production Research Centre Nitra, Hlohovecká 2, 951 41 Lužianky, Slovakia; michalickova@cvzv.sk

The objective of this study was to calculate the economic and technical efficiency of the milk production costs on 15 cattle farms (database of APRC Nitra) for the period 2006-2010. Economic efficiency was determined by individual indicators of profitability of individual cost items formed in the calculation formula and by the synthetic indicator of the total cost profitability. A nonparametric methodological approach Data Envelopment Analysis (DEA) was used to analyze the technical efficiency. When evaluating the cost-effectiveness, market price of milk with and without set-off direct subsidies was taken into account. The average profitability rate of the cost in milk production was -8% (from -23% without subsidies counted in 2009 to 8% taking subsidies into account in 2007). The highest efficiency of costs (upper limit of technical efficiency) in milk production was found in 2007 and the lowest in 2009 (without inclusion of subsidies) or in 2010 (after subsidies incorporation). Mean score of technical efficiency was 0.88 in the evaluated period. Growing demand for milk, increased price and higher milk production per cow determined the improved efficiency in milk production in 2007. The sharp fall in milk prices in 2009, the subsequent reduction in the number of cows and savings in feeds consumption resulted in the lower efficiency of milk production. These negatives were partially offset by higher value of subsidies in 2009. Including of subsidies into milk price improved profitability on average by 6 percentage points and effectiveness of costs by 3 percentage points. The biggest impact on the profitability of milk production was found for own and purchased feeds, depreciation of basic stock and of long-term assets, while the lowest importance was observed for repairs and service and for management overhead costs.

Effect of diets differing in rumen soluble nitrogen on poor quality roughage utilization by sheep
Van Niekerk, W.A., University of Pretoria, Animal and Wildlife Sciences, Lynnwood, Hillcrest, 0002, Pretora, South Africa; willem.vanniekerk@up.ac.za

The aim was to determine whether a rapid release N source (urea) can be substituted with a slow release N source (Optigen® II) when feeding sheep a poor quality hay. Five rumen cannulated wethers were used in a 5×5 Latin square design. Treatments were as follow (urea : Optigen): 100 : 0; 75 : 25; 50 : 50; 25 : 75; 0 : 100, all on an iso N, ME and mineral basis. Organic matter intake was higher (P<0.05) for the 25 : 75 than the 100 : 0, 50 : 50 and 0 : 100 treatments. No differences were recorded for DM and NDF digestibility as well as for total rumen microbial N production. Nitrogen digestibility was higher (P<0.05) for all the treatments compared to the 0 : 100 treatment. Overall, rumen pH, rumen NH3-N and rumen volatile fatty acid production did not differ, but the rumen NH3-N concentrations were lower (P<0.05) at 2 and 4 hours after infusion of N for the 0 : 100 than the 100 : 0 treatment. The N balance was the highest for the 50 : 50 and the lowest for the 0 :100 (P<0.05) treatment. No differences were recorded for both DM and NDF rumen degradability. It can be concluded that urea might be partly substituted by Optigen® II, but the higher cost /kg N of Optigen® II has to be considered.

Self-sufficiency is key to explain economic sustainability of sheep farming in marginal areas
Ripoll-Bosch, R., Joy, M. and Bernués, A., CITA, Av. Montañana 930, 50059 Zaragoza, Spain; rripoll@aragon.es

In the Euro-Mediterranean basin many sheep farming systems are located in marginal areas, often considered as High Nature Value (HNV) farmland. The analysis of sustainability of these farming systems involves the consideration of multiple variables and attributes, such as productivity, adaptive capacity or self-reliance, which are key to understand how farms might face changes in the future. Mixed cereal-sheep farms rearing a local breed (Ojinegra, Spain) were analyzed to identify key technical and economical parameters that determine their sustainability. Data regarding farm structure, management, economic and social aspects, were obtained in 2008 through direct interviews to farmers (n=30, total population 41). Principal Components Analysis and Cluster Analysis allowed to identify relationships among variables that explained the diversity of farms, and thereafter to classify them into homogeneous groups. Three main factors, explaining more than 60% of the original variance, were identified: (1) 'feed self-sufficiency'; (2) 'animal productivity'; and (3)'mixed sheep-cereal orientation'. These factors allowed identifying 4 homogeneous groups of farms. Feed self-sufficiency and reliance on natural resources greatly determined the economic profitability of the farms due to lower variable costs (i.e. low feed inputs). Animal productivity allowed for lower dependency on premiums (economic self-sufficiency), and was also related to economic profitability. However, diversity of production (lamb meat-cereals) had no relation with the economic results.

Session 44 Theatre 1

Nutrient sensing in the lingual epithelium
Margolskee, R.F., Monell Chemical Senses Center, 3500 Market Street, Philadelphia, PA 19104, USA; rmargolskee@monell.org

During the past decade beginning with the discovery in 1992 of the taste G protein gustducin many of the receptors and signalling molecules underlying taste have been identified. Confirmed and candidate receptors have been identified for the five taste qualities of sweet, bitter, salty, sour and umami (Japanese for delicious, and corresponding to the taste of glutamate and other amino acids). Type 1 taste receptors (T1rs encoded by Tas1r genes) initiate sweet (T1r2+T1r3) and umami (T1r1+T1r3). Bitter is initiated by a family of 25-30 Type 2 taste receptors (T2rs encoded by Tas2rs). The downstream components of sweet, bitter and umami are mediated by signalling elements common to all three pathways: heterotrimeric gustducin ($\beta3\cdot\gamma13\cdot\alpha$-gustducin), phospholipase $\beta2$, inositol trisphosphate receptor type 3, and the Ca^{++} gated cation channel TrpM5. Salty and sour are less well understood: the epithelial type sodium channel (ENaC) has been implicated in salty, but the sour transducer is still somewhat mysterious. We have found that many of the receptors and downstream signalling elements involved in taste detection and transduction are expressed also in intestinal hormone-producing (endocrine) cells where they underlie key chemosensory functions of the gut and pancreas. Knockout mice lacking α-gustducin or the sweet taste receptor subunit T1r3 have deficiencies in secretion of glucagon like peptide-1 (GLP-1) and in the regulation of plasma levels of insulin and glucose. In turn, taste cells of the oral cavity express GLP-1, other 'gut' hormones and the insulin receptor. Most recently, we have identified intestinal-type glucose transporters (GLUTs and SGLT1) and pancreatic-type ATP-gated K^+ channels (K_{ATP} metabolic sensors) as being present in taste cells and likely functioning in the detection of the sweet taste of sugars. In sum, these studies point out similarities in gustation and gut chemosensation and indicate the importance of 'taste cells of the gut' and 'endocrine cells of the tongue' in coordinating the body's hormone responses to regulate glucose homeostasis.

Session 44 Theatre 2

Glucose sensing in the intestinal tract; relevance to gut heath
Shirazi-Beechey, S.P., University of Liverpool, Functional and Comparative Genomics, Veterinary Sciences Building, Liverpool L69 7ZJ, United Kingdom; spsb@liverpool.ac.uk

In recent years, a number of nutrient sensors, activated by various dietary nutrients, have been identified and shown to be expressed in endocrine cells of the gut. Nutrient sensing initiates a cascade of events involving hormonal and neuronal pathways. This culminates in functional responses that ultimately regulate processes such as nutrient digestion and absorption, food intake, insulin secretion and metabolism. The sweet taste receptor, T1R2-T1R3 heterodimer, coupled to G-protein, gustducin, is expressed in the intestinal endocrine cells and functions as the luminal sweet sensor. Direct activation of T1R2-T1R3 by natural sugars and artificial sweeteners evokes secretion of a number of peptides including glucagon like peptide-2 (GLP-2). This gut hormone has multiple signalling functions including the enhancement of intestinal growth and glucose absorption. We have been interested in the role played by intestinal sweet sensing in regulation of glucose transporter, Na^+/glucose cotransporter 1, SGLT1. SGLT1 is the major route for the transport of dietary sugars from the lumen of the intestine into enterocytes. Dietary sugars and artificial sweeteners directly enhance SGLT1 expression and the capacity of the gut to absorb glucose; T1R2-T1R3 and gustducin are required for this upregulation. T1R2-T1R3 and gustducin reside in enteroendocrine cells, whereas SGLT1 is expressed in neighbouring absorptive enterocytes, thus proposing a signalling pathway between chemosensory enteroendocrine cells and absorptive enterocytes. GLP-2 receptor expressed in enteric neurons, responds to GLP-2 by initiating a neuronal signal resulting in increased functional expression of SGLT1 in absorptive enterocytes. The findings provide the molecular basis for the effect of dietary supplementation with artificial sweeteners on early weaned piglets. This practice results in enhanced intestinal tissue growth and glucose absorption promoting the health of the young animal.

The extracellular calcium sensing receptor, CaSR, as a nutrient sensor in physiology and disease
Riccardi, D., Brennan, S.C., Davies, T., Schepelmann, M., Iarova, P. and Kemp, P.J., Cardiff University, Museum Avenue, CF10 3AX, Cardiff, United Kingdom; riccardi@cf.ac.uk

From bacteria to mammals, cells possess 'sensors' which monitor nutrient availability in the environment. The extracellular Ca^{2+}-sensing receptor, CaSR, represents the first sensor of this kind capable of detecting the levels of an inorganic ion, Ca^{2+}. Genetic studies in humans have confirmed the importance of CaSR in the control of extracellular Ca^{2+} concentration. Activation of this receptor leads to the production of classic second messengers. Pharmacological studies have shown that the CaSR is activated by Ca^{2+} and other divalent cations, but also that its function is modulated by the presence of naturally occurring ligands such as L-aromatic amino acids and polyamines, extracellular pH and ionic strength. Because of its ability to act in a multimodal manner, this receptor also plays an important role outside the Ca^{2+} homeostatic system, acting as an integrator of multiple environmental signals for the regulation of many vital processes, including intercellular communication, secretion and cell fate. In the gut, where the CaSR is highly expressed, activation of this receptor leads to gastric acid secretion and to fluid reabsorption by the colon. CaSR-dependent inhibition of fluid secretion is being targeted for the treatment of secretory diarrhoea. CaSR is also required for cholecystokynin secretion in response to certain amino acids. CaSR activation also suppresses colonic crypt cell proliferation, therefore suggesting a mechanism linking low dietary Ca2+ intake and increased risk of colon cancer. The last fifteen years have seen a great expansion of our knowledge of CaSR function in health and disease. This presentation will address the role of the CaSR as a nutrient sensor, its functions in the gastrointestinal tract and CaSR-based therapeutics. Funded by the FP7 Marie Curie ITN grant ('Multifaceted CaSR,' grant 264663; http://www.multifaceted-casr.org/).

Lipid sensing in the gut mucosa; regulation of gut function and food intake
Raybould, H.E., UC Davis, School of Veterinary Medicine, 1321 Haring Hall, Davis, CA 95616, USA; heraybould@ucdavis.edu

The gastrointestinal epithelium is endowed with numerous sensory mechanisms that detect the presence of nutrients, including the three major macronutrients, carbohydrates, protein and lipid. Detection of the chemical composition of the meal by specialized cells in the gut epithelium, the enteroendocrine cells (EECs), releases peptide hormones and other mediators that act as hormones via the blood stream to influence gut and pancreatic endocrine and exocrine function. In addition, these bioactive molecules released from EECs can act via neural pathways to influence gut function and food intake. Several mechanisms have been demonstrated in EECs by which nutrients are detected. Detection of long chain triglycerides likely involves at least two mechanisms, one that involves G-protein coupled receptors free fatty acid receptor (FFAR1; also referred to as GPR-40) and a mechanism involving apolipoprotein A-IV (apo A-IV). These sensors have been shown to play a role in the release of a number of gut hormones, such as cholecystokinin (CCK), peptide YY (PYY) and the incretin hormone, glucagon-like peptide-1 (GLP-1). The gut epithelium is also richly innervated by vagal afferent fibers that express receptors for many of the products of EECs, including CCK, GLP-1 and PYY. Activation of the vagal afferent pathway results in reflex regulation of gut function but is also crucial in the regulation of food intake. In this way, the presence of lipid in the gut lumen is detected followed by appropriate physiological responses, such that the digestive and absorptive capacity of the small intestine is matched with the entry of food. Dysregulation of the systems may disrupt gastrointestinal homeostasis and may also lead to altered regulation of body weight and glucose homeostasis, such as that seen in obesity and type 2 diabetes.

G protein-coupled receptors, nutritional and therapeutic targets

Milligan, G., University of Glasgow, College of Medical, Veterinary and Life Sciences, University Avenue, G12 8QQ, Glasgow, United Kingdom; Graeme.Milligan@glasgow.ac.uk

A number of G protein-coupled receptors sense the presence of free fatty acids and hence report nutritional status. In both man and rodents Free Fatty Acid receptors FFA2 and FFA3 respond to short chain fatty acids such as acetate and propionate that are produced in the gut by microbial fermentation of non-digestible carbohydrates. When examining the bovine orthologues of these two receptors, although bovine FFA3 generated responses to the same group of short chain fatty acids as this receptor from other species, bovine FFA2 displayed very distinct pharmacology, responding effectively to fatty acids with chain length C6-C8. This allowed us to identify a number of naturally produced and synthetic ligands that were able to activate bovine FFA2 and not bovine FFA3. These will be useful to explore the specific contributions of bovine FFA2. Consideration of the sequences of bovine and human FFA2 and the production of homology models allowed mutation to convert the human orthologue of FFA2 to respond only to previously bovine selective ligands and provided novel insights into the othosteric binding pocket of the receptor. We have also recently developed a selective antagonist of FFA2 that will also be of great use in defining the physiological functions of this receptor. GPR40 and GPR120 is a further pair of G protein-coupled receptors. These respond to long chain fatty acids. Selective activation of GPR40 is a potential therapeutic treatment for diabetes as this receptor is expressed by pancreatic beta cells and stimulates the release of insulin in a glucose dependent manner. GPR120 is also considered as a potential target to regulate the release of insulin as it is reported to enhance release of the incretin GLP-1 from gut enteroendocrine cells. We have recently developed the first potent and highly selective GPR120 agonist and the effects of this ligand and whether it provides support for this idea will be discussed.

Sweet sensing by gut lactobacillus enhances its growth and population abundance

Daly, K.[1], Hall, N.[1], Bravo, D.[2] and Shirazi-Beechey, S.P.[1], [1]University of Liverpool, Functional and Comparative Genomics, Veterinary Sciences Building, Liverpool L69 7ZJ, United Kingdom, [2]Pancosma SA, Voie-des-Traz 6, Geneva 1218, Switzerland; nkd@liverpool.ac.uk

Lactobacilli have been implicated in promoting gut health via stimulation of host immunity and anti-inflammatory responses as well as protection against mucosal pathogen invasion. Lactobacillus strains are also widely used as probiotic organisms. They grow by fermenting sugars and starches and produce lactic acid as their primary metabolic product. For efficient utilization of varied carbohydrates, lactobacilli have evolved diverse sugar transport and metabolic systems; these systems can be specifically induced by their own substrates. We have employed DNA-based pyrosequencing technology to investigate changes in the intestinal microbiota of piglets weaned to a hydrolysable carbohydrate (CHO) diet supplemented with either lactose, or an artificial sweetener (SUCRAM®, Pancosma SA) consisting mainly of saccharin. Addition of either lactose or saccharin to feed dramatically increased the caecal population abundance of a specific strain of lactobacillus, with concomitant increases in luminal lactic acid concentrations. The population of this lactobacillus strain increased from 4% of the total microbiota in pigs fed the hydrolysable CHO diet to 20% in pigs fed diet supplemented with either lactose or saccharin. The increase in lactobacillus in response to lactose is expected, as lactose is a readily utilized energy source. In contrast, saccharin remains intact in the luminal contents of the intestine. It is known that artificial sweeteners are sensed by a specific receptor expressed in intestinal endocrine cells leading to enhancement of glucose transport. We hypothesise that this lactobacillus strain may also possess a sensor for recognizing saccharin, leading to upregulation of sugar transport and metabolic systems. This will provide a competitive advantage, reflected in the increased population abundance observed in response to saccharin.

Nutrient sensing by the immune system: the role of the intestinal microbiota

Bailey, M., Inman, C.F., Christoforidou, Z. and Lewis, M.C., University of Bristol, School of Veterinary Science, Langford House, Langford, Bristol BS40 5HT, United Kingdom; mick.bailey@bristol.ac.uk

Humans and animals are composed of 100 times more bacterial cells than eukaryotic cells. The commensal intestinal microbiota modifies nutrients ingested by the host: in agricultural species a high proportion of macronutrients may be derived from the microbiota. In addition, changes in nutrient supply are capable of altering the composition of the intestinal microbiota. Many physiological systems can be affected by such changes, and the mucosal immune system is critically dependant for normal function on the composition and function of the intestinal microbiota. Thus, the microbiota constitute a signalling intermediate between nutrients and the immune system. The mucosal immune system protects the intestine from potential pathogens. However, it also maintains tolerance towards harmless antigens associated with food and commensal microbiota. Young piglets mount immune responses to proteins in the weaning diet, and studies in rodents and pigs indicate that the ability to develop tolerance towards these proteins may be dependent on the process of microbial colonisation. We have used two approaches to study the effects of early colonisation on intestinal immunological function: firstly, the highly reductionist approach of caesarean-derived, gnotobiotic piglets, colonised with defined oligobiota; secondly, conventionally born piglets either left on-farm with the sow or moved into a high-security, SPF isolator system. Manipulation of the developing microbiota affects development of the mucosal immune system, both structurally and functionally. An early event in this process involves differences in the types of antigen-presenting cells (such as dendritic cells) which are recruited to the intestine, but this is followed by effects on B- and T-cells. We suggest that deliberate manipulation of early-life colonisation by microbiota using nutritional intervention may be one approach to controlling infectious and immunologically-mediated diseases in neonates.

Does calf starter composition affect unweaned dairy calves' preferences to grass-clover silage?

Vestergaard, M., Nielsen, M.F., Weisbjerg, M.R. and Kristensen, N.B., Aarhus University, Department of Animal Science, Foulum, 8830 Tjele, Denmark; mogens.vestergaard@agrsci.dk

The objective was to test if milk-fed dairy calves' preferences to four different grass-clover silages depended on the composition of the calf starter offered. A total of 14 Holstein bull calves (15 ± 2 d and 53.8 ± 10.2 kg) were used for a 38 d experiment. Calves were offered 2×2.42 kg/d of a skim-milked based milk replacer. Seven calves had access to a high starch (STARCH) and 7 to a high sugar (SUGAR) based calf starter. The composition of starters (g/kg DM, STARCH vs. SUGAR) was: CP (204 vs. 227), NDF (203 vs. 230), starch (387 vs. 116), and sugar (75 vs. 195). All calves had free access to four buckets with each of the four primary growth grass-clover silages (GCS). GCS were obtained from silage bunkers 10-11 months after harvest and wrapped in round bales and frozen at -20 °C until fed. The four GCS (S1-S4) were similar in CP (14-17%), ash (8%), NDF (37-43%), OMD (77-78%), and pH (4.1-4.4) but varied in their content (g/kg DM) of acetic acid (S1: 40.5 vs. S3: 5.3), ethanol (S2: 22 vs. S4: 3), sugar (S1 and S4: 4.5 vs. S3: 14), and glucose (S1: 0.6 vs. S3: 45.8). Total calf starter intake (681 vs. 731 g DM/d) and total silage intake (173 vs. 198 g DM/d) was not different between STARCH and SUGAR, respectively. The proportional intake of the four GCS was 0.09, 0.18, 0.59, and 0.14 for S1, S2, S3, and S4, respectively, and not affected by calf starter. The higher intake of S3 is most likely related to the high sugar/glucose and low acetic acid content. Despite the large differences in composition of the calf starter, it did not affect the individual proportional intake of the four grass-clover silages. Furthermore, total roughage intake was not dependent on the calf starter composition.

Effect of dietary free or protein-bound Lys, Thr, and Met on expression of b0,+ and myosin in pigs
Grageola, F., García, H., Morales, A., Araiza, A., Arce, N. and Cervantes, M., Universidad Autónoma de Baja California, ICA, Mexicali, BC, 21255, Mexico; miguel_cervantes@uabc.edu.mx

Pigs fed diets supplemented with free amino acids (AA) do not perform as good as those fed protein-bound AA. It is speculated that free AA are absorbed faster than protein-bound AA, which may lower the performance of pigs. Since Lys is the first limiting AA, an experiment was conducted to analyze the expression of the cationic AA transporter b0,+ in intestine, liver, and two muscles, and myosin in two muscles of pigs fed diets with either free or totally protein-bound AA (TP-AA). Twelve pigs (31.7±2.74 kg) were used. Dietary treatments (T) were: T1, wheat-based diet plus 0.59% L-Lys, 0.32% L-Thr, and 0.10% DL-Met; T2, wheat-soybean meal diet, TP-AA. Both diets contained 0.95% Lys and 0.70% Thr, and met or exceeded the requirements of the essential AA. Feed intake was restricted to 1.53 kg/d per pig. All pigs were sacrificed at the end of a 28-d trial; samples from jejunum (JE) and ileal (IL) mucosa, liver (LI), and longissimus (LM) and semitendinosus (SM) muscles were collected to analyze expression of b0,+; expression of myosin in LM and SM was also analyzed. The relative expressions (arbitrary units, mRNA Mol/18S rRNA Mol) of b0,+ for pigs on T1 and T2 were: JE, 0.041, 0.104; IL, 0.571, 0.348; LI, 0.013, 0.003; LD, 0.005, 0.006; SM, 0.001, 0.002; the expression of myosin was: LD, 0.006, 0.005; SM, 0.005, 0.005. The expression of b0,+ was higher in JE but lower in LI, in pigs fed the free-AA diet (P<0.01), and was not affected in IL, LM and SM (P>0.05); the expression of myosin was not affected in LM nor SM. There was a negative correlation in the expression of b0,+ between JE and LI (r=0.80; P=0.001). The expression of b0,+ was higher in IL as compared to JE. These data show that b0,+ is more expressed in the small intestine, and suggest that free AA are absorbed at a different rate than protein-bound AA. Also, these data suggest that the high expression of b0,+ in LI compensates for the lower expression in JE.

Genomic signatures of selection in the horse
Mickelson, J.R., Petersen, J.L., Mccue, M.E. and Valberg, S.J., University of Minnesota, College of Veterinary Medicine, 1988 Fitch Ave, St Paul, MN 55108, USA; micke001@umn.edu

Many breeds of horses have been created to uniformly exhibit particular phenotypes. We are using genome-wide SNP genotype data collected from greater than 20 horses from 33 breeds at over 54,000 loci to attempt to identify some of these genomic regions under selection. Putative loci under selection were identified using the F_{ST}-based statistic (d_i) calculated in sliding 500 kb windows. This statistic detects locus specific deviation in allele frequencies for each breed relative to the genome-wide average of pair-wise F_{ST} summed across breeds. Numerous potential targets of selection were identified, and analysis of breeds fixed for the chestnut coat color mutation demonstrated the utility of this method. Striking features of these genome scans include a 6 Mb region on ECA18 with a highly significant d_i value in the Quarter Horse and Paint Horse. Further analysis of the region revealed a ~1 Mb conserved haplotype surrounding the MSTN gene that is present in 92.8% of QH and 50% of Thoroughbreds, but rare (<1%) in all other breeds. Sequencing of MSTN identified a promoter variant due to a SINE insertion, which, along with intronic polymorphisms, are significantly associated with the conserved haplotype. Histological data from 79 horses shows a significant association of muscle fiber type proportions with the MSTN polymorphisms. The gaited breeds, including the Standardbred, Icelandic, Peruvian Paso, Paso Fino, and others, share a highly conserved haplotype on ECA23 under a strong signal of selection that contains a polymorphism demonstrated to be important in the ability to gait. Conserved haplotypes underlying signals of selection on ECA11 in the Belgian, Percheron, Shire, Clydesdale, and Miniature horse suggest the presence of a locus important in the determination of size. Mapping signatures of selection in the modern horse is the first step in the identification of genes important in the domestication and specialization of modern horse breeds.

A mutation in a novel transcription factor affects the pattern of locomotion in horses

Andersson, L.S.[1], Schwochow, D.[1], Rubin, C.[2], Arnason, T.[3], Petersen, J.L.[4], Mccue, M.E.[4], Mickelson, J.R.[4], Cothran, G.[5], Mikko, S.[1], Lindgren, G.[1] and Andersson, L.[1,2], [1]SLU, Dept of Animal Breeding and Genetics, Box 597, 75124 Uppsala, Sweden, [2]Uppsala Univ., Dept of Medical Biochemistry and Microbiology, Box 582, 75123 Uppsala, Sweden, [3]The Agricultural Univ. of Iceland, Land and Animal Resources, Hvanneyri, 311 Borgarnes, Iceland, [4]Univ. of Minnesota, College of Veterinary Medicine, 988 Fitch Ave, St Paul, MN 55108, USA, [5]Texas A&M Univ., Dept of Integrative Biosciences, TAMU4458, College Station, TX 77843, USA; Lisa.Andersson@slu.se

The three naturally occurring gaits in horses are the walk, trot and canter. Beside the basic gaits, most Icelandic horses can perform the four-beat toelt and many can also pace. A pacing horse simultaneously moves the two legs on the same side in a lateral movement. We performed a GWAS using the Illumina EquineSNP50 BeadChip in Icelandic horses with and without the ability to pace. A chi^2 test was performed for each marker separately using PLINK assuming a recessive mode of inheritance which identified a highly significant association. Whole genome resequencing of two Icelandic horses identified a single base change causing a premature stop in a transcription factor within a 438 kb IBD region. This gene is expressed in a subset of neurons in the spinal cord of the horse and the mouse. The association was verified in 190 additional Icelandic horses and horses from gaited and non-gaited breeds were screened for the mutation. Most individuals from gaited breeds were homozygous for the mutation. The mutation was also found at high frequency in trotters used for harness racing where it appears to be advantageous for performance by delaying the transition from the trot to gallop. The mutation was absent in other non-gaited breeds. Thus, this mutation appears to be permissive for the ability to perform alternate gaits, which can be either pace or four-beat ambling gaits. We hypothesize that the mutation modifies the neuronal circuitry in the horse spinal cord into a more flexible state.

Genome-wide association mapping and genomic breeding values for warmblood horses

Distl, O., Metzger, J., Schrimpf, R., Philipp, U. and Hilla, D., University of Veterinary Medicine Hannover, Institute for Animal Breeding and Genetics, Buenteweg 17p, 30559 Hannover, Germany; ottmar.distl@tiho-hannover.de

Dressage, show-jumping, riding horse points and conformation are the main breeding objectives in Hanoverian warmblood horses. We genotyped 246 Hanoverian stallions of the National State Stud of Lower Saxony using the Illumina Equine 50K beadchip in order to perform genome-wide association studies for quantitative trait loci (QTL) of performance traits, osteochondrosis (OC), radiological changes of navicular bones and stallion fertility. Mixed linear animal model technology was employed to control population stratification and the identity-by-state relationships among stallions. Traits for analysis were deregressed breeding values (BV). All the stallions under analysis had accuracies of BVs for performance traits above 0.7 and health traits above 0.5. We identified six QTL for show-jumping, twelve QTL for dressage, four QTL for riding horse points and two QTL for limb conformation. For health and fertility traits, QTL were found on several horse chromosomes. Genes involved in muscle structure, development and metabolism are crucial for elite show jumping performance whereas genes responsible for coordination, ataxia and learning aptitude might play a major role for excellent dressage performance. There was not any significant QTL overlapping among dressage and show-jumping. Some OC-QTL were located near or within QTL for performance traits. The 50K genotypes of these widely used Hanoverian stallions are the backbone for prediction of genomic BVs as their haplotypes can be found in most Hanoverian warmblood horses and many horses of many other warmblood breeds. A one step evaluation method for genomic BVs has been developed to evaluate the possible improvement in prediction of breeding values in young horses. Particularly for health and fertility traits, we expect that breeders will put more emphasis in selection on these traits.

First results on genomic selection in French show-jumping horses

Ricard, A.[1], Danvy, S.[2] and Legarra, A.[3], [1]INRA, UMR1313, 78352 Jouy-enJosas, France, [2]IFCE, Recherche et Innovation, 61310 Exmes, France, [3]INRA, UR 631, 31326 Castanet-Tolosan, France; anne.ricard@toulouse.inra.fr

Genomic selection could be highly interesting for horse breeding because it would reduce the nowadays high generation interval, at a low cost compared to the value of an animal. The aim of this study was to estimate the observed accuracies of genomic estimated breeding values in a representative set of show jumping horses. A sample of 908 stallions specialized in show jumping (71% Selle français (SF), 17% Foreign sport horses (FH), 13% Anglo Arab(AA)) were genotyped. Genotyping was performed using Illumina Equine SNP50 BeadChip and after quality tests, 44,444 SNP were retained. From whole population BLUP-based estimated breeding values and their reliability, a specific procedure was developed in order to obtain de-regressed proofs combining own performances and performances of relatives outside the genotyped sample. Two methods were used for genomic evaluation: GBLUP and Bayes Cπ, and 6 validation data sets were compared, chosen according to breeds SF+FH+AA or SF+FH, family structure (more than 3 half sibs), reliability of sires (>0.97) or sons (>0.72). In spite of a favorable genetic structure (linkage disequilibrium equal to 0.24 at 50Kbp), results showed low advantage of genomic evaluation. On the validation sample SF+FH+AA, the correlation between de-regressed proofs and GBLUP or BayesCπ predictions was: 0.39, 0.37, 0.51 according to the different validation data sets compared to 0.36, 0.33, 0.53 obtained with BLUP predictions. Correlations were much lower on the SF+FH sample. No practical applications are proposed at present. Research is pursued in order to improve the number of pairs sire-son with high number of measured progeny and to improve methodology in this context, less favorable than dairy cattle breeding.

The myostatin sequence variant g.66493737T>C detects evolution and domestication in horses

Dierks, C.[1], Mömke, S.[1], Philipp, U.[1], Lopes, M.S.[2] and Distl, O.[1], [1]University of Veterinary Medicine Hannover, Institute for Animal Breeding and Genetics, Buenteweg 17p, 30559 Hannover, Germany, [2]University of Azores, Biotechnology Centre of Azores, Rua Capitao Joao D'Avila, 9700-042 Angra do Heroísmo, Portugal; ottmar.distl@tiho-hannover.de

Myostatin (MSTN) is a negative regulator of muscle growth. Particular or complete loss of function of MSTN leads to muscle hypertrophy. We determined the presence of the sequence variant g.66493737T>C polymorphism and the promoter insertion g.66495327-[Insertion227b]-66495326 of MSTN associated with sprinting ability in thoroughbreds, donkeys and 19 different horse populations. In addition, we analyzed the haplotypes surrounding these MSTN-polymorphisms using the genotyping data from the Illumina Equine 50K beadchip. The C-allele was found in all domestic horse breeds but not in Przewalski horses and donkeys genotyped here. The frequency of the promoter insertion was quite low (0.5%) and not in linkage disequilibrium with the g.66493737T>C polymorphism. AMOVA for donkeys and the 19 horse populations showed that 25% of the variance of the g.66493737T>C SNP is due to species and population differences and 75% of the variability can be attributed to differences within species and populations. Nei's standard genetic distances among donkeys and the 19 horse populations ranged from 0 to 0.29. Genetic distances were lowest among Przewalski and Arabian, Lewitzer, Lusitano, Hanoverian, Westphalian, Exmoor and Icelandic horses (0-0.01). Genetic distances were largest among Przewalski and Sorraia as well as Black Forest horses. Cluster analysis revealed five main clusters for the C-allele frequency. The analysis of the C/T-polymorphism of the MSTN gene highlights an east to west increase of the C-allele and indicates that this MSTN mutation arose in the West European E. ferus or during domestication in western European horses. In conclusion, the g.66493737T>C polymorphism seems to be a good indicator for differentiating domestic horses due to their western or eastern origin.

Informative genomic regions for insect bite hypersensitivity in Shetland ponies in the Netherlands

Schurink, A.[1], Ducro, B.J.[1], Bastiaansen, J.W.M.[1], Frankena, K.[2] and Van Arendonk, J.A.M.[1], [1]Wageningen University, Animal Breeding and Genomics Centre, P.O. Box 338, 6700 AH Wageningen, Netherlands, [2]Wageningen University, Quantitative Veterinary Epidemiology Group, P.O. Box 338, 6700 AH Wageningen, Netherlands; bart.ducro@wur.nl

Insect bite hypersensitivity (IBH) is the most common allergic disease present in horses worldwide. It has been shown that IBH is under genetic control but knowledge on associated genes is limited. We conducted a genome-wide association study to identify and quantify genomic regions contributing to IBH in the Shetland pony population. 97 cases and 91 controls were selected and matched on withers height, coat colour and pedigree to minimize population stratification. From participating Shetland pony mares a blood sample was collected, their IBH phenotype was scored and the owner filled in a questionnaire. 40,021 SNP were fitted in a univariable logistic model fitting an additive effect. Analysis revealed no effects of population stratification. Significant associations with IBH were detected for 24 SNP on 12 chromosomes ($-\log_{10}$(P-value) >2.5). Odds ratios of allele substitution effects of the unfavourable allele were between 1.94 and 5.95. The most significant SNP was found on chromosome 27, with an odds ratio of 2.31 and with allele frequency of the unfavourable allele of 0.72 in cases and 0.53 in controls. The prevalence in the Shetland pony population, being 7.6%, could be reduced to values between 4.4 and 6.8% (depending on population attributable fraction per SNP) when the unfavourable allele would be eliminated. Genome-wide association studies in additional horse populations are desired to validate the identified associations, to identify the genes involved in IBH and to develop genomic tools to decrease IBH prevalence.

Alternative splicing of the elastin gene in horses affected with chronic progressive lymphedema

De Keyser, K.[1], Schroyen, M.[1], Oosterlinck, M.[2], Raes, E.[3], Stinckens, A.[1], Janssens, S.[1] and Buys, N.[1], [1]KU Leuven, Biosystems, Kasteelpark Arenberg 30, bus 2456, 3001 Leuven (Heverlee), Belgium, [2]Ghent University, Department of Surgery and Anaesthesiology of Domestic Animals, Salisburylaan 133, 9820 Gent (Merelbeke), Belgium, [3]Ghent University, Department of Veterinary Medical Imaging and Small Animal Orthopaedics, Salisburylaan 133, 9820 Gent (Merelbeke), Belgium; Kirsten.DeKeyser@biw.kuleuven.be

Belgian draught horses are susceptible to chronic progressive lymphedema (CPL). Main causes are a failure of the lymphatic system and its sustaining elastic network, resulting in reduced lymphatic drainage. Clinical symptoms are mainly restricted to skin deformations of the lower limbs. A genetic background for this condition is strongly suggested. In several mammalian species, there is a considerable variation in the sequence of elastin gene (ELNgene) mRNA's, due to alternative splicing of the primary gene transcript. Alternative splicing of the ELNgene mRNA occurs less in adult animals compared to juvenile ones (fetal and neonatal), except in pathological situations. Therefore, skin biopsies from three CPL-affected Belgian draught horses were examined for alternative splicing of the ELNgene primary transcript. Based on inspection and palpation, horses were classified according to the severity of CPL-associated lesions (1 mildly, 1 severely and 1 extremely affected). From each affected horse, skin biopsies were taken from the neck (clinically healthy region) and the dorsal aspect of the right fore fetlock (clinically affected region). Skin from a healthy Belgian warmblood horse served as a negative control. Messenger RNA was extracted and cDNA was produced, amplified and electrophoretically separated. Primary results indicate the existence of different splice products in affected draught horses as well as the negative control. Fragments of CPL-affected Belgian draught horses and the warmblood horse will be compared.

Genetic variation in horse breeds derived from whole genome SNP data

Mickelson, J.R., Petersen, J.L. and Mccue, M.E., University of Minnesota, College of Veterinary Medicine, 1988 Fitch Ave, St Paul, MN 55108, USA; micke001@umn.edu

Hier ben ik

Horses have been used in transportation, warfare, agriculture, and athletic competition since their domestication, and selection for desired traits and fitness has resulted in a diverse population distributed across the world. Our limited knowledge regarding the genetic diversity of horses impacts our ability to correctly define population-based issues, identify and preserve characteristics that define breeds, and decipher the history of the modern horse and the basis of numerous complex genetic traits. This report describes the use of a genome-wide set of 54,000 autosomal SNPs and a group of 814 horses from 38 populations contributed by a collaborative community of equine researchers who are working to build a comprehensive understanding of genetic diversity among equine populations across the world. The data are being utilized to quantify diversity (effective population size N_e, expected heterozygosity H_e, inbreeding coefficient F_{IS}) within breeds, and define relationships between populations (F_{ST}, STRUCTURE, parsimony analyses) to provide the first description of equine breed diversity of this magnitude. Analyses show substantial variability in genetic diversity amongst breeds; estimates of N_e range from 163-751, H_e from 0.225-0.309, and F_{IS} from 0.000-0.063. Pairwise FST values between breeds ranged from 0.01-0.16. Structure output, parsimony, neighbor joining trees, and FST analyses demonstrated relationships among the breeds that largely reflect geographic origins and known breed histories. Low levels of population divergence within and between breeds were observed in cases of both ongoing admixture and/or high levels of diversity. The results of this work better allows the elucidation of relationships among breeds, describes diversity within breeds, should stimulate studies into the origins of breeds and breed-defining traits, and guide efforts to preserve genetic diversity.

Genomic research in horses: the view of practitioners

Von Velsen-Zerweck, A. and Burger, D., European State Stud Association ESSA, Gestütshof 1, 72532 Gomadingen, Germany; Astrid.vonVelsen-Zerweck@hul.bwl.de

Genomic research is actually the overwhelming topic in the horse breeding industry. Important and successful efforts of competitive scientists meet challenged but also interested officials of breeding federations as well as frightened breeders in the context of a financially difficult and emotional market. Despite the high potential and opening of new possibilities of genomic research especially for performance and health selection, as main challenges and concerns show up the lack of communication between science and practice as well as the fact, that phenotype often considerably differs from genotype. In addition, a common sense of breeding priorities has not yet been found, creating an emotional debate. Various authors representing the practical field of horse breeding will highlight/ focus on objectives, problems, use of genomic research and expectations of modern breeders, followed by a general discussion between practitioners and scientists.

Genomic selection in the Swiss Franches-Montagnes horse breed

Signer-Hasler, H.[1], Flury, C.[1], Haase, B.[2], Burger, D.[3], Stricker, C.[4], Simianer, H.[5], Leeb, T.[2] and Rieder, S.[3], [1]Berne University of Applied Sciences, School of Agricultural, Forest and Food Science, Laenggasse 85, 3052 Zollikofen, Switzerland, [2]University of Bern, Vetsuisse Faculty, Institute of Genetics, P.O. Box 8466, 3001 Bern, Switzerland, [3]Agroscope Liebefeld-Posieux Research Station ALP, Swiss National Stud Farm, P.O. Box 191, 1580 Avenches, Switzerland, [4]agn Genetics GmbH, Boertjistrasse 8b, 7260 Davos Dorf, Switzerland, [5]Georg-August-University, Department of Animal Sciences, Albrecht-Thaer-Weg 3, 37075 Goettingen, Germany; christine.flury@bfh.ch

The Franches-Montagnes (FM) is a genetically closed and indigenous Swiss horse breed. The population consists of about 21'000 individuals. The running breeding program includes the yearly estimation of BLUP breeding values for 43 different traits (28 conformation traits, 12 riding and driving performance traits, 3 coat color traits). 1'151 FM horses were selected for genotyping with the Illumina Equine SNP50 Bead Chip. These selected FM horses represent about one-third of the active breeding population, which comprises about 3'500 animals in total. After data preparation and plausibility check we started to map traits from the running breeding program and to estimate allele effects for those. The latter were used to deduce genomic breeding values, as well as for the identification of chromosomal regions harboring genes with major influence on diverse phenotypes in the horse. The accuracies of the first genomic breeding values for conformation type, height at withers and white markings were found to be 47%, 45%, and 88%, respectively using a training data set consisting of 90% oldest and a validation data set consisting of 10% youngest horses. In addition, results from cross-validation will be presented. The average accuracy of breeding values at birth due to the information of performance from ancestors is only between 20% and 35%. Thus, genomic breeding value estimation seems to be feasible and valuable in the Franches-Montagnes horse breed.

Sub-clinical mastitis in small ruminants: prevalence, comparative aspects and prevention

Leitner, G.[1] and Silanikove, N.[2], [1]Kimron Veterinary Institute, P.O. Box 12, Bet Dagan, 50250, Israel, [2]A.R.O, The Volcani Center., P.O. Box 6, 50250, Israel; leitnerg@moag.gov.il

Mammary gland infection is a major cause of economic loss for small ruminant farmers. The infection, caused by microorganisms (mainly bacteria as well as virus and algae) can be expressed in severe clinical local and systemic symptoms leading to the complete destruction of the gland and death and/or sub-clinical infection with no symptoms. The severity of the infection depends on the causing agent and the animal's condition. Due to the nature of sub-clinical infections, the actual prevalence of infection in many herds, especially in non-dairy ones, is unknown. Studies of non-dairy herds reported infection levels of ~5%, whereas in dairy herds from 35%-60%. Various coagulase-negative staphylococci, on the skin of the udder and its surroundings, are the major types of bacteria involved in subclinical mastitis. Zoonosis agents (brucellosis, tuberculosis, etc) may also cause sub-clinical infections, which are devastating to workers and consumers. Sub-clinical infection decreases milk yield and quality. In non-dairy herds, if the dam has sufficient milk for her offspring the infection has no economical value. In dairy herds treatment strategy for sub-clinical mastitis is more complicated and controversial; frequently, the infection is ignored, especially if the animal remains profitable. In countries where somatic cell count is a payment criterion, a high level of infection reduces profit and may result in rejection by the dairies. Moreover, recent studies related to the quality of cheese and yogurt indicate that the presence of bacteria in the udder changes the composition of the milk leading to lower quantity and quality of the product. The key to lowering the level of infection is prevention. In dairy herds, hygiene, mainly during milking and teat dipping, can reduce new infection. Treatment of sub-clinical infection is problematic because in many cases the cost and availability of bacterial examination and medicine do not justify the economic cost of the treatment.

Through the colostrogenesis knowledge to the small ruminant neonate health

Castro, N., Universidad de Las Palmas de Gran Canaria, Animal Science Department, Arucas, 35413 Las Palmas, Spain; ncastro@dpat.ulpgc.es

Immune passive transfer is a critical process in newborn small ruminants, due to the lack of immunoglobulin transfer during gestation through the placenta. In addition the immune system of the neonates is unable to produce its own Ig (in an effective way) during the first month of life. Despite it has been described the presence of IgG in the Peyer's patch of goat kids at birth, the survival of these animals depends on colostrum feeding during the first hours of life, due to in the neonate kid there is only a minimum production. Colostrogenesis starts several weeks before partum which coincides with the dry period, thus the dam management during this period (induction of parturition, length of dry off period) influence the colostrum quality. Colostrum contains a mixture of different components; as well IgG is one of the principal differences between colostrum and mature milk in small ruminants there are other constituents related to immune function (complement system, Chitotriosidase enzyme). Colostrum provides antimicrobial protection but one means of the infectious diseases to newborn small ruminants is through contaminated colostrums, one effective method to prevent the transmission of infectious diseases to newborn via colostrum is pasteurization, but new advances for colostrums sanitation by using not heating methods (high pressure, biocides) are being developed. Due to the important role of colostrum in small ruminants health it's necessary to understand its synthesis mechanism and its management.

Influence of a mixture of conjugated linoleic acid on dairy performance, and milk fatty acid composi

Ghazal, S., INRA-Agroparistech, UMR MoSAR, 16 rue Claude Bernard, 75231 Paris Cedex 05, France; ghazal@agroparistech.fr

The aim of this work was to study the effect of conjugated linoleic acid (CLA) supplementation (45 g/d) on milk performance and milk fatty acid (FA) composition in dairy goats fed a diet based on corn silage and rich in concentrates The CLA consisted of 2 isomers in equal quantity; 4.5g of C18:2 cis-9, trans-11 and 4.5g of C18:2trans-10, cis-12. 24 multiparous dairy goats in early to mid-lactation were used in a 9 weeks trial with the first 2 weeks for adaptation and a 7 week experimental period. Throughout the experiment, goats were fed, a TMR with, corn silage (35%), beet pulp (20%), barley (15%), a commercial concentrate (30%) and after 2 weeks adaptation to the TMR, experimental groups were fed 45g/d of a lipid supplement either CLA or 45g/d of Ca salts of palm oil (Control) added on the top of the TMR. Individual milk production and composition (fat, protein and lactose) was recorded weekly (from week 0 to 9), and milk FA composition was analysed in weeks 1, 5 and 6. All data were evaluated using the MIXED procedure of SAS for repeated measurements. The CLA supplementation had no effect on dry matter intake, body weight, milk yield, milk protein, and lactose (yield and content) but it decreased the milk fat yield (MFY, P=0.02) and content (MFC, P=0.0002) by 8.4% and 11.8%, respectively. Further, it improved the energy balance by 6% (P=0.05). CLA treatment changed milk FA profile, it decreased the proportion of C16:0-C16:1 (P=0.05) and the sum of cis 18:1 (P=0.001) but increased the proportion of long-chain FA (>C16:0) (P=0.02) without modifying the total trans C18:1 and the proportion of FA synthesised de novo (C<16). In conclusion, CLA supplementation associated with corn silage based diet rich in concentrates decreased MFY, MFC, modified the FA composition of goat milk and improved their energy balance.

Milk ejection occurrence before teat cup attachment on milkability of ewes

Tančin, V.[1,2], Antonič, J.[1], Mačuhová, L.[2], Uhrinčať, M.[2] and Jackuliaková, L.[1], [1]Slovak University of Agriculture in Nitra, Tr. A. Hlinku 2, 94901 Nitra, Slovakia, [2]Animal Production Research Centre Nitra, Hlohovecká 2, 95141 Lužianky, Slovakia; tancin@cvzv.sk

The dairy ewes are attached without any udder pre-stimulation so milk ejection in most of ewes is induced by milking machine as compared to cows. The aim of work was to study the effect of milk ejection before cluster attachment on milkability and milk composition in two groups of ewes differed by milk flow pattern during control milking. After three pre-experimental control milking 22 dairy ewes of two breeds Tsigaj (TS, n=11, 5 with one – 1P and 6 with two milk emissions – 2P) and Improved Valachian (IV, n=11, 6 with 1P and 5 with 2P) with healthy udders were selected from flock of 400 dairy ewes. The animals were divided in a two groups, each with 11 sheep (6 TS, 5 IV, and with the same ratio of 1P / 2P). Milk flow data were recorded during three consecutive evening milkings. During the first milking the first group was treated by 5 UI i.m. of oxytocin and the second group by physiological saline 60 s before milking. During third milking the procedure was changed. Milk flow was recorded using electronic jars collecting the full milk at milking. After OT treatment milk flow curves changed: from 11 ewes in 2P group had 8 ewes 1P and 3 ewes still had 2P; 11 ewes in 1P group had 9 ewes 1P, 1 ewe PL and 1 ewe hadn´t milk flow after OT treatment. Ewes with 1P significantly increase the milk yield (0.192 ± 0.06 l vs. 0.241 ± 0.07 l) and fat content (17.02 ± 4.95 g vs. 22.91 ± 7.97 g) as compare with group of ewes with 2P where differences were not found. OT treatment reduced milking time in 2P from 54.55 ± 24.46 s to 27.00 ± 11.66 s (P<0.05), but no effect was found out in 1P group. In conclusion, we don´t expect alveolar milk ejection in ewes with only 1P milk flow, because the milkability parameters and fat content were clearly changed after treatment by OT, and therefore during milk recording the milk flow pattern could significantly influence the results.

Formic acid inactivation of caprine arthritis encephalitis virus in colostrum

Morales-Delanuez, A.[1,2,3], Trujillo, J.[1,2], Plummer, P.[2,4], Hemnani, K.[1,2], Hernández-Castellano, L.E.[3], Castro, N.[3], Nara, P.[1] and Argüello, A.[3], [1]Iowa State University, Center for Advanced Host Defense Immunobiotics and Translational Comparative Medicine, Ames, Iowa, USA, [2]Iowa State University, Department of Veterinary Microbiology and Preventive Medicine, College of Veterinary Medicine, Ames, Iowa, USA, [3]Universidad de Las Palmas de Gran Canaria, Animal Science, Arucas, 35413 Las Palmas, Spain, [4]Iowa State University, Department of Veterinary Diagnostic and Production Animal Medicine, Ames, Iowa, USA; amorales@becarios.ulpgc.es

Caprine Arthritis-Encephalitis Virus (CAEV) is a lentivirus which causes in goats goats. The primary route of CAEV transmission in goats is from dam to kid through ingestion of infected colostrum/milk. Traditionally, prevention of CAEV transmission for eradication protocols include removal of kids prior to consumption of colostrum, and the administration of heat inactived colostrum or feeding colostrum replacers and segregation. Formic Acid (FA) historically has been used in dairy calves for room temperature stabilization of milk and for its antimicrobial properties without detrimental effects on passive transfer of essential immunogical or nutritional components of colostrum. The objective of this study was to evaluate the utility of FA to inactivate CAEV in colostrum. Cell free colostrum was spiked with CAEV (10^5TCID_{50}) then treated with varying amounts of FA (10% w/vol) to acidify colostrum to pH of 3, 4, 4.5, and 5, for 15 or 30 minutes. pH was returned to 7 with NaOH (5N). Residual viral particles ($TCID_{50}$) was enumerated utilizing the virus titration assay. Acidification of CAEV spiked colostrum to a pH of 3 and 4 after a 15 and 30 min resulted in a 99.99% of reduction of infectious virus particles, Acidification of spiked colostrum to a pH 4.5 and 5 did not significantly reduce the virus infectivity. presented differences with the non-acidified colostrum. Preliminary results demonstrate that acidification of Colostrum spiked with CAEV to a pH of 4 or results in effective in inactivation of CAEV.

Sodium dodecyl sulfate to inactivation caprine arthritis encephalitis virus in DMEM
Morales-Delanuez, A.[1,2,3], Argüello, A.[3], Hartmann, S.[4], Martell-Jaizme, D.[3], Castro, N.[3], Nara, P.[1] and Trujillo, J.[1,2], [1]Iowa State University, Center for Advanced Host Defense Immunobiotics and Translational Comparative Medicine, Ames, 50011 IA, USA, [2]Iowa State University, Department of veterinary medicine and preventive medicine, Ames, 50011 IA, USA, [3]Universidad de Las Palmas de Gran Canaria, Animal Science, Arucas, 35413 Las Palmas, Spain, [4]Drexel University, Microbiology & Immunology and Obstetrics & Gynecology, New College Bldg, Philadelphia, USA; amorales@becarios.ulpgc.es

The Caprine arthritis-encephalitis Virus (CAEV) is a lentivirus of goats with worldwide distribution. In the majority of industrialized countries, CAE represents an economic problem for goat farming in several European countries. The primary route of CAEV transmission in goats is from dam to kid through ingestion of colostrum/milk containing CAEV. Traditionally, prevention of CAEV transmission includes removal of kids prior to consumption of colostrum, and the administration of heat inactived colostrum/milk. Previously it was demonstrated that the antimicrobial effects of Sodium Dodecyl Sulfate (SDS) could be efficacious in inactivation of Human Immunodeficiency virus (HIV-1) in milk. Moreover goats fed milk spiked with 1%SDS suffered no ill effects with regard to immune status. Therefore we set up to determine if varying percentages of SDS could inactivate a known amount of CAEV spiked in DMEM. DMEM was spiked with CAEV (10^5TCID$_{50}$), then the colostrum was treated with varying amounts of SDS (10% solution) to a final concentration of SDS of 1%, 0.1%, 0.01% and 0.001%. Residual viral particles (TCID$_{50}$) was enumerated utilizing the virus titration assay on following removal of SDS utilizing centrifugation. At an SDS concentration of 1% and 0.1% resulted in 99.99% reduction of the virus input titer TCID$_{50}$, while a final concentration of 0.01% and 0.001% failed to provide significant reduction. Preliminary results demonstrate that a concentration of 1% and 0.1% of SDS in Colostrum spiked with CAEV results in effective in inactivation of CAEV.

Genetic parameters of chosen udder morphology and milkability traits and somatic cell score
Margetín, M.[1,2], Apolen, D.[1], Milerski, M.[3], Debrecéni, O.[2], Bučko, O.[2], Tančin, V.[1,2] and Margetínová, J.[1], [1]Animal Production Research Centre Nitra, Hlohovecká 2, 951 41 Lužianky, Slovakia, [2]Slovak University of Agriculture Nitra, Tr. Andreja Hlinku 2, 949 76 Nitra, Slovakia, [3]Institute of Animal Science, Přátelství 815, 104 00 Praha Uhřiněves, Czech Republic; margetin@cvzv.sk

Udder morphology traits (udder depth – UD, teat position – TP, teat size – TS, udder shape – US) were assessed throughout milking period using 9 point linear scores (LS) in ewes of an experimental sheep flock. In the same ewes after linear scoring of udders chosen milkability (MA) parameters were assessed (machine milked milk – MMM; portion of machine stripped milk to total milk yield – PMSM, portion of milk milked in 30 seconds to TMY – PM30s). Milk probes were collected from each ewe for determination of somatic cell count (SCC). Genetic parameters were estimated using non-transformed data for LS traits and MA traits and transformed data for SCC (somatic cell score – SCS; 1124 records in 344 ewes for each trait). Multiple-trait models (VCE program) were used for estimation. In addition to random additive genetic effect of animal and permanent effect of ewe, the models involved fixed effects of control year (7 levels), lactation stage (4 levels), breed group (9 levels) and parity (3 levels). MMM of ewes in experimental flock was in average 319.2±167.45 ml, PMSM 27.38±15.14%, PM30s 54.0±18.24% and SCS 2.44±0.99. Heritability coefficients for MMM, PMSM and PM30s were medium high (0,204, 0,134 and 0,147 respectively). Heritability coefficients for SCS were low (0,058). Genetic correlation between the UD and PMSM was 0.267; between UD and SCS 0.346. It means that too deep and baggy udders are not suitable in terms of milkability and milk quality. The important are also findings about positive genetic correlation between TP and MMM in ewes (0.544), and negative genetic correlations between TP and PMSM (-0.278), between MMM and PMSM (-0,792) and between MMM and SCS (-0,215).

Milkability and the udder morphology traits in Tsigai, Improved Valachian, and Lacaune ewes

Mačuhová, L.[1], Uhrinčať, M.[1], Mačuhová, J.[2] and Tančin, V.[1,3], [1]Animal Production Research Centre Nitra, Hlohovecká 2, 951 41 Lužianky, Slovakia, [2]Institute for Agricultural Engineering and Animal Husbandry, Prof.-Dürrwaechter-Platz 2, 85586 Poing, Germany, [3]Slovak University of Agriculture, Trieda A. Hlinku 2, 94901 Nitra, Slovakia; tancin@cvzv.sk

The aim of this study was to compare the milkability traits and investigate the relationship between correlation udder morphology traits and milking characteristic in breeds mostly bred in Slovakia. The trail was performed with 24 ewes of three breeds: Tsigai (TS, n=8), Improved Valachian (IV, n=8;) and Lacaune (LC, n=8). Ewes were routinely milked twice a day in 1×24 milking parlour. The milkability was measured in two consecutive months June and July. Experimental milkings were performed during three successive days in the middle of both months. Udder morphology traits (teat position, cistern depth, and udder cleft) had been assessed by the use of linear scores one day before first evening milking when the milkability measurements started. During milkings, an actual milk yield was recorded in 1-second intervals using a graduated electronic milk collection jars. In total 286 measurements were recorded. The breed had significant effect on total milk yield P<0.0439). The average total milk yield was 0.44±0.02 l, 0.40±0.03 l, and 0.53±0.02 l in TS, IV, and LC; resp. Total milk yield was positively correlated with maximal milk flow rate (r=0.37; P<0.0001), machine stripping milk yield (r=0.28; P<0.0001) and machine milking time (r=0.37; P<0.0001). Positive and significant correlations were found out between teat position and cistern depth (r=0.61; P<0.0001), teat position and maximal milk flow rate (r=0.52; P<0.0001). Total milk yield decreased with increased stage of lactation. Teat position and cistern depth changed throughout lactation. The breed had significant effect on total milk yield (P<0.0033).

The goat mammary gland parenchyma: breed and milking frequency influence

Suárez-Trujillo, A.[1], Hernández-Castellano, L.E.[2], Capote, J.[3], Argüello, A.[2], Arencibia, A.[1], Castro, N.[2], Morales, J.[1] and Rivero, M.A.[1], [1]Universidad de Las Palmas de Gran Canaria, Morphology, Arucas, Gran Canaria, Spain, [2]Universidad de Las Palmas de Gran Canaria, Animal Science, Arucas, Gran Canaria, Spain, [3]Instituto Canario de Investigaciones Agrarias, La Laguna, Tenerife, Spain; lhernandezc@becarios.ulpgc.es

Tissular percentages (secretory, connective, ductal and vascular tissues), number and size of the alveoli were studied in udders of three dairy goat breeds under two milking frequencies (once vs. twice daily milking). The objectives of this study were to elucidate the influence of the breed and milking frequency on the proportion of the tissue components in the mammary gland in dairy goats;and to correlate the productive parameters (milk yield, milk composition, milk fractions and udder morphology) with the tissue parameters. Three goats of each studied breed (Majorera, Palmera and Tinerfeña), were milked during 6 weeks (mid lactation). The right half udder was milked twice daily and the left half udder was milked once daily. Moreover, during the experimental period,the productive parameters were recorded. Two samples from each gland were taken for the histological study and were analyzed using morphometric software. Macro and microscopic observations revealed a healthy mammary parenchyma, not damaged by the milking frequency. The statistical analysis revealed that tissue parameters were not influenced by the milking frequency, and the breed determined different percentages of tissue components. Correlations between udder morphological parameters and milk yield parameters determined the importance of globosity and udder cisternal compartment in milk yield of these breeds. Furthermore, it was determined that the percentage of secretory tissue in the mammary parenchyma had no correlation with the milk yield parameters in different high-production dairy breeds. In conclusion, the histological parameters were only influenced by the breed, and there were not correlations with the milk yield parameters.

Anticipating on market requirements and its changes in a modern breeding program
Van Haandel, E.B.P.G. and Huisman, A.E., Hypor, P.O. Box 30, 5830 AA Boxmeer, Netherlands;
Benny.van.haandel@hendrix-genetics.com

The cornerstone of a breeding program is good quality phenotypic measurements, allowing the breeder to select the right animal for a specific breeding goal. Typical breeding goals cover a wide range of traits: reproduction, production performance, meat and carcass quality. Modern breeding programs focus on whole chain efficiency, meeting standard requirements for meat and carcass quality for commodity markets. Within the porkchain different powers determine pig value and quality requirements. Power switches between the meatpacking industry and primary producers, dependent on shortages or surpluses in pork production. Typically when producers are empowered, they focus on 'meat per feed' or on 'kg of pork produced per surface unit'. In this scenario feed efficiency will be one of the most important determinants of the breeding goal. When meatpackers can impose their requirements, they tend to focus on packing plant efficiency; carcass quality; carcass uniformity and meat quality. The last decade developments have become more global. The US market dominates and focuses strongly on packing plant efficiency. Slaughter costs per kg have initiated the request for increased slaughter weights. This has impacted most markets worldwide. Since the latest pork crisis lasted longer than any previous pork crisis there has been a predominant influence of meatpackers on carcass value requirements. Due to this development carcass weights have risen over 100 kg in the US market and anywhere from Canada to Northern Europe and Japan carcass weights have increased by at least 5 kg. This tendency requires another type of finishing pig from breeding companies, and thus a change in breeding goal. On top of that there is a global trend towards more emphasis on animal welfare and health, and less time to take care of individual animals. Since modern breeding goals no longer focus on pureline but directly on final product performance, these changes are directly taken into account in the day-to-day breeding program.

How the transition to free farrowing systems should work
Baumgartner, J., University of Veterinary Medicine Vienna, Veterinaerplatz 1, 1210, Austria;
johannes.baumgartner@vetmeduni.ac.at

In the last 50 years crates have become the predominant farrowing environment across the world, mainly due to a reduction of costs and a decrease in piglet mortality compared to traditional free farrowing systems. More recently farrowing crates have been discussed critically by scientists and in wider society from a welfare point of view. Against this background, 32 experts met at the 'Free Farrowing Workshop Vienna 2011' in order to gather scientific and experience-based knowledge and to discuss options, obstacles and questions regarding free farrowing systems. There was good scientific agreement on the principles that make free farrowing systems work. However, the robustness of systems has to be demonstrated in large scale studies. Temporary crating may be an intermediate step towards free farrowing. The change has to tackle farmer's attitudes and beliefs before it will take place in commercial practice. Breeding and selection should aim in vital piglets that need short time to suckle after birth. Birth weight, within litter birth weight variation and thermoregulation should be given more attention compared to litter size. Reactivity to screams during crushing, passivity to neonatal litter competition and passivity to a stockperson are essential characteristics of good maternal behaviour. Mobility, udder quality, placental efficiency and longevity of sows should also be taken into account to a higher extent. This should result in a low live born piglet mortality rate which is essential in different farrowing environments. The breeding companies have to solve a number of problems when breeding for free farrowing conditions. The heritability for behavioural traits is low. The available data set is limited and data quality is expected to be poor especially concerning piglet crushing. Genetics, housing and management have to be adapted at the same time. Finally the Free Farrowing Workshop Vienna 2011 concluded that the transition to free farrowing will be an evolutionary process, but driven by some degree of ultimate urgency.

Relation between sows' aggressiveness post mixing and skin lesions recorded 10 weeks later

Tönepöhl, B.[1], Appel, A.[1], Voß, B.[2], König Von Borstel, U.[1] and Gauly, M.[1], [1]Department of Animal Sciences, Livestock Production Systems Group, Georg-August-Universität Göttingen, Albrecht-Thaer-Weg 3, 37075 Göttingen, Germany, [2]BHZP GmbH, An der Wassermühle 8, 21368 Dahlenburg-Ellringen, Germany; btoenep@gwdg.de

This study was done to assess the use of skin lesions as an indicator for the aggressiveness of sows under commercial farm conditions and the potential use of this trait for breeding docile pigs due to increased group housing of pigs. Therefore the aggressive behaviour of 112 German Landrace sows was indirectly observed for the first 3 hours after regrouping. During this period the number of agonistic interactions (aggressor and receiver) and the numbers of reciprocal fights were recorded. Seventy-one days later, the sows were washed and scored for skin lesions (1 = no lesions to 4 = very frequent and severe lesions) separately for three body sections (front, middle, rear). Mixed model analysis revealed that a skin lesion score may not be a suitable indicator for individual aggressiveness because there was no effect (P>0.1) of the number of times a sow was recorded as an aggressor on skin lesions. However, there was a tendency that the receivers had more severe skin lesions in the cranial section when compared with aggressors. The skin lesion score increased by 0.04 ± 0.02 scores per additional attack (P=0.0619). In addition our results show a significant, negative relationship between higher skin lesion scores and performance (i.e. total born piglets, live born piglets). This gives evidence that increased aggressiveness reduced animal welfare and reproductive performance of sows. The main factor explaining differences in the extent of the skin lesion was the group. A greater number of aggressive sows obviously led to more agonistic interactions and led therefore to more skin lesions within the whole group. These results suggest that a group selection strategy rather than a selection based on individuals' breeding values only might be a better possibility for breeding of calm and docile pigs.

Producers' preferences for traits important in pig production

Wallenbeck, A.[1], Ahlman, T.[1], Ljung, M.[2] and Rydhmer, L.[1], [1]Swedish University of Agricultural Sciences, Department of Animal Breeding and Genetics, P.O. Box 7023, 75007 Uppsala, Sweden, [2]Swedish University of Agricultural Sciences, Department of Urban and Rural Development, P.O. Box 7012, 75007 Uppsala, Sweden; Anna.Wallenbeck@slu.se

Traditionally both the traits included in breeding goals, and the relative importance of these traits has been determined solely by the economic return associated with the genetic gain. However, over the last decade the interest in additional non-market values, such as animal welfare and environmental impact, has increased as a response to changes in consumer preferences, societal values and political decisions. In the long run, it is the consumers' demand that determine how pig meat is produced and thus how breeding companies design their breeding programmes. Breeding companies are, however, dependent on the demand of animal material from commercial producers, who ultimately decide what animal material they want to buy. There is a lack in knowledge on producers' preferences, especially for alternative pig production systems. We investigate producers' preferences regarding traits of pigs important in their production system using an advanced web-based questionnaire. The interactive questionnaire was sent out to Swedish pig producers (519 herds) in February 2012, and will be closed in March 2012. Herdsmen are asked to state which traits they intuitively consider important in their herd, rank 15 given traits against each other, weigh traits against each other given the estimated genetic gain (calculated based on selection index theory, within the questionnaire) and indicate to what extent they consider traits to be related to productivity, environmental impact and animal welfare. The preferences are compared between different classes of herd characteristics such as herd size, herdsman's experience and production system (e.g. conventional or organic). Finally, the agreement between producers' preferences and the breeding goal for the animal material currently used will be assessed.

Improving fertility and minimizing inbreeding within the endangered pig breed 'Bunte Bentheimer'
Biermann, A.D.M.[1], Pimentel, E.C.G.[1], Tietze, M.[1,2] and König, S.[1], [1]University of Kassel, Nordbahnhofstr. 1a, 37213 Witzenhausen, Germany, [2]University of Göttingen, Albrecht-Thaer-Weg 3, 37075 Göttingen, Germany; a.biermann@agrar.uni-kassel.de

The 'Bunte Bentheimer' as a 'fat pig breed' is only marketable in a niche focusing on meat quality. However, meat quality is not included in current breeding strategies. The ultimate breeding goal is the improvement of litter size to increase the population size. The objective of this study was to analyze the population structure, to estimate genetic parameters for fertility traits, and combine aspects of inbreeding and genetic gain in a breeding strategy by applying the Optimum Genetic Contribution theory. A total of 3,004 reproduction records including number of piglets born alive (NBA) and number of piglets weaned (NW) were used. The pedigree included 1,518 individuals. Genetic parameters and breeding values for both traits were estimated with a bivariate animal model. An algorithm implemented in the GENCONT program was used to select sires and females with their optimal mating frequencies to breed the next generation. Results were compared to truncation selection (TS). Average inbreeding coefficient was 11%. Inbreeding increased on average 0.58% per generation. Despite the high level of inbreeding, there was no effect of inbreeding depression on NBA or NW. Heritabilities for NBA and NW were 0.12 and 0.10, respectively, and genetic correlation between these traits was 0.96. The OGC- concept was applied for NBA at different constrained relationships. Compared to TS, average genetic value of selected parents was lower using OGC at the same level of average relationship. However average relationship within selected sires and selected females was marginally higher using TS. Furthermore, an own computer algorithm based on OGC was developed to mate natural service boars with sows in the same herd. The flexibility of the whole concept allows to include additional traits in the future breeding goal (e.g. meat quality) while minimizing genetic relationships in the long term.

Farmers' breeding goals and requirements for PDO products: the Nustrale pig breed in Corsica
Casabianca, F.[1], Lauvie, A.[1], Muller, T.[1,2] and Maestrini, O.[1], [1]INRA, SAD, LRDE Quartier Grossetti, 20250 Corte, France, [2]AgroParisTech, Master EDTS, 16, rue Claude Bernard, 75231 Paris V, France; fca@corte.inra.fr

Nustrale pig breed is located in the mountainous part of Corsica Island. After designing the standard, creating the herd book and managing the diversity (several strains identified), official recognition of the breed is obtained in 2006. Now, the breed is involved in an application for a Protected Designation of Origin (PDO) for three products. Market positioning of the PDO is an up-market strategy, with niche characteristics in a very touristic region. So, PDO specification is quite demanding. Pure breed is compulsory. Several requirements concern age at slaughter, carcass weight, back fat and intramuscular fat. How the collective management of the breed is taking into account the PDO specification? We followed the progress of meetings of both breed association and PDO body on the long term. We analyzed how the breed management was concerned by the PDO specifications. Those data were completed through 27 interviews with breeders about their representation of a 'right' animal ie the animal fitting what should be representing the local breed. Breed management is qualifying the boars on growing speed and adiposity. Nustrale breed is considered as a slow growing pig fitting the PDO specification quite well. But PDO is requiring a level of adiposity to be maintained by the breed managers. Breeders gave 3 set of criteria for their own selection: i) aesthetic vision of the animals, morphology and color patterns, ii) ability to survive in harsh conditions, in forest and pasturelands (great variation of available feeding along the seasons) and iii) quantity of meat to process according the traditional recipes. As a discussion, we show the trade-offs farmers are making between i) the collective expectations, both for breed management and PDO requirements and ii) their own vision of the 'right' animal. Some difficulties appear for balancing conservation requirements and production expectations.

Breeding goals for niche markets: a parma ham example
Knol, E.F., TOPIGS research, R&D, Schoenaker 6, 6641 SZ Beuningen, Netherlands; egbert.knol@ipg.nl

Features of a niche market are: small, specialised, creatable, price elasticity. In pork production niche markets exist in different forms like outdoor-, antibiotic free-, high intramuscular fat-, British bacon-, and Parma ham production. Method of choice for a niche market production is to start exploiting processing variation as for British bacon and for high marbling. One step forward in the chain is purchasing appropriate slaughter pigs in terms of weight and carcass and meat quality. Second step forward is the adaptation of the production environment, low on medication, free range, outdoor etc. Last step forward is the exploitation of genetic variation. On occasion existing genetic variation is enough in terms of line differences or sorting of sires within lines. If this is not enough then a dedicated genetic program could be set up with all costs to come with it. Parma ham production is about curing heavy hams. Heavy hams are available in the normal variation of slaughter. Curing requires careful drying of ham with consequences for shape and fat layer of the ham. This niche market is high price, high value and relevant extra traits are found in marbling, absence of blood spots and fat quality. An extra complication is the legal protection of this niche market in terms of regional production and regional birth of piglets. Slow gain is part of the protocol, partly because of fat quality and partly because of regulations, animals should be older than nine months of age. Economy of the production itself is difficult, value has to come solely from the hams. The selection program is expensive because taking phenotypes in hams after curing (1) tends to be invasive and (2) is almost two years after the piglet is born (curing for at least one year); and is expensive because the niche breeding goal is quite opposite to main stream, favouring slower growing fatter animals with excellent feed conversion. Traits added to the index are waterbinding- (salting) and marbling- and fat quality- related.

Selection for number of live piglets at five-days of age increased litter size and reduced mortality
Nielsen, B.[1], Madsen, P.[2] and Henryon, M.[1], [1]Pig Research Centre, Breeding & Genetics, Axeltorv 3, 1609 Copenhagen V, Denmark, [2]Aarhus University, Department of Genetics and Biotechnology, Blichers Allé 20, 8830 Tjele, Denmark; BNi@lf.dk

Back in the early nineties market requirements was in terms of increased litter size. Different experiments were conducted to determine if selection for increased litter size was feasible. DanBred changed the dam line breeding goals accordingly in 1992. Since then, litter size has increased significantly. However, there are concerns that increased litter sizes also increase the piglet mortality. A large experiment was conducted in Danish nucleus herds to find a selection strategy that avoids the increase in number of still birth and death of piglets in the pre-suckling face. In 2004, DanBred responded the new market requirements by including the total number of live piglets at day five (TN5) in the breeding goal. The objective of this study was to investigate whether the changes in the breeding goal have affected litter size and mortality and thereby to show if the market requirements was meet. Data included records of first litter from 43432 Landrace and 34446 Yorkshire sows in Danish nucleus herds from 2004-2010. At farrowing, litter size was recorded as total number born (TNB) including number of still-births. Litter size and mortality from birth to day five after birth were analyzed using a two-trait animal model assuming normality. The results indicate that breeding for TNB has increased the litter size, but also the mortality from 1991 to 2004 as mentioned by market requirements. However the introduction of TN5 in the breeding goal reduced mortality in the following period. The annual means of BLUP values related to piglet mortality showed that the genetic gain has reduced the piglet mortality by 4%-points in Landrace and Yorkshire from 2004 to 2010. The genetics gain was confirmed by decreased phenotypic annual mortality in the breeding and multiplier herds.

Supporting adaptation of farming systems to climate change and uncertainty: a position paper
Martin, G., INRA, UMR 1248 AGIR, 31326 Castanet Tolosan, France; guillaume.martin@toulouse.inra.fr

Livestock farming systems (LFS) are faced with complex, dynamic and interrelated changes in the production context related (among other things) to climate change, volatile input and output prices and rising energy costs. The pace, scale and even the direction of such changes are hardly predictable. Farmers have no alternative than to adapt their LFS by incorporating innovations in objectives, organization and practices adapted to this changing world. In this context, scientific research is expected to provide farmers with the required knowledge and methodology to adapt their LFS. A variety of quantitative and qualitative methodologies have been developed to support this adaptation process. In this communication, I review several such methodologies relying on computer models or stakeholders' participation. I point out their respective strengths and weaknesses. My main findings are the following. Computer models are integrative and enable to assess the relevance and feasibility of possible adaptations in a variety of production contexts. However, parameterization of such models for these production contexts is data-demanding and difficult, especially when farmer's decisions and actions are modeled. Moreover, stakeholders generally see computer models as black boxes. This may compromise implementation of suggested adaptations of LFS. With participatory approaches, exploration of possible adaptations of LFS is more flexible and transparent as it relies on the creativity of stakeholders. However, identification of relevant and feasible adaptations of LFS is highly dependent on their skills and knowledge. Based on this analysis, I propose a conceptual framework to develop methodologies supporting adaptation of LFS to climate change and uncertainty that better comply with farmers' needs. The potentialities of this conceptual framework are illustrated with the example of a participatory and simulation-based methodology relying on a board game called forage rummy.

Alpacas and llamas in Peru; comparison of two production systems: how does the future look like?
Gutierrez, G.[1], Flores, E.[1], Ruiz, J.[1], Schrevens, E.[2] and Wurzinger, M.[3], [1]UNALM-Universidad Nacional Agraria La Molina, Av. La Universidad S/N, 12 Lima, Peru, [2]KU-Leuven, Celestijnenlaan 200e, 3001 Heverlee, Belgium, [3]BOKU-University of Natural Resources and Life Sciences, Gregor-Mendel-Strasse 33, 1180 Vienna, Austria; maria.wurzinger@boku.ac.at

Alpacas and llamas are kept next to each other in the central highlands of the Peruvian Andes, where crop production is not feasible due to low temperature. The aims of two studies were to document the current situation of Alpaca and llama rearing and identify the dynamics and on-going changes of the two different systems. The studies were carried out in the Department of Cerro de Pasco where a total number of 106 llama keepers and 23 Alpaca breeders were interviewed using a semi-structured questionnaire. First results show a clear distinction of both systems. Llamas are usually kept by individual smallholders. On the contrary, Alpacas are kept by farmers' cooperatives, where the animals remain property of the famers, or in large community farms, where animals are owned by the community. This is also reflected in larger Alpaca herds. Llamas are traditionally kept for meat production and are used as pack animals, whereas Alpacas are kept for fibre. Farmers put more emphasis, time and money in improving the management of Alpacas. They purchase better breeding males from different sources, whereas llama males are often replaced by animals from the own herd or are obtained from neighbours. NGOs and the local government promote a more market oriented Alpaca production and support farmers in their efforts by providing training and other inputs. On the other hand, the predicted impact of climate change for this region indicates a reduction in precipitation and as a consequence reduced pasture growth. This would actually favour llamas, which are known to be better adapted to harsh and dry conditions. There seems to be a dichotomy between the economic advantage of Alpacas in the short run and the possible more ecological sustainable llama production in the future.

Several animal species in the same farm: a system from the past or an innovation for the future?

Cournut, S.[1], Bajusz, I.[2] and Ingrand, S.[2], [1]VetAgro Sup, UMR 1273 Métafort, Campus agronomique, 89 boulevard de l'Europe, BP 35, 63370 Lempdes, France, [2]INRA, UMR 1273 Métafort, Theix, 63122 Saint-Genès-Champanelle, France; sylvie.cournut@vetagro-sup.fr

The climatic changes which regularly undergo the breeders bring the researchers to analyse the configurations of breeding systems allowing to reduce their sensibility to the disturbances of the environment. The diversity of animals constituting the herd, allowed by the association of two species of ruminant animals on the same farm, is a track of research, assuming it allows flexibility, adjustments in the use of available resources, thanks to the different and complementary capacities from every animal category, especially for the land use. The management by the breeder of competitions and complementarities existing between two herds of different species, in the particular case of our study, dairy cows and meat sheep, thus appears as particularly interesting to analyze. Inquiries with eighteen mixed farms breeding dairy cows and meat sheep of Massif Central were realized. They were used as a basis to the construction of a typology of functioning, which reports interactions between animal diversity (species and batches) and variety of the available resources, on which the breeder can act and which are a source of flexibility for his system. Four logics of functioning were identified and characterized. These logics fairly differ on the spatial configuration of grazing and on the level of intensification of surfaces and show a gradient in the interweaving of the two herds management. These different logics were associated with contrasted way of responding to climate incertainty.

Rearing two-breed-dairy-cattle to improve farms' forage self-sufficiency in relation to climate chan

Thenard, V.[1], Mihout, S.[1] and Magne, M.A.[2], [1]INRA, UMR AGIR, 31326 Castanet Tolosan, France, [2]Université de Toulouse/ENFA, UMR1248 AGIR, 31326 Castanet Tolosan, France; vincent.thenard@toulouse.inra.fr

One challenge in Livestock Farming System research is adaptation to climate change. At the farm level, source of flexibility, is the diversity of animal and forage resources. In intensive dairy farms with high level feeding, high herd turnover rate and few breed choices, animal diversity is limited. In France, though rearing two-breed-dairy-cattle is not promoted, there are such systems in less favored regions. In this study, we wanted to test if breed diversity increases farm's adaptive capacity, faced with climate change. The aims have been to define farms' diversity management and to assess the farms' forage self-sufficiency status. Two kinds of data, collected in 22 two-breed-dairy-cattle herds (Holstein breed in association with a hardier breed) have been analyzed: structural and performance variables extracted from the French Dairy Herd Information and livestock farming practices have been recorded by farmers' interviews. Data have been analyzed (MCA) and five patterns of farms' diversity management have been built by two axes characterizing respectively: the individualization of herd management practices and the valorization of the breed diversity. Forage self-sufficiency has been defined as the annual ratio between grass production and grass consumption. For each studied farms, we could characterize forage self-sufficiency status. We have studied the links between this status and the pattern of diversity management. Finally in the case of systems based on grazing feeding, we have shown that farmers, who promote the diversity, forage self-sufficiency increases in their farm. On the contrary, in the farms based on maize feeding, promoting diversity limits the forage self-sufficiency. Results suggested that breed diversity can be a lever for farmers to adapt to climate change, but increasing self sufficiency must be analyzed according to the feeding management.

Livestock farming and uncertainties: exploring tipping points and resilience with viability tools
Tichit, M.[1], Puillet, L.[2], Martin, O.[2], Douhard, F.[1,2], Friggens, N.C.[2] and Sauvant, D.[2], [1]INRA, UMR 1048 SAD-APT, 16 rue C. Bernard, 75231 Paris, France, [2]INRA, UMR 0791 MOSAR, 16 rue C. Bernard, 75231 Paris, France; muriel.tichit@agroparistech.fr

Livestock farming is experiencing a period of increasing uncertainty. This uncertainty is, in itself, nothing new. Still, the multiplicity of the parameters concerned, their interactions and the speed with which they change are impacting the decisions farmers must make in terms of herd size, composition and culling. These decisions are increasingly seen as a major source of endogenous uncertainty which can drive livestock systems toward unsustainable transitions. As complex systems, livestock systems may have critical thresholds – so-called tipping points – at which the system shifts abruptly from one state to another. Identifying indicators of, and management decisions to prevent reaching tipping points is crucial to livestock system resilience. The objective of this study was to improve our understanding of resilience in livestock herds and explore tipping points under climatic and market uncertainties. We developed a simple dynamic herd model. The model is based on Viability Theory which provides mathematical tools for the analysis of system dynamics under constraints. Using this framework we predicted viable management decisions with respect to a given set of constraints. These constraints defined critical thresholds, i.e. the limits within which the system should be maintained to remain sustainable. We studied the relative weight of exogenous (climate, market) and endogenous uncertainties (management decisions) on herd dynamics and resilience. We identified indicators that could be used as early-warning signals. Using a time of crisis algorithm, we revealed trajectories that minimize the time during which critical thresholds are violated. We discuss which major changes in herd management could reinforce the herd's ability to face a changing environment.

Climatic hazards in suckling farms: analysis of farmers needs for grassland insurance
Fourdin, S., Dockès, A.-C. and Le Floch, E., French Livestock Institute, 149 rue de Bercy, 75012 Paris, France; simon.fourdin@idele.fr

In France, feeding of suckling livestock is mainly based on grass production (grazed or harvested grass). This production is strongly exposed to climatic hazards and principally to drought. Following the CAP health check of 2008, and the potential implementation of risk management measures such as insurance schemes, the French government expects development of multi-climatic risk insurance for grassland. The French Livestock Institute has carried out a study, supported by the French Ministry of Agriculture, to evaluate needs and expectations of suckling farmers about 'grassland insurance'. Firstly, individual and collective qualitative interviews have been carried out in 6 major breeding areas to illustrate: perceptions of the climatic risk by farmers, strategies of adaptation, and motivations and barriers for contracting 'grassland insurance'. Secondly, a quantitative questionnaire was built. 181 livestock farmers, in 8 French agricultural areas were interviewed to provide statistically relevant data. According to this study, farmers change their practices on grassland in order to be less impacted by climatic hazards. In particular they introduce drought resistant species and use different techniques for harvesting and stocking fodders. However, during years with severe droughts, for example 2011, these adaptations are not sufficient to produce enough fodder stock. Farmers have to buy complements. Until now, they used to be financially supported by a National Fund for the Management of the Agricultural Risk. One third of the livestock farmers we interviewed could consider purchasing a 'grassland insurance'. They are in favor of an insurance which covers all grassland productions, harvesting and pasture. This kind of multi-climatic risk insurance on grassland already exists in Canada and Spain, it could be adapted to fit expectations of the French farmers.

Effect of farming practices and alternative land uses on greenhouse gas emissions of beef production
Nguyen, T.T.H.[1,2,3], Doreau, M.[3], Eugène, M.[3], Corson, M.S.[2] and Van Der Werf, H.M.G.[2], [1]Valorex, La Messayais, 35210 Combourtillé, France, [2]INRA/Agrocampus Ouest, UMR1069 Soil Agro and hydroSystem, 35000 Rennes, France, [3]INRA/VetAgro Sup, UMR1213 Herbivores, 63122 Saint-Genès-Champanelle, France; Thi-Tuyet-Hanh.Nguyen@rennes.inra.fr

This study assessed the effects of a change in farming practices of a suckler-beef cattle production system on greenhouse gas (GHG) emissions by using (1) a classic life cycle assessment (LCA) approach or (2) an LCA including alternative land uses, where any land becoming available due to more efficient farming practices was converted to forest (with an average 55-year cycle). The change in farming practices we examined was intensified fattening of female calves not used for replacement from 9 to 19 months with a diet based on maize silage instead of rearing them as heifers used for replacement and fattening them for 4 months until slaughter at 33 months (the reference scenario). With a classic LCA approach, this new practice resulted in a reduction relative to the reference scenario of GHG emissions and land occupation per kg of carcass weight by 3.5 and 9.3%, respectively. Land occupation per kg of carcass decreased for permanent grassland (by 4.7 m^2), temporary grassland (0.6 m^2), cereals (0.2 m^2), but increased for silage maize (1.1 m^2). As a result of this change in farming practices land previously used for cereals (0.2 m^2), temporary grassland (0.6 m^2) and permanent grassland (0.5 m^2) was used to produce silage maize. The net land area released per kg of carcass weight was 4.2 m^2 of permanent grassland. With an LCA considering forest as an alternative land use, the new practice resulted in a reduction relative to the reference scenario of GHG emissions per kg of carcass of 7.9%. Accounting for alternative land uses in the assessment of changes in farming-practices may affect estimated GHG emissions considerably, and identifying alternative land-management options may play an important role in mitigating environmental impacts of farming systems.

Controversial role of mobility faced to climatic changes in the rainfed coastal zone of Egypt
Alary, V.[1,2], Aboul-Naga, A.[3], Abdelzaher, M.[3], Hassan, F.[3], Messad, S.[2], Bonnet, P.[2] and Tourrand, J.F.[2], [1]ICARDA, 15G Radwan Ibn El Tabib Street, GIZA, 2416, Egypt, [2]CIRAD, ES, TA C-112/A, 34398 Montpellier cedex 5, France, [3]APRI/ARC, Nadi Al Said, Dokki, 12619, Egypt; monaabdelzaher@yahoo.com

Mobility is well known as a factor of flexibility and adaptation of livestock farming system in harsh conditions. By enlarging the resource potential, the mobility is considered as a way to reduce climatic risk. The objective of the study is to understand the role of mobility facing 15 drought years that have affected the North West Coastal zone of Egypt (1995-2011). A field survey has been conducted in 2011 among a sample of 120 farmers located in the rainfed zone, from El-Alamein east to Libya boarder west. The analysis is based on multiple factorial methods based on synthetic indicators related to mobility (duration, distance), supplementary feeding (during and after the transhumance) and animal performance (reproduction and mortality rates) and profitability (income/head of animal). A first typology shows that, only the large breeders with more than 400 sheep and goats practiced long transhumance during the last 15 drought years. Rate of profitability remains low due to high mortality. The other groups adapt complementary feeding (mainly concentrates and grains) according to strategy of maintaining the animal stock and maximizing reproduction rate or survival strategy by maximizing the profitability per animal during the last drought. The lowest profitability is registered for breeders that maintain short mobility (less than 7 km). The natural range did not fulfill the energy requirements during the walk. So mobility appears as a successful adaptive mechanism to drought conditions to maintain their animal stock for large breeders, at the detrimental of profitability. But it has its limitations for small breeders, with the objective of economic survival due the high rate of mortality. Only in good climatic conditions, the mobility becomes a factor of profitability.

Points of view on adaptation of local breeds to harsh conditions: the Corsican cattle breed case

Lauvie, A.[1], Casabianca, F.[1], Coquelle, C.[2] and Pretrel, M.[1,3], [1]INRA UR LRDE, Quartier Grossetti, 20250 Corte, France, [2]Corsica Vaccaghji, 7 rue du Colonel Feracci, 2050 Corte, France, [3]Student in BTSA PA, Lycée Agricole de Tours Fondettes, La plaine, 37230 Fondettes, France; anne.lauvie@corte.inra.fr

Adaptation of local breed to harsh conditions is a question of great importance, particularly in a context of climate changes. We present a part of a larger study concerning 14 local cattle breeds in the Mediterranean area. Indeed, those breeds were chosen because they are often considered as bred in quite harsh conditions. Understanding better their adaptation could be useful for other breeds facing climate change. That information will be crossed later with genomic study of those animal populations. Our aim is to precise main characteristics of the livestock farming systems for those 14 cattle breeds in southern and northern Mediterranean area and to understand how their adaptation is considered by experts. The step presented in this communication consists in building a grid to be filled in by experts in each breed area. The aim of the grid is to gather elements on various characteristics of the farming systems the breeds are involved in, but also to have elements to assess if the breed is considered as adapted by the experts, and why, and if the adaptation is managed, how? We test it in the case of the Corsican cattle breed thanks to interviews with stakeholders and practitioners. Then we analyze the information that can be obtained thanks to such a grid. We show that only a part of the information, on the diversity of situation and points of views of the farmers, can be gathered with such a grid. In a second step of our study, we plan to get further information on the way adaptation of local breed is seen and how the breed is managed, by interviewing a diversity of breeders.

Effect of using shading on sheep performance in summer season

Abedo, A.A.[1], El-Sayed, H.M.[2], El-Bordeny, N.E.[2], Hamdy, S.M.[2], Soliman, H.S.[2] and Daoud, E.N.[3], [1]National Research Center, Animal Production Dept., Dokki, Giza, 12311, Egypt, [2]Fac.of Agric, Ain Shams Univ, Animal Production Dept., Cairo, 68 Hadayek Shoubra, Egypt, [3]Alex. Copenhagen Company, Meat and Milk production, Noubarya, 12311, Egypt; abedoaa@yahoo.com

This study was conducted in summer season, during July to August to evaluate effect of using shading technique to reduce the effect of heat stress on feed intake, nutrients digestibility and rumen liqur parameters for sheep. An ambient temperature was ranged from 33 to 36 0C and the relative humidity was ranged from 52 to 56%. Twelve Barki lambs weighted in average 50 ± 1.5 kg were randomly designed among two groups; un-shaded and shaded group. Lambs were fed concentrate fed mixture at 2% of their body weight and fed sorghum grass ad lib. as a source of roughage, feeding trial extended to 8 weeks, and at the end of trial feces and rumen liquor samples were taken to analysis. The results shown that intakes of concentrate (918.8), roughage (612.1) and total dry matter (1530.9 g) were significant ($P\leq0.001$) increased for shaded animals in comparing with 872.9, 586.4 and 1459.3 g/h/day for un-shaded group. Nutrients digestibility of dry matter, organic matter, crude fiber, ether extract, nitrogen free extract and feeding value as total digestible nutrients were significant increased (78.61, 82.11, 69.0, 79.22, 70.83 and 66.86%) for shaded lambs compared with 72.69, 77.25, 53.17, 68.56, 66.92 and 60.63%, respectively for un-shaded animals, while crude protein digestibility and feeding value as digestible crude protein were insignificant increased. The ruminal pH values at 3 hrs. after feeding, total volatile fatty acids (VFA·s) at 3 and 6 hrs. and ammonia concentrations at 3 and 6 hrs. were significant higher, but pH at 0 hr. and 6 hrs., the VFA·s at 0 hr. and ammonia at 0 hr. were insignificant higher for shaded animals compared with un-shaded group.

Cognitive processes involved in the development of animal stress and welfare

Boissy, A.[1], Destrez, A.[1], Deiss, V.[1], Aubert, A.[2] and Veissier, I.[1], [1]INRA, UMR 1213 Herbivores, Centre of Theix, 63122 Saint-Genès Champanelle, France, [2]Rabelais University, Parc de Grandmont, 37200 Tours, France; alain.boissy@clermont.inra.fr

Emotions are now largely recognised as a core element in animal welfare: the welfare of an animal depends on its emotions, which may evolve into prolonged affective states. The objective of the talk is to highlight the emotions of an animal and the cognitive processes by which its welfare is built. Based on recent advances in cognitive psychology, it has been shown in sheep that emotions are triggered by a cognitive process whereby the situation is evaluated from a limited number of criteria including the novelty of the triggering situation, its predictability and controllability. The nature of the emotions the animals may feels can be thus inferred from their evaluative abilities. From such an approach it was deduced that sheep can feel emotions like fear, rage and pleasure. However, emotions are acute and transient processes while welfare refers to a prolonged affective state. The interactions between emotions and cognition can be once again helpful to understand welfare states. Indeed, emotions can in turn influence evaluative processes in a way that can prolong an emotional state. For instance, sheep subjected for one month to repeated unpredictable and uncontrollable negative events, which are known to induce negative emotions, are less prone to respond to a stimulus signalling the delivery of a positive or negative event, especially when the signal is ambiguous (i.e. between a positive and a negative signal), indicating reduced capacities for judgment. Therefore, the level of welfare of an animal depends not only on the actual environment appraised by the animal but also on the past experiences of the animal that may affect this appraisal. Studies on animal welfare must integrate the animal's own point of view of the situations in which it lives, and recommendations for improving the living conditions of farm animals shall be based on their cognitive and emotional capacities.

Cognitive bias as welfare indicator in Japanese quail

Horváthová, M.[1], Košťál, L.[1] and Pichová, K.[1,2], [1]Institute of Animal Biochemistry and Genetics, Slovak Academy of Sciences, Moyzesova 61, 90028 Ivanka pri Dunaji, Slovakia, [2]Faculty of Science, Comenius University, Mlynská dolina, 84215 Bratislava, Slovakia; maria.horvathova@savba.sk

The relationship between cognition and affective states provides a new framework for an assessment of animal emotions and welfare. Emotional state of an individual influences a number of cognitive processes. Judgement bias reflects the tendency of a subject to show behaviour indicating anticipation of either relatively positive or relatively negative outcomes in response to affectively ambiguous stimuli. Here we report our attempt to adopt the concept of cognitive (judgement) bias for the assessment of Japanese quail welfare in relation to housing conditions. Cognitive performance was measured using operant chamber with touch-screen monitor. Adult hens were first trained in a Go/NoGo task to respond by pecking to a positive stimulus (white circle) associated with a positive event (food) and to refrain from pecking at the negative stimulus (80% gray circle) to avoid a negative event (white noise). Once the birds discriminated successfully both stimuli (significant rho in 3 consecutive daily sessions), they were divided into the two groups according to housing conditions (wire cage vs. deep litter) and after the 2 weeks they were subjected to ambiguous stimuli test in 3 daily sessions. In these tests in addition to the reinforced and punished stimuli a non-reinforced intermediate cues (20%, 40% and 60% gray circles) have been presented. The proportion of birds classifying ambiguous stimuli as positive was higher in deep litter housed birds than in cage ones in both preliminary experiments (n=6 and n=12, respectively). However, after another change of housing only in one of the experiments the responses to ambiguous stimuli differed between the housing conditions. To conclude, these pilot studies suggest that cognitive bias concept is a perspective test for the assessment of affective states in poultry. This work was supported by APVV-0047-10 and VEGA-2/0192/11.

Relationship between eye temperature, heart rate and show jumping performance in horses
Bartolomé, E., Sánchez, M.J., Márquez, M., Romero, J. and Valera, M., ETSIA. Universidad de Sevilla., Ciencias Agroforestales., Ctra. Utrera km 1, 41013, Spain; ebartolome@us.es

Recent studies have shown a good potential of the infrared thermography technology (IRT) to measure eye temperature as a means to detect stress in horses during competitions. The present study goes one step further, comparing performance results and different environmental factors from a Show Jumping competition with IRT measurements and with heart rate variability. 114 Spanish Sport Horse animals, ranging from 4 to 6 years old were used for this study. Maximum eye region temperature was assessed using a portable IRT camera to collect images and a portable pulsometer was used to measure hear rate. Both physiological measures were collected during 2 Show Jumping competitions (278 participations), in three different moments within each competition: 3 hours before competition (BC), just after it (JAC) and 3 hours after it (AC). 8 environmental factors were analyzed. A GLM analysis was developed in order to assess the influence of changes in eye temperature or heart rate levels on final rankings and positive points obtained by the animal, showing significant differences in performance results for all the physiological parameters except for heart rate variability measured 3h after competition. The influence of environmental factors on eye temperature and heart rate levels was also assessed, showing significant differences in 7 of them (sex, age, trip duration, arrival time, training duration, number of previous competitions and training hours). Spearman Rank correlations between physiological traits were calculated, highlighting significant correlations (P<0.001) between IRT measurement and heart rate level of 0.42 BC, of 0.51 JAC and of 0.75 AC. When intervals between phases were considered (JAC-BC and JAC-AC), a significant correlation (P<0.01) of 0.34 was found for JAC-AC interval.

The effect of BLUP selection for socially affected traits on the rate of inbreeding: a simulation
Khaw, H.L.[1,2] and Bijma, P.[1], [1]Animal Breeding and Genomic Centre, Wageningen University, 6708 WD Wageningen, Netherlands, [2]The WorldFish Center, Jalan Batu Maung, 11960 Penang, Malaysia; h.khaw@cgiar.org

Social interactions often occur among living organisms, including aquatic animals. Empirical findings show that social interactions affect phenotypes of individuals and their group mates. A social genetic effect (indirect genetic effect) is a heritable effect of an individual on the phenotype of another individual. Selection for socially affected traits may increase selection response, but also affect rate of inbreeding (ΔF). A simulation study was used to examine the effect of the Best Linear Unbiased Predictor (BLUP) selection for socially affected traits on ΔF. A basic scenario without social effects and three alternatives with different magnitudes of social effects were simulated. Each generation, 25 sires and 50 dams were mated (eight progeny/ dam). The population was selected for 20 generations using BLUP. Each generation individuals were randomly assigned to groups of eight members, with two families per group each contributing four progeny. Heritabilities (both direct and social) of 0.1, 0.3 and 0.5, and genetic correlations between direct and social effects of -0.8, -0.4, 0, 0.4, and 0.8 were used for each scenario. The ΔF was calculated from generation 10 to 20, and averaged over 100 replicates. For basic scenario, the ΔF were 0.041, 0.028 and 0.019 for h^2 of 0.1, 0.3 and 0.5. Overall, the range for ΔF for different magnitudes of social effects was 0.022 to 0.058. A h^2 of 0.5 always yielded the lowest ΔF. At low h^2, BLUP selection favors the choice of family members which increases inbreeding. In the scenarios with social effects, ΔF was greater than in basic scenarios. Apparently, the social interactions with groups of two families increased the resemblance between relatives within a group. We conclude that selection for socially affected traits may increase the ΔF. To maintain inbreeding at an acceptable rate we should apply a selection algorithm that restricts the increase in mean kinship, such as optimum contribution selection.

Identification of QTL for behavioural reactivity in sheep using the ovineSNP50 beadchip
Hazard, D.[1], Foulquié, D.[2], Delval, E.[3], François, D.[1], Bouix, J.[1], Sallé, G.[1], Moreno, C.[1] and Boissy, A.[3], [1]INRA, UR631 SAGA, 31326 Castanet-Tolosan, France, [2]INRA, UE321 La Fage, Saint Jean et Saint Paul, 12250 Roquefort-sur-soulzon, France, [3]INRA, UMR1213 Herbivores, 63122 Saint-Gènes-Champanelle, France; dominique.hazard@toulouse.inra.fr

The evolution of sheep extensive farming systems including increased size of flock, reduced support provided by human and use of harsh environments implies that animals are required to have a greater level of behavioural autonomy to contribute for a sustainable production. In particular, social reactivity and reactivity to human are becoming relevant traits, which have previously been reported with medium to high heritability (0.2 to 0.5). A genome wide association study of behavioural traits was carried out. Male and female Romane lambs (n=933) reared outside were individually phenotyped just after weaning in two tests based on the presence or absence of flock-mates and of human. The reactivity to isolation and to human was assessed by measuring bleats, locomotion, vigilance, flight distance from humans. QTL detection was carried in 9 half-sib families (103 lambs per family): lambs were genotyped using the Illumina ovineSNP50 beadchip. Data were analysed using linkage analyses (LA), linkage disequilibrium (LD) analyses and joint analyses (LD and LA) implemented with the QTLmap software. At the 0.1% genome wide threshold, a region of the chromosomes OAR16 and 21 showed significant association with traits involved in the social reactivity. Additional SNPs associated with the social reactivity reached the 0.1% or 1% chromosome wide threshold on OAR5, 10, 24, 26 and the 5% chromosome wide threshold on OAR2, 10, 12, 13, 17, 19, 22, 24 and 25. Two regions were also found to be associated with the reactivity to human on OAR4 and 16 at the 5% chromosome wide threshold. These results will lead to a better knowledge of genetic variability of behavioural traits in sheep. SNPs polymorphisms found in this study may offer opportunities for improving behavioural adaptation in sheep.

Physiological aspects of stress and welfare
Von Borell, E., Martin-Luther-University Halle-Wittenberg, Institute of Agricultural and Nutritional Sciences, Theodor-Lieser-Str. 11, 06120 Halle, Germany; eberhard.vonborell@landw.uni-halle.de

Animal welfare comprises the physical and psychological well-being of animals. The main components include biological functioning (growth performance, health and reproduction), affective states (suffering, pleasure, pain, emotions) and the expression of (normal) species-specific behaviour. Stress is a broad term that implies a threat to which the body needs to adjust. Stress can be classified as physical, psychological, or interoceptive in nature but usually contains components of all three classifications. The adjustment to stress induces a broad range of physiological and behavioural changes that allow for a rapid recovery or adaptation to the change. Stress per se may not negatively affect welfare, and the welfare of an individual might be impaired even when signs of stress are not obviously visible. Recent research suggests that successful coping to sensory and cognitive challenges may improve farm animal welfare. On-farm animal welfare assurance programmes demand for tools to assess diseased states, injuries as well as pain and distress induced by management procedures but also need to consider coping processes that generate positive emotions. Physiological and health indices are increasingly obtained from non-invasive techniques that allow for continuous monitoring of farm animals. However, direct routine inspections of individual animals are still necessary and some diagnostic measurements from blood and milk can only be obtained and analysed from invasive blood collection and laboratory analysis. This presentation specifically reviews some of the existing stress concepts and measuring tools for the on-farm welfare assessment based on physiological and health indices.

Dairy cow welfare in commercial French farms: major issues and the influence of farm type

De Boyer Des Roches, A.[1,2], Veissier, I.[2] and Mounier, L.[1,2], [1]VetAgroSup, 1 avenue Bourgelat, 69280 Marcy L'Etoile, France, [2]INRA, UMR 1213 Herbivores, 63122 Saint-Genès-Champanelle, France; alice.deboyerdesroches@vetagro-sup.fr

The process of welfare improvement relies on identifying major problems and prioritizing corrective actions. Information on farm characteristics identified as risk indicators of poor welfare helps refining welfare plans. The objectives were to highlight positive, negative and variable aspects of dairy cow welfare in commercial farms, and to identify which farm characteristics are associated with low welfare. We assessed dairy cow welfare using the Welfare Quality® protocol in a stratified sample of 131 French farms of different sizes, locations, breeds, housing and milking systems. The protocol covers 12 welfare criteria and produces scores (0-100 value scale) that express the degree of farm compliance with these criteria. Multivariate linear models were used to analyze the influence of farms characteristics on scores. Three welfare aspects were achieved but with some variability between farms: absence of hunger (mean score±SE: 57.4±3.1) or thirst (61.5±3.3), and the possibility to express species-specific behaviors (71.7±2.6). Four aspects were critical: pain due to management procedures (25.4±0.8), resting comfort (36.4±1.5), diseases (37.2±1.4), and integument alterations which also proved variable (38.6±2.6). Farm characteristics did not impact human-animal relationship (R^2=0.001) nor pain (R^2=0.06), but hunger, thirst, resting comfort, integument alterations, positive emotional state, social and species-specific behaviors (R^2=0.24, 0.16, 0.43, 0.12, 0.13, 0.10, and 0.31). Most important effects concerned cubicles being associated with poor resting comfort (P<0.001) and integument alterations (P<0.001), Holstein breed being associated with hunger (P<0.001), and farms in lowland with thirst (P<0.001). In conclusion, to improve dairy cow welfare in France, stakeholders of the dairy industry should reduce painful procedures on all farms, and other welfare aspects in farm types identified as being at risk.

Aggressive behaviour of dairy cows at a large dairy farm; a case study

Coster, A.[1], Borrell, M.[2] and Van Zwieten, J.T.[1], [1]Dairyconsult, Rohorst 11b, 7683PE Den Ham (ov), Netherlands, [2]Gurisat, Veïnat de l'Esclet, 17244 Cassà de la Selva, Spain; albart@dairyconsult.nl

Aggressive behaviour of dairy cows towards humans is a serious problem when it occurs, and to our knowledge little is known about this phenomenon. Motivated by a sudden and specific increase of this aggressive behaviour of adolescent and adult animals at a large dairy farm, we performed a case study to investigate its occurrence and some possible causes. The behaviour will be illustrated with a video and can best be described as fighting behaviour (i.e. beginning with display of lateral threat and followed by an attack). The behaviour was not associated to parturition, and hence can not be confounded with maternal protection of a calf. In affected animals, the behaviour was initiated at an age of approximately 1 year and was mainly maintained during life. Animals of three and more lactations on the farm were not affected by this behaviour; indicating that the behaviour either was associated to the age of the affected animals or that it was due to a causal factor occurring during an early stage of life. Causal factors investigated in this study include the raising scheme of the young animals, productivity of the animals, age group, and genetic factors. Despite obvious difficulties associated to finding causality for this complex behaviour in an observation study, the results of this study enabled us to discard several putative causal factors for this behaviour and helped us to quantify the problem at this specific farm. Furthermore, since little is known about this problem, we aim to use this study to motivate a scientific discussion and study of the prevalence and prevention of this problem.

Body positions of pigs in an early stage of aggression

Ismayilova, G.[1], Oczak, M.[2], Costa, A.[1], Sonoda, L.[3], Viazzi, S.[4], Fels, M.[3], Vranken, E.[2], Hartung, J.[3], Berckmans, D.[4] and Guarino, M.[1], [1]Università degli Studi di Milano, via Celoria 10, 20133, Milan, Italy, [2]Fancom Research, Industrieterrein, 34, 5981 NK, Panningen, Netherlands, [3]University of Veterinary Medicine Hannover, Foundation, Bünteweg 17 P, 30559, Hannover, Germany, [4]Katholieke Universiteit Leuven, Kasteelpark Arenberg 30, 3001, Leuven, Belgium; gunel.ismayilova@unimi.it

The present study is the initial part of a larger analysis investigating the possibility of automatic monitoring and control of aggressive behaviour among pigs. The aim was to investigate the body positions of pigs in an early stage of aggressive actions through the video image labelling techniques. The experiment was carried out at a commercial farm on a group of 11 male pigs (23 kg average), kept in a pen of 4 × 2.5 m. A total of 8 hours of video recordings taken during the 3 days after mixing were observed and 177 aggressive interactions were identified. Each aggressive interaction was labelled according to the body positions of pigs. A total of 12 positions indicating the direction of the pigs bodies before (P1-P12) and 7 at the start of aggressive interaction (P6-P12) were identified. The time which passed from the moment when an attack could be noticed until the first aggressive contact happened was 1-3 s in 68% of the interactions and 1 s in 28%. Most common positions at the start of an attack were P3-pigs approaching each other face to face (24%) and P2-initiator pigs attacking from side(18%). In 16% of cases a P3 attack was followed by an interaction from P12-inverse parallel position. P12 was the most frequent position in the first initial phase of aggressive interactions (39.5%), P7-nose to nose, 90° covered 19.77% and P9-nose to head-13.5% of all interactions. In conclusion, since aggressive behaviours seem to start often from the same body positions, it is important to focus on these body positions in order to develop an image-based monitoring system, able to automatically identify aggressive behaviour in an early stage.

Effect of excess lysine and methionine on immune system and performance of broilers

Bouyeh, M. and Haddadi, A., Azad University, Animal Science, Rasht Branch, Rasht, 4193963115, Iran; mbouyeh@gmail.com

The present work was carried out to investigate the effects of excess dietary Lysine (Lys) and Methionine (Met) on some blood immune parameters and the performance of broiler chicks. Three hundred male Ross 308 broilers were allotted to five groups, each of which included four replicates (15 birds per replicate) in a completely randomized design. The treatment groups received the same basal diet supplemented with Lys and Met (as TSAA) in 0, 10, 20, 30 or 40% more than NRC (1994) recommendation. The collected data were analyzed by SPSS software and Duncan's test was used to compare the means on a value of P<0.05. The results indicated that the two highest levels of Lys and Met treatments (30 and 40% more than NRC recommendation) led to significant increase in blood lymphocytes and decrease in heterophyles and ratio of heterophyles to lymphocytes as a stress index (P<0.01). There was a linear increase in Newcastle antibody parallel with increasing dietary Lys and Met in 42 days of age (P<0.01) but not 21d. Carcass efficiency, breast muscle yield, heart and liver weight were also increased by the two highest levels of Lys and Met (P<0.05), whereas feed conversion ratio (FCR) was the least in these two treatment groups (P<0.05). Addition Lys and Met 40% more than NRC tend to significant decrease in body weight gain (P<0.05) but there was no significant effect of treatments on thigh and leg yield. The finding of this experiment showed that increasing Lys and Met to diets of today's broiler in excess of NRC recommendations can improve immune system functions, FCR, abdominal fat deposition, breast meat yield and carcass efficiency. Results reported here support the hypothesis that it is possible to produce more healthy and economic poultry meat by supplementation excess Lys and Met to broiler diets.

New single nucleotide polymorphisms in dbh and th genes related to behavior in beef cattle breeds
Pelayo, R.[1], Azor, P.J.[1], Membrillo, A.[1], Molina, A.[1] and Valera, M.[2], [1]University of Córdoba, Dpt. of Genetics, C.U.Rabanales. Ed. Mendel, 14071 Córdoba, Spain; [2]University of Seville, Dpt. of Agroforestry Science, Crta.Utrera, Km 1, 41013, Seville, Spain; rociopega55@hotmail.com

Behavioural traits are complex and their individual variability results from the interaction between many genes as well as interaction with the environment. Two exons from DBH gene (DBH1 and DBH3) and two exons from nervous system are very important in the modulation of behaviour in vertebrates. Dopamine and noradrenaline are catecholamine neurotransmitters that are produced by biosynthetic enzymes such as dopamine β-hydroxylase (DBH) and tyrosine hydroxylase (TH). The aim of this study was to detect new polymorphisms (SNP) in coding regions of genes of the dopaminergic pathway (DBH and TH), and look for possible association among different presented alleles TH gene (TH4-5 and TH13) were amplified and sequenced. In this study 20 animals belonging to fighting bull breed selected for their aggressive behavior, were used. In the same manner, 45 animals from different beef breeds (and therefore, selected directly or indirectly for their calm behavior) were chosen. Twenty new SNPs were observed. In order to asses association between the different polymorphisms and the breeds a Fisher's exact test was performed. As a result, 13 of 20 SNP studied were significantly associated with the different breeds. The haplotypes that had the SNPs in linkage disequilibrium were significantly associated with the different breeds for 3 of the 4 exons studied. The present results suggest that the polymorphisms of the genes encoding catecholamine biosynthetic enzymes may become important markers for examining the genetic background of behavioural characteristics in beef cattle. Further investigations might be developed in order to assess association between genetics markers and behaviour by functional studies.

Drinking behaviour of suckler cows during transition period
Mačuhová, J., Jais, C., Oppermann, P. and Wendl, G., Institute for Agricultural Engineering and Animal Husbandry, Prof.-Dürrwaechter-Platz 2, 85586 Poing, Germany; juliana.macuhova@lfl.bayern.de

The aim of this study was to evaluate drinking behaviour of German Simmental and German Yellow suckler cows (altogether n=22) during three phases of transition period (i.e. two weeks before calving (day –14 to –1), calving day (day 0) and two weeks after calving (day 1 to 14)) throughout one winter season. Cows were kept in loose housing together with other animals of suckler cow herd and few days around calving in calving box. To record the water intake, individual drinking bowls with in-line flow meters were installed in the stable and animals were fitted with electronic ear tags. For statistical data evaluation, repeated measure tests (Friedman test or one way ANOVA) with Tukey adjustment for multiple comparison at P<0.05 were used. Before calving, the day had no significant effect on any of tested parameters (daily water intake, daily duration of drinking, daily number of drinking events, daily number of drinking bouts, water intake per drinking event and per drinking bout) except drinking rate. After calving, the day had significant effect on daily duration of drinking, drinking rate, water intake per drinking event and per drinking bout. However, whereas drinking rate seems to decreased with increasing day after calving the other parameters seem to increase (mainly during first few days after calving). The phase of transition period had significant effect on all tested parameters except daily number of drinking events. Only drinking rate and daily number of drinking bouts were significantly different (both higher) at calving day than before calving. Daily water intake, daily duration of drinking and water intake per drinking event changed significantly (increased) only after calving. In conclusion, no changes in drinking behaviour were observed before calving signalising forthcoming calving. Changes in drinking behaviour after calving can be assigned to start of lactation (higher water requirement) but also to maternal behaviour.

Cognitive bias as an indicator of welfare in laying hens: designing apparatus and tests

Pichová, K.[1,2], Horváthová, M.[2] and Košťál, Ľ.[2], [1]Faculty of Science, Comenius University, Bratislava, Department of Animal Physiology and Ethology, Mlynská dolina, 842 15 Bratislava 4, Slovakia, [2]Institute of Animal Biochemistry and Genetics, Slovak Academy of Sciences, Ivanka pri Dunaji, Moyzesova 61, 900 28 Ivanka pri Dunaji, Slovakia; katarina.pichova@savba.sk

To improve the quality of farm animals life we need to know how they feel about their lives and environment. Based on the methodology of the cognitive bias measurement in rats and starlings described in the literature, and our own experience with quail, the aim of the present study was to develop the operant chamber for laying hens and to optimize both training and testing of cognitive performance in hens. We have built an in-house apparatus controlled by the Biopsychology Toolbox, the software designed for behavioural experiments management. It is used to control the touch-screen (ELO 1529L), i.e. the presentation of stimuli and collection of touch/peck data and also to open and close the automatic feeder. The automatic feeders available on the market are designed for pigeons and they are too small for hen testing. Therefore we have developed an automatic feeder of our own design using the modified CD-ROM stepper motor. An interface between the feeder and the computer is based on the I/O Warrior chip. This operant chamber represents a modern and flexible solution for the presentation of visual cues and data collectionand can be used with various behavioural paradigms (autoshaping, fixed ratio, fixed interval, Go/NoGo tasks). Our first experience shows that it is suitable for the measurement of cognitive bias as an indicator of laying hen emotions and welfare. This work was supported by APVV-0047-10 and VEGA-2/0192/11.

Radiotelemetric and behavioural monitoring of laying hens welfare in enriched cages

Bilčík, B., Cviková, M. and Košťál, Ľ., Institute of Animal Biochemistry and Genetics, Slovak Academy of Sciences, Moyzesova 61, 900 28 Ivanka pri Dunaji, Slovakia; bbilcik@gmail.com

In connection with the ban of conventional cages from 2012, enriched cages and the extent of their welfare-friendliness is often discussed. In our study we followed selected physiological and behavioural measures in groups of laying hens in enriched cages with different group size and stocking density. We used 30 adult ISA Brown laying hens, 65 wk of age, obtained from the commercial breeder. They were kept in the EU-125 enriched cages (DKG Hostivice, CZ) with the area of 7500 cm^2. Two experimental group sizes/ stocking densities were used – recommended group size (10 birds, 750 cm^2/ hen) and group size/stocking density reduced to 50% (5 hens, 1500 cm^2/hen). Hens were fed *ad libitum*. Behavioural observations (direct observations with scan sampling and video recording) aimed on the use of space and resources, exploratory behaviour, aggressive pecking and feather pecking. Physiological data were obtained using the radiotelemetric system (Data Sciences International Inc., USA), with surgically implanted transmitter TL11M2-C50-PXT. System allowed us to measure motor activity, ECG, body temperature, heart rate and blood pressure in freely moving animals. Birds from smaller group size (5 hens) displayed higher frequency of aggressive pecking as well as severe feather pecking and were more active. Birds from the larger group size (10 hens) spent more time eating and standing. The use of the nest box differed as well – 88% of eggs in smaller group were laid in the nestbox, compared to only 45% in the larger groups. There was a trend towards lower blood pressure and body temperature in the smaller group size birds. Contrary to expected results, some of the measured parameters at lower stocking density indicated lower welfare levels compared to hens kept at higher density. Currently we are collecting and analyzing additional physiological and behavioural data to validate our results. This experiment was supported by the APVV-0047-10.

New methods for long-term measuring of animal welfare: preliminary results of implantable ECG logger

Frondelius, L.[1], Saarijärvi, K.[1], Vuorela, T.[2], Hietaoja, J.[3], Mononen, J.[1] and Pastell, M.[3], [1]MTT Agrifood Research Finland, Halolantie 31 A, 71750 Maaninka, Finland, [2]Tampere University of Technology, Department of Electronics, Korkeakoulunkatu 3, 33101 Tampere, Finland, [3]University of Helsinki, Department of Agricultural Sciences, Latokartanonkaari 5, 00014 University of Helsinki, Finland; lilli.frondelius@mtt.fi

There is a need to develop new methods for long-term monitoring of animals' welfare. Many studies on different species, including cattle, show that heart rate variability (HRV) is a promising stress indicator. However measuring HRV has been dependent on ECG monitors placed on the skin and these monitors do not function properly with animals and can be used only for short periods of time. Our aim was to test a new implantable long-term ECG and activity logger for dairy cattle. Lithium-battery powered implants were able to record ECG, acceleration and temperature data continuously and store the measured data on a secure digital (SD) memory card, with sample rates of 341.3 Hz, 16Hz and 0.0625 Hz, respectively. Total of 12 cows were operated in January 2011. The implants were placed on the left side of the cow into a pocket made between the skin and subcutaneous tissue. After the surgery the cows were kept in tied stalls for a week, and then transferred to loose housing. Two weeks after the surgery the cows were exposed to physical stress, i.e. 500 m walk outside twice a day for a week. Eight out of 12 implants collected data and the operation time was 15 to 17 days. The quality of collected ECG-signal was good and plenty of 5 minute intervals, required for spectral analyses, were easily obtained. The device collected data also during physical exercise, and the data were interpretable. This is a clear improvement as compared to our previous surface ECG measurements. However, comparison of ECG data to accelerometer signals showed that during acceleration peaks ECG signal had higher noise level. In conclusion, the implant could provide a simple tool for long-term monitoring of animal welfare.

Effects of sire lineage on maintenance and abnormal behavior

Strmenova, A.[1], Juhas, P.[2] and Broucek, J.[1], [1]Animal Production Research Centre Nitra, Hlohovecka 2, 951 41 Luzianky, Slovakia, [2]Slovak University of Agriculture in Nitra, Tr. A. Hlinku 2, 949 76 Nitra, Slovakia; strmenova@cvzv.sk

Abnormal behaviors, i.e. deviations from the normal natural behavior of animals, are considered behavioral response of the organism to unfavorable conditions in the life of animals. The number of subjects displaying abnormal behavior and the time they devote to it may indicate the level of welfare in farming. The goal of this research was to monitor the daily activities occurring in fattening bulls indoors. We examined a group of 32 Holstein bulls weighing 500 kg to 550 kg, kept in a loose-housing barn. We focused on how much time during the 5 days of 10 hours observation bulls spent standing, lying, moving, eating and ruminating. Later, we described how many times during the measurement occurred comfortable, playful and aggressive behavior. We selected five bulls from the group, which showed abnormal behavior (up to 15.6% of the time). These bulls were from 3 sires. The following abnormal behavioral problems were noticed: fight – between two or more bulls, licking genitals of other bulls, jumping on other bulls and screaming as expression of the most aggressive bulls. No significant differences were found among sires in maintenance behavior. Significant differences were found between 1 and 3 days in frequency of aggressiveness and playing behavior (aggressiveness: 7.78 ± 0.66 vs. 11.22 ± 0.66, $P<0.05$; playing behavior: $0.56\pm0,588$ vs. 3.78 ± 0.59, $P<0.05$). A sire effect was manifested in the frequency of aggressive behaviour and playing. Descendants of Sire 2 showed a significantly higher frequency of aggressive behaviour compared with the descendants of the Sire 3 (8.33 ± 0.75 vs. 1.67 ± 0.75, $P<0.05$). Results of this research demonstrated significant influence of sire on abnormal behavior of fattening bulls.

In ovo injection of synbiotics influences immune system development in chickens

Sławińska, A.[1], Siwek, M.[1], Bardowski, J.[2], Żylińska, J.[2], Brzezińska, J.[1], Gulewicz, K.A.[3], Nowak, M.[4], Urbanowski, M.[5] and Bednarczyk, M.[1], [1]University of Technology and Life Sciences, Mazowiecka 28, 85-084 Bydgoszcz, Poland, [2]Institute of Biochemistry and Biophysics, PAS, Pawińskiego 5a, 02-106 Warszawa, Poland, [3]Institute of Bioorganic Chemistry, PAS, Noskowskiego 12/14, 61-704 Poznań, Poland, [4]Wroclaw University of Environmental and Life Sciences, C.K. Norwida 31, 50-375 Wrocław, Poland, [5]Vet-Trade Polska, Słoneczna 223a, 05-506 Lesznowola, Poland; slawinska@utp.edu.pl

Development of the avian immune system takes place in the embryo growing inside the egg, therefore it can be influenced before hatch. The goal of this study was to analyze impact of _in ovo_ injection of synbiotics on immune system development in two distinct genotypes of chickens: broiler chicken (Ross) and general-purpose chicken (Green-legged Partridgelike (GP)). For _in ovo_ injection at 12th day of incubation, three types of synbiotics were used: two in-house developed strains of lactic acid bacteria (LAB) in combination with lupin seed extract (RFO, raffinose family oligosaccharides) and a commercial synbiotic including two strains of LAB and lactose. The analyses comprised of measurements and histopathology of main immune organs, i.e. bursa of Fabricius, thymus and spleen in 3rd and 6th week of life. The level of antibodies against viral antigens was measured with ELISA in three time points (1-day, 3rd and 6th week). The relative weight of the spleen and bursa of Fabricius was higher in GP than in Ross and GP chickens responded better to _in ovo_ stimulation. Histological analyses showed that the bursal follicles in GP underwent lymphocytic depletion to a lesser extent than in Ross. _In ovo_ injection of synbiotics did not alter the serological profile of the chickens. Early application of the immunostimulatory substances during the embryogenesis had positive effect on the immune organs formation both in broilers and general-purpose chickens. This research project was supported by grant No. N311 623938 from the National Science Centre, Poland.

Welfare assessment in dairy herds and relationship with health and milk production

Soriani, N., Bertoni, G. and Calamari, L., Università Cattolica del Sacro Cuore, Facoltà di Agraria, istituto di Zootecnica, Via Emilia Parmense 84, 29122 Piacenza, Italy; giuseppe.bertoni@unicatt.it

The aim of this study was to assess welfare by using our model (IDSW: Integrated Diagnostic System Welfare) and blood physiological indices in dairy farms characterized by different milk yield (MY). The model, based on many welfare indices included in three clusters (environment, feeding and animal based indices), has been used in two comparable dairy cow herds (100 lactating cows each). Furthermore, for a better assessment of the real welfare status, a blood sample has been obtained from cows in early lactation for a wide metabolic profile, as well as the frequency of clinical diseases has been recorded. Average daily milk yield was 21 and 33 kg/cow in herd A and B, respectively. The welfare score obtained with IDSW was below the acceptable value (75/100) in herd A (62.5/100); conversely in herd B it was slightly above (76/100). For herd A the major concerns were about environment (barn structures, space availability, management of rest area and groups) and some based animal indices: low fertility (207 days open), poor body condition, low MY, high somatic cell (250,000 and 450,000 n/ml in bulk milk), mastitis (3.4 clinical mastitis per month), and feet and limbs lesions (particularly around calving). In herd B the clinical diseases prevalence was within acceptable range, and only fertility was not optimal (129 days open). At blood level more frequent inflammatory phenomena have been observed in herd A, with haptoglobin still high at 30 days in milk (0.58±0.28 and 0.28±0.12 g/l in A and B, respectively) and lower concentration of albumin, a negative acute phase protein (33.7±2.7 and 38.5±1.8 g/l in A and B, respectively). These results seem to confirm that good welfare conditions are possible in dairy farms and this can live together with high milk yield. Moreover, the welfare assessment obtained by using IDSW model has been confirmed by blood and health indices.

Aggresion in piglets: ways of solution by environment modification

Juhás, P., Debrecéni, O. and Vavrišinová, K., Slovak University of Agriculture in Nitra, Department of Animal Husbandry, Tr. Andreja Hlinku 2, 94976 Nitra, Slovakia; Ondrej.Debreceni@uniag.sk

We looked for possibilities to decrease the aggressive behavior in piglets after shifting to a new pen and litter mingling. We observed the behavior and recorded the number of attacks in 5 different housing conditions: a pen with concrete floor cowered by straw (CFS), the same pen divided by a temporary barrier during litters shifting and with gap stuffed with straw. The gap was step by step opened by the piglets during environment exploration and mingling was spontaneous (BAR). A third pen was with slatterred floor (SLAT), a fourth pen was similar but equipped with a tube (diameter 31.5 cm, TUBE). The tube was fixed to the floor and was intend as a shelter during attack. The last environment was a pen with a pet bottle for manipulation (PET). We have evaluated the number of the attacks during 8 h after shifting to the new environment (TSA), average number of attacks per 20 min intervals (X20) and number of attacks during 1st hour after shifting and mingling (S1H). The environments were compared by Kruskall-Wallis test. The highest number of attacks in all evaluated traits was recorded in PET (TSA = 33, X20 = 1.38 and S1H = 9.67). The smallest number of attacks in all evaluated traits was recorded in BAR (TSA = 7.01, X20 = 0.29 and S1H = 1.54). A significant difference was found only in X20. The BAR environment had the most impact on behaviour: The piglet explored the new environment in group of sibs; exploration decreased the excitment from new envirinment, so at the moment of opening the piglets were calmer; they can join the group of unfamiliar piglets when they 'desired'. The next advantage of partitioning is the possibility of escape to other side of the barrier in case of an attack. The PET environment triggered the highest number of attacks, possibly due to the competition for the pet bottle. The research was supported by VEGA 1/2717/12, ECOVA and ECOVA Plus.

The relationship between backtest and agonistic behaviour in pigs

Scheffler, K. and Krieter, J., Christian-Albrechts-University Kiel, Institute of Animal Breeding and Husbandry, Olshausenstr. 40, 24098 Kiel, Germany; kscheffler@tierzucht.uni-kiel.de

The study aimed at analysing the relationship between backtest and agonistic behaviour using video observations. The backtest characterises the behaviour of a piglet in a standardised stress situation and is a method to obtain the coping style of that animal. It was performed twice on 1383 piglets from 139 litters. At the age of 12 and 19 days, each piglet was restrained in a supine position for 1 min. The number of escape attempts (NEA), the duration of all escape attempts (DEA) and the latency to the first escape attempt (LEA) were recorded. Extremely responding piglets (the top and bottom 25% of the distribution) were classified as 'low-resistant' (LR; <2 NEA; <5 s DEA; >38 s LEA) and as 'high-resistant' (HR; >3 NEA; >16 s DEA; <7 s LEA). All remaining piglets were assigned as 'doubtful' (D). A general linear model was used to determine the influence of fixed effects: parity, batch, cross-fostering, number of penmates, test day, sex and birth weight. Only the effects of batch, test day and birth weight had a significant influence on all three parameters ($P<0.05$). At the age of 12 days, the piglets showed more NEA, a longer DEA and a smaller LEA. The effect of the birth weight revealed that lighter piglets showed an increased NEA and DEA, but a decreased LEA in comparison to heavy piglets. To analyse the relation between the coping responses, Kappa coefficients (κ), as degrees of consistency, were calculated between NEA, DEA and LEA. Kappa coefficients were lower in the first backtest (NEA*DEA κ=0.49; NEA*LEA κ=0.38; DEA*LEA κ=0.39) than in the second one (NEA*DEA κ=0.53; NEA*LEA κ=0.43; DEA*LEA κ=0.45). The effect of the test day and the higher kappa coefficients in the second test showed the habituation to the test situation. The second test was more suitable to describe the real coping behaviour of the piglets because the panic reaction could be reduced. The results from the backtest will be compared with agonistic behaviour (number and duration of fights).

Measuring relatedness and studying recent mutations

Kong, A., deCODE genetics, Statistics, Sturlugata 8, 101 Reykjavik, Iceland; Augustine.Kong@decode.is

Advances in genotyping and sequencing technology, together with newly developed methodology, allows us to measure relatedness between individuals in a manner that is a quantum leap from what could be done just a few years ago. This has an impact on virtually all genetic investigations. Using Icelandic data, we will illustrate applications that include sequence imputations, hybrid linkage analysis, study of parent of origin effects, recombinations, transmission distortions, recent mutations, heritability and population structure.

Patterns of genomic diversity among three Danish dairy cattle breeds

Guldbrandtsen, B., Lund, M.S. and Sahana, G., Aarhus University, Centre for Quantitative Genetics and Genomics, P.O. Box 50, Ørum, 8830 Tjele, Denmark; Bernt.Guldbrandtsen@agrsci.dk

For each of three Danish dairy cattle breeds, Danish Holstein, Danish Red Dairy Cattle and Danish Jersey Cattle approximately 30 key ancestors were sequenced to a depth of about 10X. These key ancestors were chosen to maximize their contributions to present day populations. The three breeds are compared with respect to differences in amount of sequence variation observed. The observed patterns in and differences between breeds are interpreted in the light of known individual and population histories. Also, patterns of occurrence of short indels are compared. Particular attention is paid to variants that occur in several of the breeds. Regions occurring in multiple breeds and suspected of harboring lethal or sub lethal sequence variants are examined for similarities across breeds in occurrence of sequence variants.

Detection of QTL affecting milk fatty acid composition in three French dairy cattle breeds

Govignon-Gion, A.[1], Fritz, S.[2], Larroque, H.[3], Brochard, M.[4], Chantry, C.[5], Lahalle, F.[4,6] and Boichard, D.[1], [1]INRA, UMR1313 GABI, Domaine de Vilvert, 78352 Jouy en Josas, France, [2]UNCEIA, 149 rue de Bercy, 75595 Paris, France, [3]INRA, UR 631 SAGA, Chemin de Borde Rouge, 31326 Castanet, France, [4]IDELE, 149 rue de Bercy, 75595 Paris, France, [5]LABOGENA, Domaine de Vilvert, 78352 Jouy en Josas, France, [6]CNIEL, 42 rue de Chateaudun, 75314 Paris, France; armelle.gion@jouy.inra.fr

Combining Mid Infra Red (MIR) prediction of fatty acid concentration in milk with high throughput SNP genotyping provides a strong opportunity to identify the main genomic regions responsible for genetic variation in bovine milk fat composition. Milk fatty acid composition was estimated from MIR spectrometry for more than 450,000 test-day from 86,458 cows in first or second parity in the three French Montbeliarde (MO), Normande (NO) and Holstein (HO) cattle breeds, within the national PhénoFinlait project funded by ANR and ApisGene. Among these cows, 8,000 cows (3,000 HO, 2,500 MO and 2500 NO) were genotyped either with the Illumina 50K Beadchip® (7,500) or with the Illumina LD Beadchip® (500) and then imputed. In a first step, analysis was relative to total saturated, unsatured, mono- and polyunsaturated, C16:0, C14:1c9, C18:1c9, and C18:2c9t11 (CLA). Individual test-day records were first adjusted for environmental effects (herd × test-day, month of calving and days in milk within year and parity) after a polygenic evaluation and then averaged per cow. In a second step, after quality control, phasing and imputation, QTL detection was carried out within breed by Linkage and Linkage Disequilibrium Analysis (LDLA) within each breed. The most important regions were found on already described regions known to affect fat components, milk production, or fatty acid desaturation, e.g. on chromosomes 14, 19, or 26. Several dozens of additional QTL were found for each trait and overlapped only very partially across breeds. Most QTL were found to affect several traits simultaneously, in agreement with the genetic correlations and the metabolic pathways.

Genome-wide association analysis of tuberculosis resistance in dairy cattle

Bermingham, M.L.[1], Bishop, S.C.[1], Woolliams, J.A.[1], Allen, A.R.[2], Mc Bride, S.H.[2], Wright, D.M.[3], Ryder, J.J.[4], Skuce, R.A.[2,3], Mc Dowell, S.W.J.[2] and Glass, E.J.[1], [1]The Roslin Institute and and R(D)SVS, University of Edinburgh, Midlothian, EH25 9RG, United Kingdom, [2]Agri-Food and Biosciences Institute, Belfast, BT4 3SD, United Kingdom, [3]Queen's University of Belfast, Belfast, BTA 3SD, United Kingdom, [4]Institute of Integrative Biology, University of Liverpool, Liverpool, L69 7ZB, United Kingdom; john.woolliams@roslin.ed.ac.uk

Mycobacterium bovis is the aetiological agent of bovine tuberculosis (TB). Diagnosis is based on the tuberculin test and abattoir surveillance. The failure of the UK and several countries to eradicate TB indicates a need to consider alternative strategies, such as genetic selection of cattle for increased resistance to TB. The aim of this study was to conduct a case/control genome wide association study using the newly available Illumina bovine high density SNP chip to identify loci associated with variation in TB resistance in the Northern Ireland Holstein-Friesian dairy cattle population. Blood samples from 3,715 cattle were collected from 464 herds between 2008 and 2009. Cases were sampled at slaughter, and were defined as animals with both a positive reaction to the tuberculin skin test and a confirmed TB lesion. Age matched controls were sampled from a subset of case herds, and were defined as animals that were negative to the tuberculin skin test. In total, 1,424 cattle were genotyped for 777,962 SNPs. After QC edits, 1,151 cattle (592 cases and 559 controls) and genotype data from 617,610 SNPs remained. Genome wide association using mixed models and regression was used to test for associations between SNPs and TB resistance. The estimated heritability for resistance to TB was 0.21 (SE 0.06). Chromosome wide significant associations detected at eight loci, and together these SNPs explained 8.2% of variance in resistance to TB in the cattle population sample investigated in this study. We aim to independently replicate these findings and understand the genetic architecture of the significant regions.

Bayesian stochastic search variable selection using phenotypes collected across research herds in EU

Veerkamp, R.F.[1], Wall, E.[2], Berry, D.P.[3], De Haas, Y.[1], Coffey, M.P.[2], Strandberg, E.[4], Bovenhuis, H.[5] and Calus, M.P.L.[1], [1]Wageningen UR Livestock Research, Animal Breeding and Genomics Centre, Vijfde Polder 1, 6708 WC Wageningen, Netherlands, [2]Scottish Agricultural College, Sustainable Livestock Systems Group, Easter Bush Campus, Edinburgh EH25 9RG, United Kingdom, [3]Teagasc, Animal & Grassland Research and Innovation Centre, Moorepark, Co. Cork, Ireland, [4]Swedish University of Agricultural Sciences, Department of Animal Breeding and Genetics, P.O. Box 702, 75007 Uppsala, Sweden, [5]Wageningen University, Animal Breeding and Genomics Centre, P.O. Box 338, 6700 AH, Wageningen, Netherlands; roel.veerkamp@wur.nl

Genome-wide association studies (GWAS) for difficult-to-measure traits are generally limited by the sample size with accurate phenotypic data and are usually carried out considering individual single nucleotide polymorphism (SNPs) one at a time. The objective of this study was to compare variance components from a linear animal model with those estimated using BSSVS. Phenotypic data on primiparous Holstein-Friesian cows from experimental farms in Ireland, the UK, the Netherlands and Sweden (FPCM, DMI, BCS and LW) were available on up to 1,629 genotyped animals (37,590 SNPs). Genetic parameters from the two analysis were comparable, but not equivalent. For some traits, SNPs explained a larger part of the total genetic variances than for other traits. This was also reflected in the genetic correlations between the traits. The variation explained by SNPs on each chromosome was related to the size of the chromosome and was relatively consistent for each trait with the possible exceptions of BTA4 for BCS, BTA7, BTA13, BTA14, BTA18 for LW, and BTA27 for DMI. For LW, BCS, DMI, and FPCM, 266, 178, 206 and 254 SNPs had a Bayes factor >3, respectively. Olfactory genes and genes involved in the sensory smell process were over represented in a 500 kb window around the significant SNPs. Potential candidate genes were involved with functions linked to insulin regulation, epidermal growth factor and tryptophan.

Assessment of the genomic variation in a cattle population by low-coverage re-sequencing

Fries, R., Pausch, H., Jansen, S., Aigner, B. and Wysocki, M., Technische Universitaet Muenchen, Lehrstuhl fuer Tierzucht, Liesel-Beckmann-Str. 1, 85354 Freising, Germany; ruedi.fries@tum.de

The results of the first phase Human 1000 Genomes Project prompted us to apply low-coverage re-sequencing to key ancestors and some contemporary animals of the German Fleckvieh (FV) population. Forty-three animals chosen for re-sequencing explain 68% of the population's gene pool. Next-generation sequencing with Illumina instruments provided an average effective coverage of 7.4 x. Multi-sample variant detection yielded genotypes at 17.3 million sites. After BEAGLE imputation, 97% of all genotypes had phred-scaled quality scores of at least 10. Sensitivity amounted to 98% and specificity to 3% after comparing the sequence-derived genotypes with high-density array genotypes. Combining BEAGLE and MiniMac facilitated the imputation of genotypes at 12 million sites of 3,600 FV animals via high- and medium-density genotypes. The imputation accuracy exceeded 90%. Imputed genotypes were used in whole-genome association studies. The association signals were more significant than those obtained from array-based markers and allowed to narrow down the positions of causal variants to a few hundred bases. It turns out that causal variants are often in non-coding regions. A compound QTL for somatic cell count, e.g., consists of causative sites in the first intron as well as in a putative distant upstream enhancer of the lactotransferrin gene. Minimal haplotypes comprising the causal variants allow for thorough characterization of QTL within and between populations and open new perspectives for using information on specific QTL in genomic selection.

Refining QTL with high-density SNP genotyping and whole genome sequence in three cattle breeds
Sahana, G., Guldbrandtsen, B. and Lund, M.S., Aarhus University, Molecular Biology and Genetics, Blichers Allé 20, 8830 Tjele, Denmark; goutam.sahana@agrsci.dk

Genome-wide association study was carried out in Nordic Holsteins, Nordic Red and Jersey breeds for functional traits using BovineHD Genotyping BeadChip (Illumina, San Diego, CA). The association analyses were carried out using both linear mixed model approach and a Bayesian variable selection method. Principal components were used to account for population structure. The QTL segregating in all three breeds were selected and a few of the most significant ones were followed in further analyses. The polymorphisms in the identified QTL regions were imputed using 90 whole genome sequences available from these three breeds. Imputations were done using IMPUTE v2.2. Association analyses with imputed polymorphisms were repeated for the targeted regions. The QTL genotypes of the sires with more than 20 sons were determined by an a posteriori granddaughter design. The concordance of sires for putative quantitative trait nucleotide was determined.

Fine mapping of a QTL region for androstenone levels on pig chromosome 6
Hidalgo, A.M.[1], Bastiaansen, J.W.M.[1], Harlizius, B.[2], Megens, H.J.[1] and Groenen, M.A.M.[1], [1]Wageningen University, Animal Breeding and Genomics Centre, De Elst 1, P.O. Box 338, 6700 AH, Wageningen, Netherlands, [2]Institute for Pig Genetics B.V., Schoenaker 6, P.O. Box 43, 6640 AA, Beuningen, Netherlands; andre.hidalgo@wur.nl

Androstenone is one of the main compounds causing boar taint in uncastrated male pigs. A genome-wide association study, using Sus scrofa Build 9 as the reference genome, unveiled an 8 Mb region on chromosome 6, associated with androstenone levels in boars. Improved map positions of markers, based on pig reference genome build 10.2, reduced the associated region on chromosome 6 from 8 to 3.75 Mb. Based on the SNPs in this region that were used in the GWAS study, we estimated haplotype effects and applied individual whole genome sequencing to further narrow the region associated with androstenone levels. SNP genotypes and androstenone phenotypes were available for 2,750 animals from 6 pure lines and 5 crosses. Haplotypes, consisting of 46 markers, were identified using Haploview and their effects on androstenone were estimated within each line via regression analysis. Two of the 10 identified haplotypes were present at moderate frequencies in all of the populations. Haplotype 1 was consistently related with high levels, and haplotype 2 with low levels of androstenone. Data from individually sequenced animals was used to identify a larger number of SNPs that differentiate the favorable and unfavorable haplotypes 1 and 2. We present an analysis of SNPs that carry alternative alleles between these two haplotypes, aimed at the identification of the genes and SNP variants that affect the development of androstenone levels in pigs.

Genetic architecture of environmental variance of somatic cell score using high density SNP data

Mulder, H.A.[1,2], Crump, R.[1], Calus, M.P.L.[1] and Veerkamp, R.F.[1], [1]Animal Breeding and Genomics Centre, Wageningen UR Livestock Research, P.O. Box 65, 8200 AB Lelystad, Netherlands, [2]Animal Breeding and Genomics Centre, Wageningen University, P.O. Box 338, 6700 AH Wageningen, Netherlands; han.mulder@wur.nl

In recent years it has been shown that not only the phenotype is under genetic control, but also the environmental variance. Very little is, however, known about the genetic architecture of environmental variance. The objective of this study was therefore to perform a genome-wide association study and to quantify the accuracy of genome-wide breeding values for somatic cell score (SCS) in dairy cows. SCS was used because previous research has shown that the environmental variance of SCS is partly under genetic control and reduction of the variance of SCS by selection is desirable. In this study we used 37,590 SNP genotypes and 44,032 test-day records of 1,563 cows at experimental research farms in four countries in Europe. Using either a pedigree or a genomic relationship matrix, we calculated for both a residual variance per cow as a proxy for the environmental variance. In addition, we calculated the within-cow mean and phenotypic variance of somatic cell score. These four traits were analysed with a Bayesian stochastic search variable selection method. No SNP were found to be associated with environmental variance of SCS, whereas 34 SNP were found to be associated with mean SCS. Based on 10-fold cross-validation, the accuracy of genome-wide breeding values was 0.4-0.5 for both environmental variance and mean SCS. Environmental variance was estimated to be affected by ~2,000-2,500 independent chromosome segments, whereas ~1,700 independent chromosome segments were affecting the mean SCS. About 25-28% of 50-SNP windows were required to explain 50% of the genetic variance of environmental variance. It is concluded that environmental variance of SCS is likely to be determined by many genes and accuracy of genomic breeding values for environmental variance is comparable to that of genomic breeding values for mean SCS.

Finding gene to genome epistatic effects

Filangi, O.[1], Bacciu, N.[1], Demeure, O.[1], Legarra, A.[2], Elsen, J.M.[2] and Le Roy, P.[1], [1]INRA, PEGASE, UMRGA INRA-Agrocampus, 65 rue de St Brieuc, 35042 Rennes, France, [2]INRA, SAGA, chemin de borde rouge, BP 52627, 31326 Castanet-Tolosan, France; elsen@toulouse.inra.fr

Gene to gene epistatic interactions and Genetic Association Interaction Networks are essential components of the genetic architecture of complex traits. SNP chips potentially provide large scale information about these interactions. Unfortunately, evidencing these interactions at a genome level is a nearly intractable question, due to its very high complexity. At the lowest level (pair wise interactions), the number of two genes epistatic effects to be estimated is proportional to n_S^2 (for n_S SNPs). However, the practical interest of such findings may be very limited as their effect on the trait variability is generally diluted by all other sources of variation. It is thus very tempting to focus only on interactions with strong effect. We propose a statistical test of gene to genome interactions. The very simple idea is, for a given candidate gene (a SNP), to compare GEBVs calculated from SNP effects estimated in subpopulations defined by the genotype at the candidate gene. The number of such interactions is only linear in n_S. Ad hoc decision rules must be defined for this test. Once a SNP is positively tested as interacting, more precise techniques can be used to identify which of the n_S SNP are concerned by the global interaction. A practical example is available.

Validation of gene networks constructed based on the 50K SNP chip using simulations
Szyda, J., Suchocki, T. and Fraszczak, M., Animal Genetics, Biostatistics Group, Wroclaw University of Environmental and Life Sciences, Kozuchowska 7, 51-631 Wroclaw, Poland; joanna.szyda@up.wroc.pl

Determination of gene functions for complex traits plays a very important role in the modern genetics. Especially interesting is knowledge about the functions of genes which effects are too small to be detected individually. The novel approach allowing for a better understanding of the genetic determination of complex traits is gene network construction. Using this method the functions of genes with both large and small effects, as well as their interactions can be derived. However, a severe drawback of a gene set approach lays in the problem of network validation. In our study, for the identification of influential genes, we used a training data set from the Polish routine genomic evaluation system. The data set consists of 2 588 bulls from the Polish Holstein-Friesian breed and 54 001 genotypes of Single Nucleotide Polymorphisms (SNP) available for each individual. Effects of the SNPs were estimated for production and fertility traits. These genes are then further used to form a network, by considering information from publicly available genomic recourses such as Gene Ontology (GO) or Kyoto Encyclopedia of Genes and Genomes (KEGG). For each of the considered traits a resulting network is validated by: a) randomly selecting from the bovine genome the same number of genes, which was selected using SNP effect estimates; b) construction of a network based on the selected genes; c) repeating steps a) and b) 100 times. Empirically derived significance is assigned: to each of the genes, Gene Ontology terms and physiological pathways represented in the original network.

A multi breed reference improves genotype imputation accuracy in Nordic Red cattle
Brøndum, R.F., Ma, P., Lund, M.S. and Su, G., Aarhus University, Department of Molecular Biology and Genetics, Blichers Alle 20, 8830 Tjele, Denmark; mogens.lund@agrsci.dk

The objective of this study was to investigate if a multi breed reference would improve genotype imputation accuracy from 50K to high density (HD) single nucleotide polymorphism (SNP) marker data in Nordic Red Dairy Cattle, compared to using only a single breed reference, and to check the subsequent effect of the imputed HD data on the reliability of genomic prediction. HD genotype data was available for 247 Danish, 210 Swedish and 249 Finnish Red bulls, and for 546 Holstein bulls. A subset 50 of bulls from each of the Nordic Red populations was selected for validation. After quality control 612,615 SNPs on chromosome 1-29 remained for analysis. Validation was done by masking markers in true HD data and imputing them using Beagle v. 3.3 and a reference group of either national Red, combined Red or combined Red and Holstein bulls. Results show a decrease in allele error rate from 2.64, 1.39 and 0.87% to 1.75, 0.59 and 0.54% for respectively Danish, Swedish and Finnish Red when going from single national reference to a combined Red reference. The larger error rate in the Danish population was caused by a subgroup of 10 animals showing a large proportion of Holstein genetics, which was not present in the reference. When adding Holstein animals to the reference, the error rate was further decreased to 1.17% for Danish Red, whereas the Swedish and Finnish Red animals were unaffected. Three Danish animals with a high proportion of genetic origin from breeds not included in the reference were also unaffected by the addition of Holstein to the reference. Results show that a multi breed reference is superior to a single breed reference when imputing from 50K to HD marker data in the Nordic Red breeds, due to sharing the information of Holstein which was previously used in Danish Red. Investigation of the reliability of genomic prediction using imputed HD data shows small improvements compared to 50K data.

Phenotype networks for lipid metabolism in pigs inferred from liver expression and SNP data

Aznárez, N.[1], Hernández, J.[1], Cánovas, A.[1], Pena, R.N.[1], Manunza, A.[2], Mercadé, A.[2], Amills, M.[2] and Quintanilla, R.[1], [1]Institute for Research and Technology in Food and Agriculture, Animal genetics & Breeding, Rovira Roure 191, 25198 Lleida, Spain, [2]Universitat Autònoma de Barcelona, Dept. Ciència Animal i dels Aliments, Edifici V, Campus UAB, 08193 Bellaterra, Spain; nitdia.aznarez@irta.cat

Lipid metabolism in pigs represents a complex system gathering traits related to animal health, carcass performance, and meat quality. In the present study, global liver expression and high-density SNP data was used to infer phenotype networks among ten lipid metabolism traits, including fatness (backfat thickness and lean percentage), serum lipid concentrations (total cholesterol, LDL, HDL and triglycerides), intramuscular fat content (IMF) and muscle fatty acid composition (MUFA, PUFA and SFA). A population of 350 Duroc castrated males was genotyped with the Illumina PorcineSNP60 BeadChip. Global liver mRNA expression levels were obtained for 104 of these individuals using GeneChip Porcine Genome arrays (Affymetrix). Transcriptomic and genomic correlation matrices between the ten traits were constructed on the basis of expression-phenotype associations (6919 probes after applying the minimum fold-change filter) and SNP genotype effects on phenotypes (37252 SNPs after quality control), respectively. The PCIT algorithm was used to filter out pair-wise correlations arising via strong correlations with a third trait. The two phenotype networks inferred from either liver expression or SNP data showed positive relationship among backfat thickness, IMF, MUFA, SFA and serum lipid concentrations, whereas PUFA and lean percentage were negatively related with them all. Interestingly, both transcriptomic and genomic association analyses yielded relatively similar phenotype network topologies, but network built on the basis of liver mRNA expression levels was notably denser and showed much higher correlation values between traits. Future studies will be performed in order to disentangle gene networks involved in the global regulation of pig lipid metabolism.

A web-interface for the detection of selection signatures: development and testing

Biscarini, F.[1], Del Corvo, M.[1], Albera, A.[2], Boettcher, P.J.[3] and Stella, A.[1], [1]PTP, Via Einstein, 26900, Lodi, Italy, [2]ANABORAPI, Str. Trinità 32/A, 12061, Carrù, Italy, [3]FAO, V. Terme di Caracalla, 00153, Rome, Italy; filippo.biscarini@tecnoparco.org

Selection signatures are of great interest and practical use both in population genetics, for the understanding and characterization of the evolution and genetic background of populations, and in animal and plant breeding, where their detection may help increase the efficiency of selection. We have been developing DeSSign (Detection of Selection Signatures), a web-interface for the on-line detection of selection signatures, based on a Fortran77 computer programme that calculates the composite log likelihoods (CLL) of the difference in allele frequency between contrasting populations along the genome. The web-interface has been developed using Php, Jquery and XML technologies, and allows for the guided upload and automatic formatting of data, prior to the launching of the statistical analysis, which will run independently on the server. It returns as output interactive graphs of CLL at each locus per chromosome, and a list with the names and positions of the most significant marker loci. It is planned to extend the output by retrieving from on-line databases the corresponding genes. DeSSign can be used with data of any diploid species. DeSSign was first tested on the cattle Hapmap dataset as in Stella *et al.*, obtaining the same results. It is now being tested with an available dataset of Piedmontese cattle genotyped with the 54k bovine SNP chip: animals affected by congenital arthrogryposis were compared to healthy animals in a cases/controls design to detect signatures for this malformation. Preliminary results did not show any significant signature of selection, thus confirming previous GWAS results that did not report significant associations for arthrogryposis (Biffani, personal communication). DeSSign will be further tested with data of more contrasting populations -dairy, beef and zebu cattle- before official release of its Beta version.

Aleutian mink disease virus infection may induce host gene shutoff
Farid, A.H.[1], Moore, S.S.[2] and Basu, U.[2], [1]Nova Scotia Agricultural College, Plant and Animal Sciences, Truro, Nova Scotia, B2N 5E3, Canada, [2]University of Alberta, Agriculture, Food and Nutritional Science, Edmonton, Alberta, T6G 2P5, Canada; hfarid@nsac.ca

The objective of this study was to investigate the effects of inoculation of mink with the Aleutian mink disease virus (AMDV) on the expression of genes in leucocytes. Four full-sib black male mink from each of three families were used. Animals were sedated and inoculated intra-nasally with $600XID_{50}$ of a spleen homogenate containing a local strain of the AMDV. Blood was collected by heart puncture from one mink from each family on days 0 (pre-inoculation), 1, 2 and 7 post-inoculation (pi). Libraries were prepared from total RNA using the Illumina TruSeq™ RNA kit, and were sequenced twice using HiScanSQ, producing 75 bp reads from each end. The number of paired-end reads were 794,241,780 in the first analysis (ranging from 7.5 M to 182 M reads per sample), and 819,437,070 in the second analysis (ranging from 13.4 to 176.2 M reads per sample). The reads were aligned to the dog genome using the Burrows-Wheeler Aligner (BWA) method. In total, 8.8% of the reads from the combined analyses were aligned to exons of the dog genes (5.6% and 11.9% for analysis 1 and 2, respectively). These transcripts were aligned to 14,845 of the 19,565 annotated genes on the dog genome, with the coverage depths of up to 3.7 M reads. Compared with the non-inoculated controls, 6 and 575 of the mapped genes were significantly up- and down-regulated, respectively, within 24 h pi. The level of expression of the six up-regulated genes remained elevated ($P>0.05$) compared with the controls until day 7 pi. There was no significant change in gene expression between controls and animals sampled on days 2 or 7 pi. Expression of 128 genes were significantly higher on day 2 pi compared with those on day 1, and expression of only one gene decreased in this period. The results may suggest an abrupt mink gene shutoff by the AMDV within 24 h after infection, parallel with the virus multiplication phase.

Application of different statisical techniques for the selection of significant SNPs on the 50K chip
Fraszczak, M. and Szyda, J., Animal Genetics, Biostatistics Group, Wroclaw University of Environmental and Life Sciences, Kozuchowska 7, 51-631 Wroclaw, Poland; magdalena.fraszczak@up.wroc.pl

A whole bunch of statistical models is applied to performing genome-wide association studies based on dense Single Nucleotide Polymorphism (SNP) chips. These vary between a single SNP models and models with effects of all the available SNPs fitted simultaneously. Our study is focused on the comparison of different approaches to select SNPs significant for complex traits based on their estimates obtained by different models. The material of this study consists of 2588 Polish Holstein-Friesian bulls genotyped by the Illumina 50K SNP chip with deregressed proofs available for production, type and conformation, udder health as well as fertility traits. Selection of SNPs is based on their effects estimated by four models: (1) a model with random effects of all the SNPs fitted simultaneously; (2) a model with random effects of all the SNPs fitted simultaneously and a random residual polygenic effect; (3) a series of models fitting a single SNP effect at a time; (4) a series of models fitting a single SNP effect at a time and a random polygenic effect. Then the following approaches are compared: (a) SNPs are selected based on the comparison of their estimated effects obtained by models (1) and (2) with the appropriate quantiles of the standard normal distribution, without correction for multiple testing; (b) SNPs are selected as above, but the possible confounding between them wich is due to LD is corrected by using step-wise multiple linear regression approach as proposed by Segrè *et al.*; (c) SNPs are selected based on P-values associated with the Wald test, based on their effects obtained by models (3) and (4), with Bonferroni correction for multiple testing; (d) SNPs are selected based on P-values associated with the Wald test, based on their effects obtained by models (3) and (4), with multiple testing correction based on multiple linear regression approach as proposed by Segrè *et al.*

Genome-wide association study of fertility and performance parameters in piglet production

Bardehle, D.[1], Preissler, R.[1], Lehmann, J.[1], Looft, H.[2] and Kemper, N.[1], [1]Institute of Agricultural and Nutritional Science, Martin-Luther-University Halle-Wittenberg, Animal Hygiene and Reproduction, Theodor-Lieser-Straße 11, 06120 Halle (Saale), Germany, [2]PIC Germany, Ratsteich 31, 24837 Schleswig, Germany; danilo.bardehle@landw.uni-halle.de

The aim of this study was to detect loci affecting fertility and performance parameters in piglet production in a genome-wide association approach. In total, 590 sows were genotyped at 62,163 single nucleotide polymorphisms (SNPs) using the Illumina PorcineSNP60 BeadChip. Quality and data control as well as statistical analysis was performed using PLINK and the specific R-package GenABEL. After quality control applying an individual callrate threshold of 0.95, a maximum missingness per SNP of 0.1 and a MAF of 0.05, 581 sows and 49,697 SNPs remained for further analysis. The number of SNPs was reduced to 45,422 as several SNPs did not have a known location. In a first approach, the number of piglets born alive (NBA), the number of stillborn (SB) and the total number born (TNB) piglets in first parity was used as phenotype. In a second approach, the mean of NBA, SB and TNB for all available litters was used. In all approaches, we accounted for the fixed effect of line in a simple score test and raw p-values were corrected using the genomic control method. Regarding sows' performance parameters, the necessity for applying birth assistance was used as a binary trait and was analysed using a score test corrected by principal components method. Applying a cut off for the corrected p-values of 0.00001, we identified significantly associated SNP on porcine chromosome (SSC) 18 for NBA within the gene ubiquitin protein ligase E3C (UBE3C). Regarding SB, loci on SSC2, SSC3 and SSC7 were found to be associated significantly. Further loci associated with TNB were located on SSC8, SSC12 and SSC18. The necessity for applying birth assistance was significantly associated with the SNP MARC0005177 on SSC12. These results are promising for further studies including confirmation and replication studies.

Efficiency, competitiveness and structure of ruminant husbandry in Eastern Europe and Western Balkan

Schroeder, E., ADT Projekt GmbH, Adenauerallee 174, 53113 Bonn, Germany; ekkehard.schroeder@adt.de

The economic transition in many countries brought significant changes into farming system during the last 2 decades, particular into the ruminant production sector. The privatisation and restructuring process and the decline of consumer purchase power in the first decade brought a dramatic decline of ruminant inventories and production. A stabilisation process has been observed during the last decade since overall economic development came up with increasing demand for food products and the modernisation of processing units increased the market chances of domestic production. Although the milk yields per cow are today usually significantly higher in these countries than two decades ago, the milk yield is in most of these countries in average still significant below 5,000 kg per cow/year. Thereby unsatisfactory productivity and efficiency of individual farms are usually not predominantly depending on farm size. Most important issues are the overall farm management, forage production, animal feeding and animal health management. The natural conditions for international competitive cow-milk production are particular in several regions of Russia, Belarus and Ukraine good. Land is principally available for cost-effective fodder production. In several Western Balkan countries ruminant husbandry is particular prevalent in hilly or mountain areas and sheep and goats husbandry with traditional breeds has a strong tradition in production and consumption. Strategies to improve livestock yields and production parameters should take into account the specific potentials of each country and region. Beside genetic improvement, organization of breeding and animal health services, research, extension services, establishment of training and information centers on advanced technologies, vocational training programmes for livestock breeders, utilization of potentials of market chains as well as specific national support programmes to introduce advanced technologies and related know-how were/are among the main subjects to optimize production parameters.

Challenges and problems of milk production sector in Russia
Turyanski, A.[1], Khmyrov, A.[2], Kulachenko, I.[1] and Dorofeev, A.[1], [1]Belgorod State Agricultural Academy, Belgorod, Russian Federation, [2]Belgorod Department for Agriculture of Belgorod Region, Belgorod, Russian Federation; irinakulachenko@mail.ru

An important challenge of food security in Russian Federation is to realize an increase in volume of non-polluted milk and dairy products, which fulfill a special place in population food allowance. According to Russian Federal Service of State Statistics, ratio of milk and dairy products consumption to recommended norm is in recent years about 80%, and share of milk suitable for intensive thermal processing in separate regions of Russia doesn't exceed 60-75%. Solution is to increase cow efficiency and decrease labor and material inputs/ton of milk. To increase efficiency, it is principally necessary to use 'intensive factors. In aggregate, these factors have impact on milk volume, but also on production of a high-quality competitive product with minimum expense of resources. Practices to improve are selection and breeding work, intensification of reproduction of herd, rational organization of food supply and high-grade feeding. It is also necessary to take care of physiological features of animals – breed, age at first calving, in heath process, calving interval, longevity, ability of dry matter intake of forages, fitness to machine milking and stability to mastitis. The organization of milk production with use of modern economic technologies, like milking techniques, and the use of "intensive factors' allows to increase production level on many farms in the region to 6,000-7,000 kg/cow/year, whereas on the average in recent years no more than 5,000 kg is achieved, while only 66 calves per 100 cows are born. Damage from mastitis and a decrease in dairy efficiency of ill cows is estimated to be equivalent to 5-8% of gross annual yield of milk. Practice shows that only at the expense of a good comfort and maintenance of laktiruyushchy cows, it is possible to have a lactation yield increase of 1000-1,800 kg milk and the highest quality milk. Thus, increase of volume of milk is possible when improving the efficiency factors mentioned.

Efficiency, cooperative aspects and prospects for dairy and beef sectors in Macedonia
Bunevski, G., Klopcic, M. and Krstevski, A., Faculty of Agriculture and Food, Ss. Cyril and Methodius University, Depart. of Cattle Breeding, Aleksandar Makedonski bb, 1000 Skopje, Macedonia; bunevski@gmail.com

Cattle breeding in the R. of Macedonia (FYROM) is in very strong competition with vineyard, vegetable and fruit production, from one side, and with small ruminant production on hill-mountain area on the other side. R. of Macedonia is 2/3 a hill-mountain country, with 6 basic agricultural regions, and 52 sub regions, with 3 kinds of climate (mediterranean, continental, high mountain). There are 11 larger ex-state dairy farms in our state, with more than 200 milking cows per farm. There are no special beef farms with larger capacity, but most of the farms are with dual-purpose production (milk and beef). Total agricultural production in GNP takes around 11%, animal production in total agriculture around 35%, and cattle production in our total animal production takes 30-35%. Because of the small average of dairy cattle per farm (3.5), low milk yield (2,880 kg/cow), with on average 3.5 lactations per cow and an unstable state subsidy policy, it is very hard to develop the dairy sector into an efficient agricultural branch. The national dairy sector provides almost 90% of our state demand. The domestic beef production for many years varies between 7-9,000 t/year, which supplies 35-50% of the total state demand of beef. Facing the development of the other agricultural branches in present situation, dairy and beef production is in a dificult situation, with a tendency of decreasing of the number of farmers, cattle and arable land, but with a slow increase in farm capacity and dairy and beef yield per animal. Agricultural cooperatives in dairy and beef sector are a part of the Federation of Farmers, but tendency is from 2012 to cooperate in special organization according the breed and species. In our country there are 86 totally registered dairy plants, but 4 of them takes 80% of the whole milk processing. For beef purposes, there are 6 slaughterhouses in our state.

Efficiency, cooperative aspects and prospects for dairy and beef sectors in Ukraine
Getya, A.[1] and Ruban, S.[2], [1]Ministry of agriculture and foodstuff of Ukraine, Khreshchatyk street 24, Kiev 01001, Ukraine, [2]Institute of animal breeding and genetic of NAAS, Pogrebnyaka street 1, Chubynske, Kiev region 08321, Ukraine; getya@ukr.net

Dairy and beef husbandry are the most important sectors of animal husbandry in Ukraine. But over the last 10 years a decrease in cow number of 2-3% per year has occurred. This decrease influenced the total volume of milk negatively and that is why the consumption of milk and dairy products calculated to milk dropped from 225.6 kg in the year 2005 to 204.9 liter per capita/year in year 2011. In Ukraine two big groups of milk producers exist being agricultural enterprises (farms with different forms of ownership) and family farms –FF, where mostly 1-2 cows per family are kept. From total number of 2.58 mln. of cows, 0.59 mln. are kept in enterprises and 1.99 mln. are on family farms. In year 2011, 11.1 mln. tons of milk was produced by all categories together, of which only 4.5 mln. tons was processed in the milk processing factories. Despite big differences in cows number between the two groups of milk producers, the amount of milk delivered for processing was the same in 2011, because 25% of milk from family farms and 90% of milk from enterprises was delivered for processing. Last year dairy cattle husbandry in the enterprises showed some improvement: an increase in total milk production was noted, even while cow population is still reducing by an higher productivity of cows which now is 4,100 kg milk/ cow/year compared to 3,975 kg in year 2010. Improvements in management of farms can be realized by introduction of modern technology of animal maintenance and optimization of feeding ration. We expect different development scenarios in family farms compared to enterprises. In enterprises the number of cows will grow, as well as milk production level. Cow number in family farms will diminish by 2-3% every year. The only possibility for family farms to develop is the creation and further development of cooperatives. The beef cattle sector in Ukraine is not very developed with 35.4 thousand beef cows bred in the whole country.

Adjustment of cattle and sheep production in Croatia to the actual economic and market environment
Ivanković, A.[1], Štoković, I.[2] and Barać, Z.[3], [1]Faculty of Agriculture, Department of animal production and technology, Svetošimunska cesta 25, 10000 Zagreb, Croatia, [2]Faculty of Veterinary Medicine, Department of Animal Husbandry, Heinzelova 55, 10000 Zagreb, Croatia, [3]Croatian Agricultural Agency, Ilica 101, 10000 Zagreb, Croatia; aivankovic@agr.hr

In last two decades, agricultural production, processing and consumption in Croatia have adjusted to new economic and market trends. During the same period, beef and milk production in Croatia have stagnated, with consumption declining by 36 and 23%, respectively. Sheep production is sufficient to satisfy domestic needs. Changes are more obvious in cattle than in sheep production. Liberalisation of genetic breeding material and animal products' markets have encouraged production and processing subject to a strategic association towards more apparent common interest. Producers are forming groups around criteria, such as primary production (milk, meat), breed of animal, region. Public services are encouraging establishment of breeding associations, which are very important for improvement of production techniques. The processing industry is non-formally split up in interest groups, imposing qualitative and quantitative standards to the primary producers in order to satisfy consumer needs. In last twenty years, the number of cattle and sheep has declined by 46 and 17%, respectively. Structurally, cattle husbandry is adjusting through production unit enlargement, introduction of modern technologies and genotypes. Decline in the number of dairy cattle had no significant effect on milk volume, but it did have a negative effect on the number of quality beef calves. Sheep production experiences small structural changes, but local breeds are in breeding sense enhanced and traditional production technologies are being standardized. Sheep milk is processed into quality cheeses, and marketed as a valued-added product. Standard models of sheep and beef meat production are available on national level, but we need standardization of the quality of production to fit into an appropriate market approach.

Various practices of subsidies for the conservation of local breeds across some countries of Europe
Kompan, D. and Klopčič, M., University of Ljubljana, Biotechnical Faculty, Dept. of Animal Science, Groblje 3, Domžale, 1230, Slovenia; drago.kompan@bf.uni-lj.si

In the ERFP (European Regional Focal Point) project 'Proper way of supports for endangered livestock breeds' we have analysed the impact of subsidy measures on population trends of breeds across countries of Europe. In the project was included formally 13 countries and additional 12 countries expressed the interest for collaboration. Altogether, 25 countries provided the data. Each country has its own system of evaluation and its own approach to conservation, depending on the state of autochthonous farm breeds. By means of an extensive questionnaire, we collected information on status of animal genetic resources in case the country did already adopt the Multiyear Program of Work in accordance with Interlaken Declaration, and when the country has a national preservation program in execution for local breeds. Furthermore, we investigated system, size and type of support or subsidies designed to preserve local breeds. We have reviewed the legal arrangements on this topic, which country programme (Action Plan) is available for the conservation of AnGR, which criteria do countries have for determining endangerment, which breed definitions are used, and which method is applied for the calculation of financial subsidies or support. The outcomes of study are very diverse between countries. Some states do not have definitions of indigenous breeds, other have. Interestingly, the definitions for indigenous breeds are quite different. We analysed the impact of the subsidy on population trend of the breeds' and how breeds are included into the environmental programs in the countries.

Efficiency, cooperative aspects and prospects for dairy and beef sectors in Belarus
Kyssa, I., Lutayeu, D., Halubets, L., Babenkov, V., Yakubets, Y. and Popov, M., International Public Association of Animal Breeders 'East-West', Kazintsa str. 88-10, 220108 Minsk, Belarus; lutayeu.dzmitry@hotmail.com

This paper gives an overview of changes in the dairy and beef sector in Belarus during 2000-2011 and main prospects. The number of livestock has decreased from 5.054 ths in 1995 to 3.924 ths in 2003. After 2003 the state policy was to make milk and beef husbandry an export sector and the number of cattle slowly increased to 4.247 ths in 2011. However, number of cows decreased in the same time from 2.137 ths with 31% to 1.477 ths in 2011. The number of cows in small private household keeping is constantly decreasing and has reduced with 14% to 149 ths during year 2011. Milk level slowly increased from 2.370 kg in 2000 to nowadays 4.400 kg kg/cow/year. The competition level among processors is quite high – there are about 50 milk processing plants today. Beef sector also got a good development to 30 ths cattle in 2011 and with average gain of 950 g/day. Nowadays milk and beef consumption is respectively 252 and 88 kg per inhabitant (year 2011). Special for the agricultural sector in Belarus is that the majority of farms are owned by the state with high subsidies level and with price regulation. Due to overproduction of milk and beef products per inhabitant, food export in Belarus became an important branch with a total value of 4 billion dollars in 2011, with Russia as main buying country. Our research also shows that most farms pay more attention to semen quality – about 90% of semen used for AI is produced by bulls which have a big % of Canadian and American Holstein cattle genes. Farmers also pay now more attention to management in feeding and reproduction – for example calving interval is about 13,2 months. We can predict that the beef and dairy sectors will slowly increase in intensity and efficiency in coming years, both because of state plans to double income from export and farmers own interest to increase efficiency due to constantly growing costs, and cooperation between farms in processing can improve as well.

An assessment of efficiency and prospects for dairy and beef sectors in Serbia

Bogdanovic, V.[1], Djedovic, R.[1], Perisic, P.[1], Stanojevic, D.[1] and Petrovic, M.D.[2], [1]University of Belgrade, Faculty of Agriculture, Institute of Animal Sciences, Nemanjina 6, 11080 Zemun-Belgrade, Serbia, [2]University of Kragujevac-Faculty of Agronomy, Cara Dusana 34, 32000 Cacak, Serbia; vlbogd@agrif.bg.ac.rs

Agriculture and food processing are the most important branches in Republic of Serbia: in accounts for 17% of GDP (10.6% agr. production and 6.4% food processing). In agr. production 70% comes from plant and 30% from livestock production. Consumption of milk in Serbia is 56 liters while consumption of meat products is 38 kilograms/capita/annum (18 kg of poultry, 16 kg of pork and 4 kg of beef). To have a more precise description of the dairy and beef sectors in Serbia, as well as obtaining data for improving zoo-technical and farm conditions in which ruminant production is organized, a survey has been conducted among farmers. 1180 questionnaires have been mailed to farmers whose farms are registered for cattle, sheep, goat or mixed production. Questionnaire was divided into 6 sections: general information about farm, technical and structural information, education and advisory services, perspectives for future activities, sanitary, veterinary and zoo-technical aspects. For more than 70% of farmers livestock production is the only source of income. One of the greatest structural problems which impedes efficient development of ruminant production in Serbia is the small scale of agricultural production. A small property with a small arable surface results also in a smaller number of animals kept. About 37% of farmers raises up to 10 cows on their farms, less than 20% farmers questioned have more than 20 sheep while goat keeping is the least represented sector. On the other hand, farmers/breeders associations are very fragmented and not able to support their associates in effective way. About 75% of farmers have expressed a positive expectation from future membership of Serbia in EU, although these expectations are not clearly defined. This paper will provide detailed analysis about structural and zoo-technical features of ruminant sectors in Serbia.

Economic aspects of suckling cow production systems in Slovakia

Koleno, A.[1], Huba, J.[2], Krupová, Z.[2] and Michaličková, M.[2], [1]RAJO a.s., Studená 35, 823 55 Bratislava, Slovakia, [2]Animal Production Research Centre Nitra, Hlohovecká 2, 951 41 Lužianky, Slovakia; andrej.koleno@gmail.com

The aim of this study was to define the optimal realization price of calves to reach a positive economic result in the production conditions of Slovakia based on real economic results achieved in the monitored farms (n=5) in 2010. The annual costs per suckler cow were at 818 € and the average realization price per 1 kg live weight of calves (LW) at 2.45 €. In the first model (VAR1) we consider the realization of production by sale of weaned calves at the end of a grazing period at about 8 months of age in LW about 280 kg, i.e. average daily gain (ADG) 1000 g/day. In the second model (VAR2) we consider the realization of production by direct sale of meat from fattened calves to consumers. The models are based on fertility of cows at 100 born calves per 100 cows. We expected only minimal losses of calves to the weaning at average 3% and low culling of cows at average 15%. It means assumption to sell 82 calves from 100 cows per year, representing a model biological maximum. With respect to the defined natural-economic parameters it is impossible to achieve positive economic profit without subsidies. Estimated loss calculated in VAR1 was -255 € per cow and year. In VAR2, in addition to grazing and suckling mother's milk, we consider the fattening of calves to early December with average daily feed ration at 8 kg of silage and 2 kg of grain feed. It causes increase of average costs by 0.3 €/day, i.e. from 2.24 € to 2.54 €. We supposed ADG of calves at about 1300 g/day, slaughtering in LW about 405 kg (about 235kg of meat with bone) and price acceptable by consumers at 6.0 €/kg of meat from young calves. For achieving zero economic profit with above defined basic natural-economic parameters (without subsidies), according to our calculations, should be in VAR1 minimal realization price at 3.56 €/kg of LW and in VAR2 at 5.33 €/kg of meat, even after including slaughtering costs on average 100 € per animal.

Impact of market prices on economic values of dairy cattle and sheep traits

Krupová, Z.[1], Wolfová, M.[2], Krupa, E.[2] and Huba, J.[1], [1]Animal Production Research Centre Nitra, Hlohovecká 2, 951 41 Lužianky, Slovakia, [2]Institute of Animal Science, P.O. Box 1, 104 01 Prague, Czech Republic; krupova@cvzv.sk

The aim of this study was to evaluate the impact of market prices of dairy products on relative economic values (REVs) of traits in the most common breeds of dairy cattle (Holstein) and sheep (Improved Valachian) in Slovakia using a bio-economic model. Dairy milk price was based on milk volume, fat and protein contents and somatic cells count, sheep milk price on volume, fat and protein contents and on cheese price. The average prices of dairy products (0.314 €/kg of dairy milk, 0.714 €/kg and 5.975 €/kg of sheep milk and cheese, resp.) varied by ±40% in the evaluated period (2004-2010). For all levels of milk prices were the most economically important milk yield, protein and fat content and somatic cell score in cattle, but only milk yield in sheep (milk traits), productive lifetime and conception rate of cows in cattle and in addition litter size in sheep (functional traits). REV of milk yield was the most sensitive to milk price, both in cattle and sheep. When changing the dairy milk price from the lowest to the highest value, the sum of REVs for all milk traits increased from 52 to 68%, but the sum of REVs for functional traits dropped from 31 to 23%. In sheep, the REV of milk yield varied from 31 to 48% when the extreme prices for both sheep commodities were applied. The REVs of growth and carcass traits (in cattle) or growth and wool traits (in sheep) had a low relative economic importance at all milk price levels (from 1 to 15% in cattle and from 0 to 9% in sheep). Taken into account the price trends in the current year (2012), the use of the average milk price for dairy milk and the maximal prices for sheep milk and cheese presented in our study can be recommended when calculating economic values of traits included in the breeding objective of Holstein cattle and dairy sheep in Slovakia.

Idiazabal cheese: a product linked to land

Ruiz, R.[1], Molina, M.[2] and Ugarte, E.[1], [1]NEIKER-Tecnalia, Animal Production, Campus Agroalimentario de Arkaute, Apdo 46, 01080 Vitoria-Gasteiz, Spain, [2]CRDO Idiazabal, Campus Agroalimentario de Arkaute, Apdo 46, 01080. Vitoria-Gasteiz, Spain; rruiz@neiker.net

Idiazabal cheese is a quality product labelled under a protected denomination of origin that protects the breed and production area. It's made with raw milk from the Latxa and Carranzana breeds in the Spanish Basque Country and Navarra. The average production is around 1400-1500 tn per year and 50% corresponds to homemade transformation. Nowadays, the cheese has international prestige and it has been recognized with different prizes. The Denomination of Origin for Idiazabal cheese was created in 1987 to define the basic regulations for the product's manufacture and to guard the quality of the product. The paper describes the work that the Denomination of Origin, created in 1987 has developed to set the basic regulations for the in relation with the transformation process (homogenization, hygienic and organoleptic quality, etc) research (breed identification, autochthonous starters, etc) and marketing(exhibitions, competitions, direct selling, local markets, etc). The Regulatory Council of the Origin Denomination, is responsible of guarding for the quality and the making processes established for Idiazabals Cheese, owns a blender team, the Official Taste Committee of the Denomination of Origin, who qualifies and value which particular characteristics the cheese have. The paper also explains financial aspects, organizational structure and working rules The creation and work of Regulatory Council of idiazabal cheese has been crucial to conserve and improve the Latxa and Carranzana breeds before the threat of high productive foreign sheep breeds in the area.

Level, quality, profitability and sustainability of the milk production of small ruminants
Kukovics, S. and Németh, T., Research Institute for Animal Breeding and Nutrition, Gesztenyés út 1, 2053 Herceghalom, Hungary; sandor.kukovics@atk.hu

Milk production has different importance in sheep and in goat in Hungary. While the income from milk can reach 75-85% of the total income in goats, its ratio is between 20 and 45% in sheep (depending on intensity of production and breed). This determines the requirements and conditions of keeping of both species. The sheep milk production has been changed significantly in the last some decades in Hungary: it reached its maximum level in 1970 (22.9 million litres). At present the sector is basically sustained by meat production and different types of subsidies. The importance of wool production is infinitesimal (2-3%), while milk production (1%) reaches the 20-45% of the total income only on milking farms. Currently only 5-6% of the total number of ewes are under milking, while their ratio was 75, 33 and 12% in the top years (1970, 1985, 2003) of milk production. The importance of goat milk production is very limited in Hungary, because of the dispersion of sector and characteristics of breeds. Contrary to the 967 000 ewes, the number of does was just about 16 000 in the official register in 2010, while the total population was 44 000 by Hungarian Central Statistical Office. The quantity and composition (fat, protein, lactose and fat-free dry matter), quality (total number of bacteria, somatic cell count, specific gravity, titre, pH value, and freezing-point) of sheep and goat milk and their changes during lactation were analyzed. Based on production costs and buying up prices the profitability and sustainability of milk production of both species was analyzed. As conclusion besides increasing of individual milk production increasing of number of milked animals per farms is also necessary. The intensity of production affects yields, but the effect of breed is not neglectable. Over a certain cost level the milk production cannot be profitable. Sale of own produced and processed products could increase incomes, but the available market has basic effect on sustainability.

Impacts of once-daily milking for Rocamadour on-farm cheese-makers
Dutot, S.[1], Durand, G.[2], Gaudru, M.L.[2], Martin, B.[3], Pomiès, D.[3], Hulin, S.[4] and Marnet, P.G.[1], [1]Agrocampus Ouest INRA UMR 1348 Pegase, 65 Rue de Saint Brieuc, 35000 Rennes, France, [2]Syndicat des Producteurs de Fromages Rocamadour, 430 Avenue Jean Jaurès, 46004 Cahors, France, [3]INRA UR 1213 Herbivores, Theix, 63122 Saint Genès Champanelle, France, [4]Pole AOP Fromager Massif Central, 20 Côte de Reyne, 15000 Aurillac, France; solene.dutot@wanadoo.fr

An experiment was conducted with 6 on-farm cheese-makers from the Rocamadour Protected Designation of Origin (PDO). The objectives were to analyse zootechnical effects on goat milk production and to assess composition and sensory quality of cheeses made with their technology. The farmers were divided into 2 groups and managed their herd over two 6 weeks periods of twice daily-milking (TDM) and once-daily milking (ODM), inverted during the lactation season. For each milking frequency, milk quality and quantity were measured and 12 groups of cheeses were tasted by 2 specialized panels. Data were analysed with Proc Mixed Anova procedure for repeated measurements. Otherwise, the GLM model was used. ODM leaded to a reduction of milking time of 45 min/d for 100 goats and a milk yield reduction from about 22% calculated at the change of milking frequency. About milk quality, we noticed a significant increase of +1.5 and +2.0 g/kg of fat and protein contents, included casein. If curds were not affected by milking frequency, ripened cheeses were richer in dry matter and had a significantly lower fat / dry matter percentage (-1.35%). The PDO panel of judges confirmed this with a texture score and a total score significantly lower during ODM than TDM periods (2.76 vs. 2.97 and 11.48 vs. 11.96). The sensory analysis showed only 9 differences among the 37 descriptors, a lower development of the cheese covering moulds and a texture judged dryer during ODM period. These modifications were in PDO classical range and linked to the richer milk composition. Regarding the results, dairy goats of Rocamadour PDO adopt easily ODM management without impairing on cheese quality production.

Effects of udder morphological characteristics on the quality of sheep milk

Malá, G.[1], Milerski, M.[1], Švejcarová, M.[2] and Knížek, J.[1], [1]Institute of Animal Science, Přátelství 815, 104 00 Prague 10 – Uhříněves, Czech Republic, [2]Milcom a.s., Ke Dvoru 12a, 160 00 Prague 6 – Vokovice, Czech Republic; milerski.michal@vuzv.cz

Udder morphological characteristics and especially the teat position (angle) in dairy sheep are very important parameters in term of machine milking, which influences the milking process and welfare of ewes during it. The aim of this study was to determine the influence of chosen udder size and shape characterictics on the quality of sheep milk. Totally 141 purebred Eastfriesian and crossbred Lacaune ewes were included into investigation. Udder width, udder depth, teat lenght and teat angle were mesured. Milk samples were collected twice per lactation, i.e. 25 days after weaning of lambs (102 ± 10 days after lambing) and at the end of lactation (177 ± 12 days after lambing). The selected milk composition characteristics (fat %, protein %, casein % and lactose %) and microbiological parameters (total bacteria count – TBC, coliform bacteria – CB, somatic cells count – SCC) were determined in the individual milk samples. The highest correlation (r=-0.52) was found between fat content and udder depth. This relationship, however, was conditioned by high correlations between milk yield and udder depth (r=0.79) on one side and a negative correlation between milk yield and fat content (r=-0.58) on the other. Low correlations were found between udder depth and logTCB (r=0.23) and logCB (r=0.21). Teat angle showed no significant correlation relations to monitored milk quality traits. The study was supported by NAZV QH72286.

Genetic parameter estimation for major fatty acids in French dairy goats

Maroteau, C.[1,2,3], Palhière, I.[2], Larroque, H.[2], Clément, V.[4], Tosser-Klopp, G.[1] and Rupp, R.[2], [1]INRA, UMR444 Génétique Cellulaire, chemin de Borde-Rouge, 31326 Castanet-Tolosan, France, [2]INRA, UR631, SAGA, chemin de Borde-Rouge, 31326 Castanet-Tolosan, France, [3]UNCEIA, Service Génétique, 149 rue de Bercy, 75595 Paris Cedex 12, France, [4]Institut de l'Elevage, Site de toulouse, 31326 Castanet-Tolosan, France; cyrielle.maroteau@unceia.fr

Fatty acids (FA) are well-known for their importance on human nutrition. In France, since 2008, an important research and development project on phenotyping and genotyping the milk composition (FA and proteins) of cattle, sheep and goat has been carried out ('PhénoFinlait'). The project was based on a large scale on-farm phenotyping scheme for milk components allowed by the use of mid infrared (MIR) spectra. In the present study, genetic parameters for twenty fatty acids and milk production traits were estimated by restricted maximum likelihood with an animal model, using 45,259 testday records from 13,677 first lactation of Alpine and Saanen goats. Heritability estimates ranged from 0.19 to 0.51 for fatty acids and were highest (0.21 to 0.37) for short and medium chain fatty acids which are beneficial to human health, i.e. C6:0 to C14:0. High positive correlations (>0.50) were found between fatty acids of the same origin: short and medium chain fatty acids, i.e. from C6:0 to C14:0, synthesised de novo in the mammary gland and long chain unsaturated fatty acids coming from the diet and biohydrogenate in the rumen. In both the Saanen and Alpine breeds, no significant genetic correlation was found between C16:0 and fat content, whereas positive correlations (0.17 to 0.87) were found between fat content and specific goat fatty acids, i.e. C6:0 to C10:0. This result suggests that selection on fat content will not be correlated with undesirable changes in FA profile for human health. A sample of 2 254 goats was genotyped with the 50K goat SNP beadchip (Illumina) for QTL detection. This study was funded by ANR, Apis-Gène, CASDAR, CNIEL, FranceAgriMer, France Génétique Elevage and Ministry of Agriculture.

The quality of Slovak ewe milk based on fatty acids health affecting compounds
Soják, L.[1], Kubinec, R.[1], Blaško, J.[1], Margetín, M.[2] and Apolen, D.[2], [1]Comenius University, Faculty of Natural Sciences, Mlynská dolina, 842 15 Bratislava, Slovakia, [2]Animal Production Research Centre Nitra, Hlohovecká 2, 951 41 Lužianky, Slovakia; apolen@cvzv.sk

Our objective was to determine the content of 70 C4-C24 FA in fat of Slovak ewe milk from pasture and winter seasons. The ewe milk samples of grazed ewes were collected at 4 farms. The effect of individuality, breed, lactation stage and parity of grazed ewes is investigated on 330 Tsigai, Valachian and Lacaune ewes, milk samples were taken on the same day. The content of individual FA in milk samples as well as in pasture plants during whole pasture season was determined by GC, unseparated CLA isomers were resoluted by chemometric deconvolution. Determined content of FA was evaluated using a one-way ANOVA statistical package. FA content in milk fat of grazed ewes is up to 4-fold higher for CLA, 3-fold for TVA, and 2-fold for α-linolenic acid (ALA) than in the fat during winter diet. The content of CLA (3.2%) and TVA (7.2%) in milk fat of grazed ewes is higher than that in milk of pasture grazed ewes or cows published so far. Both CLA and TVA content in milk fat decreased from May to mid-summer and then increased until mid-September, ewe milk in mid-September has a similar FA composition to that in May. The variations in CLA content in milk fat during pasture season are primarily determined by the ALA content in pasture. The FA milk profile of cows grazing on Alpine pastures is rather similar to Slovak milk profile of grazing ewes despite different forage and altitude. The second–most abundant CLA t11c13 isomer in cows' milk was proposed as an indicator of milk products of Alpine origin. We observed identical result for milk of ewes grazing on pastures in altitude range of 250-800 m. Further improvement in quality of ewe milk can be achieved by selecting individual ewes with higher CLA milk fat content (up to 5-fold inter-individual variation) and milk yields (up to 12-fold) and by oversowing pastures. This work was supported by the Slovak RDA No. APVV-0458-10.

Adaptation of livestock to climatic stress
Mitchell, M.A., SAC, Animal and Veterinary Science, The Roslin Institute Building, Easter Bush, Midlothian, EH25 9RG, United Kingdom; malcolm.mitchell@sac.ac.uk

It is now recognised that climate change constitutes significant threat to the efficiency of animal production and the welfare of all livestock, The climate scenarios proposed indicate that even in the short term (e.g. by 2030) alterations in average, maximum/minimum temperatures and frequency of extreme events will require changes in production strategies and operations. Adaptations may be regarded as the interventions required by the producer industries and/or the physiological responses exhibited by animals in the face of imposed environmental challenges. Recognition of the importance of both these operational definitions is essential to underpin an integrated approach to the development of future livestock production strategies. Thermal challenge in terms of heat and/or cold stress constitutes the biggest risk to animal production efficiency and welfare, thus:- (a) Intensive production of housed animals e.g. poultry and pigs is most vulnerable to chronic heat stress due to increased average temperatures and heat stress during extreme events (b) Intensive production has, however, more adaptations or strategies available in the face of thermal challenge and technological solutions may be possible but the cost and effectiveness of these measures requires further investigations and evaluation (c) Animal transport for all species is a major concern: poultry and pigs transported to slaughter are the main concerns (d) The transportation of young animals is a major concern (e) The risks to production efficiency and welfare of animals in outdoor extensive productions are very significant (cattle and sheep). These sectors may be considered to be at less risk than those above in the short to medium term. Incorporation of an understanding of the physiological adaptive and thermoregulatory capacities of each species of livestock and at appropriate ages and physiological conditions must underpin the design and development of adaptation strategies in each of the key areas highlighted above in relation to the various climate change scenarios.

Assessing heat stress effects on production traits of Holsteins in a temperate region

Hammami, H.[1,2], Bormann, J.[3] and Gengler, N.[2], [1]National Fund for Scientific Research, 5, rue d'Egmont, 1000 Brussels, Belgium, [2]University of Liège, Gembloux Agro-Bio Tech, Animal Science Unit, 2, Passage des Déportés, 5030 Gembloux, Belgium, [3]Administration des Services Techniques de l'Agriculture, 16, route d'Esch BP 1904, 1019 Luxembourg, Luxembourg; hedi.hammami@ulg.ac.be

Heat stress impaired productive, reproductive and animal behaviour. Examination of milk production loss due to heat stress was mainly evaluated in USA and few tropical countries using the temperature humidity index (THI1) developed in the 1950's. The main objective of this study was to evaluate the effects of six new environmental stress indices on production traits in a temperate region. These indices include solar radiation and wind speed in addition to the temperature and humidity only defining THI1. For that purpose, 530820 milk, fat, protein, and somatic cell count first-lactation test-day records (TD) collected between 2000 and 2011 in Luxembourg were used. TD records were merged with meteorological data from 14 public weather stations. Each TD was assigned to the average daily thermal index of the 3 days before. Firstly, broken-line regression models were applied to identify the threshold of heat stress (THR) for each of the thermal indices. The decline of daily production of the different traits above THR point was assessed by mixed linear models including regression on unit of each of the thermal index. Estimates of THR were specific to each thermal index and trait. Significant decrease of milk, fat and protein yields above THR was observed for all studied thermal predictors. Whereas, an increase of the somatic cell score was observed above the specific THR point. THR estimates for THI1 and THI6 were of 68 and 80 and 62 and 78 for milk and protein yields respectively. Respective rate of decline per unit were of 0.243 and 0.188 kg and 0.0045 and 0.0051 kg respectively. Results from this phenotypic analysis show that evaluating heat tolerance based on bio-meteorological indicators is promising and phenotypic and genetic implications should be investigated.

Using fibrolite enzymes to reduce effect of heat stress on Holstein dairy cows

El-Sayed, H.M.[1], Abedo, A.A.[2], El-Bordeny, N.E.[1], Soliman, H.S.[1], Hamdy, S.M.[1] and Daoud, E.N.[3], [1]Ain Shams Univ., Animal Production Department, Cairo, 68 Hadayek Shoubra, Egypt, [2]National Research Center, Animal Production Dept., Dokki, 12311, Egypt, [3]Alex. Copenhagen Company, Meat and Milk production, Noubarya, 12311, Egypt; abedoaa@yahoo.com

This study aimed to evaluate the effect of using fibrolytic enzymes (Fibrozyme) on reducing the effect of heat stress on performance of lactating Holstein Friesian cows. An ambient temperature was ranged from 33 to 36 0C and the relative humidity was ranged from 52 to 56%. Ninety-four cows were randomly designed among two groups; the first fed without Fibrozyme (control), the second group fed diet supplemented with 15 g Fibrozyme /h/d. The results show that value of rectal temperature and respiration rate for control group were 40.50C and 82.6 and was 40.420C and 81.8 breaths/minutes, respectively for Fibrozyme group, those differences were not significant. Milk production, fat corrected milk (FCM), fat and total solids yields were significantly increased (29.37, 28.67, 1.13 and 3.67 kg) for cows fed Fibrozyme diet compared with 26.04, 23.40, 0.87 and 3.10 kg/h/day for cows fed control diet, respectively, while yields of protein, lactose, and solids not fat were insignificant increased. Milk composition was not significantly affected by adding Fibrozyme, except milk fat percentage was significantly ($P \leq 0.001$) increased. Feed conversion as kg dry DM, CP, TDN and NEL per kg FCM were significantly improved for cows fed Fibrozyme diet in comparing with control group. Fibrozyme supplementation caused significant increase in serum total protein, glucose, Triiodothyronine and total bilirubin, but contents of albumin, ALT, AST, and urea were not significant different. It is conclude that Fibrozyme supplement improved productivity of lactating cow in summer season.

Genetics show that adaptation to drought periods means less lambs for young ewes and more for old
Rose, G.[1], Mulder, H.A.[1], Thompson, A.N.[2], Ferguson, M.[2], Van Der Werf, J.H.J.[3] and Arendonk, J.A.M.[1], [1]Wageningen University, Animal Breeding and Genomics Centre, P.O. Box 338, 6700 AH Wageningen, Netherlands, [2]Department of Agriculture and Food Western Australia, 3 Baron-Hay Court, 6151 South Perth, Australia, [3]University of New England, School of Environmental and Rural Science, 2351 Armidale NSW, Australia; gus.rose@wur.nl

Sheep in Australia can be bred to be more resilient to drought periods. Resilient sheep lose less weight when grazing poor quality pasture. However, we do not know if breeding for weight change affects reproduction. So we estimated genetic correlations between weight change and reproduction. We used ~6800 fully pedigreed Merino ewes managed at Katanning in Western Australia. Weight change was measured during mating (42 days) on poor pasture and during pregnancy (90 days) on poor and medium quality pasture on ~1900 2 year old ewes, ~1500 aged 3 and ~1100 aged 4. Reproduction traits were number of lambs and total weight of lambs born and weaned measured on ~5300 2 year old ewes, ~4900 aged 3 and ~3600 aged 4. We estimated genetic correlations (r_g) between weight change and reproduction within age. Two year old ewes that gain weight during mating gave birth to more lambs (r_g=0.37±0.18) and weaned more lambs (r_g=0.62±0.19). Two year old ewes that lost more weight during pregnancy weaned more lambs (r_g=-0.45±0.19). In contrast, three year old ewes that gained weight during pregnancy weaned more lambs (r_g=0.65±0.30) and had a higher total birth weight (r_g=0.42±0.17). All other correlations were low or had high standard errors. We concluded that young, immature ewes were more sensitive to weight change during mating because they had more difficulty getting pregnant compared to older, more fertile, ewes. Older ewes also had higher litter sizes that require more energy at lambing and lactation to produce healthy lambs. This means weight gain during pregnancy is more important for older ewes. In conclusion, breeding for resilience to drought periods improves reproduction in old ewes.

Mild heat load alters some aspects of nutrient partitioning and gene expression in sheep
Digiacomo, K., Dunshea, F.R. and Leury, B.J., The University Of Melbourne, Department of Agriculture and Food Systems, Parkville, Vic, 3010, Australia; kristyd@unimelb.edu.au

In southern Australia sheep commonly experience mild heat stress, although the effects of mild heat load on intake, nutrient partitioning and tissue heat shock protein (HSP) responses in sheep are not well documented. This experiment examined the effects of a mild heat load on metabolism in sheep utilizing a glucose challenge (IVGTT) and tissue biopsies to examine HSP gene expression in adipose tissue and skeletal muscle. Six castrated male and six female cross bred sheep (34±0.5 kg) were fed *ad libitum* and exposed to 24 days of either thermoneutral (TN) (17-24 °C) or diurnal mild heat conditions (17-36 °C). Physiological responses (respiration rate (RR), skin (T_S) and rectal (T_R) temperature) were observed at 09:00, 13:00 and 17:00 hours on days 0-20 and blood samples obtained on days 7 and 14 at 09:00 and 15:00 hours. The glucose challenge was conducted on day 21 and tissue biopsies obtained on day 24. The RR, T_S and T_R responses were greatest at 1300 compared with 09:00 hours (P<0.001) and RR and T_S at 13:00 hours was higher in heated compared to TN sheep (P<0.001). Plasma prolactin concentrations were higher at 15:00 compared to 09:00 hours (163 vs 77 ng/ml, P<0.001) and mild heat increased plasma prolactin concentrations at 15:00 hours (83 vs 273 ng/ml, P=0.003). During the IVGTT mild heat increased the plasma NEFA $AUC_{0-10min}$ (P=0.03) and tended to decrease the mean time to reach a minimum insulin concentration (195 to 155 mins, P=0.08). The clearance rate of plasma $NEFA_{6-10mins}$ was faster for TN animals (sed 0.04, P=0.06). Mild heat exposure did not alter HSP expression in either tissue type, while the expression of all measured genes was greater in adipose compared to muscle tissue (P<0.001). Thus, in the present experiment mild heat was sufficient to influence some aspects of physiology, hormone secretion and nutrient partitioning in sheep.

Pig adaptation to cold environment enhances oxidative and glycolytic Longissimus muscle metabolism
Faure, J.[1,2], Lebret, B.[1,2], Bonhomme, N.[1,2], Ecolan, P.[1,2], Kouba, M.[1,2] and Lefaucheur, L.[1,2], [1]INRA, UMR 1348, 35590 Saint Gilles, France, [2]Agrocampus Ouest, UMR 1348, 35000 Rennes, France; justine.faure@rennes.inra.fr

Environmental temperature influences energy intake utilization and tissue deposition in pig. However, physiological mechanisms involved in muscle metabolic adaptation to cold temperature are unclear. This study aimed at evaluating the effects of a cold temperature applied during the post-weaning (PW) and growing-finishing (GF) periods of pigs on growth performance, body composition and Longissimus muscle (LM) energy metabolism, with emphasis on the AMP-protein kinase (AMPK) pathway involved in homeostasis regulation. A total of 84 Large White × (Large White × Landrace) piglets (castrated males and females) fed *ad libitum* were submitted to either Cold (C: from 23 °C to 15 °C) or Thermoneutrality (T: from 28 °C to 23 °C) temperature during PW. 12 C and 12 T piglets were slaughtered at 25 kg and the remaining piglets were reared at 12 °C (CC) or 23 °C (TT), up to 115 kg. During PW, cold exposure induced strong responses on growth performance. Despite a higher average feed intake (AFI, + 50 g/d, P<0.01), C piglets exhibited a lower daily gain (ADG) and carcass adiposity (P<0.01) compared to T piglets. During GF period, CC pigs still exhibited a higher AFI (+ 394 g/d, P<0.001) than TT pigs but ADG and carcass adiposity did not differ between CC and TT pigs. Whereas a cold PW exposure reduced glycogen content (P=0.01), a long term GF adaptation stimulated oxidative and glycolytic muscle metabolism by increasing glycogen and lipid contents (P<0.001), and lactate dehydrogenase, citrate synthase and β-hydroxy-acyl-CoA dehydrogenase activities (P<0.001). Evaluated by an increase in total AMPK level without change in AMPK phosphorylation level in CC pigs, homeostasis potential was preserved in a cold environment. LM adaptations to cold temperature impaired technological meat quality (higher drip loss and lower ultimate pH), but higher lipid content could positively modulate sensory quality.

Do the genes involved in longevity interact with in farm temperature in rabbit females?
Sanchez, J.P. and Piles, M., IRTA, Animal Genetics and Improvement, Avd Rovira Roure 191, 25198, Spain; juanpablo.sanchez@irta.cat

The aim of this study was to assess the magnitude of the interaction between additive genetic effects on longevity and in farm temperature. To this end 6,743 longevity records from females belonging to a line selected for post-weaning growth since 1983 were used, being right censored 34% of the records. The model for the analysis was a piecewise constant hazard model including as time-dependent covariates the year-season, the ordinal of positive pregnancy test, the doe physiological status and the number born alive; in addition two additive genetic effects were included, associated to the intercepts and slopes of animal-specific regressions on weakly average of daily mean temperatures, i.e. the regressor was a time-dependent covariate, with changes every weeks. In these animal specific regressions intercepts and slopes were assumed to be independent and, for each one, a (co)variance structure proportional to the additive genetic relationship matrix was assumed. The model was fitted using an MCMC Bayesian procedure, performing the sampling of the appropriate conditional distribution by Adaptative Rejection Sampling. The posterior means (posterior standard deviations) of the additive genetic variance for the intercept and the slope were 0.76 (0.10) and 9.5E-5 (3.5E-5), respectively. Thus, for the range of temperature usually observed in the farm, 15-25 °C, a nearly flat effective heritability pattern was estimated, from 0.44 to 0.45. Therefore, the interaction between genes and the environmental temperature seems not to be a relevant factor in the determination of longevity. However, this result must be interpreted with caution since the independence between intercept and slope in the current model it is an important constraint, and in fact when the regression of weakly temperatures was not considered the effective heritability dropped to 0.38. Further research it is needed to relax model assumptions.

Effect of thermic stress on nitrogen retention in pigs

Brestenský, M., Nitrayová, S., Patráš, P. and Heger, J., Animal Production Research Centre Nitra, Hlohovecká 2, 95141 Lužianky, Slovakia; m_brestensky@cvzv.sk

We used 7 gilts (initial BW 50.5 kg±1.7 kg) for evaluation the effect of environmental heat stress on nitrogen retention in body. Animals were housed in metabolism balance cages. The whole experiment last for 28 days. Animals were housed at thermo neutral temperature – NT (20.61±0.1 °C) for 14 days. After this period followed second 14 days period during which the animals were housed at high environmental temperature – HT (30.41±0.4 °C). Animals were fed with standard diet twice a day in two equal doses at daily rate of 90 g.kg $^{-0.75}$. Water was offered *ad libitum*. On day 6 and 13 of each period we carried out 24 hours balance periods during which urine and faeces were collected separately using bladder catheters. There were 4 balance periods all together through the whole experiment – two balance periods at NT and two periods at HT. At the end of the experiment we pooled the samples for each 14 days period. The remnants of diets were collected daily, dried and N intakes were corrected on dry matter basis. Based on analyzed content of nitrogen in samples diets, urine and faeces we calculated nitrogen retention Nitrogen retention was lower by 20.6% (P=0.001) at HT in comparison with NT. The nitrogen retention was 32.6 g/d at environmental temperature 20.61±0.1 °C and 25.8 g/d at environmental temperature 30.41±0.4 °C. Nitrogen utilization decreased from 56.4% at NT to 46.4% at HT (P=0.003). Through the continuous long term effect of high environmental temperatures the nitrogen retention and utilization of nitrogen in body decreased as a result of decline utilization of metabolisable energy during respiration. This article was written during realization of the project 'BELNUZ 26220120052' supported by the Operational Programme Research and Development funded from the European Regional Development Fund.

Hematological and blood biochemical parameters of horses during winter in Latvia

Caune, I.N.T.A. and Ilgaza, A.I.J.A., Latvian university of Agriculture, Preclinical institute, Kr. Helmana 8, 3004, Latvia; intacaune@inbox.lv

The typical winter conditions have been in Latvia last two years. We kept horses in warm stalls, open stalls or in a paddock with shed. The aim of the research is to find out how to the weather in typical for winter time, includes enviromental conditions and physiological status (pregnancy) affected at horses blood biochemical and haematological paremeters. Research objectives: to research what are the changes in haematological and blood biochemical parameters for horses in early winter, mid-term and at the end, to analyze what are the changes in haematological and blood biochemical parameters in different keeping conditions and to compare how different are haematological and blood biochemical parameters in pregnant and non-pregnant animals for winter period. The research includes 20 adult warmblood horses, who is dividend into two groups: pregnant mares and hobby horses. During the daylight horses were into paddock, but in the dark time they were in different conditions – 10 horses in warm stalls and 10 in open stalls. In each group had five pregnant and five non-pregnant animals. We made a clinical investigation of horses, collected data of their history, as well as blood samples for laboratory examination. In our researh we included only clinically healthy horses. The research is continue, but at present getting data show that the haematological and blood biochemical parameters in pregnant and non-pregnant horses are different and change dependence keeping in winter climat and conditions.In winter mid-term for all pit housed horses have seen dehydration, as manifested by increases hemoglobin, hematocrit increases and changes in blood biochemistry. At the end of winter the blood biochemical analyses of horses will reduce amount of total protein, especially for pregnant mares. The research continue, our aim is to find the optimal housing conditions for pregnant and non-pregnant horses in various conditions in winter.

Sheep and wool production in Central and Eastern Europe

Niżnikowski, R. and Strzelec, E., Warsaw University of Life Sciences – WULS, Sheep and Goat Breeding, Ciszewskiego 8, 02-786 Warsaw, Poland; roman_niznikowski@acn.waw.pl

Changes in wool production in the countries of Central and Eastern Europe (CEEC) have processed decisively in response to the declining sheep population, which began in the region after 1990. This was due to the adoption of a new management system and the abandonment of subsidizing of wool production, which was usually carried out in the aim of maintaining employment in the textile industry. Sheep population at this time dropped from about 55 million heads in 1990 to 35 million heads in 2010. In the same time, the wool production has fallen from around 60 thousand tons in 1990 to 30 thousand tons in 2010. The reduction of sheep population in each country at that time was also progressed and the largest reduction was observed in Poland (94% less) whereas the smallest was presented in Hungary and Romania (41 and 42% less, respectively). Similar decrease in production of greasy wool was observed on the market of Central and Eastern European Countries, excluding Romania and Bulgaria where the greasy wool markets were maintained. In other countries of the region the serious problems with the management of sheep wool exist and generally, wool is used mainly in artistic handicrafts or production of various types of thermal insulation products, e.g. insulation mats. This has a significant impact on the profitability of sheep production. The economic yield from the sale of wool is now about 5-10% of total income from sheep production, depending on the state of the region. This situation in sheep wool production leads to the search for a new development of this resource, which leads to a return to traditional methods of processing wool for self-supplying or the supplying of local markets due to the promotion of specific handicraft products typical for the individual regions. Therefore as the result, the wool processing is most often localized in local farms focused on tourism and agro-tourism, which is presented quite widely in mountainous regions of CEEC.

Current considerations regarding the use of goat fibers in Europe

Zamfirescu, S., 'Ovidius' University, Agriculture, Mamaia Blvd., No. 124, 900527 Constanta, Romania; zamfirescustela@yahoo.com

The return to 'natural products' represents a growing trend in the wool industry. At international level, the use of fibers from rare animals has gained dimension for the past few years. These fibers are also known as 'noble, precious fibers', due to their specific characteristics: superior capacity for thermal insulation, soft, nappy touch, increased volume, natural sheen, pleasant aspect, increased degree of comfort. In the Central and East European countries, the textile industry registered a significant decline after 1989, which culminated with the partial or total abolition of the factories that processed natural animal fibers. Now, the current policies rethink methods for the revival of the industries of goat productions. Since natural fibers have a more important role in the manufacturing of ecological textiles, the international efforts are focused towards the accomplishment and application of economic strategies to support the animal breeders that produce rare fibers (Angora, Cashgora, Cashmere), to support the processing industry and the research activity in the field, through an adequate economical policy with the purpose of increasing the production of items with content of rare fibers with improved esthetic and functional characteristics. At European level, the number of goats exceeds 15911631 (source: Eurostat, 2007), with an ascending evolution in some countries from central and eastern Europe (Romania, Hungary) and a marked decrease in western Europe (France, Italy, Great Britain). The Angora, Cashmere and Cashgora breeds or other small ruminants (alpaca, yak) exploited for rare fibers have reduced numbers, under 500000 heads. This paper presents current data regarding the breeding of fiber-producing goats and their exploitation manner in Europe in general, and in the Central and East European countries in particular.

Variation in wool fibre characteristics in sheep differing in genotype

Moreno Martinez, L., Mcdonald, A.J.S. and Galbraith, H., University of Aberdeen, School of Biological Sciences, 23 St Machar Drive, AB24 3RY, Aberdeen, United Kingdom; h.galbraith@abdn.ac.uk

Hair fibre 'wool' is a biological product synthesised in hair follicles in the skin of mammals. It is produced from specialised epidermal keratinocytes and is influenced by internal chemical signalling and nutrient supply. Sheep wool production in EU27 is substantial at approximately 186,000 tonnes annually (Faostat). Its commercial value is inconsistent and determined by physical properties of fibres and structure and function of individual follicles. The aim of this work was to investigate physical properties and variability. Samples of wool from 11 EU breeds were studied in addition to fibres from three cross-breed meat genotypes from the same local flock. Individual samples of wool were divided into lots of finest and coarsest according to diameter. Mean diameters (5 values at 1.0cm intervals) of 30 fibres from each lot were recorded. Separated lots of North Ronaldsay wool contained the finest and coarsest fibres with values (μm, mean, (SD)) of diameter of 21.2 (3.34) and 100 (14.2) respectively. Mean values (μm, (SD)) for finest fibres in fleece samples for three cross breeds in the same flock were 44.1 (4.17), 37.8 (5.21) and 33.5 (4.1). Values for means (with SD) of coefficient of variation (CV%) for diameter between individual fibres were, for coarsest fibres, greatest for North Ronaldsay wool (19.70 (14.33)) and least for Hebridean (3.18, (2.80)). Similarly for finest fibres, values were greatest for Charollais (5.39 (2.62)) and least for North Ronaldsay (1.90 (1.18)). Comparisons, for finest fibres, for measurements of cuticle scale height (μm, mean, (SD)) were greatest for Greyface (20.8 (4.0)) and least for Suffolk (9.9 (1.1). These results describe variation in physical properties of wool between and within breeds, within samples of fibres and within individual fibres. Improvements in uniformity will depend on better understanding of follicle characteristics and application in breeding programmes to enhance quality of fibre product.

Changes in wool classes proportions in the most numerous sheep breeds in the Czech Republic

Milerski, M., Institute of Animal Science, Přátelství 815, 104 00 Prague 10 – Uhříněves, Czech Republic; milerski.michal@vuzv.cz

The only indicator of wool performance surveyed in the Czech Republic in the last 20 years in a frame of official recording scheme is the wool class subjectively assessed by experienced classifiers and mainly based on fineness of wool. The aim of this study was to evaluate changes in the wool class proportions in the chosen sheep breeds during the last decade. Totally 51025 animals of 10 sheep breeds (Charollais, Romney, Merinolandschaf, Oxford Down, Romanov, Šumava sheep, Suffolk, Texel, Wallachian sheep and Eastfriesian) were included into evaluated dataset. Wool class assessments were carried out in young animals aged 6-18 months. In the majority of breeds the proportions of wool classes have not changed, but in Romanov, Suffolk, Texel and Wallachian sheep there is a systematic increase in proportion of coarse wool. This is probably due to the fact that these breeds are increasingly being kept on pasture year-round and breeders often select animals with a rather coarse wool. The largest shift can be observed in Wallachian sheep which since 2000 has increased the average diameter of hairs by almost 9μm (from 43,4 μm in animals born in 2000 to 52,1 μm in animals born in 2010). In this rustic breed is an effort to reconstruct the original very coarse, mixed character of fleece.

Fleece structure of Italian dairy and meat sheep

Gubbiotti, M.[1], Pazzaglia, I.[2], Lebboroni, G.[2], Antonini, M.[2] and Renieri, C.[2], [1]University Guglielmo Marconi, Via Plinio 44, 00193 Roma, Italy, [2]University of Camerino, School of Environmental Sciences, Via Gentile III da Varano s/n, 62032 Camerino, Italy; carlo.renieri@unicam.it

Fleece structure from 33 Sarda, 20 Comisana, 20 Delle Langhe, 20 Fabrianese and 20 Bergamasca adult ewes were investigated. Fibers were processed through a projection microscope at x500 magnification according the ASTM D2130-90 method. The mean fiber diameter (MFD) was 38.45 (s=19.19), 41.32 (s=22.74), 42.05 (s=23.45), 34.71 (s=9.05) and 38.41 (s=13.07) μm, with a coefficient of variation of 49.9%, 55.02%, 55.76%, 26.1% and 34.03%, respectively. Fibers without medulla (wool) accounted for 81.31%, 77.9%, 69.9%, 86.9% and 81.3%, respectively. MFD was 32.54 (s=14.22), 32.07 (=10.0), 31.87 (s=10.55), 32.31 (s=8.9) and 10.44 (s=10.44) μm, with a CV of 43.69%, 30.34%, 33.09%, 27.54% and 31.00%, respectively. Fibers with medulla (hair, kemp and heterotype hair) accounted 18.69%, 22.1%, 31.1%, 13.1% and 18.7%, respectively. MFD was 64.14 (s=16.69), 70.75 (s=29.57), 65.74 (s=27.72), 62.47 (s=25.70) and 67.13 (s=27.55), with a CV of 26.02%, 41.79%, 42.16%, 41.28% and 41.04%, respectively. The comparison of general, fibers without medulla and fibers with medulla MDF within and among breeds was highly significant (P≤0.001). Skewness and kurtosis for general, fibers without medulla and fibers with medulla diameter distributions was highly significant. The examined breeds retain a double coated, as a result of lack of selection for fiber diameter.

Phenotypic and genetic variation of cashmere production on Chinese Alashan white goats

Valbonesi, A.[1], Lou, Y.[2], Antonini, M.[1], Sun, Z.[2], Luan, W.[2], Tang, P.[3], Ma, N.[2] and Renieri, C.[1], [1]University of Camerino, School of Environmental Sciences, Via Gentile da Varano s/n, 62032 Camerino, Italy, [2]Jilin Agricultural University, Xin Cheng 2888, 130118 Chang Chun, China, [3]Station for Livestock Improvement of Alashan, Alashan Zoogi, 75030 Inner Mongolia Autonomous Region, China; carlo.renieri@unicam.it

Six hundred and nine one-year-old white cashmere goats (340 females and 269 males) bred at the Station of Livestock Improvement of Alashan (Left Banner, Inner Mongolia, P.R. China) were chosen for an investigation of phenotypic and genetic variation of down weight (DW), down fiber diameter (DFD) and its coefficient of variation (CVDFD). A one-way ANOVA was performed for testing differences between males and females. Phenotypic correlations were calculated by Pearson's coefficient. Hereditability was estimated on the base of a multiple trait animal model, using MTDFREML program. Sex was included as a fixed effect for each parameter. The average DW was 501.4 gr (s=120.6 gr) and 436.9 gr (120.6 gr) for males and females, respectively. The analysis of variance indicates a significant variance component due to the sex (F=42.97; P<0.0001). DW is significantly correlated, at the 0.05 level, to the DFD (r=0.095), whereas it's not correlated with CVDFD (r=0.005). Hereditability was very low (0.02). The average DFD was 14.63 μm (s=0.94 μm) and 14.49 μm (s=1.10 μm) for males and females, respectively. The analysis of variance indicates no evidence of a significant variance component due to the sex (F=2.72; P≤0.10). DFD resulted positively and significantly correlated, at the 0.01 level, with CVDFD. Hereditability was insubstantial. The average CVDFD was 27.09% (s=3.78%) and 27.65% (s=1.10%) for males and females, respectively. The analysis of variance indicates no evidence of a significant variance component due to the sex (F=1.83; P=0.18). Hereditability was low (0.15).

Economically weighting fibre and morphological traits in an alpaca breeding program

Gutiérrez, J.P.[1], Cervantes, I.[1], Pérez-Cabal, M.A.[1], Pun, A.[1], Burgos, A.[2] and Morante, R.[2], [1]UCM, Puerta de Hierro s/n, 28035 Madrid, Spain, [2]Pacomarca S.A., Miguel Forga 348, Arequipa, Spain; gutgar@vet.ucm.es

Nowadays, the fiber diameter is considered as the main selection objective in alpaca populations all over the world. ICAR recommendations define the fiber diameter (FD) and its coefficient of variation (CV) as the most important traits to be considered in breeding programs for this specie. Using these two traits as selection criteria can be accompanied by desired or undesired genetic responses in other highly related and economically important traits for the industry such as comfort factor (CF) or standard deviation (SD) as well as to other less important traits being selection objectives such as these morphological traits: density (DE), crimp (CR) or lock structure (LS) for respectively Huacaya (HU) and Suri (SU) breeds, head (HE), coverage (CO) and balance (BA). The goal of this study was to study the expected correlated genetic trends by considering different alternative procedures of weighting all the involved traits. Heritabilities and genetic and phenotypic correlations estimated from the data set belonging to the PACOMARCA experimental farm for SU and HU were used. Three approaches were used to check the consequences of a set of subjective economic weights essayed. The coefficients of selection indexes were obtained for the set of economic weights, and equivalent economic weights obtained when applied those values as coefficients of hypothetical selection indexes directly on phenotypes were drawn. Also relative expected genetic responses were computed in different cases. Results showed that almost in all cases for both breeds, the weight applied to CF should be surprisingly negative. Concerning genetic responses, only CO was compromised in some cases for the HU breed, but morphological traits could be negatively modified in many cases for SU breed. If selection is focused in CV and FD, morphological traits will be penalized only in the SU breed.

Author index

A

Aass, L.	253	Andersson, L.S.	323
Abd-Elsamee, M.O.	159	André, G.	48
Abdelzaher, M.	339	Angón, E.	183, 316
Abedo, A.A.	158, 340, 368	Anskiene, L.	101
Abeni, F.	94, 160	Antalíková, J.	285
Aboul-Naga, A.	339	Antler, A.	194, 195
Abraham, K.J.	310	Anton, I.	250
Åby, B.A.	253	Antonič, J.	329
Acero, R.	183, 316	Antonini, M.	374, 374
Acuña, O.	220	Apolen, D.	108, 262, 264, 269, 330, 367
Aditia, E.	175	Appel, A.	333
Aerts, M.	33	Appel, A.K.	89
Agabriel, J.	143	Araiza, A.	322
Agena, D.	199	Arana, A.	251
Agovino, M.	114, 115	Arce, N.	322
Aguilar, I.	135, 304	Archibald, A.L.	73
Ahlman, T.	333	Ardiyanti, A.	247
Ahmadzadeh Bazzaz, B.	61, 62, 121	Arencibia, A.	331
Aigner, B.	353	Argüello, A.	106, 329, 330, 331
Aihara, M.	201	Arias, G.	297
Ait-Ali, T.	73	Arnason, T.	323
Akdag, F.	82	Arnesson, A.	259
Akit, H.	211	Arranz, J.	120, 191
Al-Allak, Z.S.	241	Arranz, J.J.	309
Alarslan, E.	100	Arts, D.	40
Alary, V.	339	Árvayová, M.	292
Albanell, E.	53	Asgarijafarabadi, G.	70
Albayrak, S.	31, 243	Asheim, L.J.	183
Albera, A.	357	Ashfield, A.	141
Albertí, P.	251	Ashtiani, R.	236
Alexandridou, M.	166	Ashworth, C.J.	175
Alfaia, C.P.M.	75	Aso, H.	278, 279
Ali, M.	56, 57	Astruc, J.M.	132
Alkholder, H.	136	Atzori, A.S.	301
Allahyarkhankhorasani, D.	70	Aubert, A.	341
Allais, S.	8, 231	Auclair, E.	213, 252
Allen, A.R.	352	Aviles, C.	80, 81
Allison, G.G.	54, 155	Avino, V.	158
Almeida, A.	106	Aygün, T.	100, 265
Alòs, N.	222	Aznárez, N.	357
Aloulou, R.	180	Azor, P.J.	32, 81, 346
Alstrup, L.	113, 296		
Altarriba, J.	233, 305	**B**	
Altenhofer, C.	202	Babenkov, V.	362
Aluwé, M.	276	Babilliot, J.M.	187
Amatiste, S.	23	Bacciu, N.	355
Amazan, D.	46	Bachmann, H.	215
Amills, M.	357	Baes, C.	89
Ampe, B.	243	Bagheri, M.	236
Ampuero, S.	89	Bahelka, I.	218, 219
Anderson, F.	96, 301, 302	Bahr, C.	194, 195, 195
Andersson, L.	323	Bailey, M.	321

Bailoni, L.	192, 255	Beriain, M.J.	251
Bajusz, I.	337	Bermingham, M.L.	352
Baldi, A.	67	Bernard, L.	283
Baldi, G.	115	Bernard-Capel, C.	231
Baldinger, L.	117	Bernhard, V.	257
Ball, A.J.	301	Bernués, A.	140, 317
Baloche, G.	8, 132	Berry, D.P.	92, 127, 353
Banchero, G.	77	Bertolini, F.	276
Baneh, H.	261	Bertoni, G.	349
Bannink, A.	54, 55, 296	Bessa, R.J.B.	75, 76
Banos, G.	92, 169	Bessong Ojong, W.	31
Bar, D.	194	Bidanel, J.	272
Barać, Z.	266, 361	Bidanel, J.P.	240, 276
Barahona, M.	80, 81	Bielański, P.	235, 235
Barbey, S.	228	Biermann, A.D.M.	334
Barbotte, L.	187	Biffani, S.	130
Bardehle, D.	45, 359	Bijma, P.	88, 342
Bardowski, J.	287, 349	Bilčík, B.	206, 347
Bareille, N.	51	Billon, Y.	276
Barillet, F.	8, 132	Binggeli, M.	103
Barlowska, J.	245	Bionaz, M.	63
Barna, B.	254, 254	Bíreš, J.	270
Baro, J.A.	187	Birǵele, E.	83
Barragan, C.	10	Biscarini, F.	95, 357
Bartolomé, E.	39, 41, 342	Bishop, S.C.	73, 352
Barton, L.	80	Bittante, G.	151
Bassami, M.	66	Blair, H.	6
Bastiaansen, J.W.M.	52, 230, 325, 354	Blaj, V.A.	159
Bastin, C.	15, 17, 18, 78, 91, 92, 93	Blasco, I.	266
Basu, U.	358	Blaško, J.	262, 367
Bauer, J.	133	Blazkova, K.	198
Bauer, M.	252, 269	Bleumer, E.J.B.	48
Baumgartner, J.	332	Blok, M.C.	190, 191
Baumrucker, C.R.	122, 126	Blonk, H.	225
Baur, A.	188	Blouin, C.	36
Bavčević, L.	283	Bluyssen, H.	287
Beatson, P.	307	Boddicker, R.L.	43
Beck, R.	109	Bodin, L.	177, 187
Becker, A.	274	Bodó, I.	97
Becker, A.-C.	35	Bodo, S.	227
Bednarczyk, M.	287, 349	Boettcher, P.J.	357
Bee, G.	89	Boga, M.	59, 163
Beghelli, D.	21	Bogdanovic, V.	363
Belanche, A.	54, 155	Boichard, D.	188, 228, 308, 352
Beldman, A.	196, 314	Boissy, A.	110, 207, 341, 343
Beldman, A.C.G.	196, 310	Boivin, X.	110
Bell, M.J.	200	Bokkers, E.A.M.	195
Ben Othman, H.	180	Boldižár, M.	105
Benmansour, A.	207	Bonhomme, N.	370
Bennedsgaard, T.W.	49	Bonnet, P.	339
Bennewitz, J.	184, 260	Bonnot, A.	179
Benoit, M.B.	279	Bono, C.	181
Berckmans, D.	194, 195, 195, 345	Borka, G.	111
Bereta, A.	215, 237	Borling, J.	213
Beretta, V.	118, 249	Bormann, J.	368
Bergfeld, U.	172	Borrell, M.	344

Ertugrul, M.	69, 102	Formelová, Z.	56, 164
Escribano, A.J.	30, 282	Forni, S.	130
Escribano, M.	30, 282	Fortin, F.	273
Esquivelzeta, C.	135, 264	Foskolos, A.	53, 112
Estany, J.	212	Fouilloux, M.N.	308
Estelle, J.	240	Foulquié, D.	343
Esteve-Garcia, E.	210	Fourdin, S.	148, 338
Eugène, M.	339	Fourichon, C.	51
Evans, A.C.O.	185, 198	France, J.	54, 55, 296
		Francois, D.	249, 259
F		Francois, J.	197
Fabre, S.	177	Frankena, K.	325
Facchin, F.	178	Frankič, T.	164
Facciolongo, A.M.	269	Franzén, J.	91
Fahey, A.	168	Fraszczak, M.	356, 358
Fahey, A.G.	38	Fremaut, D.	225
Fahri, F.T.	211	French, P.	16, 146
Fair, M.D.	234	Fréret, S.	178
Fair, T.	185	Frey, H.J.	281
Falchi, L.	178	Frieden, L.	274
Falcucci, A.	224	Friedrichs, P.	64
Falo, F.	266	Fries, R.	353
Fanning, J.	109	Friggens, N.C.	201, 338
Fantuz, F.	21, 22	Fritz, S.	188, 308, 352
Farid, A.H.	358	Froidmont, E.	161
Farkas, J.	271, 273	Frondelius, L.	348
Farmer, C.	2	Frydas, I.	72
Fatet, A.	178, 299	Fuentes, E.	117
Faure, J.	276, 370	Fuerst, C.	49, 171
Faverdin, P.	115, 143, 298	Fuerst-Waltl, B.	49, 171
Felleki, M.	90	Furtado, S.	297
Fels, M.	345	Furukawa, T.	238
Ferencakovic, M.	12	Fusi, E.	67
Ferguson, M.	284, 369	Füssel, A.-E.	103
Fernandez, A.	10		
Fernández, J.	11, 41	**G**	
Ferraro, S.	22	Gabler, N.K.	43
Ferret, A.	112	Gabory, A.	2
Fiems, L.O.	190	Gagliardi, D.	258
Fievez, V.	117	Gaiato, A.P.R.	7
Fikse, F.W.	226	Galbraith, H.	206, 373
Fikse, W.F.	90, 239	Galik, R.	227
Filangi, O.	355	Galio, L.	2
Fina, M.	264	Galisteo, A.M.	32
Finocchiaro, R.	130	Gallard, Y.	228
Fiorelli, C.	24	Gallus, M.	178
Fischer, R.	172, 306	Gamba, D.	118
Fitzsimons, C.	168	García, A.	183, 316
Florek, M.	79	García, H.	322
Flores, E.	336	Garcia, J.F.	9, 94, 132
Flury, C.	148, 229, 268, 327	Garcia, S.	158
Flysjö, A.	223	Garcia-Gamez, E.	309
Foertter, F.	232	García-Rodríguez, A.	120, 191
Foltys, V.	125	Gardner, G.	96, 301
Font-I-Furnols, M.	210	Gardner, G.E.	302
Fontes, C.M.G.A.	75, 76	Gaspar, P.	30, 282

Gaspardo, B.	253	Groeneveld, E.	137, 172, 306
Gatellier, P.	276	Groot, J.C.J.	142
Gaudru, M.L.	365	Gross, J.J.	65
Gauly, M.	89, 175, 291, 292, 333	Große-Brinkhaus, C.	274
Gautier, J.M.	105	Guadagnin, M.	192
Gay, K.D.	315	Guarino, M.	345
Gbur, P.	311	Guatteo, R.	51, 122
Gebre, K.T.	29	Gubbiotti, M.	374
Geers, R.	217	Güemez, H.	220
Geertsema, H.G.	171	Guerci, M.	149
Gellrich, K.	199	Guerra, M.H.	297
Gengler, N.	15, 17, 18, 18, 78, 91, 93, 131,	Guiatti, D.	139
	173, 309, 368	Guillaume, F.	308
Gerber, P.J.	224	Guillouet, P.	276
Gergátz, E.	179	Guinard-Fament, J.	116
Gerjets, I.	287	Guinard-Flament, J.	122
Gernand, E.	248	Gulbe, G.	20
Geršiová, J.	219	Guldbrandtsen, B.	351, 354
Getya, A.	361	Gulewicz, K.A.	349
Ghazal, S.	328	Gulyás, L.	247
Ghita, E.	159, 262	Gupta, S.	4
Gianni, G.	224	Gutierrez, G.	336
Gilbert, H.	240, 272, 276	Gutiérrez, J.P.	34, 39, 236, 375
Gilliland, T.J.	182	Gutierrez-Gil, B.	309
Ginneberge, C.	225	Gutzwiller, A.	215
Giral, B.	132	Guzzo, N.	255
Gispert, M.	210	Gyökér, E.	179
Gizaw, S.	29		
Glass, E.J.	352	**H**	
Głażewska, I.	70, 71	Haas, L.S.	184
Goddard, M.	131	Haase, B.	327
Goddard, M.E.	128	Hadaš, Z.	219
Gomes, E.	286	Haddadi, A.	345
Gómez, I.	251	Hadjipavlou, G.	267
Gómez, M.D.	32, 32, 34, 41	Haesaert, G.	225
Gómez Castro, G.	126	Haeussermann, A.	47, 193
Gómez-Cortés, P.	126	Hahn, A.	35
Gomez-Raya, L.	188	Haile, A.	29
Gomez-Romano, F.	10	Halachmi, I.	194, 195, 195, 299
González, M.	74	Hall, N.	320
González-Rodríguez, A.	233	Haltia, S.	271
Gorgulu, M.	59, 163	Halubets, L.	362
Gorjanc, G.	134	Haman, J.	133
Gorniak, T.	58	Hamann, H.	261
Götz, K.-U.	229	Hamdy, S.M.	158, 340, 368
Götz, K.U.	261	Hammami, H.	368
Govignon-Gion, A.	352	Han, Y.-K.	160
Goyache, F.	236	Hancock, B.	301
Gradassi, S.	23	Hanotte, O.	8
Grageola, F.	322	Hansen Axelsson, H.	226
Gras, M.	262	Hanus, A.	292
Gredler, B.	129	Hanusová, E.	219, 292
Green, M.H.	126	Haque, M.N.	115, 116
Gremmen, B.	143	Harlizius, B.	354
Grigoletto, L.	151	Hartmann, S.	330
Groenen, M.A.M.	354	Hartung, E.	47, 193

Hartung, J.	345	Horváthová, M.	341, 347
Harty, D.	38	Hostiou, N.	24, 24, 26
Hasan, E.F.	241	Hoze, C.	308
Haščík, P.	217, 250	Højsgaard, S.	201
Hassan, F.	339	Hrdlicova, A.	68
Hassan, H.M.A.	159, 166	Hristov, A.N.	57
Häussler, S.	6	Huba, J.	312, 316, 363, 364
Hayes, B.J.	128, 167, 200	Hue, I.	299
Hazard, D.	343	Huhtanen, P.	147
Hazuchová, E.	246, 246	Huisman, A.E.	332
Hedger, M.	284	Hulin, S.	365
Heetkamp, M.J.W.	248	Hulsegge, I.	11
Heger, J.	371	Humphreys, J.	150
Heinz, J.	6	Huquet, B.	242
Helander, C.	259, 288	Hurtaud, C.	299
Hellwing, A.L.F.	294, 296	Hüttinger, K.	202
Hemnani, K.	329	Huţu, I.	119, 120
Hendriks, W.H.	56, 57	Huws, S.A.	119
Henne, H.	272	Hvelplund, T.	113
Hennessy, D.	182	Hynd, P.	109
Henningsen, A.	260	Hyslop, J.	86
Henry, C.	124		
Henry, M.L.E.	108	**I**	
Henryon, M.	335	Iannuccelli, N.	240
Henseler, S.	260	Iarova, P.	319
Herholz, C.	103	Ibáñez, M.B.	236
Hernández, J.	357	Ibañez-Escriche, N.	222
Hernandez, L.L.	123	Ibi, T.	174
Hernández-Castellano, L.E.	106, 329, 331	Ikauniece, D.	20
Herold, P.	261	Ilgaža, A.	83
Heuer, C.	305	Ilgaza, A.I.J.A.	371
Heugebaert, S.	44	Ilieski, V.	209
Heuven, H.C.M.	20	Illmann, G.	208
Heuze, V.	190	Ilska, J.J.	308
Hickey, J.M.	232	Ingrand, S.	281, 337
Hidalgo, A.M.	354	Ingvartsen, K.L.	49
Hietaoja, J.	348	Inman, C.F.	321
Hilla, D.	323	Insausti, K.	251
Hille, K.	175	Ismayilova, G.	345
Hirayama, T.	174, 247	Istasse, L.	185, 244, 244, 245, 255, 300
Hnisová, J.	165	Ivanković, A.	12, 361
Hochstuhl, D.	134		
Hocquette, J.F.	87, 231, 302	**J**	
Hoedemaker, M.	199	Jaayid, T.A.	13, 241
Hofer, A.	89	Jack, M.C.	169
Hoffman, L.C.	263	Jackuliaková, L.	329
Hofmanova, B.	40, 68, 98, 99	Jafarikia, M.	273
Hofstetter, P.	281	Jais, C.	346
Holló, G.	250, 251, 254, 254	Jakopović, T.	75
Holló, I.	250, 251	Jalč, D.	161
Homolka, P.	161, 165	Jammes, H.	2
Hoofs, A.I.J.	146	Jančík, F.	161, 165
Hopster, H.	209, 290	Jankauskas, I.	312
Horcada, A.	32	Jankowski, P.	79
Hornick, J.L.	185, 244, 244, 245, 255, 300	Jansen, S.	353
Horovská, Ľ.	285	Janssens, S.	33, 103, 276, 325

López-Mazz, C.	30
Lopez-Villalobos, N.	138
Lou, Y.	374
Louvandini, H.	286
Løvendahl, P.	86, 113, 172
Luan, W.	374
Lukač, D.	22
Lukic, B.	274
Lund, M.S.	351, 354, 356
Lund, P.	113, 155, 157, 294
Lundeheim, N.	44, 45
Lunden, A.	230
Lunnan, T.	183
Lutayeu, D.	362
Luther, H.	89
Lyberg, K.	213

M

Ma, N.	374
Ma, P.	356
Machpesh, G.	157
Macijauskiene, V.	275
Maciuc, V.	315
Macleay, C.	284
Macleod, M.	224
Mačuhová, J.	331, 346
Mačuhová, L.	329, 331
Madelrieux, S.	24, 27
Madsen, J.	86
Madsen, P.	40, 335
Maestrini, O.	334
Maeztu, F.	152
Magne, M.A.	337
Mahmoodi, Z.	167, 285
Mahmoudi, M.	66
Mahrer, D.D.	148
Maignel, L.	273
Majzlik, I.	40, 68, 98, 99
Mäki-Tanila, A.	315
Malá, G.	366
Malak-Rawlikowska, A.	313
Malavolta, M.	21
Malinowski, A.	10
Malovrh, S.	271
Maltar, Z.	274
Maltz, E.	194, 195, 195
Mandaluniz, N.	120, 191
Mangale, M.	216
Mani, V.	43
Mantovani, R.	255
Mäntysaari, E.A.	170
Mäntysaari, P.	170
Manunza, A.	357
Manzi, G.M.	263
Marchitelli, C.	189
Marcon, D.	259

Marett, L.	200
Margetín, M.	218, 250, 262, 264, 330, 367
Margetínová, J.	330
Margolskee, R.F.	318
Mariani, P.	22
Marichal, M.D.E.J.	297
Marie, M.	144, 280
Marnet, P.G.	365
Maroteau, C.	366
Márquez, M.	342
Marriott, D.T.	212
Martell-Jaizme, D.	330
Marti, E.	103
Martin, B.	365
Martin, C.	295
Martin, G.	93, 284, 336
Martin, O.	139, 338
Martin, P.	124
Martín-Collado, D.	74, 310
Martínez Marín, A.L.	126
Martínez-Camblor, P.	187
Marton, J.	79
Martyniuk, E.	177
Masar, J.	241
Masoero, F.	160
Masuda, Y.	307
Matheson, S.M.	107
Mathur, P.	72
Matkovic, K.	27
Mattei, S.	89
Mattsson, B.	214
Mattsson, P.	214
Maura, L.	178
Maurice - Van Eijndhoven, M.H.T.	16
Mavrogenis, A.P.	267
Mc Bride, S.H.	352
Mc Dowell, S.W.J.	352
Mc Parland, S.	230
McCartney, C.A.	295
McCoard, S.	6
McCue, M.E.	322, 323, 326
McDonald, A.J.S.	373
McEwan, J.C.	9
McGee, M.	168, 168
McManus, C.	286
McParland, S.	92, 93, 169, 173
Megens, H.J.	354
Mehrzad, J.	64, 66
Mele, M.	5
Melfsen, A.	193
Melišová, M.	208
Meloni, G.	178
Melzer, N.	88, 95
Membrillo, A.	346
Mendel, C.	261
Mendizabal, J.A.	251

Nara, P.	329, 330	**O**	
Nascimento, O.F.D.	49	Ober, U.	10
Naserian, A.	58	Öberg, M.	62
Näsholm, A.	36	Obregon, J.F.	220
Nath, M.	107	Oczak, M.	345
Nauwynck, H.J.	72	Oczkowicz, M.	277
Nazemi, S.	123	ODoherty, J.	290
Ndila, M.	8	O'Donovan, M.	182
Necpalova, M.	150	Oficialdegui, M.	152
Nedellec, C.	299	Ogink, N.	186, 226
Negrao, J.A.	7, 7	Oguey, C.	153
Negussie, E.	170	Oikawa, T.	174
Nel, F.	157	Oldenbroek, J.K.	68
Nemeš, Ž.	22	Olfaz, M.	96
Németh, T.	111, 179, 365	O'Mahony, M.J.	16
Nennich, T.D.	315	Onder, H.	96
Neser, F.W.C.	82	Ondruška, Ľ.	164
Neuhoff, C.	274	Oomen, G.J.M.	142
Neumeister, D.	148	Oosterlinck, M.	325
Neuner, S.	229	Oosting, S.J.	25, 28
Neves, H.	9	Opio, C.	224
Nevez, H.H.R.	132	Oppermann, P.	346
Nevrkla, P.	219	Oravcová, M.	218, 264
Newbold, C.J.	54, 155, 252, 294	Ortigues-Marty, I.	114, 154
Ng, C.	108	Ortiz, A.	191
Ngo, T.T.	44	Otwinowska-Mindur, A.	240
Nguyen, M.	2	Ouellet, D.R.	57
Nguyen, T.L.T.	294	Özel, D.	265
Nguyen, T.T.H.	295, 339		
Nicastro, A.	269	**P**	
Nicastro, F.	269	Paboeuf, F.	44
Nicholas, K.R.	3	Pace, V.	119
Nicolazzi, E.L.	95	Pacheco, D.	6
Nielsen, B.	335	Pailleux, J.Y.	281
Nielsen, M.F.	321	Paim, T.P.	286
Nieuwhof, G.	131	Paladini, I.	23
Nimmo, S.B.	300	Palhière, I.	366
Nir (markusfeld), O.	193	Palkovičová, Z.	228
Nishiura, A.	201	Palme, R.	184
Nitrayová, S.	47, 371	Palmer, E.	257, 258
Nižnikowski, R.	372	Panis, P.	132
Noblet, J.	276	Pannier, L.	96, 302
Noguera, J.L.	222	Papachristoforou, C.	267
Nollet, L.	121	Papoutsoglou, E.S.	166
Norberg, E.	40	Papoutsoglou, S.E.	166, 282
Norouzi Ebdalabadi, M.	61, 62	Parés-Casanova, P.M.	110, 111
Noskovičová, J.	105	Park, C.	299
Nourozi Ebdalabadi, M.	121	Parkanyi, V.	241
Novák, K.	9	Pastell, M.	348
Noval, E.	117	Patoux, S.	255
Novotný, F.	105	Patráš, P.	47, 371
Nowak, M.	349	Paulaitiene, J.	151
Nowak, Z.	177	Pausch, H.	353
Nozière, P.	118	Pavlík, I.	246, 246
Nozières, M.O.	27	Paweska, J.	104
Nørgaard, P.	259, 288	Pazzaglia, I.	374

Pearce, M.C.	104	Ploegaert, T.C.W.	52
Pecceu, K.	122	Plummer, P.	329
Peeters, L.M.	103	Poccard-Chapuis, R.	27
Peharda, M.	283	Pogran, Š.	304
Pelayo, R.	81, 346	Poláčiková, M.	56, 164
Pellicer-Rubio, M.-T.	178	Polak, G.M.	98
Pelmus, R.	262	Polák, P.	252, 262, 264, 316
Peña, F.	80, 81	Polkinghorne, R.	302
Pena, R.N.	357	Pollott, G.E.	230, 231
Perea, J.	183, 316	Pomar, J.	222
Pereira, O.G.	163	Pomiès, D.	365
Pérez Alba, L.M.	126	Pongrácz, L.	247
Pérez Hernández, M.	126	Poole, C.A.	5
Perez O'brien, A.M.	94, 132	Popov, M.	362
Pérez-Cabal, M.A.	39, 375	Portier, B.	197
Perez-Vendrell, A.	213	Pošivák, J.	105
Perisic, P.	363	Pospišilová, D.	292
Perler, L.	103	Posta, J.	35
Permentier, L.	217	Potočnik, K.	97, 134, 256
Persani, L.	177	Pottier, E.	110
Peškovičová, D.	219, 292, 312	Prates, J.A.M.	75, 76
Peters, T.L.	123	Preisegolaviciute-Mozuraitiene, D.	312
Petersen, J.L.	322, 323, 326	Preissler, R.	45, 359
Petersson, K.-J.	226	Pretrel, M.	340
Pethick, D.	302	Preuss, S.	260
Pethick, D.W.	96, 301, 302	Prezelj, K.	196
Petrák, J.	217, 218	Pribyl, J.	99
Petrášková, E.	165	Přibylová, J.	133
Petrera, F.	94	Prins, B.	310
Petrovic, M.D.	363	Pröll, M.	274
Peyraud, J.L.	115, 143	Proskina, L.	162
Phelan, P.	150	Prusak, B.	71
Philipp, U.	323, 324	Pryce, J.E.	167, 200
Philipsson, J.	33, 36	Pszczola, M.	84, 85, 92
Phocas, F.	308	Ptak, E.	240
Phyn, C.V.C.	4	Puillet, L.	143, 338
Piaggio, L.	297	Pulido, A.F.	30, 282
Piasentier, E.	31	Pulido, F.	30
Piazza, A.	23	Pun, A.	39, 375
Picard, B.	302	Purroy, A.	251
Pichová, K.	341, 347	Puskur, R.	28
Piedrafita, J.	264		
Pierni, E.	258	**Q**	
Pijl, R.	50	Queiroz, M.A.A.	293
Piles, M.	370	Quesnel, H.	42
Pilmane, M.	162	Quiniou, N.	44, 211
Piloni, R.	22	Quintanilla, R.	357
Pimentel, E.C.G.	85, 236, 334	Quintans, G.	30, 77
Piñeiro, C.	46		
Pinho, L.	51	**R**	
Pinloche, E.	252	Rad, M.	66
Pinotti, L.	67	Radeski, M.	209
Piórkowska, K.	277, 277	Radisic, Z.	274
Piórkowska, M.	235	Radovčić, A.	75
Pires, V.M.R.	75	Raes, E.	325
Pirlo, G.	160, 297	Rafay, J.	241

Ragona, G.	23	Rodriguez-Ramilo, S.T.	310
Rahimi, A.	58	Roehe, R.	86
Rame, C.	177	Roepstorff, L.	36
Ramljak, J.	12	Rofiq, M.N.	78
Raoul, J.	8, 179, 187	Rogel-Gaillard, C.	240
Rapey, H.	27	Rognon, X.	69
Rashidi, H.	72	Rogosic, J.	109
Rasmussen, M.D.	49	Roh, S.-G.	247
Rauw, W.M.	188, 270	Romanini, C.E.B.	195
Ray, B.	33	Romanini, E.	195
Raybould, H.E.	319	Romero, J.	342
Razmaitė, V.	107	Romnee, J.M.	'18, 161
Rebel, J.M.J.	212	Rondia, P.	161
Rebucci, R.	67	Rönnegård, L.	90, 239
Reckmann, K.	280	Rooke, J.	86
Reecy, J.	299	Rooni, K.	99
Reents, R.	136, 136	Ropka-Molik, K.	215, 237, 277
Rehak, D.	80	Ropota, M.	159, 262
Reichstadt, M.	299	Rosa, G.J.M.	232
Reichstädterová, T.	304	Rosales, C.	284
Reinhardt, F.	136, 136, 184, 199	Roschinsky, R.	28
Reinsch, N.	34, 95	Rose, G.	369
Reis, A.P.	257, 258	Rosetti, R.	177
Rekik, M.	180, 180	Rosner, F.	50, 93
Relun, A.	51	Ross, D.	86
Renand, G.	231	Ross, J.W.	43
Renieri, C.	21, 374, 374	Rossi, C.	115
Reolon, E.	158	Rossignol, M.N.	231
Repsilber, D.	88, 95	Rossing, W.A.H.	142
Reverchon, M.	177	Roughsedge, T.	129
Revilla, R.	266	Rousset, S.	231
Rey, A.I.	46	Rovere, G.	40
Reynolds, C.K.	52	Rowland, R.R.R.	71
Rezayazdi, K.	61, 63	Royer, E.	289, 293
Ribaud, D.	110	Różycki, M.	215, 237
Ribikauskas, V.	275	Røjen, B.A.	116
Ribikauskiene, D.	275	Ruban, S.	361
Ricard, A.	37, 324	Rubin, C.	323
Ricard, E.	259	Ruis, M.A.W.	209
Riccardi, D.	319	Ruiz, J.	336
Ricci, P.	86	Ruiz, R.	120, 140, 191, 364
Rieder, S.	229, 327	Rulquin, H.	115, 116
Rios, F.	220	Rupp, R.	366
Ripoll-Bosch, R.	266, 317	Russo, V.M.	55, 154
Rischkowsky, B.	29	Rustas, B.O.	62
Rius, A.G.	4	Rutten, M.J.M.	19
Rivero, M.A.	331	Růžička, Z.	133
Robaye, V.	244, 244, 245, 255, 300	Ryder, J.J.	352
Robert-Granié, C.	132	Rydhmer, L.	226, 333
Robin, C.	110	Ryshawy, J.	27
Roche, J.R.	4		
Rodrigues, A.D.	7	**S**	
Rodrigues, R.T.S.	60, 163, 293	Saarijärvi, K.	348
Rodriguez, M.C.	10	Sabaté, J.	110, 111
Rodríguez De Ledesma, A.	30	Sabatier, R.	140
Rodríguez-López, J.M.	154	Sabbagh, M.	37